T0336192

ADVANCED ENGINEERING MATHEMATICS

with Modeling Applications

ADVANCED ENGINEERING MATHEMATICS

with Modeling Applications

S. Graham Kelly

CRC Press is an imprint of the
Taylor & Francis Group, an **informa** business

CRC Press
Taylor & Francis Group
6000 Broken Sound Parkway NW, Suite 300
Boca Raton, FL 33487-2742

© 2009 by Taylor & Francis Group, LLC
CRC Press is an imprint of Taylor & Francis Group, an Informa business

No claim to original U.S. Government works

ISBN-13: 978-0-8493-9533-8 (hbk)

This book contains information obtained from authentic and highly regarded sources. Reasonable efforts have been made to publish reliable data and information, but the author and publisher cannot assume responsibility for the validity of all materials or the consequences of their use. The authors and publishers have attempted to trace the copyright holders of all material reproduced in this publication and apologize to copyright holders if permission to publish in this form has not been obtained. If any copyright material has not been acknowledged please write and let us know so we may rectify in any future reprint.

Except as permitted under U.S. Copyright Law, no part of this book may be reprinted, reproduced, transmitted, or utilized in any form by any electronic, mechanical, or other means, now known or hereafter invented, including photocopying, microfilming, and recording, or in any information storage or retrieval system, without written permission from the publishers.

For permission to photocopy or use material electronically from this work, please access www.copyright.com (http://www.copyright.com/) or contact the Copyright Clearance Center, Inc. (CCC), 222 Rosewood Drive, Danvers, MA 01923, 978-750-8400. CCC is a not-for-profit organization that provides licenses and registration for a variety of users. For organizations that have been granted a photocopy license by the CCC, a separate system of payment has been arranged.

Trademark Notice: Product or corporate names may be trademarks or registered trademarks, and are used only for identification and explanation without intent to infringe.

Visit the Taylor & Francis Web site at
http://www.taylorandfrancis.com

and the CRC Press Web site at
http://www.crcpress.com

Dedication

In memory of my parents

Contents

Preface

This book springs from class notes, I developed for a course called Engineering Analysis which I have taught every other fall semester since 1983 at the University of Akron. The course is targeted to students who are beginning graduate study in engineering. The students enrolled are first- and second-year graduate students in the Department of Mechanical Engineering, although I have taught students from other engineering disciplines.

At the beginning of the first class, I tell my students that I have two objectives in teaching the class. The first is to prepare them for subsequent graduate courses in engineering by teaching mathematical methods that are used in major graduate classes. The second objective is to prepare students to read engineering literature, such as journals and monographs. Students engaged in thesis and dissertation research need a foundation in mathematical terminology and methods to understand previous work in their research area. The title of the course, Engineering Analysis, is vague, but it essentially means "Advanced Engineering Mathematics with Applications." The course content and the content of this book is exactly that. Contrary to many engineering mathematics courses, the applications are emphasized as well as the physics behind the applications.

The applications are directed toward problems encountered in graduate engineering classes as well as in emerging areas of practice. An understanding of the modeling methods used to derive the mathematical equations as well as the underlying physics of the problem is usually essential to developing a method to solve the mathematical equations. For this reason, issues of modeling and scaling are discussed along with the analysis methods.

The motivation for the course and this book is to provide students and readers an experience of the marriage of engineering and applied mathematics. As in a marriage, both are equal and complement one another.

A unique feature of this book is the foundation laid for study. Books and courses on real analysis and functional analysis concentrate on theory rather than applications. Graduate engineering courses in areas, such as vibration, stress analysis, fluid mechanics, and heat transfer, concentrate on applications and use mathematical methods such as eigenvalue analysis and separation of variables as tools without much explanation of the mathematical theory underlying why they can be applied. An engineering student should understand the underlying mathematics that explains why these methods work, when they can be applied, and what are their limitations, but does not need to understand how to prove every theorem. This book takes such a view.

An underlying foundation is developed using the language of vector spaces and linear algebra. Basic results are derived, such as the existence of energy inner products for self-adjoint operators, the eigenvector expansion theorem, and the Fredholm alternative, and applied to problems for discrete and continuous systems. Yes, the differences between finite and infinite dimensional spaces are addressed, especially the issues of convergence. For this study, a discussion of completeness of eigenvectors of self-adjoint operators and heuristic proofs of convergence with respect to energy norms is sufficient, and only a limited discussion of pointwise convergence of the trigonometric Fourier and Fourier-Bessel series is presented. Because the focus is on applied mathematics for beginning graduate students, topics such as continuous spectra of eigenvalues are omitted.

This book is different from other advanced engineering mathematics books in many ways, some of which are listed below:

- Applications are presented to provide motivation for the mathematics, whereas most existing books illustrate applications after developing the mathematics.
- Linear algebra is used to provide a foundation for analysis of discrete and continuous systems.
- Rigor is used in development of concepts and is used in proving theorems for which the proofs themselves are instructive.
- The view is taken that a general understanding of theory is necessary to develop applications.
- Applications from emerging technologies are presented.

Theorems and proofs are presented, but without the detail found in many mathematics books. It is not intended to have the development of the theory obscure its application to engineering problems. On the other hand, a full understanding of the solution is not possible unless the theory from which it is developed is understood.

Acknowledgement is due to students who have taken Engineering Analysis over the years. Their questions and suggestions helped refine the book. I gratefully acknowledge my wife, Seala Fletcher-Kelly, not just for her support, but for significant help in preparing the figures. I am grateful to B.J. Clark, formerly of Taylor & Francis, for his efforts in developing the project, as well as to Michael Slaughter and Jonathan Plant at CRC Press for their continued support during the project. I would also like to express appreciation for the work of Amber Donley, project coordinator, at CRC and Glenon Butler, project editor at Taylor & Francis and others at CRC and Taylor & Francis who aided in the publication of this work.

S. Graham Kelly

Author

S. Graham Kelly received a BS in Engineering Science and Mechanics from Virginia Tech and a MS and PhD in Engineering Mechanics from Virginia Tech in 1977 and 1979, respectively. He served on the faculty at the University of Notre Dame from 1979 to 1982 and at The University of Akron since, in addition to his academic work, Dr. Kelly served in several administrative positions, including associate provost and dean of Engineering. Dr. Kelly is the author of three previous texts, *Fundamentals of Mechanical Vibrations, System Dynamics and Response,* and *Advanced Vibration Analysis* as well as a *Schaum's Outline of Mechanical Vibration*. He is a member of the American Society of Mechanical Engineers, the American Physical Society, the Society for Industrial and Applied Mathematics, the Society for Engineering Science, and the American Institute for Astronautics and Aeronautics. In addition to writing projects, Dr. Kelly currently enjoys teaching, research, working with students, and writing at The University of Akron.

Chapter 1

Foundations of mathematical modeling

1.1 Engineering analysis

Engineering analysis, mathematical modeling, engineering mathematics, and applied mathematics are all related terms. Let us start with engineering analysis, which can simply be defined as the analysis used by engineers during design and in applications. The adjective "engineering" implies that a practical reason exists for the analysis and that physical sciences are used in problem development. However the term "analysis" is vague as to the method of analysis used. Engineers use a variety of analytical techniques, including empirical methods, statistical methods, graphical methods, numerical and computational methods, and mathematical methods. "Engineering analysis" can refer to the application of any of these methods.

Mathematical modeling is a procedure in which a system of mathematical equations is developed to simulate the behavior of a physical system. The procedure usually includes a method to solve the resulting equations. Engineering mathematics refers to the mathematical tools that engineers use to solve the equations derived during modeling. Engineering mathematics is a subset of applied mathematics.

This study focuses on a subset of "engineering analysis," engineering mathematics and its interrelationship with mathematical modeling. Indeed, the two are intertwined as the knowledge of the available mathematical tools may drive the modeling process, and the engineering knowledge may drive the solution process. In this context, engineering analysis begins with mathematical modeling of an engineering system. The modeling leads to the formulation of a mathematical problem. The mathematical problem is nondimensionalized (Section 1.4) to understand the scaling of physical parameters and to suggest solution techniques. A mathematical method is applied to obtain the solution of the mathematical problem. The mathematical solution is then used to solve the engineering problem.

The above is a simplified overview of engineering analysis. Perhaps an alternate definition is the rational application of basic laws of nature and constitutive equations or equations of state to derive a mathematical model for an engineering system and then to apply the methods of applied mathematics.

Both the mathematical modeling and the resulting solution pose challenges. Often these are not independent.

Knowledge of the principles and techniques of applied mathematics can aid in the development of a mathematical model of a physical problem, and often the mathematical solution is guided by the physics. Physics and mathematics are married. It is no coincidence that the discoverers of the basic principles of physics, such as Sir Isaac Newton, are also the developers of modern mathematical analysis.

One cannot formulate an exact statement of how to develop a mathematical model. While generalities can be examined, the exact nature of mathematical modeling is discipline-specific. The first step is to identify the problem. Problem identification often requires abstraction of the system to be modeled from larger systems. Problem identification also requires specification of a goal. Consider, for example, the heating of the sphere shown in Figure 1.1. One goal of the modeling may be simply to analyze the unsteady state temperature at the center of the sphere or to determine the time required for this temperature to reach a certain value. Another possible goal is to determine a material such that the temperature at the center reaches a certain value in a specified time. Still another goal is to determine how the sphere can be heated such that the minimum energy is used for the center of the sphere to reach the specified temperature. Thus possible objectives of mathematical modeling include analysis, design, and optimization.

The abstraction of the system also leads to identification of system inputs (what is feeding the system, usually assumed to be specified) and system outputs (what is to be determined). Independent variables, the variables on which the output depends, are identified. Time and spatial coordinates referred to a fixed frame are common examples of independent variables. Dependent variables are the variables which specify the output.

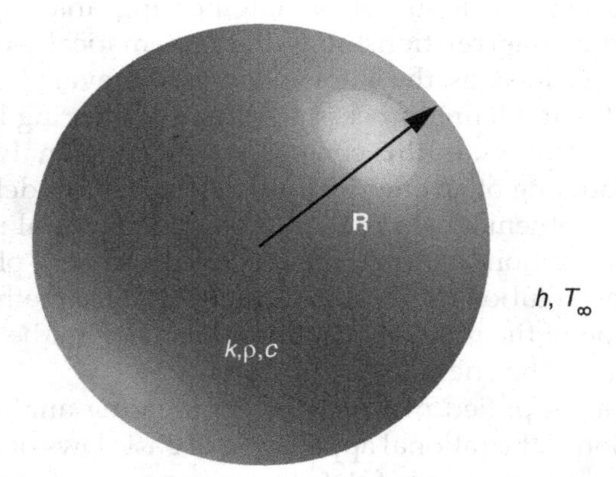

Figure 1.1 An objective of the mathematical modeling of heat transfer processes leading to the determination of the temperature distribution in the sphere may be analysis, design, or synthesis.

A necessary step in problem identification is to specify assumptions or identify possible assumptions. It may not be possible to list all reasonable assumptions until a preliminary model is completed. Assumptions can be modified during the modeling process. Scaling issues may lead to modification of assumptions.

Implicit assumptions are those that are taken for granted and rarely stated. Examples include that the earth is a Newtonian reference frame, that the acceleration of gravity on the surface of the earth is 9.81 m/s^2, and that nuclear reactions are not occurring within the system. It should be explicitly stated if a particular implicit assumption is not appropriate.

Implicit assumptions are often questioned in extreme cases. A common implicit assumption is that a continuum model can be used. This assumption implies that properties of matter are continuous. However, as engineering problems emerge at the nanoscale, the validity of the continuum assumption at this scale has been questioned. The radius of a carbon atom is approximately 0.34 nm. Can the continuum assumption be used to model nanowires and nanotubes? Answers lie in empirical results and scaling. That is, the continuum assumption is apparently valid if its application leads to results that compare favorably with empirical results. In addition, the continuum model is valid if length scales are greater than those associated with the vibration of molecules.

The alternative to the continuum assumption is to assume that each atom is a discrete particle and to apply conservation laws to each particle. A limitation to such an approach is that it leads to large numbers of simultaneous equations whose solution requires long computational times. In addition, the motion of individual atoms is not as deterministic as in continuous systems. Often Monte Carlo methods are used for molecular dynamics simulations.

Explicit assumptions are those that are specific to a particular modeling problem. Explicit assumptions are made to simplify the modeling or the subsequent mathematical solution of the resulting model. Many explicit assumptions are made for convenience. For example, when aerodynamic drag and other forms of damping are assumed to be small, modeling of the system shown in Figure 1.2 leads to a linear mathematical model. However, the resulting model is not physically practical because it leads to prediction of perpetual motion for any initial energy input.

Assumptions affect the identification of independent and dependent variables. Four independent variables can be identified for the modeling of the temperature distribution of the sphere shown in Figure 1.1. The sphere is heated until the temperature of the center reaches a certain value. The temperature clearly changes with time. The temperature of a particle in the sphere depends on its location in the sphere, which is described by three independent spatial coordinates: a radial coordinate r, a circumferential angular coordinate θ, and an azimuthal angular coordinate ϕ. Thus $T = T(r,\theta,\phi,t)$. If an assumption is made that the sphere is uniformly heated over its surface and that the sphere is at uniform temperature before heating, then the temperature is independent of θ and ϕ, and $T = T(r,t)$.

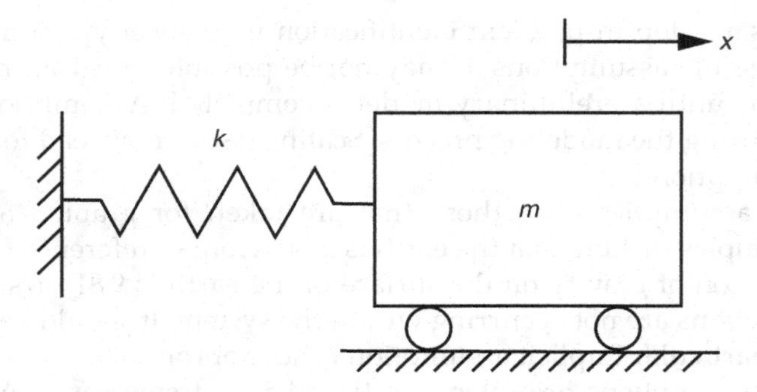

Figure 1.2 An assumption of no friction, including sliding friction and aerodynamic drag, is often used to linearize the system. While the resulting model does provide valuable information regarding the system, it leads to predicting physically impossible perpetual motion.

Five dependent variables identified for modeling of a fluid flow problem are the density ρ, the pressure *p*, and three components of the velocity vector **v**. When the flow is assumed to be incompressible, the density is constant, reducing the number of independent variables to four. If the system involves flow in a circular pipe, the velocity vector is often assumed to be directed along the axis of the pipe, reducing the number of dependent variables by two. In addition, if the pipe is of uniform diameter, then the axial component of velocity depends only on the radial coordinate and time.

Assumptions are often made regarding material behavior. Solids are usually assumed to be linearly elastic, air is assumed to be an ideal gas, thermal properties are assumed constant with temperature, and liquids are assumed incompressible. Specifications of material behavior may be in the form of a constitutive equation such as Hooke's law or Newton's viscosity law, an equation of state such as the ideal gas law, or an empirical law such as Newton's law of cooling. These laws introduce parameters into the model. Common examples of parameters include spring stiffness, viscous damping coefficient, thermal conductivity, and heat transfer coefficient. Less common parameters include Poisson's ratio and coefficient of thermal expansion.

Additional parameters are introduced from geometry. These include lengths and angular measures. Other parameters are introduced through system inputs such as velocity of a free-stream flow or frequency and amplitude of a harmonic input.

Once the problem has been identified, with independent variables, dependent variables, and parameters specified, the basic conservation laws which govern the system are identified. Conservation laws can be applied macroscopically to the entire system or a finite part of the system, or microscopically to an infinitesimal volume in the system. Their application usually requires an illustrative diagram. The result of the application of basic conservation laws and perhaps some algebraic manipulation leads to the mathematical model for the system.

The mathematics used to solve the mathematical model is the focus of the remaining chapters. The purpose of the mathematics is to provide a representation of the dependent variables in terms of the independent variables and the parameters which can be used to meet the goals of the modeling.

Mathematics and physics are interrelated. Physical explanations exist for mathematical paradoxes, and vice versa. Some examples of interaction between physics and mathematics are listed below:

- The partial differential equation governing the solution of the temperature distribution of the sphere shown in Figure 1.1 is second-order in the radial coordinate r. A second-order differential equation usually requires application of two boundary conditions to determine a unique solution; however, the only boundary condition formulated is $r=R$, the outer radius of the sphere. It is shown in Chapter 5 that a second boundary condition is not necessary due to the form of the differential equation, but the physics requires that the solution be finite everywhere in the sphere.
- Rayleigh-Ritz and finite-element approximations minimize the difference in energies between the exact solution and the approximate solution.
- The trigonometric Fourier series which is used to represent periodic functions in many applications has a theoretical development which is a result of the solution of an eigenvalue problem.
- Green's functions are derived as the response of a system to a singular input using variation of parameters.
- Solvability conditions for the existence of solutions of nonhomogeneous problems when a nontrivial solution to the corresponding homogeneous problem exists have physical interpretations.
- Eigenvalues equal to zero correspond to rigid-body motion in dynamic systems.

The focus of this study is on the interaction of the modeling process and the mathematical solution of the resulting equations. Applications are used from all fields of engineering, but mainly from mechanical engineering. Two applications are emphasized.

The temperature distribution in a solid body is modeled through application of conservation of energy to a differential volume within the body. The resulting mathematical problem depends on the assumptions used in the modeling. If the temperature distribution changes with time, a parabolic partial differential equation, the diffusion equation, is obtained. If the temperature is independent of time, an elliptical partial differential equation, Laplace's equation, is obtained. Assumptions may also reduce the number of spatial variables on which the temperature depends. A steady-state one-dimensional problem is governed by an ordinary differential equation. This differential equation has constant coefficients if the area over which conduction heat transfer occurs is constant, and has variable coefficients if the area varies over the range of the dependent variable. Heat transfer problems can be formulated for bars, cylinders, spheres, and other common volumes. In all problems, the differential equations are second-order in spatial variables and first-order in time. Such problems and their variations are considered throughout the study.

Vibrations of a beam are governed by a partial differential equation which is fourth-order in a spatial coordinate and second-order in time. A normal-mode solution is assumed, to determine the free response leading to a fourth-order ordinary differential equation. A nontrivial solution of the equation occurs only for certain values of the normal-mode parameter, the natural frequency. The problem is equivalent to an eigenvalue problem in which the natural frequencies are the square roots of the eigenvalues and the mode shapes are the corresponding eigenvectors. The mode-shape vectors satisfy an orthogonality condition. Variations of the basic beam problem studied include stretched beams, beams with transverse loading, beams on an elastic foundation, and sets of elastically connected beams.

Heat transfer problems and vibration problems describe different physical phenomena, but the underlying mathematics used in solving the problems is similar. The normal-mode solution used in the vibration problem is a form of separation of variables, which is the method used to solve the heat transfer problem. Product solutions in the heat transfer problem satisfy orthogonality conditions similar to those satisfied by the mode shapes in the vibration problem. In both cases, nonhomogeneous problems are solved using expansions in terms of homogeneous solutions. Such expansions are called eigenvector expansions or Fourier series expansions.

The remainder of this chapter focuses on mathematical modeling, including nondimensionalization and scaling. The theory of vector spaces and linear operators, presented in Chapter 2, provides a mathematical foundation on which solution methods can be based. The general theory of ordinary differential equations and methods of solution are presented in Chapter 3. Variational methods, approximate solution techniques based on energy methods, including the fundamentals of the finite-element method, are developed in Chapter 4. A comprehensive theory of eigenvalue problems for discrete and continuous systems is presented in Chapter 5. The study concludes in Chapter 6 with a study of partial differential equations.

1.2 *Conservation laws and mathematical modeling*

A finite number of basic conservation laws are used in the mathematical modeling of physical problems. Listed by name only, they include:

- Conservation of mass
- Conservation of linear momentum
- Conservation of angular momentum
- Conservation of energy

- The second law of thermodynamics
- Maxwell's equations

Each law has a specific meaning, but its manifestation may be different for different problems depending on the assumptions, the viewpoint, the materials used, and whether it is applied macroscopically or microscopically.

There are two viewpoints used in mathematical modeling of dynamic systems. For both viewpoints, imagine yourself as an observer viewing the system. Using the Lagrangian viewpoint, you fix yourself to a particle and observe the motion of the particle and other properties as you travel with the particle. The representation of the motion of the particle is in terms of time and the initial position of the particle. The Lagrangian viewpoint is best used for systems involving motion of discrete particles or for systems in which a mass can be identified and tracked. The motion of a single particle or rigid body is modeled using the Lagrangian formulation. As illustrated in Figure 1.3, the position vector for a particle as the particle moves within a Cartesian reference frame is tracked as a function of time. The Lagrangian viewpoint is also used to model the time-dependent motion of the simply supported beam shown in Figure 1.4. Let x be the distance from the fixed end to a particle along the beam's neutral axis when the beam is undeflected. The resulting displacement of the particle is a function of time. The deflection of the beam is described by $w(x,t)$, where x represents the initial position of a particle.

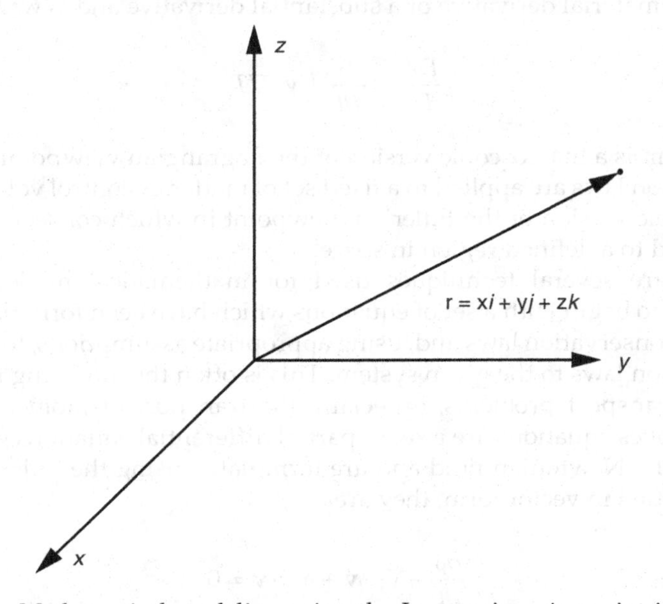

Figure 1.3 Mathematical modeling using the Lagrangian viewpoint is based on tracking the position vector for a particle as it changes with time. The independent variables used in a Lagrangian formulation are the initial position of the particle and time.

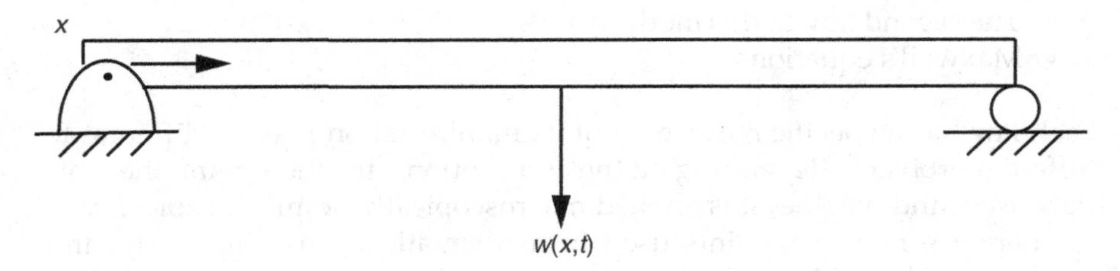

Figure 1.4 The Lagrangian viewpoint is used for modeling the vibrations of a beam. The vibrations are described using time and x, the initial distance of a particle along the neutral axis of the beam from its left end, as independent variables.

Using the Eulerian viewpoint, you fix yourself to a point in space and observe the time-dependent properties of the material surrounding you as particles of mass move past you. The Eulerian or field viewpoint is used in modeling systems in which a transport process occurs, such as the flow of a fluid. As you observe from a fixed point in space, particles of matter move by you with a velocity $\mathbf{v}$. You observe changes in properties such as temperature caused by changes occurring locally in the field, as well as changes occurring because of changes in temperature of the fluid particles that pass by you. The total rate of change of temperature is the sum of the local rate of change, $\partial T/\partial t$, and the convective rate of change, $\mathbf{v}\cdot\nabla T$. This form of the rate of change is called a material derivative or a substantial derivative and is written as:

$$\frac{DT}{Dt}=\frac{\partial T}{\partial t}+\mathbf{v}\cdot\nabla T \tag{1.1}$$

A system is a macroscopic version of the Lagrangian viewpoint in which conservation laws are applied to a fixed set of matter. A control volume is the macroscopic version of the Eulerian viewpoint in which conservation laws are applied to a defined region in space.

There are several techniques used for mathematical modeling. One method is to begin with a set of equations which have been formulated from the basic conservation laws and, using appropriate assumptions, to apply the conservation laws to the given system. This is often the modeling technique used in transport problems, especially the transport of momentum. The Navier-Stokes equations are a set of partial differential equations governing the flow of a Newtonian fluid and are formulated using the Eulerian viewpoint. Written in vector form, they are

$$\frac{\partial \rho}{\partial t}+\nabla\rho\cdot\mathbf{v}+\rho\nabla\cdot\mathbf{v}=0 \tag{1.2}$$

$$\rho\frac{\partial \mathbf{v}}{\partial \mathbf{t}}+\rho(\mathbf{v}\cdot\nabla)\mathbf{v}=\mathbf{F}_b-\nabla p+\mu\nabla^2\mathbf{v} \tag{1.3}$$

Equation 1.2 is called the continuity equation and is a statement of conservation of mass. Equation 1.3 is the vector form of conservation of linear momentum. The dependent variables in Equation 1.2 and Equation 1.3 are ρ, the mass density of the fluid, $\mathbf{v}$, the three-dimensional velocity vector, and p, the pressure. The body forces are represented by $\mathbf{F}_b$, and μ is the dynamic coefficient of viscosity. There are five dependent variables, but Equation 1.2 and Equation 1.3 provide only four scalar equations. If the flow is incompressible, then conservation of mass implies that $\nabla \cdot \mathbf{v} = 0$, and Equation 1.2 implies that the density is constant. In other cases, an equation of state relating pressure and density is used along with Equation 1.2 and Equation 1.3. Equations of state often involve temperature as a variable. If temperature varies, then an energy equation is used in conjunction with Equation 1.2 and Equation 1.3.

A second method of modeling is to apply basic forms of the conservation laws directly to the system. If time is an independent variable in the problem formulation, the conservation laws are applied at an arbitrary instant. In all cases, the conservation laws are applied for arbitrary values of the dependent variables. Two of the basic laws are conservation of momentum and conservation of energy.

Conservation of momentum is often applied to a free-body diagram of an infinitesimal particle drawn at an arbitrary instant. Conservation of momentum, Newton's second law as applied to a particle, is formulated as:

$$\sum \mathbf{F} = m\mathbf{a} \tag{1.4}$$

where $\sum \mathbf{F}$ represents the resultant force acting on a free-body diagram of the particle drawn at an arbitrary instant and $\mathbf{a}$ is the acceleration vector. Conservation of angular momentum is an appropriate form of the moment equation.

Conservation of energy as applied to a control volume takes the form that the rate of energy accumulation within the control volume is equal to the rate at which energy is transferred into the control volume through its boundaries.

A complete mathematical formulation often requires that the basic conservation laws be supplemented by laws specific to the system being modeled. For example, conservation of energy can be supplemented by Newton's law of cooling and Fourier's conduction law. The Navier-Stokes equations are written for a viscous fluid which satisfies Newton's viscosity law. Constitutive equations are used to represent stress-strain relations for many solids.

The method of application of mathematical modeling is different for different types of systems, yet the concepts are the same for all systems. The problem is first identified. Basic questions are answered. Why is a mathematical model necessary? What level of abstraction is required? What assumptions are necessary for successful modeling? What are the independent and dependent variables? What parameters are used in the modeling? Are numerical values of these parameters necessary? Based on answers to these questions, the problem is refined. Basic laws of physics, in some form, are applied to the abstracted system. Empirical laws specific to the system are applied as well as geometrical constraints.

Finally, a set of equations is developed. Before a solution method is applied, an assessment is necessary. Will solving the equations enable the objectives of the modeling to be met? Can additional physical insights be gained from examining the equations? Can these physical insights be used to aid in the solution of the equations? Answers to the previous two questions require that the mathematical equations be written in a nondimensional form.

A solution method is chosen and applied, resulting in mathematical equations defining the state of the system in terms of the independent variables. Another assessment is then required. Can these equations be used to meet the objectives of the modeling? Does the solution contradict any of the assumptions? If all appears to be well with the solution, it is then used to satisfy the objectives of the modeling.

1.3 *Problem formulation*

This section provides examples of mathematical modeling of physical systems. The differential equations are derived, but the methods of solution are found in later chapters.

Example 1.1 A fin or extended surface, as illustrated in Figure 1.5, is attached to a body to enhance the rate of heat transfer to or from the body. The base of the fin is maintained at the surface temperature, while its tip is insulated. The surface of the fin is surrounded by an ambient medium which is at a constant temperature T_∞. The convective heat transfer coefficient between the fin and the ambient is h. The fin is made of a material of thermal conductivity k. The thickness of the fin is small enough that conduction can be assumed negligible across the thickness of the fin and a one-dimensional model can be used.

A differential approach is used to derive a differential equation governing the steady-state temperature distribution in the fin. Apply the model to (a) a straight fin of constant thickness; (b) an annular fin of constant thickness; and (c) a straight fin of triangular profile. In each case, show how to determine the total heat transfer at the base of the fin.

Solution The approach used is to apply conservation of energy to a control volume of a differential element drawn at an arbitrary instant. The same basic model can be used for a straight fin and an annular fin. First, consider a straight fin. Let x represent the perpendicular distance from the base of the fin to a point on the fin. For a straight fin, assume that heat conduction occurs only in the x-direction. Let $b(x)$ represent the thickness of the fin, and let w be the width of the fin at its base.

Consider a differential slice of the fin as illustrated in Figure 1.6. Heat is transferred into the slice at a distance x from the wall by conduction at a rate of qA, where q is the rate of conductive heat transfer per unit area and A is the area of the face of the slice. Heat is transferred from the slice by conduction at

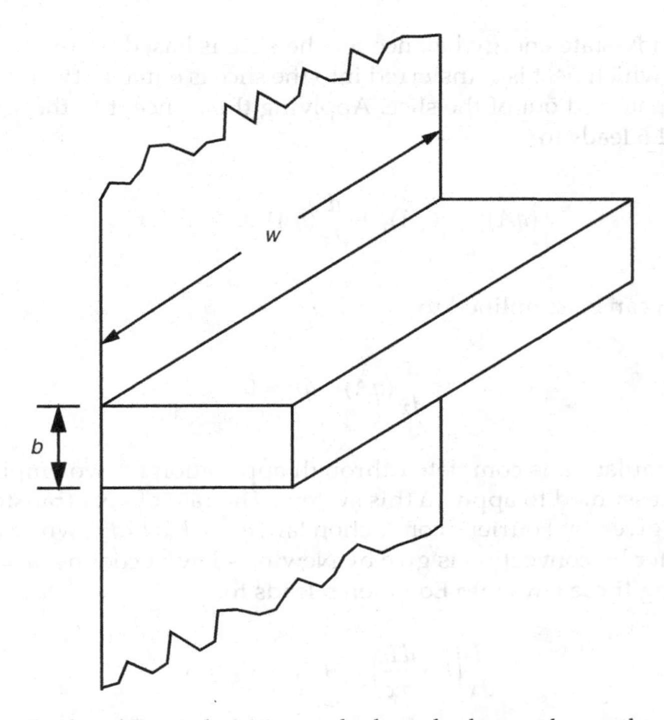

Figure 1.5 The fin of Example 1.1 is attached to a body to enhance the rate of heat transfer to or from the body. A mathematical model is formulated for the temperature distribution in the fin and the rate of heat transfer from its base.

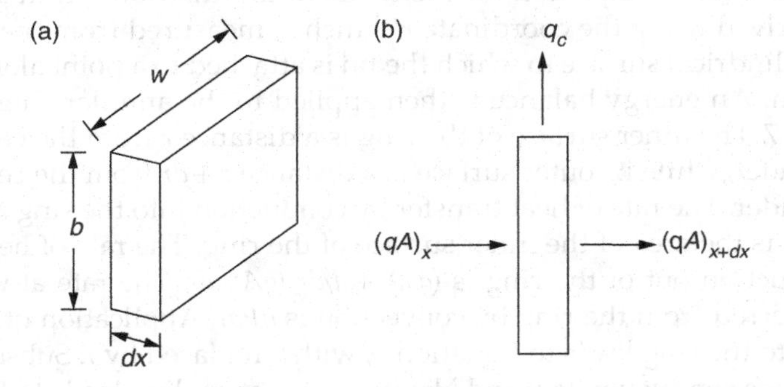

Figure 1.6 (a) Conservation of energy is applied to a differential slice of the fin. The temperature is assumed to vary only along the axis of the fin, so the area of the element is constant over the length of the fin. (b) Graphical illustration of the energy balance shows heat transfer by conduction into and out of the element through its internal faces and heat transfer by convection out of the element over its perimeter.

a distance $x+dx$ at a rate of $(qA)_{x+dx}=(qA)_x+[d(qA)_x/dx]dx$. Heat is transferred from the slice by convection at its surface. Let P be the perimeter of the surface and $\hat{q}$ the rate of convective heat transfer per unit area. The total rate of heat transfer from the surface of the element by convection is therefore $\hat{q}Pdx$.

The steady-state energy balance for the slice is based on the concept that the rate at which heat is transferred into the slice is equal to the rate at which heat is transferred out of the slice. Applying this concept to the slice shown in Figure 1.6 leads to

$$(qA)_x = \left[(qA)_x + \frac{d}{dx}(qA)_x dx\right] + \hat{q}Pdx \tag{a}$$

Equation a can be simplified to

$$\frac{d}{dx}(qA) + \hat{q}P = 0 \tag{b}$$

The formulation is completed through application of two empirical laws which are assumed to apply to this system. The rate of heat transfer by conduction is given by Fourier's conduction law, $q = -k(dT/dx)$, while the rate of heat transfer by convection is give by Newton's law of cooling, $\hat{q} = h(T - T_\infty)$. Substituting these laws into Equation b leads to:

$$\frac{d}{dx}\left(kA\frac{dT}{dx}\right) - hP(T - T_\infty) = 0 \tag{c}$$

Equation c governs the steady-state temperature distribution over an extended surface.

The differential equation for the temperature distribution in an annular fin is derived using the coordinate r, which is measured from the centerline of the cylindrical surface to which the fin is attached to a point along the axis of the fin. An energy balance is then applied to the annular ring shown in Figure 1.7. The inner surface of the ring is a distance r from the centerline of the cylinder, while its outer surface is a distance $r + dr$ from the centerline of the cylinder. The rate of heat transfer by conduction into the ring at r is $(qA)_r$, where A is the area of the inner surface of the ring. The rate of heat transfer by conduction out of the ring is $(qA)_r + d/dr\ (qA)_r\ dr$. The rate at which heat is transferred from the ring by convection is $\hat{q}Pdr$. Application of an energy balance to the ring leads to Equation b, with x replaced by r. Subsequent use of Fourier's conduction law and Newton's law of cooling leads to Equation c, with x replaced by r.

Equation c can be used to model one-dimensional steady-state heat transfer over an extended surface when A is interpreted as the area through which heat conduction occurs and P is the perimeter over which heat transfer occurs.

(a) The conductive area of a straight fin of constant thickness b is $A = bw$. The perimeter over which convection occurs is $P = 2w + 2b$. However, the thickness of the fin is assumed to be small compared to its width, such that P can be approximated by $P = 2w$. Substitution into Equation c leads to

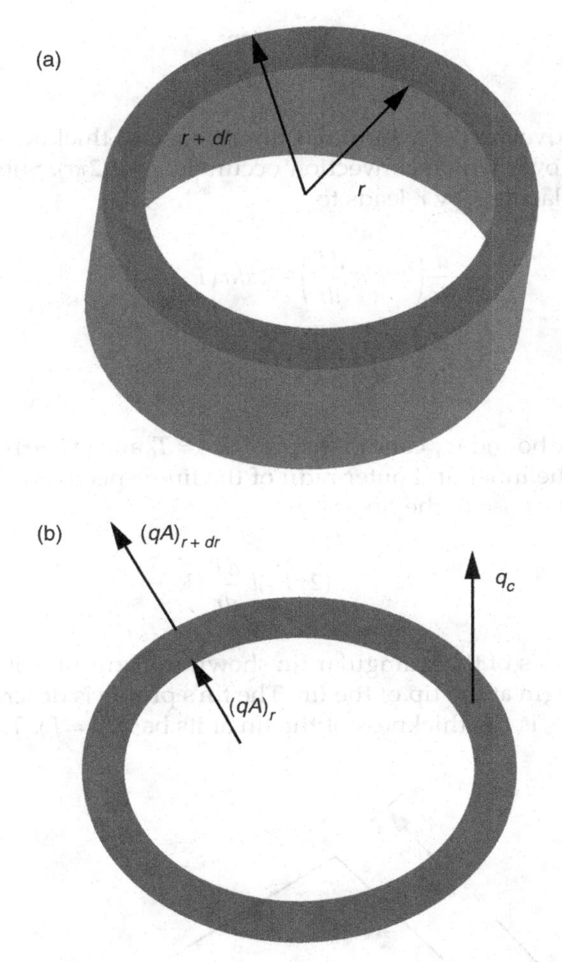

Figure 1.7 (a) The differential element used in the modeling of the temperature distribution in an annular fin is an annulus of inner radius r and thickness dr. The height of the annulus is the constant thickness of the fin. (b) Heat is transferred by conduction through the area of the annulus and out of the annulus though convection over its top and bottom perimeter.

$$\frac{d}{dx}\left(kbw\frac{dT}{dx}\right) - h(2w)(T - T_\infty) = 0$$

$$\frac{d^2T}{dx^2} - \frac{2h}{kb}(T - T_\infty) \tag{d}$$

The appropriate boundary conditions are $T(0) = T_0$ and $dT/dx\ (L) = 0$, where L is the length of the fin. The total heat transfer from the base of the fin is

$$Q_b = -kwb\frac{dT}{dx}(0) \tag{e}$$

(b) The conductive area of an annular fin of constant thickness b is $A = 2\pi rb$. The perimeter over which convection occurs is $P = 2(2\pi r)$. Substitution into Equation c, replacing x by r, leads to

$$\begin{aligned} &\frac{d}{dr}\left(2\pi kbr\frac{dT}{dr}\right) - 4\pi hr(T - T_\infty) \\ &\frac{d}{dr}\left(r\frac{dT}{dr}\right) - \frac{2h}{kb}r(T - T_\infty) = 0 \end{aligned} \tag{f}$$

The appropriate boundary conditions are $T(R_1) = T_0$ and $(dT/dr)\,(R_2) = 0$, where R_1 and R_2 are the inner and outer radii of the fin respectively. The total heat transfer from the base of the fin is:

$$Q_b = -k(2\pi R_1)b\frac{dT}{dr}(R_1) \tag{g}$$

(c) For the analysis of the triangular fin shown in Figure 1.8, it is convenient to locate the origin at the tip of the fin. The fin's profile is described by $b(x) = (b_0/L)x$, where b_0 is the thickness of the fin at its base ($x = L$). The conductive

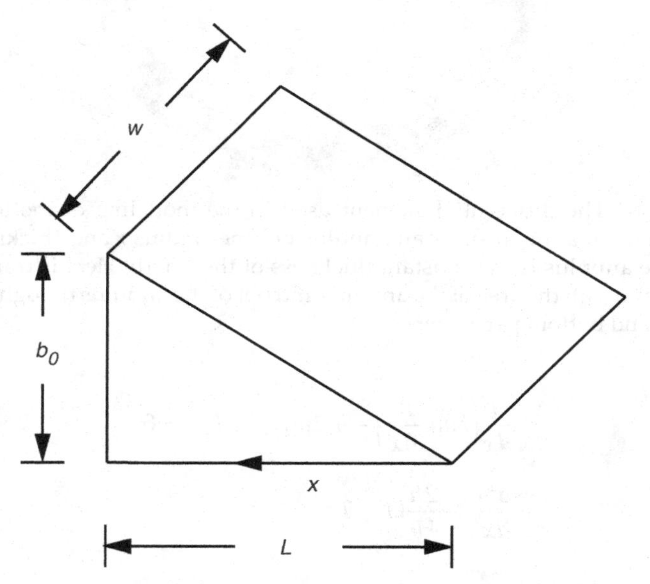

Figure 1.8 The area for heat transfer at the tip of the triangular fin is zero; thus it is convenient to measure x from the fin's tip.

area is $A=(b_0w/L)x$, while the convection perimeter is $P=2w+2b(x)$. Assuming $b_0 \ll w$, the perimeter can be approximated by $P=2w$. Substitution into Equation c leads to:

$$\frac{d}{dx}\left(k\frac{b_0}{L}x\frac{dT}{dx}\right)-2wh(T-T_\infty)=0$$

$$\frac{d}{dx}\left(x\frac{dT}{dx}\right)-\frac{2hL}{kb_0}(T-T_\infty)=0 \tag{h}$$

The area at the tip of the fin is zero. Thus there is no heat transfer from the tip as long as the temperature is finite. The only boundary condition is at the base, $T(L)=T_0$. The total heat transfer from the base of the fin is

$$Q_b=-kb_0w\frac{dT}{dx}(L) \tag{i}$$

Example 1.2 Derive the governing equation for the heat transfer in the annular fin of Example 1.1 if the temperature at the base of the fin varies over its circumference and changes with time.

Solution Define θ as a circumferential coordinate measured counterclockwise from a reference line, $0 \le \theta \le 2\pi$, as illustrated in Figure 1.9. The temperature at

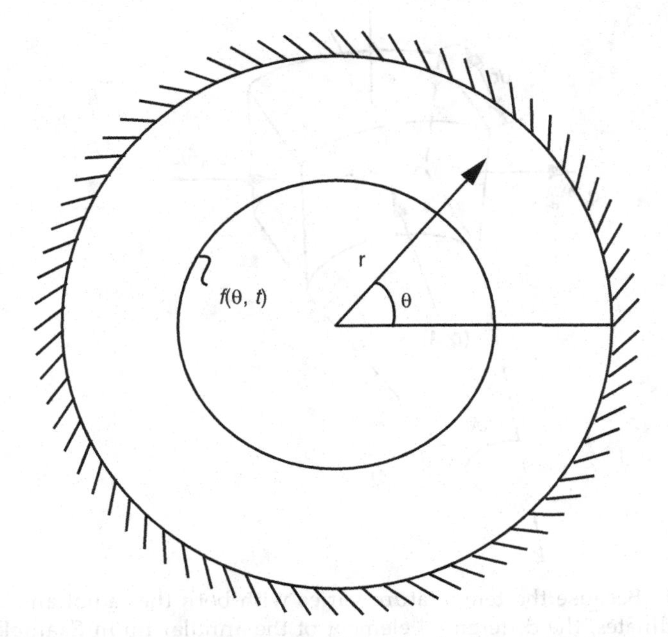

Figure 1.9 The annular fin of Example 1.2 is attached to a cylindrical surface. The temperature varies over the surface and with time.

the base of the extended surface is a function of θ as well as time, $f(\theta,t)$. Let r be a coordinate measured radially from the center of the cylinder, $R_1 \leq r \leq R_2$. Assuming that the extended surface is thin and that conduction can be neglected in the direction transverse to the extended surface, then the temperature of a particle in the extended surface is dependent on its location described by r and θ as well as on time, $T=T(r,\theta,t)$.

Consider a differential segment of the cylinder shown in Figure 1.10. The segment is a portion of the annular ring of Example 1.1, but is defined by an angle $d\theta$. Heat is transferred by conduction into the differential segment on the surfaces defined by constant values of r and θ, while heat is transferred out of the segment on surfaces defined by constant values of $r+dr$ and $\theta+d\theta$. A concern may be that since T varies with θ, the rate of heat transfer q_r may vary across the surface defined by r. This is true, but consider the total heat transfer over the surface at constant r from θ to $\theta+d\theta$,

$$Q_r = \int_{\theta}^{\theta+d\theta} q_r(r,\theta) b r d\theta \tag{a}$$

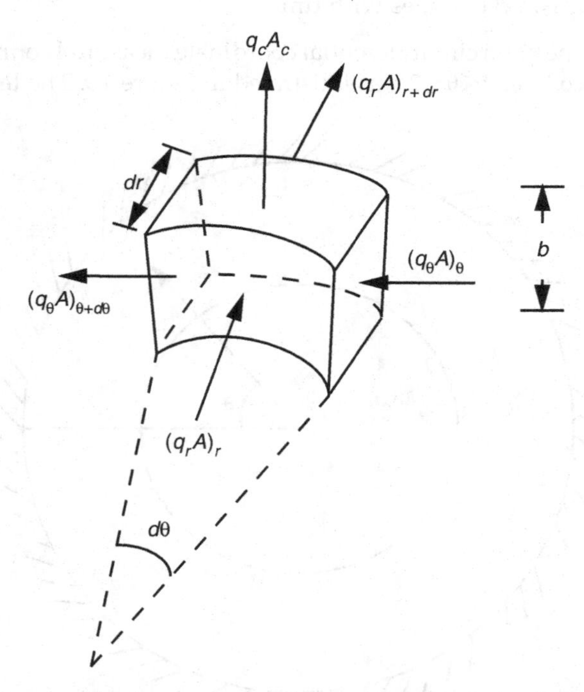

Figure 1.10 Because the temperature varies with both the radial and circumferential coordinates, the differential element of the annular fin in Example 1.2 is an infinitesimal segment of the annular ring in Figure 1.7.

The mean value theorem implies that there exists θ^*, $\theta \leq \theta^* \leq \theta + d\theta$ such that

$$Q_r = q_r(r, \theta^*) b r d\theta \tag{b}$$

Noting that the area of the surface is $brd\theta$, a Taylor series expansion of $q_r(r, \theta^*)$ about (r,θ) shows that a correction to account for the variation of Q_r with θ over the surface is of order $d\theta$ and will be smaller than all other terms in the resulting equation. Thus the rate of heat transfer can reasonably be assumed to be constant over each surface.

Heat is transferred from the element by convection from its upper and lower surfaces. Internal energy is stored in the differential element. The total internal energy is $U = \rho u dV$, where u is the specific internal energy and the differential volume is $dV = 2\pi b r dr d\theta$. The appropriate energy balance is that the rate at which energy is stored in the element is equal to the rate at which heat is transferred into the element minus the rate at which heat is transferred out of the element. Application to the differential element shown in Figure 1.10 leads to

$$\frac{\partial}{\partial t}(\rho u dV) = q_r A_r + q_\theta A_\theta - \left[q_r A_r + \frac{\partial}{\partial r}(q_r A_r) dr \right] - \left[q_\theta A_\theta + \frac{\partial}{\partial \theta}(q_\theta A_\theta) d\theta \right] - \hat{q} P dr \tag{c}$$

Equation c can be simplified to

$$\frac{\partial u}{\partial t}(\rho b r dr d\theta) = -\frac{\partial}{\partial r}(r q_r) b d\theta dr - \frac{\partial}{\partial \theta}(q_\theta) b dr d\theta - \hat{q} dr (2 r d\theta) \tag{d}$$

Equation d then reduces to

$$\rho r \frac{\partial u}{\partial t} = -\frac{\partial}{\partial r}(r q_r) - \frac{\partial q_\theta}{\partial \theta} - \frac{\hat{q} 2r}{b} \tag{e}$$

The rate of convective heat transfer per unit area is given by Newton's law of cooling, $\hat{q} = h(T - T_\infty)$. Fourier's conduction law is used to determine the rate of heat transfer per unit area on a surface: $q_n = -k \nabla T \cdot \mathbf{n}$, where q_n is the heat transfer per unit area into the surface and $\mathbf{n}$ is a unit vector normal to the surface. Noting that in polar coordinates, $\nabla T = (\partial T / \partial r)\, \mathbf{e}_r + 1/r\, (\partial T / \partial \theta)\, \mathbf{e}_\theta$, then $q_r = -k(\partial T / \partial r)$ and $q_\theta = -(k/r)\,(\partial T / \partial \theta)$.

The specific internal energy is related to temperature by $u = cT$, where c is the specific heat of the body. Substitution into Equation e, assuming that k and c are constants, results in

$$\rho c r \frac{\partial T}{\partial t} = k \frac{\partial}{\partial r}\left(r \frac{\partial T}{\partial r} \right) + k \frac{1}{r} \frac{\partial^2 T}{\partial \theta^2} - \frac{2hr}{b}(T - T_\infty) \tag{f}$$

Dividing by kr equation f can be rewritten as

$$\frac{\rho c}{k}\frac{\partial T}{\partial t}=\left[\frac{1}{r}\frac{\partial}{\partial r}\left(r\frac{\partial T}{\partial r}\right)+\frac{1}{r^2}\frac{\partial^2 T}{\partial \theta^2}\right]-\frac{2h}{bk}(T-T_\infty) \tag{g}$$

The Laplacian operator in polar coordinates is $\nabla^2 T = 1/r\,[\partial/\partial r\,(r\cdot\partial T/\partial r)] + 1/r^2\,(\partial^2 T/\partial\theta^2)$. Thus Equation g becomes

$$\frac{\rho c}{k}\frac{\partial T}{\partial t}=\nabla^2 T-\frac{2h}{bk}(T-T_\infty) \tag{h}$$

Example 1.3 A rigid mass m is attached to the end of an elastic bar is illustrated in Figure 1.11. A time-dependent force is applied to the mass. If m is much greater than the mass of the bar, a single-degree-of-freedom model is used in which the inertial effects of the bar are neglected and the bar is modeled as a linear spring of stiffness EA/L. This lumped-parameter model is illustrated in Figure 1.12. However, if the mass of the bar is comparable to m, then the distributed-parameter model in Figure 1.13 is used.

(a) Derive the differential equation governing the motion for the lumped-parameter model; (b) Derive the differential equation governing the motion for the distributed-parameter model.

Solution (a) Let $y(t)$ represent the displacement of the mass. The spring is assumed to be linear, and all forms of friction are neglected. Application of Newton's second law to a free-body diagram of the block, drawn at an arbitrary instant and illustrated in Figure 1.14, leads to:

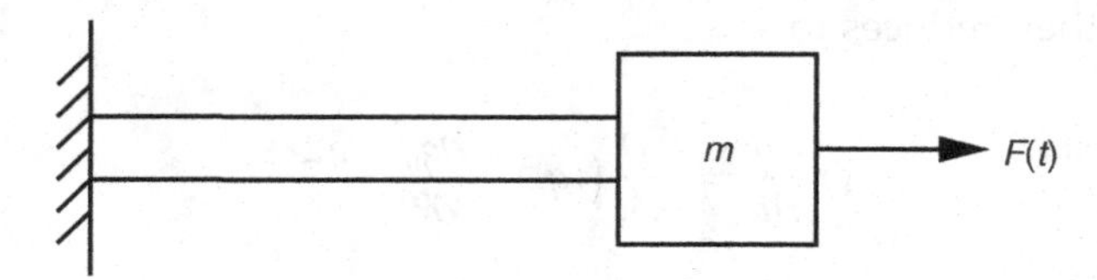

Figure 1.11 The longitudinal motion of a discrete particle of mass m attached to a flexible bar is modeled in Example 1.3.

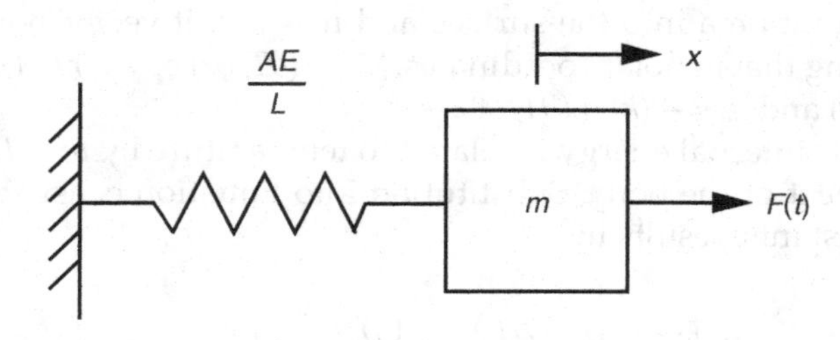

Figure 1.12 A lumped-parameter model is used if the mass of the particle is much greater than the mass of the bar.

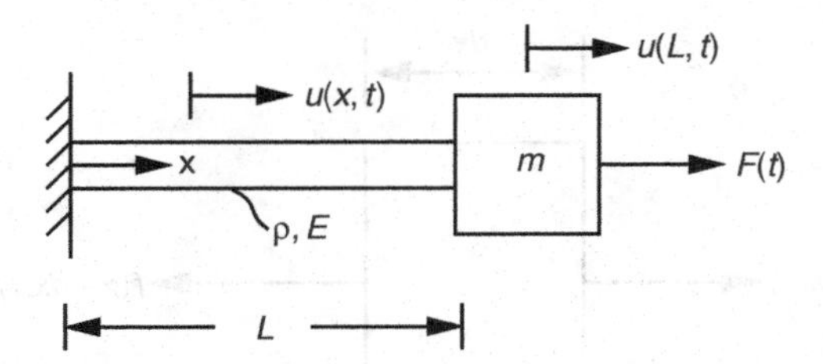

Figure 1.13 A distributed-parameter model is used if the mass of the bar is comparable to or greater than the mass of the particle.

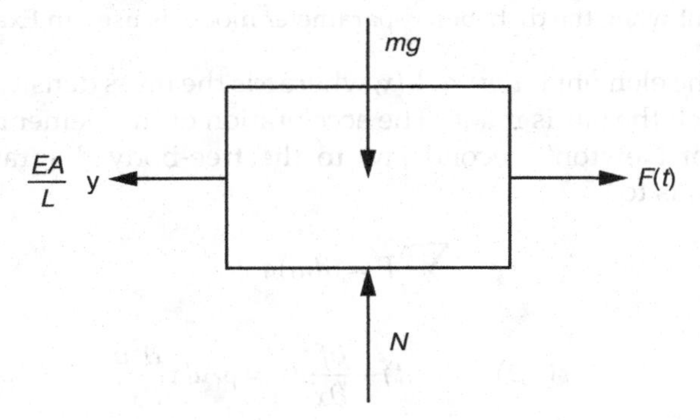

Figure 1.14 Newton's second law is applied to the free-body diagram of the particle of Example 1.3, drawn at an arbitrary instant, when the lumped-parameter model is applied.

$$\sum F = ma$$

$$F(t) - \frac{EA}{L} y = m \frac{d^2 y}{dt^2}$$

$$m \frac{d^2 y}{dt^2} + \frac{EA}{L} y = F(t) \tag{a}$$

Equation a is supplemented by initial conditions specifying the initial displacement and the initial velocity of the block.

(b) Let x, $0 \leq x \leq L$, be a coordinate along the axis of the bar. Note that the displacement of the rigid mass is $y(t) = u(L,t)$. Consider a differential segment of the bar of length dx. The area of the segment is A, the cross-sectional area of the bar. A free-body diagram of the segment, drawn at an arbitrary instant, is illustrated in Figure 1.15. The force acting on each face of the segment is the resultant of the normal stress distribution $\sigma(x,t)$, which is assumed uniform over each face. If the force acting normal to the element at x is $f(x)$, then the force acting normal to the element at $x + dx$ is $f(x+dx,t) = f(x,t) + \partial f / \partial x(dx)$.

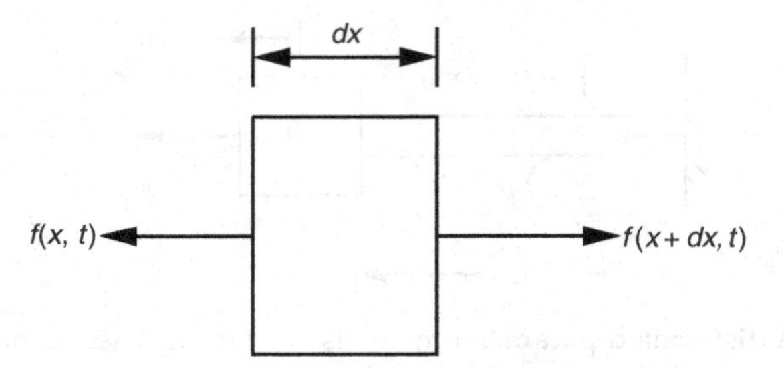

Figure 1.15 A free-body diagram of a differential element of the bar is drawn at an arbitrary instant when the distributed-parameter model is used in Example 1.3.

The mass of the element is $dm = \rho A dx$, where ρ is the mass density of the material from which the bar is made. The acceleration of the element $a = \partial^2 u/\partial t^2$. Application of Newton's second law to the free-body diagram shown in Figure 1.15 leads to

$$\sum F = (dm)a$$

$$-f(x,t) + f(x,t) + \frac{\partial f}{\partial x}dx = \rho A dx \frac{\partial^2 u}{\partial x^2}$$

$$\frac{\partial f}{\partial x} = \rho A \frac{\partial^2 u}{\partial t^2} \tag{b}$$

If the normal stress is less than the material's yield stress, then the system behaves elastically, and the normal stress is related to the normal strain ε through Hooke's law, $\sigma = E\varepsilon$, where E is the modulus of elasticity or Young's modulus. By definition, the normal strain is the change in displacement per change in length, $\varepsilon = \partial u/\partial x$. Thus the resultant force acting normal to the differential element at x is $f = EA(\partial u/\partial x)$. This force can be substituted into Equation b, leading to

$$\frac{\partial}{\partial x}\left(EA\frac{\partial u}{\partial x}\right) = \rho A \frac{\partial^2 u}{\partial t^2} \tag{c}$$

In the case of a uniform bar, Equation c reduces to

$$\frac{\partial^2 u}{\partial x^2} = \frac{\rho}{E}\frac{\partial^2 u}{\partial t^2} \tag{d}$$

The form of Equation d is that of the wave equation.

The bar is fixed at $x = 0$, thus

$$u(0,t) = 0 \tag{e}$$

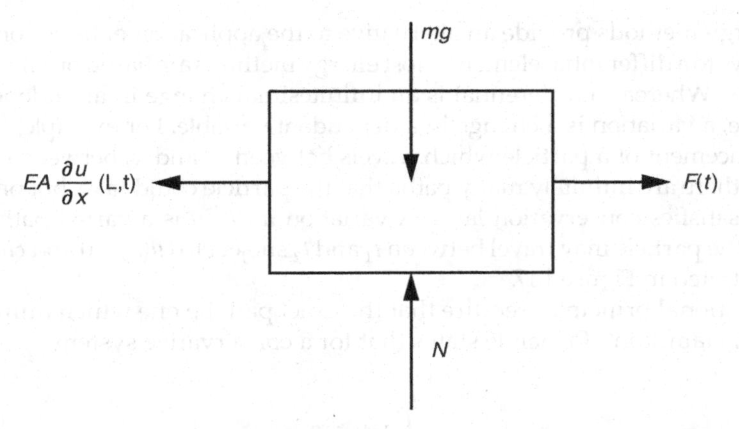

Figure 1.16 The boundary condition at $x = L$ for the distributed-parameter model of Example 1.3 is obtained through application of Newton's second law to a free-body diagram of the particle drawn at an arbitrary instant.

A free-body diagram of a rigid block, abstracted from the bar at an arbitrary instant, is illustrated in Figure 1.16. Application of Newton's second law to this free-body diagram leads to

$$-EA\frac{\partial u}{\partial x}(L,t) + F(t) = m\frac{\partial^2 u}{\partial t^2}(L,t) \tag{f}$$

Equation d, Equation e, and Equation f are supplemented by two initial conditions.

Equation 1.2 and Equation 1.3, in general, represent four scalar equations in five dependent variables. For an incompressible flow, the rate of change of volume of a constant mass of fluid is zero, leading to the incompressibility condition

$$\nabla \cdot \mathbf{v} = 0 \tag{1.5}$$

The continuity equation then leads to the conclusion that the density is constant. The incompressible form of the momentum equation becomes

$$\frac{\partial \mathbf{v}}{\partial \mathbf{t}} + (\mathbf{v} \cdot \nabla)\mathbf{v} = \frac{1}{\rho}\mathbf{F_b} - \frac{\nabla p}{\rho} + \nu \nabla^2 \mathbf{v} \tag{1.6}$$

where $\nu = \mu/\rho$ is the fluid's kinematic viscosity. Equation 1.5 and Equation 1.6 are used to develop models for incompressible flow problems.

The basic conservation laws are used to derive principles which are useful in problem formulation. For example, integration of Newton's second law along a particle trajectory leads to an energy principle.

Energy methods provide an alternative to the application of basic conservation laws to a differential element. Most energy methods are based on variational calculus. Whereas a differential is an infinitesimal change in an independent variable, a variation is a change in a dependent variable. For example, if *x(t)* is a displacement of a particle which travels between x_1 and x_2 between times t_1 and t_2, there are infinitely many paths that the particle could take, but only one which satisfies conservation laws. A variation in u, δu is a varied path along which the particle may travel between t_1 and t_2 subject to $\delta u(t_1)=0$ and $\delta u(t_2)=0$, as illustrated in Figure 1.17.

Variational principles require that the exact path be one which minimizes energy. Hamilton's Principle states that for a conservative system,

$$\delta \int_{t_1}^{t_2} L dt = 0 \tag{1.7}$$

where L, the lagrangian, is

$$L = T - V \tag{1.8}$$

in which T is the system's kinetic energy and V is the system's potential energy.

Defining $x_1, x_2, \ldots, x_n$ as a system of generalized coordinates for an n-degree-of-freedom discrete system, the potential energy is a function of these coordinates, $V = V(x_1, x_2, \ldots, x_n)$, and the kinetic energy is a function of the generalized coordinates and their time derivatives, $T = T(x_1, x_2, \ldots, x_n, \dot{x}_1, \dot{x}_2, \ldots, \dot{x}_n)$.

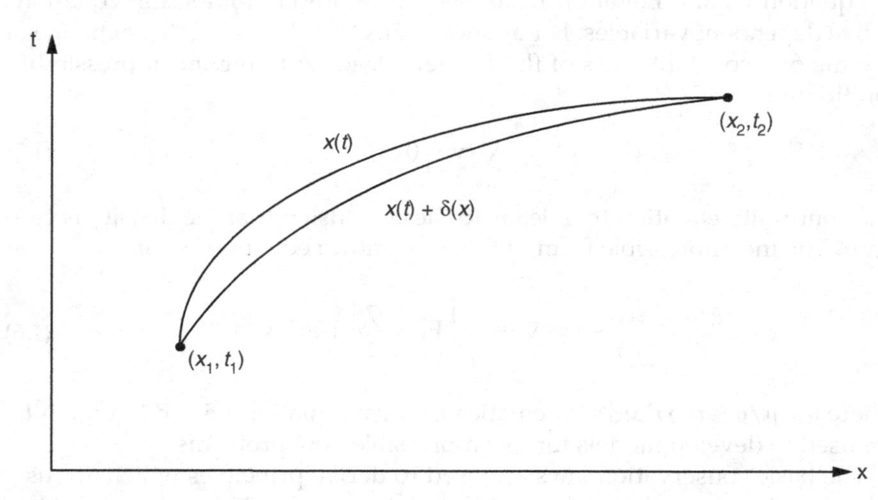

Figure 1.17 The true path taken by a particle from t_1 to t_2 is *x(t)*. The variable path $x(t) + \delta x$ is an alternate path between t_1 and t_2.

Application of Hamilton's Principle to a conservative discrete system leads to Lagrange's equations,

$$\frac{d}{dt}\left(\frac{\partial L}{\partial \dot{x}_i}\right) - \frac{\partial L}{\partial x_i} = 0 \quad i = 1, 2, \ldots, n \tag{1.9}$$

Lagrange's equations can be applied to derive the differential equations governing the motion of a discrete system.

Example 1.4 Use Lagrange's equations to derive the differential equations governing the motion of the system shown in Figure 1.18.

Solution The kinetic energy of the system at an arbitrary instant is

$$T = \frac{1}{2}m_1\dot{x}_1^2 + \frac{1}{2}m_2\dot{x}_2^2 + \frac{1}{2}m_3\dot{x}_3^2 \tag{a}$$

The potential energy in the system at the same instant is

$$V = \frac{1}{2}k_1x_1^2 + \frac{1}{2}k_2(x_2 - x_1)^2 + \frac{1}{2}k_3(x_3 - x_2)^2 \tag{b}$$

The Lagrangian for the system is

$$L = T - V = \frac{1}{2}m_1\dot{x}_1^2 + \frac{1}{2}m_2\dot{x}_2^2 + \frac{1}{2}m_3\dot{x}_3^2$$

$$-\left[\frac{1}{2}k_1x_1^2 + \frac{1}{2}k_2(x_2 - x_1)^2 + \frac{1}{2}k_3(x_3 - x_2)^2\right] \tag{c}$$

Application of Lagrange's equations to Equation c gives

$$\frac{d}{dt}\left(\frac{\partial L}{\partial \dot{x}_1}\right) - \frac{\partial L}{\partial x_1} = 0$$

$$\frac{d}{dt}(m_1\dot{x}_1) - [-k_1x_1 - k_2(x_2 - x_1)(-1)] = 0$$

$$m_1\ddot{x}_1 + (k_2 + k_1)x_1 - k_2x_2 = 0 \tag{d}$$

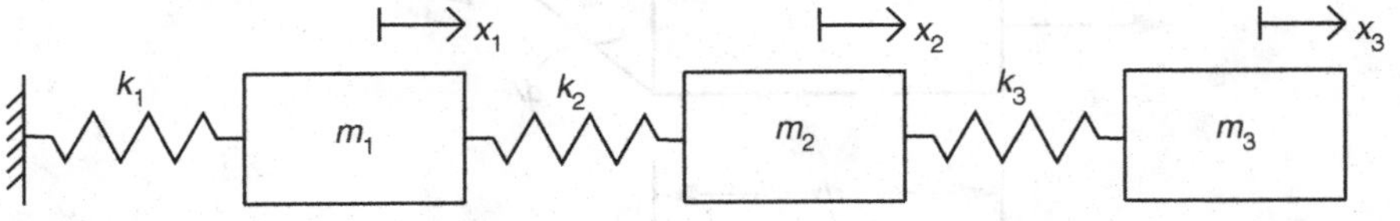

Figure 1.18 Lagrange's equations are used to derive the differential equations governing the motion of the mechanical system.

$$\frac{d}{dt}\left(\frac{\partial L}{\partial \dot{x}_2}\right)-\frac{\partial L}{\partial x_2}=0$$

$$\frac{d}{dt}(m_2\dot{x}_2)-[-k_2(x_2-x_1)-k_3(x_3-x_2)(-1)]=0$$

$$m_2\ddot{x}_2-k_1x_1+(k_2+k_3)x_2-k_3x_3=0 \tag{e}$$

$$\frac{d}{dt}\left(\frac{\partial L}{\partial \dot{x}_3}\right)-\frac{\partial L}{\partial x_3}=0$$

$$\frac{d}{dt}(m_3\dot{x}_3)-[-k_3(x_3-x_2)]=0$$

$$m_3\ddot{x}_3-k_3x_2+k_3x_3=0 \tag{f}$$

Equation d, Equation e, and Equation f are the differential equations governing the dynamic response of the system shown in Figure 1.18.

Example 1.5 Use Hamilton's Principle to derive the differential equation for the longitudinal oscillations of a particle of mass m attached to the end of an elastic bar, as illustrated in Figure 1.11.

Solution Consider a differential element of thickness dx and area dA as illustrated in Figure 1.19. The strain energy stored in this element is

$$dV=\frac{\sigma^2}{2E}dAdx \tag{a}$$

where σ is the normal stress. Assuming elastic behavior, $\sigma = E\varepsilon$, and for small displacements, the normal strain is $\varepsilon = \partial u/\partial x$, where $u(x,t)$ is the displacement

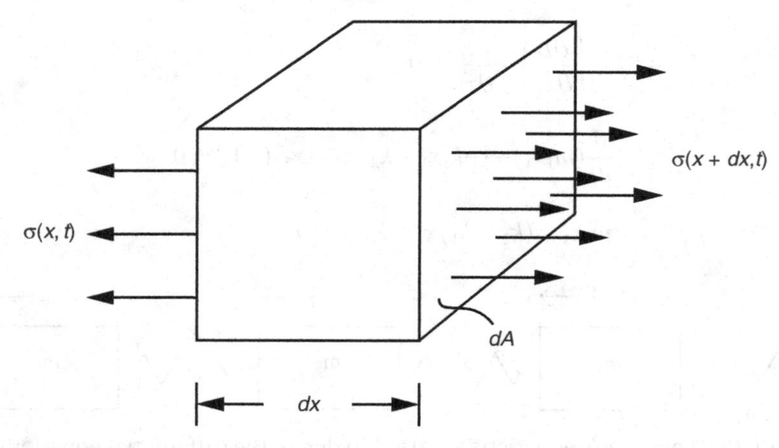

Figure 1.19 The normal stress in the elastic bar of Figure 1.11 is uniform across the face of the differential element, but varies across the length of the element.

of a particle along the longitudinal axis of the bar. The total potential energy is obtained by integrating Equation a over the volume of the bar:

$$V = \int_0^L \int_A \frac{1}{2E}\left(E\frac{\partial u}{\partial x}\right)^2 dAdx$$

$$= \int_0^L \frac{EA}{2}\left(\frac{\partial u}{\partial x}\right)^2 dx \tag{b}$$

The kinetic energy of the element is

$$dT = \frac{1}{2}(\rho A dx)\left(\frac{\partial u}{\partial t}\right)^2 \tag{c}$$

The total kinetic energy is the kinetic energy of the bar plus the kinetic energy of the particle:

$$T = \int_0^L \frac{\rho A}{2}\left(\frac{\partial u}{\partial t}\right)^2 dx + \frac{1}{2}m\left[\frac{\partial u}{\partial t}(L,t)\right]^2 \tag{d}$$

Substitution of Equation c and Equation d into Hamilton's Principle leads to

$$\delta\int_{t_1}^{t_2}\left\{\int_0^L \frac{\rho A}{2}\left(\frac{\partial u}{\partial t}\right)^2 dx + \frac{1}{2}m\left[\frac{\partial u}{\partial t}(L,t)\right]^2 - \int_0^L \frac{EA}{2}\left(\frac{\partial u}{\partial x}\right)^2 dx\right\}dt = 0 \tag{e}$$

The calculus of variations shows that it is allowable to interchange the order of the variation and the integration, that is, $\delta(\partial u/\partial t)^2 = 2(\partial u/\partial t)[\delta(\partial u/\partial t)]$, and $\delta(\partial u/\partial t) = \partial/\partial t(\delta u)$. Thus, the first integral in Equation e becomes

$$\delta\int_{t_1}^{t_2}\int_0^L \frac{\rho A}{2}\left(\frac{\partial u}{\partial t}\right)^2 dxdt = \int_{t_1}^{t_2}\int_0^L \rho A\frac{\partial u}{\partial t}\frac{\partial}{\partial t}(\delta u)dxdt \tag{f}$$

Interchanging the order of integration and applying integration by parts, $\int f dg = fg - \int g df$ with $f = \partial u/\partial t$ and $dg = \partial/\partial t(\delta u)dt$, leading to

$$\delta\int_{t_1}^{t_2}\int_0^L \frac{\rho A}{2}\left(\frac{\partial u}{\partial t}\right)^2 dxdt = \int_0^L \rho A\left\{\frac{\partial u}{\partial t}\delta u\Big|_{t=t_1}^{t=t_2} - \int_{t_1}^{t_2}\left(\frac{\partial^2 u}{\partial t^2}\delta u dt\right)\right\}dx \tag{g}$$

Remembering that $\delta u(t_1) = 0$ and $\delta u(t_2) = 0$ and again interchanging the order of integration, Equation g becomes

$$\delta \int_{t_1}^{t_2} \int_0^L \frac{\rho A}{2} \left(\frac{\partial u}{\partial t} \right)^2 dxdt = -\int_{t_1}^{t_2} \int_0^L \rho A \frac{\partial^2 u}{\partial t^2} \delta u dxdt \tag{h}$$

Next, consider the contribution of the kinetic energy of the discrete particle. Using the properties mentioned earlier, from the calculus of variations and integration by parts, this term becomes:

$$\delta \int_{t_1}^{t_2} \frac{1}{2} m \left[\frac{\partial u}{\partial t}(L,t) \right]^2 dt = \int_{t_1}^{t_2} m \frac{\partial^2 u}{\partial t^2}(L,t) \delta u(L,t) dt \tag{i}$$

The potential-energy term can be written as

$$\delta \int_{t_1}^{t_2} \int_0^L \frac{EA}{2} \left(\frac{\partial u}{\partial x} \right)^2 dxdt = \int_{t_1}^{t_2} \int_0^L EA \frac{\partial u}{\partial x} \frac{\partial}{\partial x} (\delta u) dxdt \tag{j}$$

Application of integration by parts to the inner integral of the right-hand side of Equation j with $f = \partial u / \partial x$ and $dg = (\partial / \partial x)(\delta u) dx$ leads to $[\partial(\delta u)/\partial x] dx$

$$\delta \int_{t_1}^{t_2} \int_0^L \frac{EA}{2} \left(\frac{\partial u}{\partial x} \right)^2 dxdt = \int_{t_1}^{t_2} \left\{ EA \frac{\partial u}{\partial x}(L,t) \delta u(L,t) - EA \frac{\partial u}{\partial x}(0,t) \delta u(0,t) \right.$$

$$\left. - \int_0^L EA \frac{\partial^2 u}{\partial x^2} \delta u dx \right\} dt \tag{k}$$

Combining Equation e, Equation h, Equation i, and Equation k leads to

$$\int_{t_1}^{t_2} \left\{ - \left[EA \frac{\partial u}{\partial x}(L,t) + m \frac{\partial^2 u}{\partial t^2}(L,t) \right] \delta u(L,t) + EA \frac{\partial u}{\partial x}(0,t) \delta u(L,t) \right.$$

$$\left. + \int_0^L \left[EA \frac{\partial^2 u}{\partial x^2} - \rho A \frac{\partial^2 u}{\partial t^2} \right] \delta u \right\} dt = 0 \tag{l}$$

Equation l must be true for any varied path (for all possible δu). This implies that

$$EA \frac{\partial^2 u}{\partial x^2} - \rho A \frac{\partial^2 u}{\partial t^2} = 0 \tag{m}$$

Equation l also implies that either $EA(\partial u/\partial t)(0,t)=0$ or $\delta u(0,t)=0$ and that either $EA(\partial u/\partial x)(L,t)+m(\partial^2 u/\partial t^2)(L,t)=0$ or $\delta u(L,t)=0$. These choices specify possible boundary conditions. For a bar fixed at $x=0$ and with the discrete mass attached at $x=L$, the appropriate choices are:

$$u(0,t)=0 \tag{n}$$

$$EA\frac{\partial u}{\partial x}(L,t)+m\frac{\partial^2 u}{\partial t^2}(L,t)=0 \tag{o}$$

1.4 Nondimensionalization

The motion of the one-degree-of-freedom mass-spring and viscous-damper system shown in Figure 1.20 is governed by the differential equation

$$m\ddot{x}+c\dot{x}+kx=F_0\sin(\omega t) \tag{1.10}$$

The general solution of Equation 1.10 is the sum of a homogeneous $x_h(t)$, the solution of Equation 1.10 obtained if $F_0=0$, and a particular solution, $x_p(t)$, the solution corresponding to the specific function on the right-hand side of Equation 1.10. The general solution is:

$$x(t)=e^{-\zeta\omega_n t}\left[C_1\cos\left(\omega_n\sqrt{1-\zeta^2}t\right)+C_2\sin\left(\omega_n\sqrt{1-\zeta^2}t\right)\right]+X\sin(\omega t+\phi) \tag{1.11}$$

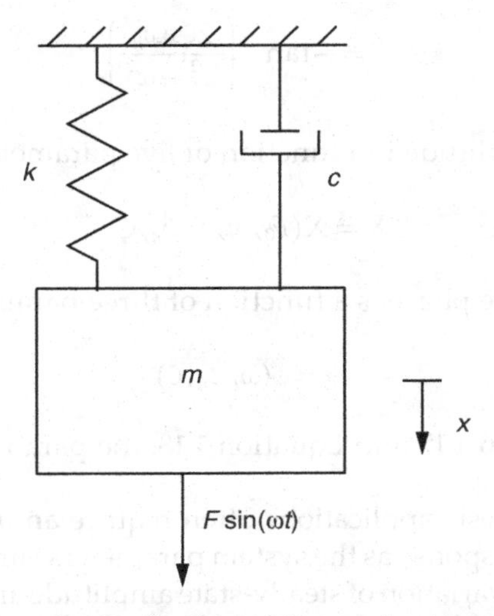

Figure 1.20 The steady-state amplitude and the steady-state phase angle of the mass-spring-viscous-damper system are each a function of five dimensional parameters.

where C_1 and C_2 are constants of integration to be determined through application of initial conditions,

$$\omega_n = \sqrt{\frac{k}{m}} \tag{1.12}$$

is the system's natural frequency, and

$$\zeta = \frac{c}{2m\omega_n} \tag{1.13}$$

is the system's damping ratio. Equation 1.11 is derived assuming $0<\zeta<1$. As t grows large, the particular solution becomes exponentially small and negligible compared with the particular solution. The steady-state response is defined as

$$\begin{aligned} x_{ss}(t) &= \lim_{t\to\infty} x(t) \\ &= X\sin(\omega t + \phi) \end{aligned} \tag{1.14}$$

where the steady-state amplitude is determined as

$$X = \frac{F_0}{m\sqrt{\left(\omega_n^2 - \omega^2\right)^2 + (2\zeta\omega\omega_n)^2}} \tag{1.15}$$

and the steady-state phase is

$$\phi = -\tan^{-1}\left(\frac{2\zeta\omega\omega_n}{\omega_n^2 - \omega^2}\right) \tag{1.16}$$

The steady-state amplitude is a function of five parameters,

$$X = X(F_0, m, \omega, \omega_n, \zeta) \tag{1.17}$$

while the steady-state phase is a function of three parameters,

$$\phi = \phi(\omega, \omega_n, \zeta) \tag{1.18}$$

Note that in Equation 1.17 and Equation 1.18, the parameters ω_n and ζ could be replaced by k and c.

Design and synthesis applications often require an understanding of the behavior of system response as the system parameters vary. The term *frequency response* refers to the variation of steady-state amplitude and steady-state phase with ω, the frequency of the excitation. It is usually easier to understand these variations when the relations are written in a nondimensional form. To this end, the frequency ratio is defined as

$$r = \frac{\omega}{\omega_n} \tag{1.19}$$

Use of Equation 1.19 in the form of $\omega = r\omega_n$ in Equation 1.15 and Equation 1.16 leads to:

$$\frac{m\omega_n^2 X}{F_0} = \frac{1}{\sqrt{(1-r^2)^2 + (2\zeta r)^2}} \tag{1.20}$$

$$\phi = -\tan^{-1}\left(\frac{2\zeta r}{1-r^2}\right) \tag{1.21}$$

The left-hand side of Equation 1.20 is a nondimensional parameter, often called the magnification factor, and is a function of the nondimensional parameters r and ζ:

$$\frac{m\omega_n^2 X}{F_0} = M(r,\zeta)$$

$$= \frac{1}{\sqrt{(1-r^2)^2 + (2\zeta r)^2}} \tag{1.22}$$

Noting that for a given system, the steady-state amplitude is proportional to the magnification factor, Equation 1.22 can be used to understand the frequency response. Equation 1.22 can be used algebraically to determine the limits of the steady-state amplitude for small and large frequency ratios, to determine the maximum value of X for a specific value of ζ, and to develop the plot shown in Figure 1.21, which summarizes the frequency response using a single set of curves. Using Figure 1.21 and Equation 1.22, the following can be determined regarding the frequency response:

- $M(0, \zeta) = 1$
- $M(r, \zeta)$ is asymptotic to $1/r^2$ large r
- $M(r, \zeta)$ has a maximum value of $1/\left(2\zeta\sqrt{1-\zeta^2}\right)$ at $r = \sqrt{1-2\zeta^2}$ if $\zeta < 1/\sqrt{2}$
- $M(r, \zeta)$ has no extrema if $\zeta > 1/\sqrt{2}$

The variation of the steady-state phase with frequency ratio is plotted in Figure 1.22. The following can be determined from Equation 1.21 and Figure 1.22:

- $\tan\phi < 0$ and $-\pi < \phi < -\pi/2$ for $r < 1$
- $\phi = -\pi/2$ for $r = 1$

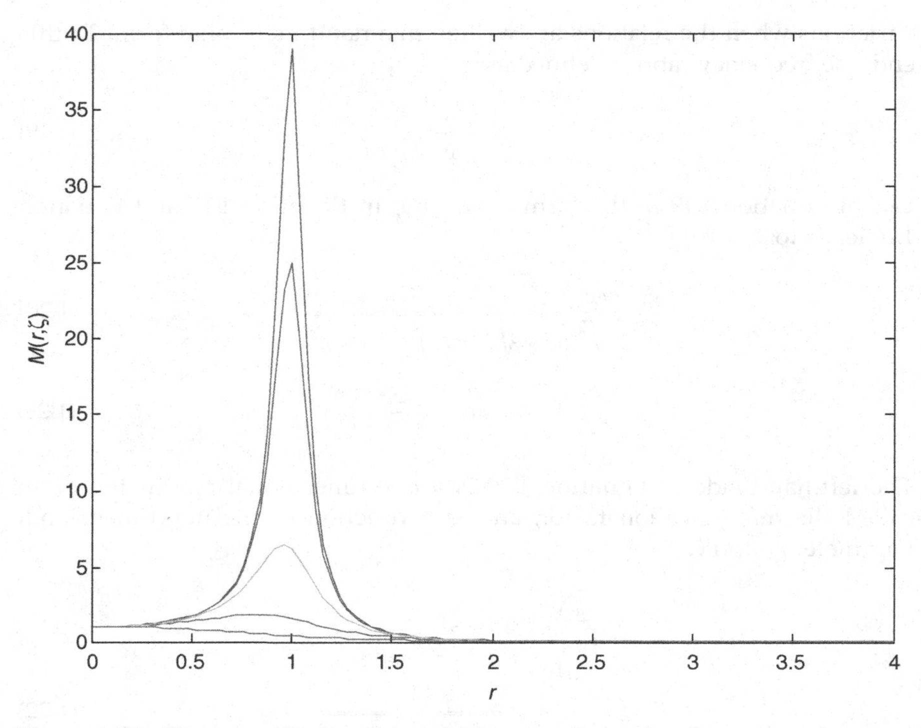

Figure 1.21 The magnification factor M is a nondimensional parameter proportional to the steady-state amplitude of a mass-spring-viscous-damper system. The magnification factor is a function of two dimensionless parameters, the frequency ratio r and the damping ratio ζ.

- $\tan\phi > 0$ and $-\pi/2 < \phi < 0$ for $r > 1$
- $\phi \to 0$ for large r

The frequency response of a mechanical system is an example where it is convenient to nondimensionalize a derived relationship between an output variable and the system parameters. In the above example, the steady-state amplitude and steady-state phase are the output parameters. It is assumed that the dimensional parameters (natural frequency, input frequency, excitation amplitude, mass, and damping ratio) vary independently of one another. An objective of nondimensionalization is to determine a set of nondimensional parameters such that a nondimensional parameter involving the output variable is represented as a function of independent dimensionless parameters. This is easy to do for the frequency response of a mechanical system when a dimensional relation between variables has been derived.

Dimensional analysis is used to determine empirically the relationship between a dependent parameter and the independent parameters on which it depends. A familiar example is the determination of the variation of a drag

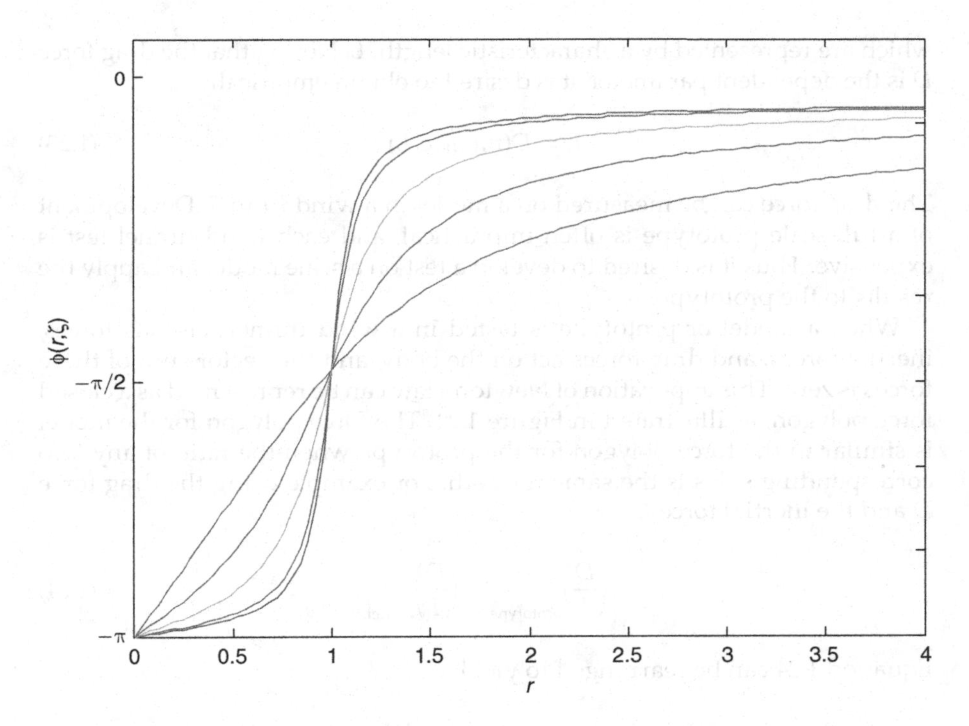

Figure 1.22 The steady-state phase angle is itself nondimensional and is a function of the nondimensional frequency ratio and the nondimensional damping ratio.

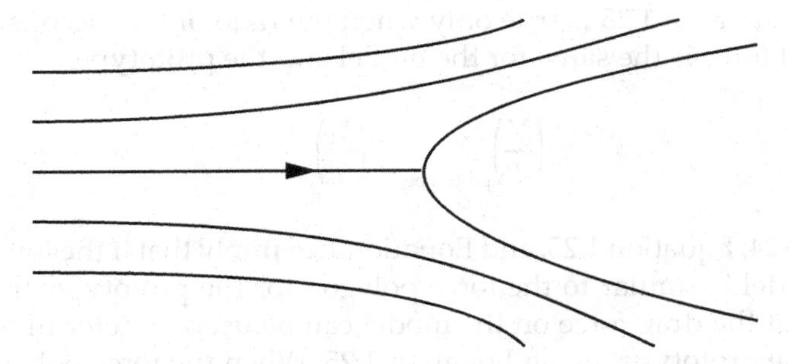

Figure 1.23 The drag force acting on a bluff body placed in a uniform free stream is dependent on the free-stream velocity, the density and dynamic viscosity of the fluid, the speed of sound in the fluid, and a length scale. The relationship can be reformulated nondimensionally as the drag coefficient as a function of Mach number and Reynolds number.

force on a bluff body, as shown in Figure 1.23, with parameters of the fluid flow and geometric parameters of the body. These parameters may include the velocity of the flow, v, the mass density of the fluid, ρ, the dynamic viscosity of the fluid, μ, the speed of sound in the fluid, c, and the geometric quantities which are represented by a characteristic length, L. Noting that the drag force D is the dependent parameter, it is desired to obtain empirically

$$D = D(v, \rho, \mu, c, L) \tag{1.23}$$

The drag force can be measured on a model in a wind tunnel. Development of a full-scale prototype is often impractical, and each wind tunnel test is expensive. Thus it is desired to develop a test on a scale model and apply the results to the prototype.

When a model or prototype is tested in a wind tunnel, viscous forces, inertial forces, and drag forces act on the body, and the vector sum of these forces is zero. This application of Newton's law can be represented as a closed force polygon, as illustrated in Figure 1.24. The force polygon for the model is similar to the force polygon for the prototype when the ratio of any two corresponding sides is the same for both. For example using the drag force D and the inertial force I,

$$\left(\frac{D}{I}\right)_{\text{prototype}} = \left(\frac{D}{I}\right)_{\text{model}} \tag{1.24}$$

Equation 1.24 can be rearranged to yield

$$D_{\text{prototype}} = I_{\text{prototype}}\left(\frac{D}{I}\right)_{\text{model}} \tag{1.25}$$

However, Equation 1.25 is true only when the ratio of the viscous force V to the inertial force is the same for the model and the prototype:

$$\left(\frac{V}{I}\right)_{\text{prototype}} = \left(\frac{V}{I}\right)_{\text{model}} \tag{1.26}$$

Equation 1.24, Equation 1.25, and Equation 1.26 imply that if the force polygon for the model is similar to the force polygon for the prototype, then a measurement of the drag force on the model can be used to determine the drag force on the prototype, using Equation 1.25. When the force polygons of the

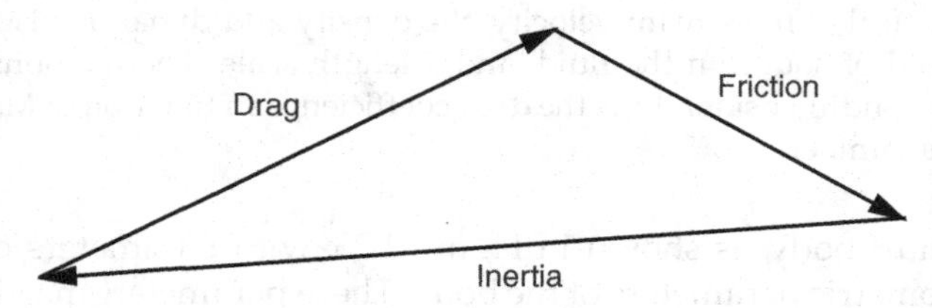

Figure 1.24 Vector addition of the drag force, the friction force, and the inertia force acting on a bluff body forms a closed polygon.

model and the prototype are similar, the model and prototype are said to exhibit dynamic similitude.

When running an experiment on a model that is used to predict the performance of the prototype, kinematic similitude must also be satisfied; meaning that the ratio of all time scales for the model and for the prototype must be the same. Kinematic similitude is usually enforced by requiring velocity ratios to be equivalent:

$$\left(\frac{v}{c}\right)_{\text{model}} = \left(\frac{v}{c}\right)_{\text{prototype}} \tag{1.27}$$

Equation 1.24, Equation 1.25, Equation 1.26, and Equation 1.27 lead to the conclusion that

$$\frac{D}{V} = f\left(\frac{V}{I}, \frac{v}{c}\right) \tag{1.28}$$

A nondimensional form of Equation 1.28 is

$$C_\text{D} = f(\text{Re}, M) \tag{1.29}$$

where C_D, the drag coefficient, is the ratio of the drag force to the inertial force,

$$C_\text{D} = \frac{D}{\frac{1}{2}\rho v^2 L^2} \tag{1.30}$$

Re, the Reynolds number, is the ratio of the viscous forces to the inertial forces,

$$\text{Re} = \frac{\rho v L}{\mu} \tag{1.31}$$

and M, the Mach number, is

$$M = \frac{v}{c} \tag{1.32}$$

The previous discussions illustrate the value of nondimensionalization. A derived mathematical relation between the steady-state amplitude and system parameters was nondimensionalized to provide a better understanding of the steady-state behavior. The fluid flow problem was nondimensionalized so that empirical results on a model could be used to predict the performance of a prototype. The functional relation between the nondimensional dependent variables and the nondimensional independent variables is determined from empirical results.

The process of mathematical modeling is used to develop a mathematical problem relating output variables to independent variables in terms of parameters. Such problems can be nondimensionalized through introduction of nondimensional independent and dependent variables. The process leads to determination of a set of nondimensional parameters. The physical meanings of the parameters are obtained from the nondimensional problem. The choice of nondimensional variables and nondimensional parameters is not unique. The process of nondimensionalizing a mathematical problem is illustrated in the following examples.

Example 1.6 The differential equation governing the motion of a one-degree-of-freedom mass-spring and viscous damper system is Equation 1.10, repeated following:

$$m\ddot{x}+c\dot{x}+kx=F_0\sin(\omega t) \tag{a}$$

Nondimensionalize Equation a using the following nondimensional variables:

$$x^*=\frac{x}{F_0/k} \tag{b}$$

$$t^*=\omega_n t=\sqrt{\frac{k}{m}}\,t \tag{c}$$

Solution Note that $x^*$and t^* are nondimensional. The choices of $x^*$and t^* are not unique. For example, an alternative choice to nondimensionalize time is $t^*=\omega t$.

Derivatives with respect to dimensional independent variables are converted to derivatives with respect to nondimensional independent variables through the chain rule:

$$\frac{d}{dt}=\frac{d}{dt^{t}}\frac{dt^*}{dt}=\omega_n\frac{d}{dt^*} \tag{d}$$

$$\frac{d^2}{dt^2}=\frac{d}{dt}\left(\frac{d}{dt}\right)=\omega_n\frac{d}{dt^*}\left(\omega_n\frac{d}{dt^*}\right)=\omega_n^2\frac{d^2}{dt^{*2}} \tag{e}$$

Substituting Equation b, Equation c, Equation d, and Equation e into Equation a leads to

$$m\omega_n^2\frac{d^2}{dt^{*2}}\left(\frac{F_0}{k}x^*\right)+c\omega_n\frac{d}{dt^*}\left(\frac{F_0}{k}x^*\right)+k\left(\frac{F_0}{k}x^*\right)=F_0\sin\left(\omega\frac{t^*}{\omega_n}\right) \tag{f}$$

Simplification of Equation f, noting that from Equation (c), $\omega_n=\sqrt{k/m}$ leads to

$$\frac{d^2x^*}{dt^{*2}}+\left(\frac{c\omega_n}{k}\right)\frac{dx^*}{dt^*}+x^*=\sin\left(\frac{\omega}{\omega_n}t^*\right) \tag{g}$$

Note that

$$\frac{c\omega_n}{k} = \frac{c}{\sqrt{km}} = 2\zeta \tag{h}$$

and that $r = \omega/\omega_n$, where ζ and r are the nondimensional frequency and damping ratios. Thus Equation f can be rewritten as:

$$\frac{d^2x}{dt^2} + 2\zeta\frac{dx}{dt} + x = \sin(rt) \tag{i}$$

It is conventional to drop the superscript * from dimensionless quantities and understand that all variables and parameters are henceforth dimensionless.

When solved, Equation i gives a mathematical representation of the nondimensional displacement in terms of two nondimensional parameters, the damping ratio and the frequency ratio. The steady-state amplitude is obtained by taking the limit $t \to \infty$. This leads to $X^* = f(r, \zeta)$. However, from Equation b, $X^* = kX/F_0 = m\omega_n^2/F_0 = M$. Hence the relation among the magnification factor, the damping ratio, and the frequency ratio can be obtained using a priori nondimensionalization of the governing differential equation.

The choices for defining nondimensional variables are not unique. For example, time could be nondimensionalized using the input frequency rather than the natural frequency, $t^* = \omega t$, in which case the nondimensional differential equation becomes:

$$\ddot{x} + \frac{c}{m\omega}\dot{x} + \frac{k}{m\omega^2}x = \frac{k}{m\omega^2}\sin(t) \tag{j}$$

where all variables are nondimensional. Equation j suggests the definition of two nondimensional parameters, $\pi_1 = k/m\omega^2$ and $\pi_2 = c/m\omega$. It should be noted that $\pi_1 = 1/r^2$ and $\pi_2 = 2\zeta/r$. The two sets of dimensionless parameters are related. Equation j is obtained by dividing the differential equation obtained after changing from dimensional to nondimensional variables by the coefficient of the inertial term. This suggests that π_2 is the ratio of the damping forces to the inertial forces. It is, but on a different time scale than ζ.

Example 1.7 A sphere of radius R is initially at a uniform temperature T_0 when it is plunged into a bath of temperature T_1, as illustrated in Figure 1.25. The sphere is made of a material of specific heat c, mass density ρ, and thermal conductivity k. The heat transfer coefficient between the sphere and the bath is h. The temperature distribution is assumed to be axisymmetric and thus to depend only on the distance from the center of the sphere, r, and on time, t; $T(r,t)$. The partial differential equation governing the unsteady state heat conduction in the sphere is:

$$\rho c\frac{\partial T}{\partial t} = \frac{k}{r^2}\frac{\partial}{\partial r}\left(r^2\frac{\partial T}{\partial r}\right) \tag{a}$$

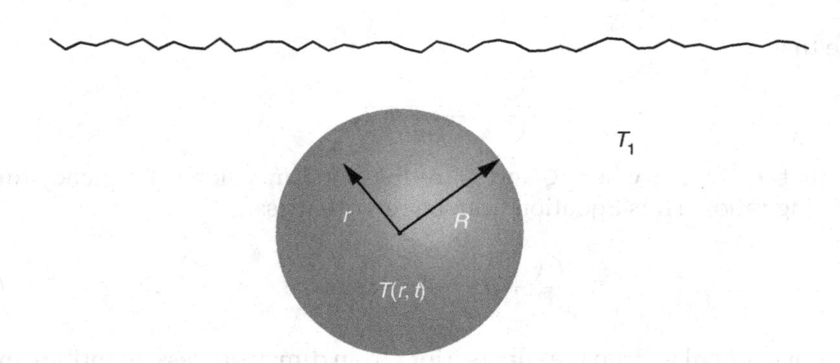

Figure 1.25 The sphere is at a uniform temperature when immersed in a bath at a different temperature. Heat transfer by convection through the surface of the sphere leads to a unsteady state temperature distribution throughout the sphere. The nondimensional temperature is dependent upon the nondimensional Biot number, which is a ratio of the rate of heat transfer by convection to the rate of heat transfer by conduction.

The initial condition is:

$$T(r,0) = T_0 \tag{b}$$

The boundary condition at the outer radius is obtained, using an energy balance on the surface of the sphere, as:

$$k\frac{\partial T}{\partial r}(R,t) = -h[T(R,t) - T_1] \tag{c}$$

The radial coordinate varies over the range $0 \le r \le R$. The temperature is required to remain finite at $r = 0$.

Introduce nondimensional variables to rewrite Equation a, Equation b, and Equation c in nondimensional forms. Identify relevant nondimensional parameters.

Solution The radius of the sphere is a characteristic length, and thus a nondimensional radial coordinate can be defined as

$$r^* = \frac{r}{R} \tag{d}$$

There is no parameter with the time dimension, and it is not clear what an appropriate nondimensionalization of the time variable would be. Let A be a combination of parameters with the time dimension, and define

$$t^* = \frac{t}{A} \tag{e}$$

The form of A will be chosen for convenience.

Temperature is a thermodynamic potential. Heat transfer occurs because of differences in temperature. Therefore, in nondimensionalizing the dependent variable of temperature, it is convenient to use a ratio of temperature differences. To this end, a nondimensional temperature, $\Theta(r^*,t^*)$, can be defined as

$$\Theta = \frac{T - T_1}{T_0 - T_1} \tag{f}$$

The chain rule is used to obtain

$$\frac{\partial}{\partial r} = \frac{\partial}{\partial r^*}\frac{dr^*}{dr} = \frac{1}{R}\frac{\partial}{\partial r^*} \tag{g}$$

$$\frac{\partial}{\partial t} = \frac{\partial}{\partial t^*}\frac{dt^*}{dt} = \frac{1}{A}\frac{\partial}{\partial t^*} \tag{h}$$

Substitution of Equation d, Equation e, Equation f, Equation g, and Equation h into Equation a leads to

$$\frac{\rho c}{A}\frac{\partial}{\partial t^*}[(T_0 - T_1)\Theta + T_1] = \frac{k}{R^2 r^{*2}}\frac{1}{R}\frac{\partial}{\partial r^*}\left\{R^2 r^{*2}\frac{1}{R}\frac{\partial}{\partial r^*}[(T_0 - T_1)\Theta + T_1]\right\} \tag{i}$$

Simplification of Equation i results in

$$\frac{\partial \Theta}{\partial t^*} = \frac{kA}{\rho c R^2}\frac{1}{r^{*2}}\frac{\partial}{\partial r^*}\left(r^{*2}\frac{\partial \Theta}{\partial r^*}\right) \tag{j}$$

For convenience, choose A such that

$$\frac{kA}{\rho c R^2} = 1$$

$$A = \frac{\rho c R^2}{k} \tag{k}$$

Substituting Equation k into Equation j and dropping the *s from nondimensional variables leads to

$$\frac{\partial \Theta}{\partial t} = \frac{1}{r^2}\frac{\partial}{\partial r}\left(r^2\frac{\partial \Theta}{\partial r}\right) \tag{l}$$

Substitution of Equation f into initial condition b leads to

$$\Theta(r,0) = 1 \tag{m}$$

Noting that $r=R$ corresponds to $r^*=1$, substitution of Equation d, Equation e, Equation f, Equation g, and Equation h into boundary condition c gives

$$\frac{(T_0-T_1)k}{R}\frac{\partial\Theta}{\partial r}(1,t)=-h\Theta(1,t)(T_0-T_1)$$

$$\frac{\partial\Theta}{\partial r}(1,t)=-Bi\Theta(1,t) \tag{n}$$

where the Biot number is defined as

$$Bi=\frac{hR}{k} \tag{o}$$

and is the ratio of the rate of heat transfer by convection to the rate of heat transfer by conduction on the surface of the sphere.

The nondimensional formulation for the unsteady state heat transfer in the sphere is the partial differential equation, Equation l, the initial condition, Equation m, and the boundary condition, Equation o. In each of these equations, it is understood that the variables are dimensionless.

Example 1.8 One continuum model of multi-walled carbon nanotubes is that of concentric Euler-Bernoulli beams connected by elastic layers which model the van der Waals forces between the layers of atoms. Manufacturing often requires carbon nanotubes to be subject to a tensile force and surrounded by a polymer gel. Consider a double-walled nanotube modeled by concentric Euler-Bernoulli beams. The tubes are each of length L. Define x such that $0\le x\le L$ is a coordinate along the centerline of the tubes, as illustrated in Figure 1.26. Using the subscript "1" to refer to properties of the inner tube and the subscript "2" to refer to properties of the outer tube, the partial differential equations governing the transverse displacements of the tubes, $w_1(x,t)$ and $w_2(x,t)$ are

$$E_1I_1\frac{\partial^4 w_1}{\partial x^4}-P\frac{\partial^2 w_1}{\partial x^2}+k_1(w_1-w_2)+\rho_1A_1\frac{\partial^2 w_1}{\partial t^2} \tag{a}$$

$$E_2I_2\frac{\partial^4 w_2}{\partial x^4}-P\frac{\partial^2 w_2}{\partial x^2}+k_1(w_2-w_1)+k_2w_2+\rho_2A_2\frac{\partial^2 w_2}{\partial t^2} \tag{b}$$

where E is the elastic modulus of the tube (usually 1 TPa), ρ is the mass density of the nanotube (approximately 1.3 g/cm^3), A is the cross-sectional area (the radius of a carbon atom is 0.34 nm), I is the cross-sectional moment of inertia, P is the tensile force, k_1 is the stiffness per unit length used to model the van der Waals forces, and k_2 is the stiffness per unit length of the surrounding polymer gel.

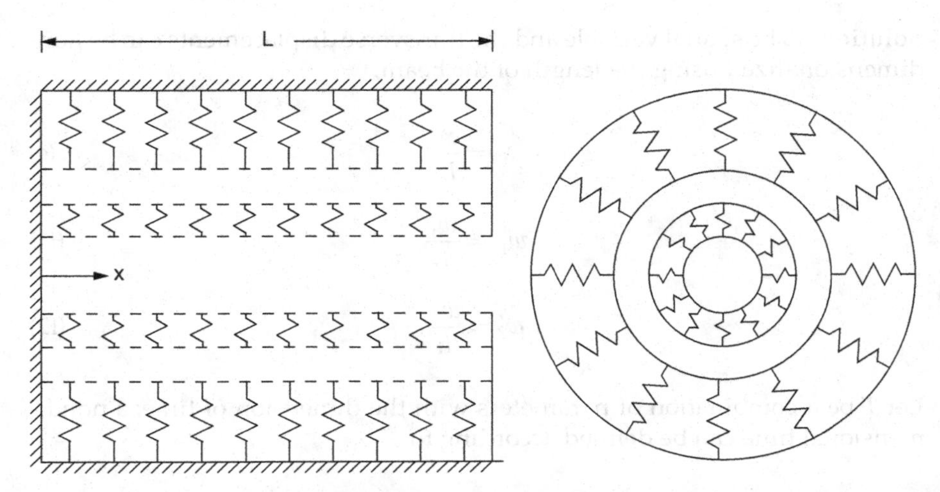

Figure 1.26 A continuum model for the displacement of carbon nanotubes uses concentric Euler-Bernoulli beams connected by elastic layers which model the interatomic van der Waals forces.

If the tubes are each fixed at $x=0$ and free at $x=L$, the appropriate boundary conditions are:

$$w_1(0,t)=0 \tag{c1}$$

$$\frac{\partial w_1}{\partial x}(0,t)=0 \tag{c2}$$

$$\frac{\partial^2 w_1}{\partial x^2}(L,t)=0 \tag{c3}$$

$$E_1 I_1 \frac{\partial^3 w_1}{\partial x^3}(L,t)+P\frac{\partial w_1}{\partial x}(L,t)=0 \tag{c4}$$

and

$$w_2(0,t)=0 \tag{d1}$$

$$\frac{\partial w_2}{\partial x}(0,t)=0 \tag{d2}$$

$$\frac{\partial^2 w_2}{\partial x^2}(L,t)=0 \tag{d3}$$

$$E_2 I_2 \frac{\partial^3 w_2}{\partial x^3}(L,t)+P\frac{\partial w_2}{\partial x}(L,t)=0 \tag{d4}$$

Nondimensionalize Equation a, Equation b, Equation c, and Equation d.

Solution The spatial variable and the transverse displacements can be nondimensionalized using the length of the beam,

$$x^* = \frac{x}{L} \tag{e}$$

$$w_1^* = \frac{w_1}{L} \tag{f1}$$

$$w_2^* = \frac{w_2}{L} \tag{f2}$$

Let T be a combination of parameters with the dimension of time; a nondimensional time can be defined according to

$$t^* = \frac{t}{T} \tag{g}$$

Substitution of Equation e, Equation f, and Equation g into Equation a and Equation b gives

$$\frac{E_1I_1}{L^4}\frac{\partial^4}{\partial x^{*4}}(Lw_1^*) - \frac{P}{L^2}\frac{\partial^2}{\partial x^{*2}}(Lw_1^*) + k_1L(w_1^* - w_2^*) + \frac{\rho_1A_1}{T^2}\frac{\partial^2}{\partial t^{*2}}(Lw_1^*) = 0 \tag{h}$$

and

$$\frac{E_2I_2}{L^4}\frac{\partial^4}{\partial x^{*4}}(Lw_2^*) - \frac{P}{L^2}\frac{\partial^2}{\partial x^{*2}}(Lw_2^*) + k_1L(w_2^* - w_1^*)$$
$$+ k_2Lw_2^* + \frac{\rho_2A_2}{T^2}\frac{\partial^2}{\partial t^{*2}}(Lw_2^*) = 0 \tag{i}$$

Rearranging Equation h leads to

$$\frac{\partial^4 w_1^*}{\partial x^{*4}} - \frac{PL^2}{E_1I_1}\frac{\partial^2 w_1^*}{\partial x^{*2}} + \frac{k_1L^4}{E_1I_1}(w_1^* - w_2^*) + \frac{\rho_1A_1L^4}{E_1I_1T^2}\frac{\partial^2 w_1^*}{\partial t^{*2}} = 0 \tag{j}$$

For convenience, choose

$$\frac{\rho_1A_1L^4}{E_1I_1T^2} = 1$$

$$T = L^2\sqrt{\frac{\rho_1A_1}{E_1I_1}} \tag{k}$$

and define

$$\varepsilon = \frac{PL^2}{E_1 I_1} \tag{l}$$

$$\eta_1 = \frac{k_1 L^4}{E_1 I_1} \tag{m}$$

Using Equation k, Equation l, and Equation m, Equation h becomes

$$\frac{\partial^4 w_1}{\partial x^4} - \varepsilon \frac{\partial^2 w_1}{\partial x^2} + \eta_1 (w_1 - w_2) + \frac{\partial^2 w_1}{\partial t^2} = 0 \tag{n}$$

where the * has been dropped from nondimensional variables. Substitution of Equation k, Equation l, and Equation m into Equation i leads to

$$\frac{E_2 I_2}{L^4} \frac{\partial^4}{\partial x^{*4}} (L w_2^*) - \varepsilon \frac{E_1 I_1}{L^4} \frac{\partial^2}{\partial x^{*2}} (L w_2^*) + \eta_1 \frac{E_1 I_1}{L^3} (w_2^* - w_1^*) + k_2 L w_2^*$$
$$+ \frac{\rho_2 A_2 E_1 I_1}{L^4 \rho_1 A_1} \frac{\partial^2}{\partial t^{*2}} (L w_2^*) = 0 \tag{o}$$

Multiplication of Equation o by $L^3 / E_1 I_1$ and dropping the * from nondimensional variables leads to

$$\mu \frac{\partial^4 w_2}{\partial x^4} - \varepsilon \frac{\partial^2}{\partial x^2} (w_2) + \eta_1 (w_2 - w_1) + \eta_2 w_2 + \beta \frac{\partial^2 w_2}{\partial t^2} = 0 \tag{p}$$

where

$$\mu = \frac{E_2 I_1}{E_1 I_1} \tag{q}$$

$$\beta = \frac{\rho_2 A_2}{\rho_1 A_1} \tag{r}$$

$$\eta_2 = \frac{k_2 L^4}{E_1 I_1} \tag{s}$$

Substitution of Equation e, Equation f, and Equation l into Equation c4 leads to

$$\frac{E_1 I_1}{L^2} \frac{\partial^3 w_1^*}{\partial x^{*3}} (1,t) + P \frac{\partial w_1^*}{\partial x^*} (1,t) = 0$$
$$\frac{\partial^3 w_1}{\partial x^3} (1,t) + \varepsilon \frac{\partial w_1}{\partial x} (1,t) = 0 \tag{t}$$

where the * has been dropped from nondimensional variables. Substitution of Equation e, Equation f, Equation l, and Equation q into Equation d4 leads to

$$\mu\frac{\partial^3 w_2}{\partial x^3}(1,t)+\varepsilon\frac{\partial w_2}{\partial x}(1,t)=0 \tag{u}$$

The nondimensional boundary conditions can be summarized by

$$w_1(0,t)=0 \tag{v1}$$

$$\frac{\partial w_1}{\partial x}(0,t)=0 \tag{v2}$$

$$\frac{\partial^2 w_1}{\partial x^2}(1,t)=0 \tag{v3}$$

$$\frac{\partial^3 w_1}{\partial x^3}(1,t)+\varepsilon\frac{\partial w_1}{\partial x}(1,t)=0 \tag{v4}$$

$$w_2(0,t)=0 \tag{w1}$$

$$\frac{\partial w_2}{\partial x}(0,t)=0 \tag{w2}$$

$$\frac{\partial^2 w_2}{\partial x^2}(1,t)=0 \tag{w3}$$

$$\mu\frac{\partial^3 w_2}{\partial x^3}(1,t)+\varepsilon\frac{\partial w_2}{\partial x}(1,t)=0 \tag{w4}$$

The nondimensional problem is comprised of the partial differential equations, Equation n and Equation p, and the eight boundary conditions given by Equation v and Equation w.

One value of nondimensionalization is to determine the order of magnitude of physical effects. The fourth-order spatial derivatives arise from normal bending stresses, the second-order spatial derivatives occur because of the normal stress due to the axial load, the second-order time derivatives are the inertial terms, and the zeroth-order derivatives occur because of the elastic layer between the tubes and surrounding. Since the parameter ε is obtained as the coefficient of the second-order spatial derivative terms after dividing by the coefficient of the fourth-order spatial derivative, it represents the ratio of the normal stresses due to axial load to the normal stresses due to bending. If $\varepsilon \ll 1$, the system is "lightly stretched." If $\varepsilon \gg 1$, the beams are highly stretched.

Similarly, if $\eta \ll 1$, the beams are lightly coupled. Asymptotic solutions as expansions in terms of powers of these independent dimensionless parameters can be used to approximate the natural frequencies and mode shapes.

Analysis of the highly stretched case is best performed with a different nondimensionalization, as illustrated in the Example 1.9.

Example 1.9 The differential equation for the transverse displacement of a nonuniform beam of length L subject to an axial load P, as illustrated in Figure 1.27, is

$$\frac{\partial^2}{\partial x^2}\left(EI(x)\frac{\partial^2 w}{\partial x^2}\right) - P\frac{\partial^2 w}{\partial x^2} + \rho A(x)\frac{\partial^2 w}{\partial t^2} = 0 \tag{a}$$

(a) Define nondimensional variables such that the nondimensional equation for the transverse displacement is

$$\varepsilon\frac{\partial^2}{\partial x^2}\left(\alpha(x)\frac{\partial^2 w}{\partial x^2}\right) - \frac{\partial^2 w}{\partial x^2} + \beta(x)\frac{\partial^2 w}{\partial t^2} = 0 \tag{b}$$

(b) A normal-mode solution of Equation b is of the form

$$w(x,t) = u(x)e^{i\omega t} \tag{c}$$

where ω represents a natural frequency of the system and $u(x)$ is its corresponding mode shape. Substitute Equation c into Equation b to obtain an ordinary differential equation for $u(x)$ with ω as a parameter.

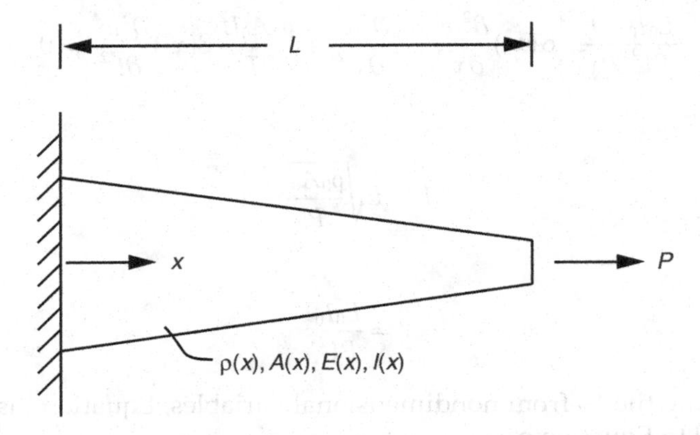

Figure 1.27 The nonuniform beam is subject to a constant axial load. When the governing equation is nondimensionalized, the highest-order spatial derivative is multiplied by a parameter defined as the ratio of the normal stresses due to axial load to the normal stresses due to bending.

Solution Define the nondimensional variables,

$$w^* = \frac{w}{L} \tag{d}$$

$$x^* = \frac{x}{L} \tag{e}$$

$$t^* = \frac{t}{T} \tag{f}$$

where T is a characteristic time whose form is to be determined. Using the subscript "0" to refer to the value of a function at $x=0$, the following nondimensional functions can be defined:

$$\alpha(x^*) = \frac{E(x)I(x)}{E_0 I_0} \tag{g}$$

$$\beta(x^*) = \frac{\rho(x)A(x)}{\rho_0 A_0} \tag{h}$$

Substitution of Equation d, Equation e, Equation f, Equation g, and Equation h into Equation a leads to

$$\frac{1}{L^2}\frac{\partial^2}{\partial x^{*2}}\left[E_0 I_0 \alpha(x^*)\frac{1}{L^2}\frac{\partial^2}{\partial x^{*2}}(Lw^*)\right] - P\frac{1}{L^2}\frac{\partial^2}{\partial x^{*2}}(Lw^*) + \frac{\rho_0 A_0}{T^2}\beta(x^*)\frac{\partial^2}{\partial t^{*2}}(Lw^*) = 0 \tag{i}$$

Multiplying by L^2/P, Equation i can be rearranged to give

$$\frac{E_0 I_0}{PL^2}\frac{\partial^2}{\partial x^{*2}}\left[\alpha(x^*)\frac{\partial^2 w^*}{\partial x^{*2}}\right] - \frac{\partial^2 w^*}{\partial x^{*2}} + \frac{\rho_0 A_0 L^2}{PT^2}\beta(x^*)\frac{\partial^2 w^*}{\partial t^{*2}} = 0 \tag{j}$$

Choosing

$$T = L\sqrt{\frac{\rho_0 A_0}{P}} \tag{k}$$

defining

$$\varepsilon = \frac{E_0 I_0}{PL^2} \tag{l}$$

and dropping the *'s from nondimensional variables, Equation i is found to be identical to Equation b.

(b) Substitution of Equation c into Equation b gives

$$\varepsilon\frac{d^2}{dx^2}\left(\alpha\frac{d^2 u}{dx^2}\right) - \frac{d^2 u}{dx^2} - \beta(x)\omega^2 u = 0 \tag{m}$$

1.5 Scaling

"Small" and "large" are relative terms. The microscale is large compared to the nanoscale, but small compared to the macroscale. When modeling physical systems, one often cannot determine whether an effect is small and potentially negligible until it is compared to other effects in the system. Even then it is not possible simply to compare dimensional quantities, and therefore the equation is nondimensionalized and nondimensional parameters introduced. It is often possible, by comparing magnitudes of nondimensional parameters, to neglect specific effects when modeling a physical system. Even this approach does not always work, because it is possible that a seemingly small parameter will be multiplied by large derivatives such that the product is of the same order of magnitude as other terms.

The above discussion suggests the following approach: model, nondimensionalize, and compare magnitudes of dimensionless parameters. Such a procedure can also be used, not only to refine a model, but also to determine the best way to solve mathematically the equations derived during the mathematical modeling exercise.

The differential equation governing the unsteady state temperature distribution over the extended surface shown in Figure 1.28a is derived by applying an energy balance to the differential volume shown in Figure 1.28b. The result is

$$k\left(\frac{\partial^2 T}{\partial x^2}+\frac{\partial^2 T}{\partial y^2}\right)=\rho c\frac{\partial T}{\partial t} \tag{1.33}$$

The boundary conditions when the surface at $x=0$ is maintained at a constant temperature, the surface at $x=L$ is insulated, and heat transfer with the ambient medium occurs over the top and bottom of the fin are represented by $y=0$ and $y=w$. The mathematical formulations of the boundary conditions are

$$T(0,y,t)=T_0 \tag{1.34}$$

$$\frac{\partial T}{\partial x}(L,y,t)=0 \tag{1.35}$$

$$k\frac{\partial T}{\partial y}(x,0,t)-h[T(x,0,t)-T_\infty]=0 \tag{1.36}$$

$$k\frac{\partial T}{\partial y}(x,w,t)-h[T(x,w,t)-T_\infty]=0 \tag{1.37}$$

The following nondimensional variables are introduced:

$$x^*=\frac{x}{L}\quad y^*=\frac{y}{w}\quad t^*=\frac{k}{\rho c L^2}t\quad \theta=\frac{T-T_\infty}{T_0-T_\infty},$$

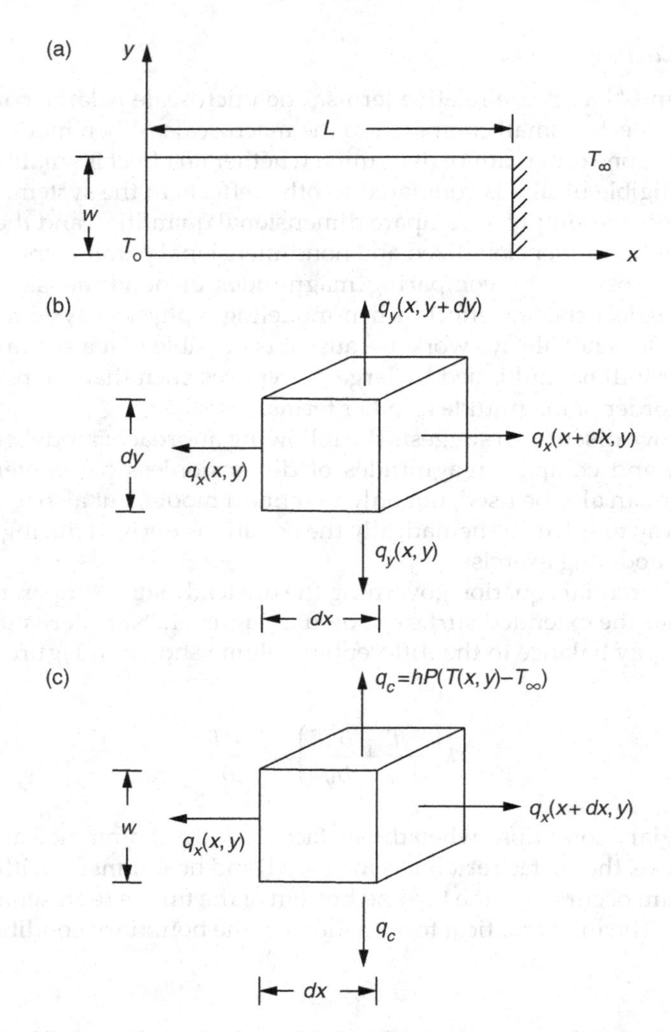

Figure 1.28 (a) A mathematical model for the temperature distribution in the extended surface is developed using several different sets of assumptions. (b) The temperature distribution is two-dimensional, a function of x and y, when L/w is O(1). (c) The temperature distribution is assumed to be a function of x only. The effect of convective heat transfer from the surface is lumped.

which, when substituted into the partial differential equation and boundary conditions, lead to

$$\frac{\partial^2\theta}{\partial x^2}+\left(\frac{L}{w}\right)^2\frac{\partial^2\theta}{\partial y^2}=\frac{\partial\theta}{\partial t} \tag{1.38}$$

$$\theta(0,y,t)=0 \tag{1.39}$$

$$\frac{\partial\theta}{\partial x}(1,y,t)=0 \tag{1.40}$$

$$\frac{\partial\theta}{\partial y}(x,0,t)-\left(\frac{hw}{k}\right)\theta(x,0,t)=0 \tag{1.41}$$

$$\frac{\partial\theta}{\partial y}(x,1,t)+\left(\frac{hw}{k}\right)\theta(x,1,t)=0 \tag{1.42}$$

If $L \gg w$, then L/w is a large quantity. Thus, for all terms in Equation 1.38 to be of the same order of magnitude, $\partial^2\theta/\partial y^2$ must be small. If this term is taken to be zero, solutions of Equation 1.38 are of the form $\theta(x,y,t)=A(x,t)+B(x,t)y$. Application of the boundary conditions at $y=0$ and $y=w$ leads to $B=0$. Then the temperature is a function only of x and t. However, simplifying the partial differential equation to $\partial^2\theta/\partial x^2=\partial\theta/\partial t$ implies that the temperature is independent of the heat transfer coefficient h. This does not seem reasonable. Instead, the derivation of the governing equation can be modified to account for this assumption. An energy balance is applied to the differential segment of Figure 1.28c which includes heat transferred by convection through the upper and lower faces of the surface. The resulting model is

$$kw\frac{\partial^2 T}{\partial x^2}-2h(T-T_\infty)=\rho cw\frac{\partial T}{\partial t} \tag{1.43}$$

Substitution of the nondimensional variables into Equation 1.43 leads to

$$\frac{\partial^2\theta}{\partial x^2}-Bi\theta=\frac{\partial\theta}{\partial t} \tag{1.44}$$

where $Bi=2hL^2/kw$ is the Biot number.

Equation 1.44 subject to the boundary conditions given in Equation 1.39 and Equation 1.40 is not difficult to solve and is solved in Chapter 6. Suppose, though, that the surface is of nonuniform cross section, $w=w(x)$. The governing equation then becomes

$$\frac{\partial}{\partial x}\left(\alpha(x)\frac{\partial\theta}{\partial x}\right)-Bi\theta=\alpha(x)\frac{\partial w}{\partial t} \tag{1.45}$$

where $\alpha(x)$ is a nondimensional function representing the nonuniform cross section. Equation 1.45 may have exact solutions for some forms of $\alpha(x)$, but even when it exists, the exact solution is difficult to determine. If the Biot number is small, the rate of heat transfer by convection is small compared with the rate of heat transfer by conduction, and then Equation 1.45 can perhaps be approximated by

$$\frac{\partial}{\partial x}\left(\alpha(x)\frac{\partial\theta}{\partial x}\right)=\alpha(x)\frac{\partial w}{\partial t} \tag{1.46}$$

If Equation 1.46 is used, it provides a first-order approximation of the temperature distribution. It is the first term, $\theta_0(x)$, in the asymptotic expansion of $\theta(x)$ in terms of the Biot number,

$$\theta(x) = \theta_0(x) + Bi\theta_1(x) + \ldots \tag{1.47}$$

The decision to assume that the temperature is independent of y is based on the scaling of the variables. The decision to expand the nondimensional temperature distribution into an asymptotic expansion in terms of the Biot number can be made only when the problem is written in nondimensional variables. Asymptotic expansions can be made only by using an independent nondimensional variable as the expansion parameter.

Consider the system shown in Figure 1.11, in which a particle of mass m is attached to the end of an elastic bar of length L and cross-sectional area A. The bar is made from a material of elastic modulus E and mass density ρ. The particle is subjected to a time-dependent force, $F(t) = F_0 \sin(\omega t)$. The system is to be modeled to determine its vibrational properties and its frequency response. Assume that all forms of friction are neglected and that normal stresses developed in the bar remain less than the material's yield stress.

An important decision in modeling the system of Example 1.3 as illustrated in Figure 1.3 is whether to use a discrete model or a distributed-parameter model for the bar. A discrete model is appropriate if the inertial effects of the bar are small compared with the inertial effects of the particle. However, this is an ambiguous statement which is not easily quantified. If inertial effects are neglected, the bar can be modeled by a massless linear spring of stiffness $k = EA/L$. The resulting mathematical model is Equation a of Example 1.3. Many vibration texts suggest that if the mass of the bar is not negligible, but not too large, the inertial effects of the bar can be approximated by imagining that a particle whose mass is one-third of that of the bar is added to the particle, giving a better model

$$\hat{m}\frac{d^2y}{dt^2} + \frac{AE}{L}y = F_0 \sin(\omega t) \tag{1.48}$$

where

$$\hat{m} = m + \frac{1}{3}\rho AL \tag{1.49}$$

If the inertial effects of the bar are not neglected, the appropriate model is the partial differential equation given by Equation d of Example 1.3, subject to the boundary conditions of Equation e and Equation f.

Most readers are able to solve the ordinary differential equation derived for the discrete model, but are not able to solve the partial differential equation of the distributed parameter model, although all should be able to do so after completing the study of this book. For this reason, it is instructive to determine under what conditions the discrete model yields accurate results. This knowledge provides guidance in developing subsequent models.

The following nondimensional variables can be introduced into Equation d, Equation e, and Equation f of Example 1.3:

$$x^* = \frac{x}{L} \tag{1.50a}$$

$$u^* = \frac{u}{L} \tag{1.50b}$$

$$t^* = \frac{1}{L}\sqrt{\frac{E}{\rho}}t \tag{1.50c}$$

which leads to

$$\frac{\partial^2 u}{\partial x^2} = \frac{\partial^2 u}{\partial t^2} \tag{1.51}$$

$$u(0,t) = 0 \tag{1.52}$$

$$-\frac{\partial u}{\partial x}(1,t) + \frac{F_0 L^2}{EA} f(t) = \frac{m}{\rho AL}\frac{\partial^2 u}{\partial t}^2(1,t) \tag{1.53}$$

The current discussion intends to determine the conditions under which the inertia of the bar can be neglected or approximated by an equivalent mass given by Equation b and the use of a one-degree-of-freedom assumption in modeling the system. One fundamental difference between the two models is that the one-degree-of-freedom model leads to the prediction of only one natural frequency and only one mode of free vibration, whereas the continuous model predicts an infinite number of natural frequencies and corresponding mode shapes. Thus, since the one-degree-of-freedom model is limited in this regard and can be used only when the lowest natural frequency is required, one means of comparison is to determine the lowest natural frequency obtained using both models.

The natural frequency obtained using the one-degree-of-freedom model is

$$\omega_1 = \sqrt{\frac{EA}{L\left(m + \frac{1}{3}\rho AL\right)}}$$

$$= \sqrt{\frac{EA}{\rho AL^2\left(\frac{m}{\rho AL} + \frac{1}{3}\right)}}$$

$$= \frac{1}{L}\sqrt{\frac{E}{\rho}}\sqrt{\frac{3}{3\beta + 1}} \tag{1.54}$$

where

$$\beta = \frac{m}{\rho A L} \tag{1.55}$$

is the ratio of the mass of the particle to the mass of the bar.

The natural frequency obtained from solution of Equation 1.51, Equation 1.52 and Equation 1.53 is a nondimensional frequency, $\omega^* = \omega(1/L\sqrt{E/\rho})$. Then the comparison can be made to determine the values of β such that the lowest nondimensional frequency obtained from solving Equation 1.51 is close to $\sqrt{3/(3\beta+1)}$.

Solution of Equation 1.51 subject to Equation 1.52 and Equation 1.53 using the methods developed in Chapters 5 and 6 reveals that the lowest natural frequency for a given value of β is the smallest positive solution of the transcendental equation,

$$\tan \omega = \frac{1}{\beta\omega} \tag{1.56}$$

Figure 1.29 shows a comparison between the values of ω calculated from Equation k and the values of $\sqrt{3/(3\beta+1)}$, whereas Figure 1.30 illustrates the percent error in using a one-degree-of-freedom model to determine the lowest natural frequency.

The above example illustrates the appropriate use of approximations. Development of a mathematical solution to a problem obtained through mathematical modeling often requires an analysis of each term in the equations to determine whether or not its effect is as large as other effects considered in the model. In comparing two terms, it is often convenient simply to look at the magnitudes of the parameters used to multiply each term. If two parameters, say η and v, are comparable in magnitude, then $\eta = O(\nu)$, which is read, "η is big 'O' of v." The formal definition of the big "O" symbol is that $f(\nu) = O(g(\nu))$ if $\lim_{\nu \to 0} |f(\nu)/g(\nu) =| = A$ such that $0 < A < \infty$.

This analysis is used to identify the largest or dominant terms in the model. Obtaining a mathematical solution using only the dominant terms is viewed as a first approximation to the solution. This approximation may be improved by using an asymptotic expansion in terms of a small dimensionless parameter. The problem involving only the dominant terms is often significantly simpler than most problems.

Consider again the system of Example 1.8 in which two stretched beams are connected through an elastic layer. The differential equations governing the beam displacements, Equation n and Equation p of Example 1.8, are repeated

$$\frac{\partial^4 w_1}{\partial x^4} - \varepsilon\frac{\partial^2 w_1}{\partial x^2} + \eta_1(w_1 - w_2) + \frac{\partial^2 w_1}{\partial t^2} = 0 \tag{1.57}$$

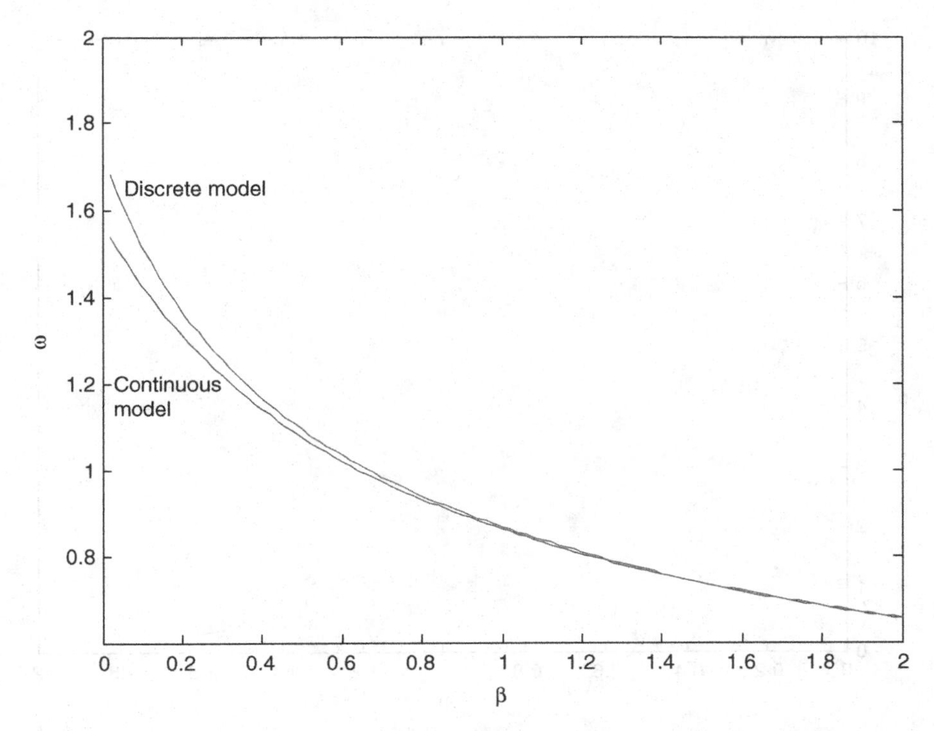

Figure 1.29 The agreement between the lowest natural frequency calculated for the system shown in Figure 1.11 using a discrete model and using a continuous model increases as the mass ratio increases.

$$\mu\frac{\partial^4 w_2}{\partial x^4}-\varepsilon\frac{\partial^2 w_2}{\partial x^2}+\eta_1(w_2-w_1)+\eta_2 w_2+\beta\frac{\partial^2 w_2}{\partial t^2}=0 \tag{1.58}$$

The nondimensional parameters can be recognized as follows:

- ε is the ratio of the normal stress due to axial load to the normal stress due to bending.
- η_1 is the ratio of the stiffness of the elastic layer connecting the beams to the stiffness of the first beam.
- η_2 is the ratio of the stiffness of the elastic layer between the second beam and the fixed support to the stiffness of the first beam.
- μ is the ratio of the stiffness of the second beam to the stiffness of the first beam.
- β is the ratio of the inertia of the second beam to the inertia of the first beam.

Equation n and Equation p are coupled through the elastic layer. If $\eta_1 \ll O(1)$, then the beams are lightly coupled. A first approximation, obtained by

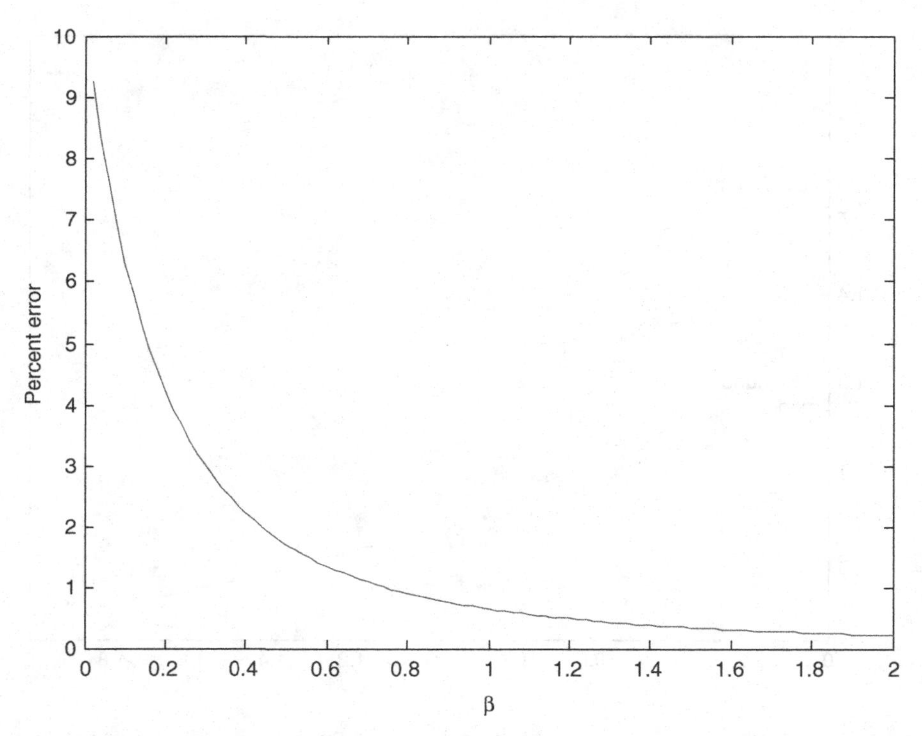

Figure 1.30 The percent error in calculation of the lowest natural frequency of the system shown in Figure 1.11 when using a discrete model is only 10 percent for a small mass ratio and less than 1 percent for a mass ratio of 0.6.

neglecting the coupling terms, leads to a set of uncoupled equations which is much easier to solve. The coupling is then a second-order effect, and its effects are included through an asymptotic expansion in terms of powers of η_1.

If ε is small, then the beams are lightly stretched. A first approximation, obtained by neglecting the stretching terms, is that of coupled Euler-Bernoulli beams. It is much easier to solve for the deflections of coupled Euler-Bernoulli beams than for deflections of coupled stretched beams.

A problem which clearly illustrates the necessity of understanding the effects of scaling is that of a uniform flow impinging on a flat plate, as shown in Figure 1.31. Assuming steady-state two-dimensional flow of an incompressible fluid, the Navier-Stokes equations are:

$$\frac{\partial u}{\partial x}+\frac{\partial v}{\partial y}=0 \tag{1.59}$$

$$\rho\left(u\frac{\partial u}{\partial x}+v\frac{\partial u}{\partial y}\right)=-\frac{\partial p}{\partial x}+\mu\left(\frac{\partial^2 u}{\partial x^2}+\frac{\partial^2 u}{\partial y^2}\right) \tag{1.60}$$

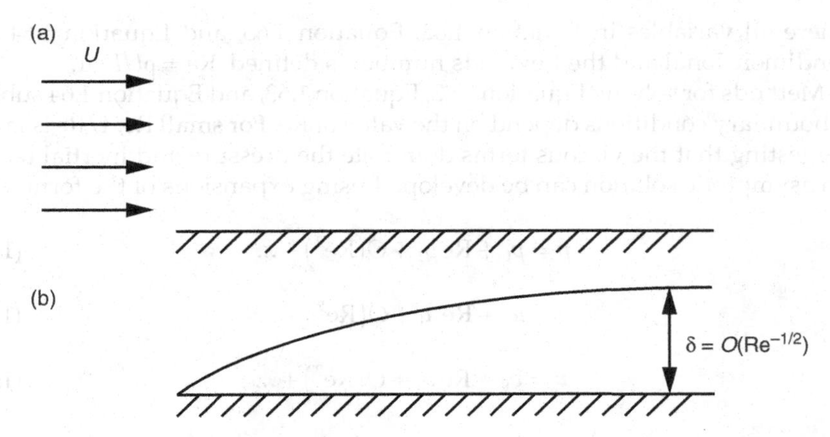

Figure 1.31 (a) A uniform flow impinging on a flat plate is analyzed using the Navier-Stokes equation. (b) When the governing equations are nondimensionalized, the highest-order derivatives are multiplied by a small parameter. This leads to the hypothesis of a boundary layer which grows along the plate. Further scaling shows that the boundary-layer thickness is inversely proportional to the square root of the Reynolds number.

$$\rho\left(u\frac{\partial v}{\partial x}+v\frac{\partial v}{\partial y}\right)=-\frac{\partial p}{\partial y}+\mu\left(\frac{\partial^2 v}{\partial x^2}+\frac{\partial^2 v}{\partial y^2}\right) \tag{1.61}$$

Equation 1.59 is a statement of conservation of mass, while Equation 1.60 and Equation 1.61 are components of the conservation-of-momentum equation. The terms on the left-hand side of the latter equations are convective acceleration or inertia terms. The terms on the right-hand sides represent pressure and viscous forces.

Equation 1.59, Equation 1.60, and Equation 1.61 must be supplemented by boundary conditions. The no-slip condition requires that $u=0$ on the surface of the plate, while the no-penetration condition requires that $v=0$ on the surface of the plate. Far away from the plate, u approaches the free-stream velocity U, while v approaches zero.

Equation 1.59, Equation 1.60, and Equation 1.61 can be nondimensionalized using $x^*=x/L, y^*=y/L, u^*=u/U, v^*=v/U$ and $p^*=p/\rho U^2$, where L is any characteristic length. Use of nondimensional variables leads to:

$$\frac{\partial u}{\partial x}+\frac{\partial v}{\partial y}=0 \tag{1.62}$$

$$u\frac{\partial u}{\partial x}+v\frac{\partial u}{\partial y}=-\frac{\partial p}{\partial x}+\frac{1}{\text{Re}}\left(\frac{\partial^2 u}{\partial x^2}+\frac{\partial^2 u}{\partial y^2}\right) \tag{1.63}$$

$$u\frac{\partial v}{\partial x}+v\frac{\partial v}{\partial y}=-\frac{\partial p}{\partial y}+\frac{1}{\text{Re}}\left(\frac{\partial^2 v}{\partial x^2}+\frac{\partial^2 v}{\partial y^2}\right) \tag{1.64}$$

where all variables in Equation 1.62, Equation 1.63, and Equation 1.64 are nondimensional and the Reynolds number is defined $\mathrm{Re} = \rho UL/\mu$.

Methods for solving Equation 1.62, Equation 1.63, and Equation 1.64 subject to boundary conditions depend on the value of Re. For small Re, 1/Re is large, suggesting that the viscous terms dominate the pressure and inertial terms. An asymptotic solution can be developed using expansions of the form:

$$p = p_0 + \mathrm{Re}\, p_1 + O(\mathrm{Re}^2) + \ldots \tag{1.65}$$

$$u = u_0 + \mathrm{Re}\, u_1 + O(\mathrm{Re}^2) + \ldots \tag{1.66}$$

$$v = v_0 + \mathrm{Re}\, v_1 + O(\mathrm{Re}^2) + \ldots \tag{1.67}$$

For large Re, 1/Re is small, suggesting that the inertial and pressure forces dominate the viscous forces. A first attempt to approximate the solutions might be to try asymptotic expansions of the form:

$$p = p_0 + \frac{1}{\mathrm{Re}} p_1 + O\left(\frac{1}{\mathrm{Re}^2}\right) + \ldots \tag{1.68}$$

$$u = u_0 + \frac{1}{\mathrm{Re}} u_1 + O\left(\frac{1}{\mathrm{Re}^2}\right) + \ldots \tag{1.69}$$

$$v = v_0 + \frac{1}{\mathrm{Re}} v_1 + O\left(\frac{1}{\mathrm{Re}^2}\right) + \ldots \tag{1.70}$$

The hierarchical equations obtained at the lowest order do not include viscous forces. However, such an attempt changes the nature of the differential equations at the lowest order. Equations containing the viscous forces are second-order in x and y, whereas equations resulting from neglecting the viscous forces are first-order in both x and y. Mathematically, the no-slip condition cannot be satisfied by this approximation.

Physically, the shear stresses are largest near the surface of the plate. Even though 1/Re is small, terms such as $\partial^2 u/\partial y^2$ are as large as Re near the surface of the plate and must be included in solving the equations near the surface. However these terms are small far from the plate and can then be neglected.

The above analysis leads to the hypothesis of a boundary layer, a region near the surface of the plate in which the viscous forces are as large as the pressure and inertial forces. This suggests rescaling the y-coordinate within the boundary layer as

$$y^{**} = \frac{y^*}{\delta} \tag{1.71}$$

where δ is the boundary-layer thickness. Substituting Equation 1.71 into Equation 1.62, Equation 1.63, and Equation 1.64 leads to:

$$\frac{\partial u}{\partial x}+\frac{1}{\delta}\frac{\partial v}{\partial y}=0 \tag{1.72}$$

$$u\frac{\partial u}{\partial x}+\frac{v}{\delta}\frac{\partial u}{\partial y}=-\frac{\partial p}{\partial x}+\frac{1}{\mathrm{Re}}\left(\frac{\partial^2 u}{\partial x^2}+\frac{1}{\delta^2}\frac{\partial^2 u}{\partial y^2}\right) \tag{1.73}$$

$$u\frac{\partial v}{\partial x}+\frac{v}{\delta}\frac{\partial v}{\partial y}=-\frac{1}{\delta}\frac{\partial p}{\partial y}+\frac{1}{\mathrm{Re}}\left(\frac{\partial^2 v}{\partial x^2}+\frac{1}{\delta^2}\frac{\partial^2 v}{\partial y^2}\right) \tag{1.74}$$

where the *'s have again been dropped from nondimensional variables

The viscous terms should be as large as the inertial and pressure terms within the boundary layer. However, conservation of mass must also be satisfied. Because δ is a small dimensionless quantity, the two terms in Equation 1.72 are of the same order of magnitude only if $v=O(\delta)$. Thus the y-component of velocity can be rescaled by introducing $v^{**}=v/\delta$ such that $v^{**}=O(1)$. Substitution into Equation 1.72, Equation 1.73, and Equation 1.74 leads to:

$$\frac{\partial u}{\partial x}+\frac{\partial v}{\partial y}=0 \tag{1.75}$$

$$u\frac{\partial u}{\partial x}+v\frac{\partial u}{\partial y}=-\frac{\partial p}{\partial x}+\frac{1}{\mathrm{Re}}\left(\frac{\partial^2 u}{\partial x^2}+\frac{1}{\delta^2}\frac{\partial^2 u}{\partial y^2}\right) \tag{1.76}$$

$$u\delta\frac{\partial v}{\partial x}+v\delta\frac{\partial v}{\partial y}=-\frac{1}{\delta}\frac{\partial p}{\partial y}+\frac{1}{\mathrm{Re}}\left(\delta\frac{\partial^2 v}{\partial x^2}+\frac{1}{\delta}\frac{\partial^2 v}{\partial y^2}\right) \tag{1.77}$$

where the *'s have again been dropped. That is, the variables in Equation 1.75, Equation 1.76, and Equation 1.77 are boundary-layer variables.

The order of magnitude of the boundary-layer thickness is determined by requiring that the viscous forces balance with the inertial and pressure forces in Equation 1.76. Invoking this requirement leads to:

$$\delta=O\left(\frac{1}{\sqrt{\mathrm{Re}}}\right) \tag{1.78}$$

Selection of the boundary-layer thickness according to Equation 1.78 makes the largest term in the y-momentum equation, Equation 1.77, equal to $\partial p/\partial y$. Thus the y-component of the momentum equation is satisfied to the lowest order in the boundary layer by requiring that the pressure be constant across the boundary layer.

A similar analysis can be used for the problem of Example 1.9, a fixed free beam with an axial load. The differential equation governing the displacement of the beam is Equation m of Example 1.9 which is repeated below:

$$\varepsilon\frac{d^2}{dx^2}\left(\alpha\frac{d^2u}{dx^2}\right)-\frac{d^2u}{dx^2}-\beta(x)\omega^2u=0 \tag{1.79}$$

If ε is small, and if an asymptotic expansion were attempted for the displacement, it would fail because the hierarchical equation obtained at the lowest order is a second-order differential equation which cannot satisfy all four required boundary conditions. Thus there must be a region where $d^2[\alpha(d^2u/dx^2)]/dy^2$ is large enough that the bending term balances with at least one other term. This region is also a boundary layer, or due to the physics of the problem, a bending layer. Actually, there are two bending layers, one at each end of the beam. The parameter ω is a natural frequency and has an infinite but countable number of values. Consider first the case when $\omega^2=O(1)$. Let δ be the boundary-layer thickness, and define the boundary-layer variable as $x^{**}=x/\delta$. Equation m can be rewritten in terms of the boundary-layer variable as:

$$\frac{\varepsilon}{\delta^4}\frac{d^2}{dx^2}\left(\alpha\frac{d^2u}{dx^2}\right)-\frac{1}{\delta^2}\frac{d^2u}{dx^2}-\beta(x)\omega^2u=0 \tag{1.80}$$

If $\delta=O(\varepsilon^{1/2})$, then both the bending and stretching terms are $O(\varepsilon^{-1})$, while the inertia term is O(1). Thus the bending and stretching terms interact in the boundary layers.

Suppose $\omega^2=O(\varepsilon^{-3})=\kappa/\varepsilon^3$. Then Equation 1.80 written in boundary-layer variables becomes

$$\frac{\varepsilon}{\delta^4}\frac{d^2}{dx^2}\left(\alpha\frac{d^2u}{dx^2}\right)-\frac{1}{\delta^2}\frac{d^2u}{dx^2}-\beta(x)\frac{\kappa}{\varepsilon^3}u=0 \tag{1.81}$$

If $\delta=O(\varepsilon)$, then both the bending and inertial terms are $O(\varepsilon^{-3})$, while the stretching term is $O(\varepsilon^{-2})$. Thus the bending and inertial terms interact in the boundary layer for larger frequencies.

The previous two problems are said to have "singularities" such that straightforward asymptotic expansions of the form of Equation 1.68, Equation 1.69, and Equation 1.70 are unsuccessful in obtaining a solution. The nondimensionalization and the physics of the problem lead to the discovery of boundary layers.

Problems

1.1. An elastic bar has a static displacement due to an applied axial load *F*. The force is suddenly removed, resulting in free oscillations of the bar. Derive the expression governing the displacement $u(x,t)$ of a particle along the axis of the bar.

a. Specify the governing differential equation as well as all boundary and initial conditions necessary to solve for $u(x,t)$.
b. Nondimensionalize the problem by introducing $x^* = x/L$ and $t^* = t/T$, where T is to be chosen for convenience.

1.2. The elastic bar of Figure P1.2 is fixed at $x=0$ and has a particle of mass m attached at $x=L$. The particle is attached to a spring in parallel with a viscous damper. The differential equation governing the displacement of a particle along the axis of the bar is:

$$EA\frac{\partial^2 u}{\partial x^2} = \rho A\frac{\partial^2 u}{\partial t^2} \tag{a}$$

a. Specify the boundary conditions at $x=0$ and $x=L$.
b. The particle of mass m is displaced a distance δ to the right, held in this position, and released. Specify the initial conditions for the resulting vibrations.
c. Introduce nondimensional variables, $x^* = x/L$, $u^* = u/L$ and $t^* = t/T$. Substitute the nondimensional variables into Equation a, the boundary conditions, and the initial conditions. Choose T such that the resulting partial differential equation is:

$$\frac{\partial^2 u}{\partial x^2} = \frac{\partial^2 u}{\partial t^2} \tag{b}$$

where all variables in Equation b are nondimensional.

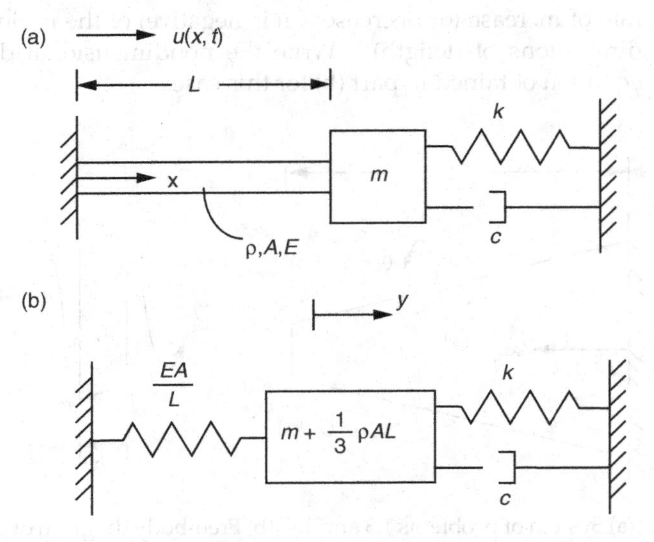

Figure P1.2 (a) System for problem 1.2 (b) Discrete system model for problem 1.2(e).

d. Identify and physically define all nondimensional parameters which appear in the nondimensional formulation of the boundary conditions.
e. If the inertia of the elastic bar is small compared to the inertia of the particle, the system shown in Figure P1.2a can be modeled by the discrete system shown in Figure P1.2b. The inertia effects of the bar can be approximated by adding a particle of mass $1/3\rho AL$ to the existing particle. Derive the differential equation governing the displacement $y(t)$ of the particle. Express the appropriate initial conditions when the particle is displaced as in part (b).
f. Nondimensionalize the differential equation obtained in part (e).
g. Discuss a strategy to determine the accuracy of the approximation using the mass ratio $m/\rho AL$.

1.3. The shaft shown in Figure P1.3a is nonuniform such that its polar moment of inertia varies with x, $J=J(x)$. Let $\theta(x,t)$ represent the angular displacement of the plane a distance x from the left support. A differential element of thickness dx is illustrated in Figure P1.3b.
a. Use the free-body diagram to derive the partial differential equation governing the torsional oscillations of the shaft. Specify appropriate boundary conditions.
b. Introduce nondimensional variables $x^* = x/L$ and $t^* = t/T$ and a nondimensional function $\alpha(x) = J(x)/J_0$, where J_0 is the polar moment of inertia of the shaft at $x=0$. Substitute these variables into the differential equation and choose T such that the governing differential equation contains no parameters.
c. The shaft is circular with a radius which varies according to $r(x) = r_0(1+\mu x)$, where r_0 is the shaft radius at $x=0$ and μ is the rate of increase (or decrease if μ is negative) of the radius and has dimensions of $(\text{length})^{-1}$. Write the nondimensional differential equation obtained in part (b) for this case.

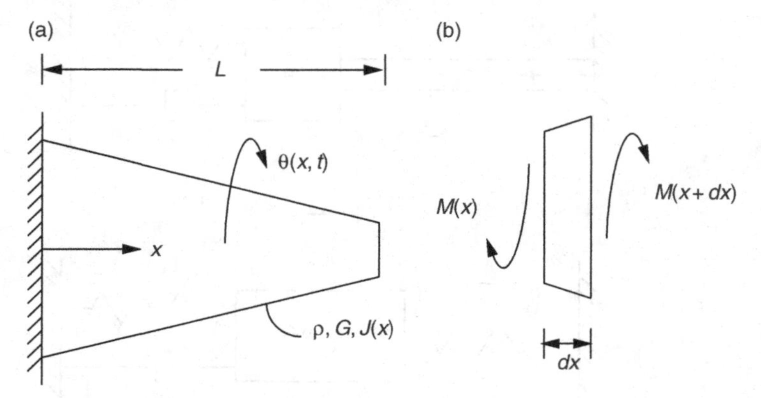

Figure P1.3 (a) System of problems 1.3 and 1.4. (b) Free-body diagram of differential element at an arbitrary instant.

1.4. Consider the shaft shown in Figure P1.3a when it is subject to a harmonic torque $M(t) = M_0 \sin(\omega t)$ applied at its free end. The radius of the shaft varies linearly as described in problem 1.3c. Let Θ represent the steady-state amplitude of the end of the shaft. The amplitude is a function of the parameters $\rho, G, L, r_0, \mu, M_0$ and ω. That is, $\Theta = \Theta(\rho, G, L, r_0, \mu, M_0, \omega)$. Determine a set of nondimensional parameters π_1, π_2, π_3 and π_4 for a nondimensional formulation of the relationship between the parameters, $\Theta = \Theta(\pi_1, \pi_2, \pi_3, \pi_4)$.

1.5. The steady-state amplitude of the mass attached to the bar shown in Figure P1.5 is:

$$X = \frac{F_0 L}{A\omega L\sqrt{E\rho}\cos\left(\omega L\sqrt{\frac{\rho}{E}}\right) - m\omega^2 L\sin\left(\omega L\sqrt{\frac{\rho}{E}}\right)} \tag{a}$$

Equation a shows $X = X(F_0, A, L, m, \omega, E, \rho)$.

a. Determine a set of nondimensional parameters which can be used to formulate Equation a in a nondimensional form.
b. Rewrite Equation a in terms of these nondimensional parameters.

1.6. The temperature distribution over the extended surface of Figure P1.6 is:

$$T(x) = T_\infty + (T_1 - T_\infty)\left[\cosh\left(\frac{2h}{kw}x\right) - \tanh\left(\frac{2h}{kw}L\right)\sinh\left(\frac{2h}{kw}x\right)\right] \tag{a}$$

The rate of heat transfer from the base of the extended surface is

$$Q = -kw\ell\frac{dT}{dx}(0) \tag{b}$$

Develop a nondimensional relationship between the heat transfer at the base and appropriate nondimensional parameters. Speculate on the physical meanings of the nondimensional parameters.

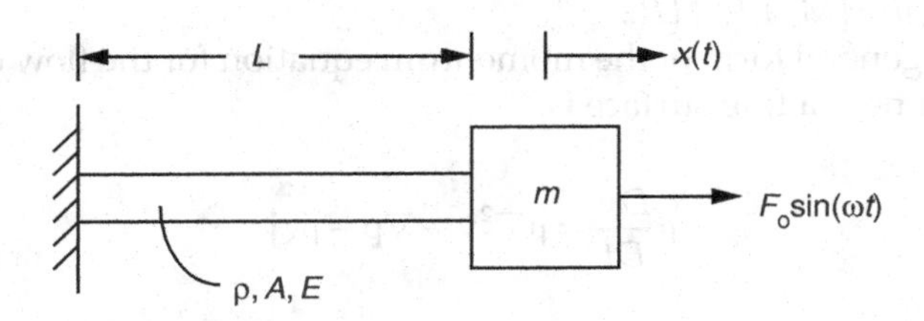

Figure P1.5 System of problem 1.5.

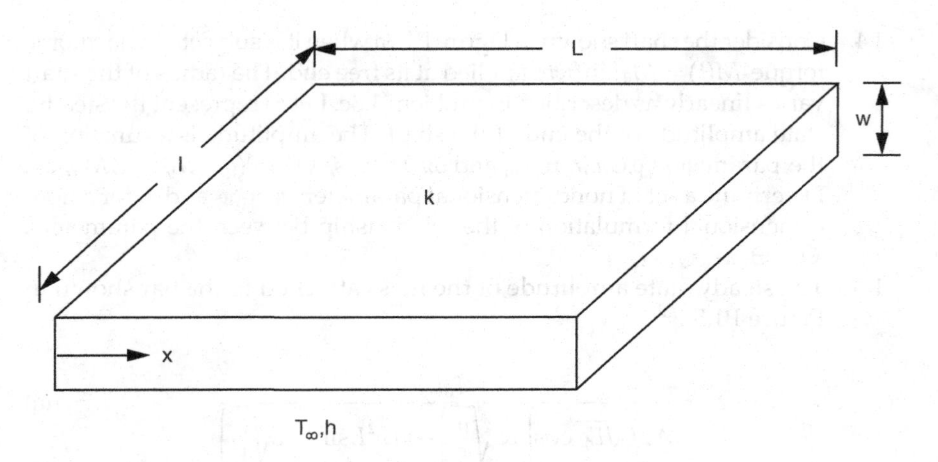

Figure P1.6 System of problem 1.6.

1.7. The differential equation governing the transverse deflection of a pipe with an inviscid fluid flowing with a velocity U is:

$$EI\frac{\partial^4 w}{\partial x^4}+MU^2\frac{\partial^2 w}{\partial x^2}+2MU\frac{\partial^2 w}{\partial x\partial t}+(M+m)g\frac{\partial w}{\partial x}+(M+m)\frac{\partial^2 w}{\partial t^2}=0 \quad \text{(a)}$$

where E is the elastic modulus of the pipe, I is the area moment of inertia of the pipe's cross section, M is the mass of the fluid in the pipe per unit length, and m is the mass of the pipe per unit length. Define the following nondimensional variables:

$$x^* = \frac{x}{L} \quad w^* = \frac{w}{L} \quad t^* = \left[\frac{EI}{M+m}\right]^{\frac{1}{2}}\frac{t}{L^2} \quad \text{(b)}$$

Nondimensionalize Equation a through introduction of the nondimensional variables of. Write the resulting nondimensional equation in terms of the nondimensional parameters, $\beta = M/(M+m)$ and $u=[M/EI]^{1/2}UL$.

1.8. The general form of the momentum equation for the flow of a viscous fluid near a free surface is:

$$\rho\frac{D\mathbf{v}}{Dt}=\mu\nabla^2\mathbf{v}-\nabla\mathbf{p}+\rho g\mathbf{j} \quad \text{(a)}$$

where $\mathbf{v} = u\mathbf{i} + v\mathbf{j} + w\mathbf{k}$ is the velocity vector, ρ is the mass density of the fluid, p is the fluid pressure, μ is the dynamic viscosity, and g is the acceleration due to gravity. Let L be a characteristic length of flow, V be a characteristic velocity, and p_∞ be a characteristic pressure, and define nondimensional variables as:

$$\mathbf{v}^* = \frac{\mathbf{v}}{V} \tag{b}$$

$$t^* = t\frac{V}{L} \tag{c}$$

$$p^* = \frac{p - p_\infty}{\rho V^2} \tag{d}$$

$$x^* = \frac{x}{L},\ y^* = \frac{y}{L},\ z^* = \frac{z}{L} \tag{e}$$

a. Substitute Equation b, Equation c, Equation d, and Equation e into Equation a to obtain

$$\frac{D\mathbf{v}}{Dt} = -\nabla p + \frac{1}{\mathrm{Re}}\nabla^2\mathbf{v} + \frac{1}{Fr}\mathbf{j} \tag{f}$$

where Re is the Reynolds number, Fr is the Froude number, and all variables are taken as nondimensional.

b. Define the Froude number mathematically and physically.

c. Write the component equations represented by Equation f.

1.9. The momentum equation for the flow of a fluid with free convection is:

$$\rho\frac{D\mathbf{v}}{Dt} = \mu\nabla^2\mathbf{v} - \rho\beta(T - T_\infty)g\mathbf{j} \tag{a}$$

where, in addition to the variables defined in problem 1.8, β is the coefficient of thermal expansion and T is temperature. The nondimensional variables defined in problem 1.8 are used as well $\Theta = (T - T_\infty)/(T_1 - T_\infty)$ where T_1 and T_∞ are reference temperatures.

a. Nondimensionalize Equation a.

b. The nondimensional formulation of Equation a contains two nondimensional parameters, Re and Gr, the Grashopf number. Define the Grashopf number both mathematically and physically.

1.10. Use Lagrange's equations to derive the differential equations governing the motion of the system shown in Figure P1.10.

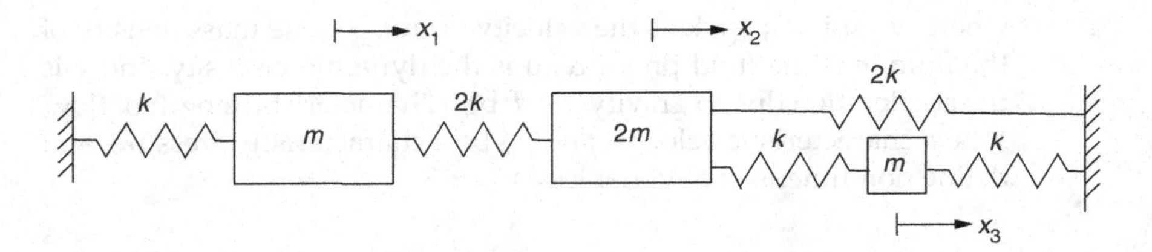

Figure P1.10 System of problem 1.10.

1.11. A non-uniform bar of length L is in a medium of ambient temperature T_∞. One end of the bar is insulated, while the other end is maintained at a temperature of T_0. The bar is made of a smart material, which produces an internal heat generation proportional to the gradient of the temperature. Assuming one-dimensional heat generation along the length of the bar and defining x as a coordinate along the length of the bar, the differential equation governing the temperature distribution in the bar is

$$k\frac{d}{dx}\left(A(x)\frac{dT}{dx}\right)+\eta\frac{dT}{dx}+hP(x)(T-T_\infty)=0 \tag{a}$$

where k is the thermal conductivity of the bar, h is the heat transfer coefficient between the bar and the ambient, and η is a parameter related to the internal heat generation. The appropriate boundary conditions are

$$T(0)=T_0 \tag{b}$$

$$\frac{dT}{dx}(L)=0 \tag{c}$$

(a) Non-dimensionalize Equation a, Equation b, and Equation c through introduction of appropriate non-dimensional variables.
(b) Identify mathematically and physically all parameters appearing in the non-dimensional formulation.
(c) Write a non-dimensional formulation of the problem when

$$A(x)=A_0\left(1-\mu\frac{x}{L}\right)^2 \tag{d}$$

and

$$P(x)=2w+2b_0\left(1-\mu\frac{x}{L}\right) \tag{e}$$

(d) Suppose the ratio of the rate of conduction to the rate of convection is assumed to be small, $1/\mathrm{Bi} = O(\varepsilon)$, where Bi is the Biot number and ε is a small non-dimensional parameter. Speculate whether a boundary layer exists. If so, determine the thickness of the boundary layer and which terms physically interact within the boundary layer.

Chapter 2

Linear algebra

2.1 Introduction

Linear algebra is the algebra used for analysis of linear systems. It provides a foundation on which solutions to mathematical problems can be developed. Success in obtaining a solution to a mathematical problem requires finding the specific solution among a possible set of solutions, the solution space. An understanding of the solution space and the properties of elements of the solution space leads to the development of solution techniques. It is in this spirit that this review of linear algebra is presented.

An exact solution of a problem is a solution for the dependent variables which satisfies without error the mathematical problem for all possible values of the independent variables. An exact solution, while desirable, is not always possible. Approximate solutions are sought when an exact solution is not available. Approximate solutions are of two types. Variational methods are used to determine continuous functions of the independent variables which provide in some sense the "best approximation," chosen from a specified set, to the exact solution. Numerical solutions provide an approximation to the exact solution only at discrete values of independent variables. Linear algebra provides a framework in which these approximate solutions can be developed and in which the error between the exact solution and an approximate solution can be estimated.

Linear algebra provides a framework for developing solutions to linear problems. Modeling of engineering systems often leads to nonlinear mathematical problems. Exact solutions exist for only a few nonlinear problems. Often assumptions are made such that the nonlinear problem can be approximated by a linear problem. Even if the assumptions that linearize the problem are not valid, some understanding of the solution can be obtained by studying the linearized problem. Indeed, knowledge of the behavior of the linearized system is necessary to understand the effect of nonlinearities on the system behavior. Numerical methods developed to approximate solutions of nonlinear problems are based upon numerical methods used to develop approximate solutions for linear problems. Approximate solutions of nonlinear problems are often assumed to be perturbations of the linear solution. Thus knowledge of linear solutions is necessary to develop approximate solutions for nonlinear problems.

This chapter provides a review of linear algebra and linear operator theory. Theorems are presented, with proofs when these proofs are themselves instructive.

2.2 *Three-dimensional space*

Every point in three-dimensional space has a unique set of *x-y-z* coordinates defined in a fixed Cartesian reference frame, as shown in Figure 2.1. The location of a particle at any instant of time is defined by the Cartesian coordinates of the point it occupies in space. A position vector, **r**, defining the location of the particle is a line segment drawn from the origin (0,0,0) of the Cartesian system to the particle. The position vector has a unique direction with respect to the Cartesian frame and a calculable length denoted by $|\mathbf{r}|$.

Let $\mathbf{r}_1$ and $\mathbf{r}_2$ represent two vectors defined in a Cartesian system. Vector addition in three-dimensional space is defined geometrically, as illustrated in Figure 2.2a. A vector parallel to **v** is drawn such that its tail coincides with the head of the second vector **u**. If $\mathbf{w}=\mathbf{r}_1+\mathbf{r}_2$, then **w** is the vector drawn from the origin to the head of the vector parallel to $\mathbf{r}_2$. Figure 2.2b shows the process repeated, but with a vector parallel to $\mathbf{r}_1$ drawn with its tail coinciding with the head of **v**. The resulting sum is the same as that obtained in Figure 2.2a. Thus vector addition is commutative,

$$\mathbf{r}_1+\mathbf{r}_2=\mathbf{r}_2+\mathbf{r}_1 \tag{2.1}$$

Figure 2.2 illustrates the triangle rule for vector addition. The vectors being added and the resultant (sum) are depicted as sides of a triangle. Since the

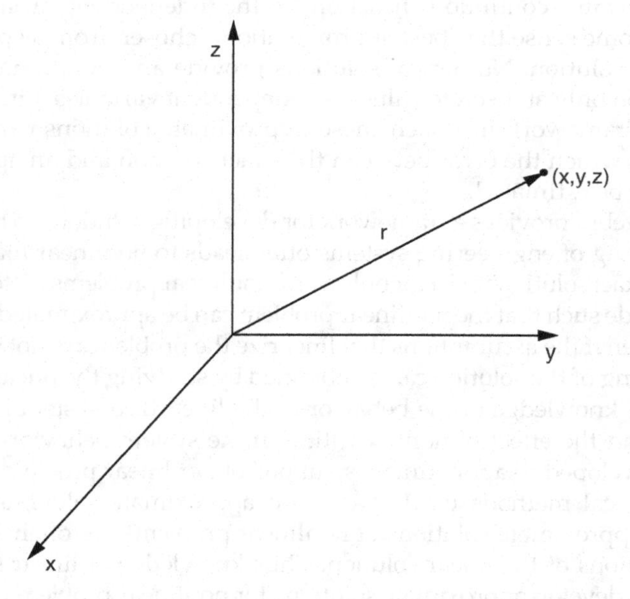

Figure 2.1 A point in a Cartesian coordinate system is referenced by three coordinates *(x,y,z)*. A position vector is drawn from the origin to the point.

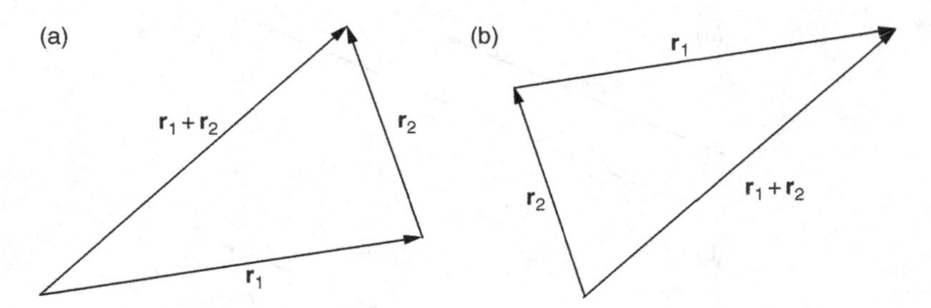

Figure **2.2** (a) Vector addition is illustrated using the triangle rule. The resultant is drawn from the tip of $\mathbf{r}_1$to the head of $\mathbf{r}_2$. (b) Vector addition is commutative.

length of any side of a triangle must be less than the sum of the lengths of the other two sides,

$$|\mathbf{r}_1 + \mathbf{r}_2| \leq |\mathbf{r}_1| + |\mathbf{r}_2| \tag{2.2}$$

Equation 2.2 is called the triangle inequality.

The concept of multiplication of a vector by a scalar is illustrated in Figure 2.3. Let α be any real value. The vector $\alpha\mathbf{r}$ is the vector parallel to $\mathbf{r}$ whose length is α times the length of $\mathbf{r}$. If α is positive, then the vector $\alpha\mathbf{r}$ lies in the same direction as $\mathbf{r}$. If α is negative, the vector $\alpha\mathbf{r}$ lies in the direction opposite that of $\mathbf{r}$. If α equals zero, then $\alpha\mathbf{r}$ is a vector whose length is zero and is called the zero vector, $\mathbf{0}$.

The following properties follow from the definitions of vector addition and multiplication of a vector by a scalar:

i. Associative law of addition: $(\mathbf{r}_1 + \mathbf{r}_2) + \mathbf{r}_3 = \mathbf{r}_1 + (\mathbf{r}_2 + \mathbf{r}_3)$
ii. Commutative law of addition: $\mathbf{r}_1 + \mathbf{r}_2 = \mathbf{r}_2 + \mathbf{r}_1$
iii. Addition of zero vector: $\mathbf{0} + \mathbf{r}_1 = \mathbf{r}_1$
iv. Multiplication by one: $(1)\mathbf{r}_1 = \mathbf{r}_1$
v. Negative vector: $-\mathbf{r}_1 = (-1)\mathbf{r}_1 \quad \mathbf{r}_1 + (-\mathbf{r}_1) = \mathbf{0}$
vi. Distributive law of scalar multiplication: $(\alpha + \beta)\mathbf{r}_1 = \alpha\mathbf{r}_1 + \beta\mathbf{r}_1$
vii. Associative law of scalar multiplication: $(\alpha\beta)\mathbf{r}_1 = \alpha(\beta\mathbf{r}_1)$
viii. Distributive law of addition: $\alpha(\mathbf{r}_1 + \mathbf{r}_2) = \alpha\mathbf{r}_1 + \alpha\mathbf{r}_2$

A unit vector is a vector whose length is one. Let $\mathbf{i}$, $\mathbf{j}$, and $\mathbf{k}$ be a set of unit vectors parallel to the x, y, and z coordinate axes respectively. Using the definitions of vector addition and multiplication of a vector by a scalar, the position vector for the particle can be written in terms of the unit vectors as:

$$\mathbf{r} = x\mathbf{i} + y\mathbf{j} + z\mathbf{k} \tag{2.3}$$

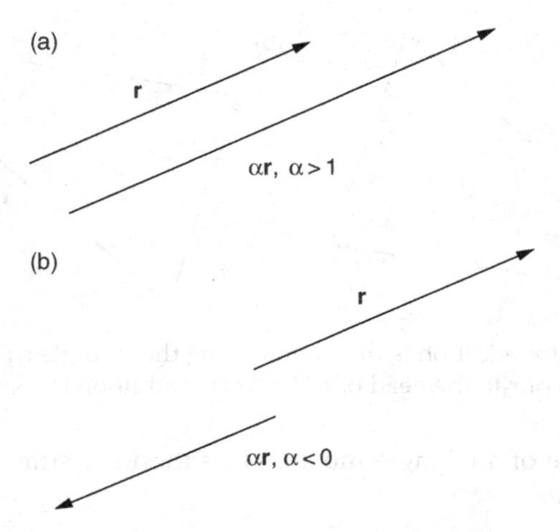

Figure 2.3 (a) Multiplication of a vector by a scalar leads to a vector lying in the same direction as the vector, but whose length is multiplied by the scalar. (b) If the scalar is negative, the product vector opposes the original vector.

Any vector in three-dimensional space may be written as a linear combination of the trio of unit vectors. The vectors **i**, **j**, and **k** form a basis for three-dimensional space.

The Pythagorean theorem is used to determine the length of the position vector as

$$|\mathbf{r}| = \sqrt{x^2 + y^2 + z^2} \tag{2.4}$$

A scalar function of two vectors, called the dot product, is defined by

$$\mathbf{r}_1 \cdot \mathbf{r}_2 = |\mathbf{r}_1||\mathbf{r}_2|\cos\theta \tag{2.5}$$

where θ is the angle made between $\mathbf{r}_1$ and $\mathbf{r}_2$. The dot product has the geometric interpretation that it is equal to the length of the projection of the vector $\mathbf{r}_1$ onto $\mathbf{r}_2$. This leads to the calculation of the dot product as:

$$\mathbf{r}_1 \cdot \mathbf{r}_2 = x_1x_2 + y_1y_2 + z_1z_2 \tag{2.6}$$

It should be noted that when computed using Equation 2.6, the dot product has the following properties:

Commutative property: $\mathbf{r}_1 \cdot \mathbf{r}_2 = \mathbf{r}_2 \cdot \mathbf{r}_1$
Multiplication by scalar: $(\alpha\mathbf{r}_1) \cdot \mathbf{r}_2 = \alpha(\mathbf{r}_1 \cdot \mathbf{r}_2)$
Distributive property: $(\mathbf{r}_1 + \mathbf{r}_2) \cdot \mathbf{r}_3 = \mathbf{r}_1 \cdot \mathbf{r}_3 + \mathbf{r}_2 \cdot \mathbf{r}_3$
Non-negative property: $\mathbf{r}_1 \cdot \mathbf{r}_1 \geq 0$ and $\mathbf{r}_1 \cdot \mathbf{r}_1 = 0$ if and only if $\mathbf{r}_1 = \mathbf{0}$

It should also be noted that from Equation 2.5 and Equation 2.6,

$$\mathbf{r} \cdot \mathbf{r} = |\mathbf{r}|^2 \tag{2.7}$$

Three-dimensional space, the definition of a vector, and the operations of vector addition and multiplication by a scalar are generalized in section 2.3 into the concept of vector spaces. The properties of vectors in a general vector space are defined in the same way as the properties satisfied by vectors in three-dimensional space. The scalar function of the dot product is generalized in section 2.5 into the concept of inner products.

Example 2.1 A force, **F**, acting on a particle is represented as a vector in three-dimensional space. The force has a magnitude $|\mathbf{F}|$ and lies in the direction of a unit vector **e**. The resultant force acting on a particle is the vector sum of all forces acting on the particle. A free-body diagram of a particle at a given instant is shown in Figure 2.4. Determine the resultant force acting on the particle at this instant. Specify its magnitude and a unit vector in the direction of the resultant.

Solution The forces are represented as:

$$\mathbf{F}_1 = -300\mathbf{j} \text{ N} \tag{a}$$

$$\mathbf{F}_2 = 600\mathbf{e}_2 \text{ N} \tag{b}$$

$$\mathbf{F}_3 = 500\mathbf{e}_3 \text{ N} \tag{c}$$

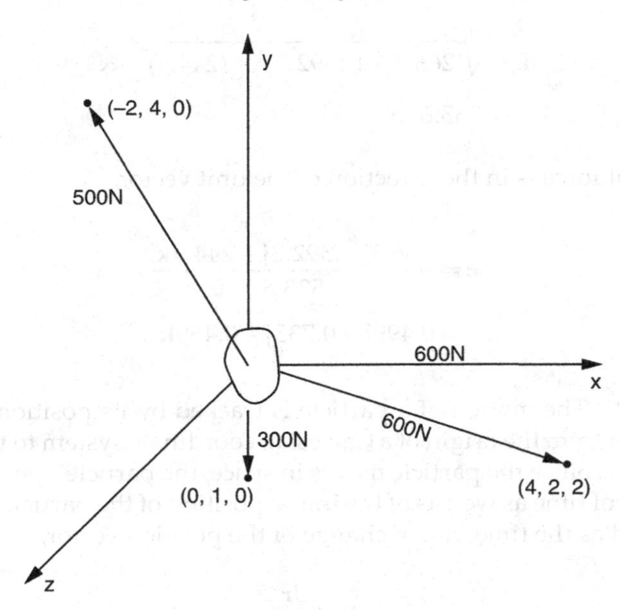

Figure 2.4 Forces on a free-body diagram are drawn in the direction of their line of action with magnitude shown. The forces are thus modeled as vectors.

where the unit vectors are

$$\mathbf{e}_2 = \frac{4\mathbf{i}+2\mathbf{j}+2\mathbf{k}}{\sqrt{(4)^2+(2)^2+(2)^2}} = \frac{\sqrt{6}}{3}\mathbf{i}+\frac{\sqrt{6}}{6}\mathbf{j}+\frac{\sqrt{6}}{6}\mathbf{k} \tag{d}$$

$$\mathbf{e}_3 = \frac{-2\mathbf{i}+4\mathbf{j}}{\sqrt{(-2)^2+(4)^2}} = -\frac{\sqrt{5}}{5}\mathbf{i}+\frac{2\sqrt{5}}{5}\mathbf{j} \tag{e}$$

The resultant force acting on the particle is

$$\begin{aligned}\mathbf{F} &= \mathbf{F}_1+\mathbf{F}_2+\mathbf{F}_3\\ &= \left[\left(200\sqrt{6}-100\sqrt{5}\right)\mathbf{i}+\left(-300+100\sqrt{6}+200\sqrt{5}\right)\mathbf{j}+100\sqrt{6}\mathbf{k}\right]\ \text{N}\\ &= [266.3\mathbf{i}+392.2\mathbf{j}+244.9\mathbf{k}]\ \text{N}\end{aligned} \tag{f}$$

The magnitude of the resultant force is

$$|\mathbf{F}| = \sqrt{(266.3)^2+(392.2)^2+(244.9)^2}\ \text{N} = 533.6\ \text{N} \tag{g}$$

The resultant force is in the direction of the unit vector,

$$\mathbf{e} = \frac{266.3\mathbf{i}+392.2\mathbf{j}+244.9\mathbf{k}}{533.6} = 0.499\mathbf{i}+0.735\mathbf{j}+0.459\mathbf{k} \tag{h}$$

Example 2.2 The motion of a particle is tracked by its position vector $\mathbf{r}$, a vector drawn from the origin of a Cartesian coordinate system to the position of the particle. Since the particle moves in space, the particle's position vector is a function of time as well as of the initial position of the particle. The velocity is defined as the time rate of change of the position vector,

$$\mathbf{v} = \frac{d\mathbf{r}}{dt} \tag{a}$$

By definition, the velocity vector is always tangent to the spatial path traversed by the particle. The acceleration is defined as the time rate of change of the velocity vector,

$$\mathbf{a} = \frac{d\mathbf{v}}{dt} = \frac{d^2\mathbf{r}}{dt^2} \tag{b}$$

Equation a and Equation b represent the velocity and acceleration of the particle as seen by an observer attached to the particle. Modeling of a system using the views of observers attached to particles is called a Lagrangian formulation.

The position vector of a particle moving in space, referenced to a Cartesian frame, is determined to be

$$\mathbf{r} = [\cos(2t)\mathbf{i} + \sin(2t)\mathbf{j} + 0.5t\mathbf{k}] \text{ m} \tag{c}$$

where t is in seconds.

Determine a unit vector tangent to the path of motion at any instant of time; (b) Determine the component of the acceleration normal to the path of motion.

Solution (a) The velocity vector is determined by applying Equation a to the position vector of Equation c, leading to

$$\mathbf{v} = [-2\sin(2t)\mathbf{i} + 2\cos(2t)\mathbf{j} + 0.5\mathbf{k}] \ \frac{\text{m}}{\text{s}} \tag{d}$$

Since the velocity vector is always tangent to the path of motion, it can be expressed as

$$\mathbf{v} = |\mathbf{v}|\mathbf{e}_t \tag{e}$$

where e_t is a unit vector tangent to the path. To this end,

$$\begin{aligned} |\mathbf{v}| &= \sqrt{[-2\sin(2t)]^2 + [2\cos(2t)]^2 + (0.5)^2} \ \frac{\text{m}}{\text{s}} \\ &= 2.87 \ \frac{\text{m}}{\text{s}} \end{aligned} \tag{f}$$

Using Equation d and Equation f, Equation e can be rearranged to give

$$\begin{aligned} \mathbf{e}_t &= \frac{\mathbf{v}}{|\mathbf{v}|} \\ &= -0.693\sin(2t)\mathbf{i} + 0.693\cos(2t)\mathbf{j} + 0.174\mathbf{k} \end{aligned} \tag{g}$$

The acceleration vector can be calculated from Equation b and Equation d as

$$\mathbf{a} = [-4\cos(2t)\mathbf{i} - 4\sin(2t)\mathbf{j}]\ \frac{\text{m}}{\text{s}^2} \tag{h}$$

The tangential component of the acceleration is

$$a_t = \mathbf{a} \cdot \mathbf{e}_t \tag{i}$$

which using Equation g and Equation h, becomes

$$\begin{aligned} a_t &= [-4\cos(2t)\mathbf{i} - 4\sin(2t)\mathbf{j}] \bullet [-0.693\sin(2t)\mathbf{i} + 0.693\cos(2t)\mathbf{j} + 0.174\mathbf{k}] \\ &= 0 \end{aligned} \tag{j}$$

Thus, since the tangential component of the acceleration is zero, the acceleration is in a direction normal to the surface.

Example 2.3 A Eulerian formulation is usually used in modeling the flow of a fluid. In a Eulerian formulation, the properties used are those observed at a fixed location in the flow field as different fluid particles pass through the point. In a Eulerian formulation, the properties are functions of spatial coordinates and time. If $\mathbf{v}(x,y,z,t)$ is the Eulerian velocity vector at a point (x,y,z) in the flow field, then the acceleration is comprised of two terms. The local acceleration is the rate of change of the velocity and is simply calculated $\mathbf{a}_l = \partial \mathbf{v}/\partial t$. The convective acceleration is the rate of change of the velocity at the point (x,y,z) due to the rate at which fluid particles pass through the point. The convective acceleration is calculated as $\mathbf{a}_c = \mathbf{v} \cdot \nabla \mathbf{v}$, where

$$\begin{aligned} \mathbf{v} \cdot \nabla \mathbf{v} &= \left(v_x \frac{\partial v_x}{\partial x} + v_y \frac{\partial v_x}{\partial y} + v_z \frac{\partial v_x}{\partial z} \right)\mathbf{i} + \left(v_x \frac{\partial v_y}{\partial x} + v_y \frac{\partial v_y}{\partial y} + v_z \frac{\partial v_y}{\partial z} \right)\mathbf{j} \\ &\quad + \left(v_x \frac{\partial v_z}{\partial x} + v_y \frac{\partial v_z}{\partial y} + v_z \frac{\partial v_z}{\partial z} \right)\mathbf{k} \end{aligned}$$

The velocity vector for a flow field is

$$\mathbf{v} = [xye^{-2t}\mathbf{i} + y^2 e^{-2t}\mathbf{j}]\ \frac{\text{m}}{\text{s}} \tag{a}$$

(a) Determine the acceleration vector for the flow
(b) The flow rate Q through a surface area is calculated as

$$Q = \int_A \mathbf{v} \cdot \mathbf{n} dA \tag{b}$$

where $\mathbf{n}$ is a unit vector normal to the surface. Determine the flow rate through a square whose vertices are (0,2,0),(0,2,– 2), (2,2,– 2), and (2,2,0).

Solution (a) Noting that the total acceleration is

$$\mathbf{a}=\frac{\partial \mathbf{v}}{\partial t}+\left(v_x\frac{\partial v_x}{\partial x}+v_y\frac{\partial v_x}{\partial y}+\right)\mathbf{i}+\left(v_x\frac{\partial v_y}{\partial x}+v_y\frac{\partial v_y}{\partial y}+\right)\mathbf{j}$$

$$=\{[-2xye^{-2t}\mathbf{i}-2y^2e^{-2t}\mathbf{j}]+e^{-4t}[xy(y)+y^2(x)]\mathbf{i}+e^{-4t}[xy(0)+y^2(2y)]\mathbf{j}\}\frac{\mathrm{m}}{\mathrm{s}^2} \quad \text{(c)}$$

$$=[(-2xye^{-2t}+2xy^2e^{-4t})\mathbf{i}+(-2y^2e^{-2t}+2y^3e^{-4t})\mathbf{j}]\ \frac{\mathrm{m}}{\mathrm{s}^2}$$

(b) The unit normal to the square is $\mathbf{n}=\mathbf{j}$, and everywhere on the square, $y=2$. Thus

$$Q=\int_{-2}^{0}\int_{0}^{2}2^2e^{-2t}dxdz$$
$$=16e^{-2t}\ \frac{\mathrm{m}^3}{s} \quad \text{(d)}$$

2.3 *Vector spaces*

Vectors in three-dimensional space are defined along with a definition of vector addition and multiplication by a scalar. These operations yield another vector in three-dimensional space. The definitions of these operations, along with the implicit definitions of addition and multiplication of real numbers, lead to eight properties that apply to all vectors in three-dimensional space.

Three-dimensional space is illustrated geometrically. However, this is only an example of the more general and abstract concept of a vector space.

Definition 2.1 A **vector space** is a collection of objects called **vectors**, together with defined operations of vector addition and scalar multiplication that satisfy a set of 10 axioms. Let the collection of objects be collectively called V, and let **u**, **v**, and **w** be arbitrary elements of V. Let α and β be arbitrary elements of an associated scalar field. Then if V is a vector space,

(i) V is closed under addition, that is, $\mathbf{u}+\mathbf{v}$ is an element of V,
(ii) V is closed under scalar multiplication, that is, $\alpha\mathbf{u}$ is an element of V,
(iii) The associative law of addition, $(\mathbf{u}+\mathbf{v}) = \mathbf{w} = \mathbf{u} + (\mathbf{v}+\mathbf{w})$, holds,
(iv) The commutative law of addition, $(\mathbf{u}+\mathbf{v}) = (\mathbf{v}+\mathbf{u})$, holds,
(v) There exists a vector in V, called the zero vector **0**, such that $\mathbf{u}+\mathbf{0}=\mathbf{u}$,
(vi) There exists a vector in V, $-\mathbf{u}$, such that $-\mathbf{u}+\mathbf{u}=\mathbf{0}$,
(vii) For the scalar 1, $(1)\mathbf{u}=\mathbf{u}$,
(viii) The distributive law of scalar multiplication, $(\alpha+\beta)\mathbf{u}=\alpha\mathbf{u}+\beta\mathbf{u}$, holds,
(ix) The associative law of scalar multiplication, $(\alpha\beta)\mathbf{u}=\alpha(\beta\mathbf{u})$, holds, and
(x) The distributive law of addition, $\alpha(\mathbf{u}+\mathbf{v})=\alpha\mathbf{u}+\alpha\mathbf{v}$, holds.

If the associated scalar field is the set of all real numbers, then the vector space is said to be a **real vector space.** If the associated scalar field is the set of complex numbers, then the vector space is called a **complex vector space.**

Example 2.4 Define R^n as the set of all ordered n-tuples of real numbers. Vectors **u** and **v** in R^n are represented as

$$\mathbf{u} = \begin{bmatrix} u_1 \\ u_2 \\ \vdots \\ u_n \end{bmatrix} \quad \mathbf{v} = \begin{bmatrix} v_1 \\ v_2 \\ \vdots \\ v_n \end{bmatrix} \tag{a}$$

The operations of vector addition and scalar multiplication are defined such that

$$\mathbf{u}+\mathbf{v} = \begin{bmatrix} u_1+v_1 \\ u_2+v_2 \\ \vdots \\ u_n+v_n \end{bmatrix} \quad \alpha\mathbf{u} = \begin{bmatrix} \alpha u_1 \\ \alpha u_2 \\ \vdots \\ \alpha u_n \end{bmatrix} \tag{b}$$

Show that R^n is a vector space under these definitions of vector addition and multiplication by a scalar.

Solution Certainly R^n is closed under vector addition and scalar multiplication. The zero vector is the vector whose components are all zero, and $-\mathbf{u}$ is defined as $(-1)\mathbf{u}$. It is easy to show that the other axioms defining a vector space hold for R^n under these definitions of vector addition and scalar multiplication. For example, consider the distributive law of scalar multiplication,

$$(\alpha+\beta)\mathbf{u} = \begin{bmatrix} (\alpha+\beta)u_1 \\ (\alpha+\beta)u_2 \\ \vdots \\ (\alpha+\beta)u_n \end{bmatrix} \tag{c}$$

However, the distributive law holds for scalar multiplication of real numbers, that is, $(\alpha+\beta)u_i = \alpha u_i + \beta u_i$. Thus

$$(\alpha+\beta)\mathbf{u} = \begin{bmatrix} \alpha u_1+\beta u_1 \\ \alpha u_2+\beta u_2 \\ \vdots \\ \alpha u_n+\beta u_n \end{bmatrix} \tag{d}$$

which using the definition of vector addition can be written as

$$(\alpha+\beta)\mathbf{u}=\begin{bmatrix}\alpha u_1\\ \alpha u_2\\ \vdots\\ \alpha u_n\end{bmatrix}+\begin{bmatrix}\beta u_1\\ \beta u_2\\ \vdots\\ \beta u_n\end{bmatrix} \tag{e}$$

$$=\alpha\mathbf{u}+\beta\mathbf{u}$$

Since the ten axioms of definition 2.1 hold, R^n is a vector space.

The vector space R^n is a generalization of the three-dimensional space R^3 discussed in section 2.2. However, vectors in R^n lack the geometric representation of vectors in R^3. Equation a of Example 2.4 provides an alternate way to Equation 2.3 to write a vector in R^3.

Example 2.5 Let C^n[a,b] represent the set of functions of a real variable, call it x, that are n times differentiable on the real number line within the interval, $a \le x \le b$. If $f(x)$ is an element of C^n[a,b], then for any x, $a \le x \le b$, $f(x)$ is a real number. If $f(x)$ and $g(x)$ belong to C^n[a,b] and α is a real scalar, then for any x, $a \le x \le b$, $f(x)+g(x)$ is the real number that is the scalar sum of $f(x)+g(x)$, and $\alpha f(x)$ is the real number defined by the scalar multiplication of α and $f(x)$. Show that C^n[a,b] is a vector space under the defined operations of vector addition and scalar multiplication.

Solution Clearly, if $f(x)$ and $g(x)$ are n times differentiable, then $f(x)+g(x)$ and $\alpha f(x)$ are also n times differentiable. The zero function is defined as $f(x)=0$ for all x, $a \le x \le b$ and $-f(x)$ is defined as $(-1)f(x)$. It is easy to show that the remaining axioms of definition 2.1 are satisfied and that C^n[a,b] is a vector space under the given definitions of vector addition and scalar multiplication.

R^n and C^n[a,b] are the two vector spaces used most extensively in this study. Other vector spaces used in applications in this book include the following.

P^n[a,b] is the set of all polynomials of degree n or less defined on the interval $a \le x \le b$. The definitions of vector addition and scalar multiplication for vectors in P^n[a,b] are the same as used in the definition of C^n[a,b].

$C^n(\Re)$ is the set of all functions which are n times continuously differentiable in the spatial volume defined by $\Re$. Vectors in $C^n(\Re)$ are functions of three spatial variables, perhaps of the form $f(x,y,z)$, or if cylindrical coordinates are used, $f(r,\theta,z)$. Since $C^n(\Re)$is really a generalization of $C^n[a,b]$, their definitions of vector addition and scalar multiplication are similar. Addition of two vectors in $C^n(\Re)$ is performed at each point in the spatial volume defined by $\Re$.

$S = R^k X C^n[a,b]$ is the set of elements of the form

$$\begin{bmatrix} f_1(x) \\ f_2(x) \\ \vdots \\ f_k(x) \end{bmatrix}$$

where $f_1(x), f_2(x), \ldots, f_k(x)$ are each in $C^n[a,b]$. Definitions of vector addition and scalar multiplication are given by:

$$\mathbf{f}+\mathbf{g} = \begin{bmatrix} f_1(x) \\ f_2(x) \\ \vdots \\ f_k(x) \end{bmatrix} + \begin{bmatrix} g_1(x) \\ g_2(x) \\ \vdots \\ g_k(x) \end{bmatrix} = \begin{bmatrix} f_1(x)+g_1(x) \\ f_2(x)+g_2(x) \\ \vdots \\ f_k(x)+g_k(x) \end{bmatrix} \qquad \alpha\mathbf{f} = \begin{bmatrix} \alpha f_1(x) \\ \alpha f_2(x) \\ \vdots \\ \alpha f_k(x) \end{bmatrix}$$

Definition 2.2 Let S be a set of vectors contained in a vector space V. Then S is a **subspace** of V if S is a vector space in its own right.

To prove that S is a subspace of V, it must be shown that the 10 axioms of definition 2.1 are satisfied by the vectors in S. However, because all vectors in S are also in V and V is a vector space, then by hypothesis, axioms (iii), (iv), (vii), (viii), (ix), and (x) are satisfied. Hence it is only necessary to show that S is closed under vector addition, S is closed under scalar multiplication, the zero vector is in S, and for every **u** in s, −**u** is also in S.

Example 2.6 Determine whether each set of vectors is a subspace of R^3.

(a) Let S be the set of vectors in R^3 whose components sum to zero. That is, if **u** is in S, then $u_1 + u_2 + u_3 = 0$.

(b) Let S be the set of vectors in R^3 whose components sum to 1, $u_1 + u_2 + u_3 = 1$.

Solution (a) If **u** and **v** are in S, then

$$\mathbf{u}+\mathbf{v} = \begin{bmatrix} u_1+v_1 \\ u_2+v_2 \\ u_3+v_3 \end{bmatrix} \tag{a}$$

and the sum of the components of $\mathbf{u}+\mathbf{v}$ is

$$\begin{aligned}(u_1+v_1)+(u_2+v_2)+(u_3+v_3) &= (u_1+u_2+u_3)+(v_1+v_2+v_3) \\ &= 0+0=0\end{aligned} \tag{b}$$

Thus S is closed under vector addition.

The sum of the components of $\alpha\mathbf{u}$ is $(\alpha u_1)+(\alpha u_2)+(\alpha u_3)=\alpha(u_1+u_2+u_3)$ $=(\alpha)0=0$. Thus S is closed under scalar multiplication. The zero vector is in S, as is $-\mathbf{u}$. Thus since V is a vector space and axioms (i), (ii), (v), and (vi) of definition 2.1 are satisfied, then S is a subspace of V.

(b) S is not a subspace of V. S is not closed under either vector addition or under scalar multiplication. If **u** and **v** are in S, then $u_1+u_2+u_3=1$ and $v_1+v_2+v_3=1$. Then if $\mathbf{w}=\mathbf{u}+\mathbf{v}$, $w_1+w_2+w_3=(u_1+v_1)+(u_2+v_2)+(u_3+v_3)=2$. Thus, by definition of S, **w** is not in S.

Example 2.7 The boundary-value problem for the nondimensional temperature distribution in one dimension, $\Theta(x)$, in a wall with both sides fixed at the same temperature and with a nonuniform internal heat generation, is of the form

$$\frac{d^2\Theta}{dx^2}=u(x)$$

$$\Theta(0)=0$$

$$\Theta(1)=0$$

The solution must be twice differentiable and satisfy both boundary conditions. Let S be the set of functions in $C^2[0,1]$ that satisfy the conditions $f(0)=0$ and $f(1)=0$. Is S, the set of possible solutions to the boundary value problem, a subspace of $C^2[0,1]$?

Solution If $f(x)$ and $g(x)$ are in S, then $f(0)=0$, $f(1)=0$, $g(0)=0$, and $g(1)=0$. Hence $f(0)+g(0)=0$ and $f(1)+g(1)=0$. Thus S is closed under addition. Moreover, S is closed under scalar multiplication, because $\alpha f(0)=0$. The function $f(x)=0$ satisfies $f(0)=0$ and $f(1)=0$, and hence the zero vector is in S. Also, $-f(0)=0$ and $-f(1)=0$, thus $-f(x)$ is also in S. Hence since axioms (i), (ii), (v), and (vi) of definition 2.1 are satisfied, then S is a subspace of $C^2[0,1]$.

Example 2.8 Let $V=P^4[0,1]\cap S$, where S is the subspace of $C^2[0,1]$ defined in Example 2.7. Show that V is a subspace of S.

Solution An element of V is of the form $p(x)$, a polynomial of degree four or less which satisfies $p(0)=0$ and $p(1)=0$. A polynomial is by definition in $C^2[0,1]$. Clearly the sum of two such polynomials is also a polynomial of degree four or less, and using an argument similar to that of Example 2.7, must also be zero at $x=0$ and $x=1$.Thus V is closed under addition. It is also clear that V is closed under scalar multiplication. Furthermore, $p(x)=0$ is in V, and $-p(x)+p(x)=0$. Hence V is a subspace of S (as well as of $P^4[0,1]$).

2.4 *Linear independence*

The mathematical solutions to problems involving discrete dynamic systems reside in a subspace of the vector space R^n. The mathematical solutions to problems involving continuous- or distributed-parameter systems reside in a subspace of the vector space C^n[a,b]. To facilitate solutions to these problems, it is imperative to understand properties of vector spaces and of sets of vectors within these vector spaces. Approximate solutions are constructed from vectors in a subspace of a vector space in which the exact solution resides. To be able to develop such solutions, it is imperative to understand properties of vector spaces and subspaces. The definitions and theorems in this section provide the foundation for these solutions.

Definition 2.3 A set of vectors is said to be **linearly independent** if when a linear combination of the vectors is set equal to zero, all coefficients in the linear combination must be zero. If the vectors are not linearly independent, they are said to be **linearly dependent.**

Definition 2.3 states that if a set of vectors, $\mathbf{u}_1,\mathbf{u}_2,\ldots,\mathbf{u}_k$ is linearly independent, then

$$\sum_{i=1}^{k} c_i \mathbf{u}_i = 0 \tag{2.8}$$

implies that $c_1 = c_2 = \ldots = c_k = 0$. This implies that no vector in the set can be written as a linear combination of the other vectors.

Application of Equation 2.8 for a set of vectors in R^n leads to a set of n equations summarized by the matrix system

$$\mathbf{Uc} = \mathbf{0} \tag{2.9}$$

where $\mathbf{U}$ is a matrix with n rows and k columns whose jth column is the vector $\mathbf{u}_j$, and $\mathbf{c}$ is a column vector of k rows whose ith component is c_i. The vectors are linearly independent if and only if a nontrivial solution to the set of linear equations exists. This leads to the following theorem, presented without proof.

Theorem 2.1 Let $\mathbf{u}_1,\mathbf{u}_2,\ldots,\mathbf{u}_k$ be a set of vectors in R^n. Let $\mathbf{U}$ be the $n \times k$ matrix whose ith column is $\mathbf{u}_i$. If $k > n$, a nontrivial solution of the system exists, and the set of vectors is linearly dependent. If $k = n$, the vectors are linearly dependent if and only if $\mathbf{U}$ is singular. If $k < n$, the vectors are linearly dependent only if every square submatrix of $\mathbf{U}$ is singular.

Application of Equation 2.8 for a set of functions in C[a,b], $f_1(x), f_2(x), \ldots, f_k(x)$, leads to

$$c_1 f_1(x) + c_2 f_2(x) + \ldots c_k f_k(x) = 0 \tag{2.10}$$

For a function to be identically zero on the interval $a \le x \le b$, it must be zero at every value of x. Differentiating Equation 2.10 k-1 times with respect to x leads to

$$\begin{aligned}
&c_1 f_1'(x) + c_2 f_2'(x) + \ldots + c_k f_k'(x) = 0 \\
&c_1 f_1''(x) + c_2 f_2''(x) + \ldots + c_k f_k''(x) = 0 \\
&\qquad \vdots \\
&c_1 f_1^{(k-1)}(x) + c_2 f_2^{(k-1)}(x) + \cdots + c_k f_k^{(k-1)}(x) = 0
\end{aligned} \tag{2.11}$$

Equation 2.10 and Equation 2.11 can be summarized in matrix form as

$$\mathbf{Wc} = \mathbf{0} \tag{2.12}$$

where $\mathbf{W}$ is a $k \times k$ matrix called the Wronskian, defined by

$$\mathbf{W} = \begin{bmatrix} f_1(x) & f_2(x) & f_3(x) & \cdots & f_k(x) \\ f_1'(x) & f_2'(x) & f_3'(x) & \cdots & f_k'(x) \\ f_1''(x) & f_2''(x) & f_3''(x) & \cdots & f_k''(x) \\ \vdots & \vdots & \vdots & \ddots & \vdots \\ f_1^{(k-1)}(x) & f_2^{(k-1)}(x) & f_3^{(k-1)}(x) & \cdots & f_k^{(k-1)}(x) \end{bmatrix} \tag{2.13}$$

and $\mathbf{c}$ is a column vector of k rows whose ith component is c_i. Equation 2.12 has a nontrivial solution if and only if $\mathbf{W}$ is singular for every value of x, $a \le x \le b$. If there exists at least one value of x such that $\mathbf{W}(x)$ is nonsingular, then the set of vectors is linearly independent. This leads to the following theorem, which is presented without formal proof.

Theorem 2.2 A set of functions in C[a,b], $f_1(x), f_2(x), \ldots, f_k(x)$ is linearly dependent if and only if the Wronskian for the set of functions, defined by Equation 2.13 is singular for all x, $a \le x \le b$.

Example 2.9 Show that the set of functions,

$$f_1(x) = 1 \quad f_2(x) = x \quad f_3(x) = x^2$$

is linearly independent in C[0,1].

Solution The Wronskian of the set can be determined to be

$$\mathbf{W} = \begin{bmatrix} 1 & x & x^2 \\ 0 & 1 & 2x \\ 0 & 0 & 2 \end{bmatrix}$$

The determinant of the Wronskian is

$$|\mathbf{W}| = 2$$

Since the Wronskian is nonsingular for at least one x (actually it is nonsingular for all x), the functions are linearly independent.

Example 2.10 The general form of a polynomial of order n, an element of the vector space $P^n[a,b]$, is

$$p(x) = a_1x^n + a_2x^{n-1} + \ldots + a_nx + a_{n+1} \tag{a}$$

A polynomial of order n is uniquely defined by its coefficients, $a_1, a_2, \ldots, a_{n+1}$, which define a vector in R^{n+1} of the form

$$\mathbf{a} = \begin{bmatrix} a_1 \\ a_2 \\ \vdots \\ a_n \\ a_{n+1} \end{bmatrix} \tag{b}$$

Use theorem 2.1 to determine whether or not the polenomials $p_1(x) = x^2 - 1$, $p_2(x) = 2x^2 - 4x + 2$ and $p_3(x) = x^2 - 4x - 3$, are linearly independent.

Solution These quadratic polynomials can be represented as vectors in R^3 as

$$p_1 = \begin{bmatrix} 1 \\ 0 \\ -1 \end{bmatrix} \quad p_2 = \begin{bmatrix} 2 \\ -4 \\ 2 \end{bmatrix} \quad p_3 = \begin{bmatrix} 1 \\ -4 \\ -3 \end{bmatrix} \tag{c}$$

The matrix **U** as defined in theorem 2.1 is

$$\mathbf{U} = \begin{bmatrix} 1 & 2 & 1 \\ 0 & -4 & -4 \\ -1 & 2 & -3 \end{bmatrix} \tag{d}$$

The determinant of **U** is calculated as 24. Thus the matrix is nonsingular, and the vectors are linearly independent.

2.5 *Basis and dimension*

A linear combination of a set of linearly independent vectors in a vector space V, because of closure, is an element of V. A change in any of the coefficients in the linear combination leads to a different element of V.

Definition 2.4 A set of vectors in a vector space V is said to **span** V if every vector in V can be written as a linear combination of the vectors in the set.

Theorem 2.3 The span of a set of vectors in a vector space V is a subspace of V.

Proof Let U = {$\mathbf{u}_1, \mathbf{u}_2, \cdots, \mathbf{u}_k$} be a set of vectors in V. A vector in S = span{U} is written as

$$a = \sum_{i=1}^{k} a_i \mathbf{u}_i$$

Since S is in V, and V is vector space axioms (iii), (iv), (vii), (viii), (ix), and (x) of definition 2.1 are true. Consider the sum of two vectors in the span of S:

$$\mathbf{a} + \mathbf{b} = \sum_{i=1}^{n} a_i \mathbf{u}_i + \sum_{i=1}^{n} b_i \mathbf{u}_i = \sum_{i=1}^{n} (a_i + b_i) \mathbf{u}_i$$

Hence $\mathbf{a} + \mathbf{b}$ is in S because it can be written as a linear combination of the elements of U. Thus U is closed under vector addition. Closure under scalar multiplication can be similarly shown. The zero vector is obtained by setting each $a_i = 0$, $i = 1, 2, \ldots, n$. The vector $-\mathbf{a}$ is obtained as

$$-\mathbf{a} = \sum_{i=1}^{n} (-a_i) \mathbf{u}_i$$

Hence all ten axioms of definition 2.1 hold, and S is a subspace of V.

Example 2.10 $P^2[0,1]$ is the space of all polynomials of degree two or less defined on the interval $0 \le x \le 1$. It is easy to show that $P^2[0,1]$ is a vector space and is a subspace of $C^n[0,1]$. Does the set

$$P = \{p_1(x), p_2(x), p_3(x)\} \quad p_1(x) = 1, \; p_2(x) = x, \; p_3(x) = x^2 \tag{a}$$

span $P^2[0,1]$?

Solution A linear combination of the set of vectors is

$$p(x) = a + bx + cx^2 \tag{b}$$

Every polynomial of degree two or less can be obtained by varying the coefficients in the linear combination. Hence the set spans $P^2[0,1]$.

The span of a vector space is not unique. Indeed, the number of vectors in a span is also not unique. If the set P of Example 2.10 is augmented to $p_4(x) = x^2 + 2x + 3$ then the augmented set also spans $P^2[0,1]$. It is shown in Example 2.10 that the vectors in P are linearly independent. However, since $p_4(x) = 3p_1(x) + 2p_2(x) + p_3(x)$, the augmented set is not a set of linearly independent vectors.

Definition 2.5 A set of linearly independent vectors that spans a vector space V is called a **basis** for V.

The choice of basis vectors is not unique. However, the number of vectors in the basis is unique. The unit vectors **i**, **j**, and **k** defined as being parallel to the coordinate axes in three-dimensional space form a basis for R^3. An alternate choice for a basis is a set of vectors parallel to the coordinate axes obtained by rotating the *x-y* axes about the *z* axis through a counterclockwise angle θ, as illustrated in Figure 2.5. These vectors are related to **i**, **j**, and **k** by $\mathbf{i}=\mathbf{i}'\sin\theta-\mathbf{j}'\cos\theta$, $\mathbf{j}=\mathbf{i}'\cos\theta+\mathbf{j}'\sin\theta$, and $\mathbf{k}'=\mathbf{k}$. The vectors $\mathbf{i}', \mathbf{j}'$, and $\mathbf{k}'$ are also a basis for R^3.

By definition, the set P, defined in Example 2.10, spans $P^2[0,1]$. Indeed, any set of three linearly independent polynomials of degree two or less constitutes a basis for $P^2[0,1]$. Not all vectors in $P^2[0,1]$ can be written as a linear combination of only two linearly independent polynomials. A set of four polynomials of degree two or less is not linearly independent. Thus the number of elements in the basis for $P^2[0,1]$ is exactly three.

Definition 2.6 The number of vectors in the basis of a vector space V is called the **dimension** of V.

The vector space R^n has dimension n. The space $P^n[a, b]$ has dimension $n+1$. These are examples of **finite-dimensional vector spaces**.

It is not possible to express every element in $C^2[0,1]$ by a linear combination of a finite number of elements of $C^2[0,1]$. A basis for $C^2[0,1]$ contains an infinite, number of elements. This is an example of an **infinite-dimensional vector space**. Any set of vectors from which any element of an infinite-dimensional vector space can be written is said to be **complete** in that space.

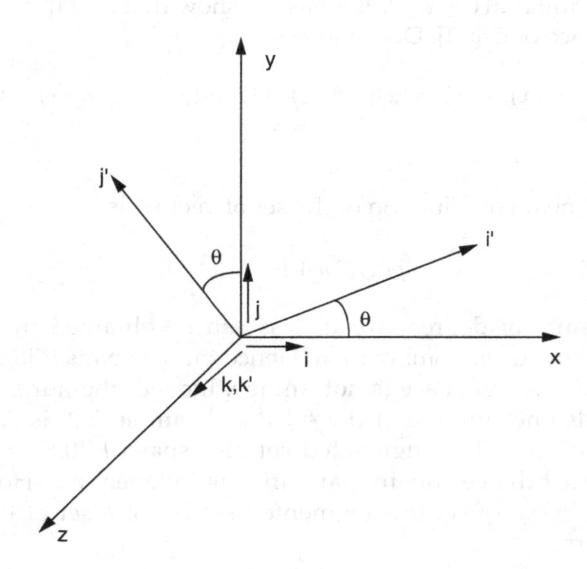

Figure 2.5 The choice of a basis is not unique. The set of vectors **i',j',k'** is a basis for R^3, as is the set of vectors **i,j,k**.

The subspace of Example 2.7 is also an infinite-dimensional vector space. It is important to determine a complete set of vectors for such a space.

The expression of a vector in an infinite-dimensional vector space in terms of members of a complete set (or basis) is an infinite series. Thus questions of convergence arise. Such questions include whether the series converges at every point within the interval [a,b], and if so, then to what value does it converge. While these questions are interesting, they are not within the scope of this study. Theorems regarding convergence of series will be presented without proof and subsequently used. Occasionally questions of convergence will not be considered. This does not mean that they are not important, only that their consideration does not improve the understanding of the material being presented. Such a case arises in the proof of theorem 2.4. If the number of vectors in U is not finite, then questions about series convergence arise when considering the closure of S under vector addition and scalar multiplication. However, it is assumed that the series for **a** and **b** converge, and hence their sum converges.

Example 2.11 The determination of the natural frequencies and mode shapes for a nonuniform beam requires the solution of a variable-coefficient differential equation of the form

$$\frac{d^2}{dx^2}\left(EI(x)\frac{d^2w}{dx^2}\right)+\rho A(x)\omega^2 w=0 \tag{a}$$

where x, the independent variable, is a coordinate along the neutral axis of the beam, measured from its left support, $w(x)$, the dependent variable, is the transverse deflection of the beam, E is the elastic modulus of the material from which the beam is made, ρ is the mass density of the material, $I(x)$ is the area moment of inertia of the nonuniform cross-section, $A(x)$ is the area of the cross-section, and ω is the natural frequency which is to be determined. After nondimensionalization, Equation a becomes

$$\frac{d^2}{dx^2}\left(\alpha(x)\frac{d^2w}{dx^2}\right)+\beta(x)\omega^2 w=0 \tag{b}$$

The nondimensional boundary conditions for a fixed-fixed beam are

$$\begin{aligned} w(0)&=0 \qquad \frac{dw}{dx}(0)=0 \\ w(1)&=0 \qquad \frac{dw}{dx}(1)=0 \end{aligned} \tag{c}$$

Since the beam is nonuniform, determination of an exact solution for the mode shapes and natural frequencies is difficult. Thus an approximate solution is sought. The exact solution resides in a subspace of $C^4[0,1]$, call it S, such that if *f(x)* is in S, then

$$f(0)=0 \quad \frac{df}{dx}(0)=0$$
$$f(1)=0 \quad \frac{df}{dx}(1)=0 \tag{d}$$

An approximate solution is sought from $P^6[0,1]$, the space of polynomials of degree six or less. To satisfy the boundary conditions, the approximate solution must also lie in S. Thus it is desired to seek an approximate solution from the subspace of S and $P^6[0,1]$ defined as $Q = S \cap P^6[0,1]$.

Determine the dimension of Q and determine a basis for Q.

Solution Let *f(x)* be an element of Q. Since *f(x)* is in $P^6[0,1]$, then it is of the form

$$f(x)=a_6x^6+a_5x^5+a_4x^4+a_3x^3+a_2x^2+a_1x+a_0 \tag{e}$$

Since *f(x)* is also in S, it must satisfy the boundary conditions. To this end,

$$f(0)=0=a_0 \tag{f}$$

$$\frac{df}{dx}(0)=0=a_1 \tag{g}$$

$$f(1)=0=a_6+a_5+a_4+a_3+a_2+a_1+a_0 \tag{h}$$

$$\frac{df}{dx}(1)=0=6a_6+5a_5+4a_4+3a_3+2a_2+a_1 \tag{i}$$

Equation f and Equation g obviously imply that $a_0=0$ and $a_1=0$. Equation h and Equation i can be manipulated to give

$$a_2=a_4+2a_5+3a_6 \tag{j}$$

$$a_3=-2a_4-3a_5-4a_6 \tag{k}$$

Thus an arbitrary element of Q is of the form

$$f(x)=a_6x^6+a_5x^5+a_4x^4-(2a_4+3a_5+4a_6)x^3+(a_4+2a_5+3a_6)x^2 \tag{l}$$

$$f(x)=a_6(x^6-4x^3+3x^2)+a_5(x^5-3x^3+2x^2)+a_4(x^4-2x^3+x^2) \tag{m}$$

Since $f(x)$ is a linear combination of three linearly independent functions, the dimension of Q is three, and a basis for Q is

$$f_1(x) = x^6 - 4x^3 + 3x^2 \tag{n}$$

$$f_2(x) = x^5 - 3x^3 + 2x^2 \tag{o}$$

$$f_3(x) = x^4 - 2x^3 + x^2 \tag{p}$$

Example 2.12 The nondimensional partial differential equation governing the steady-state temperature distribution $\Theta(x,y)$ in a thin square slab is

$$\frac{\partial^2\Theta}{\partial x^2} + \frac{\partial^2\Theta}{\partial y^2} = u(x,y) \tag{a}$$

where $u(x,y)$ is a function describing nondimensional internal heat generation. Consider a slab in which the boundary conditions are

$$\Theta(0,y) = 0 \tag{b1}$$

$$\frac{\partial\Theta}{\partial x}(1,y) = 0 \tag{b2}$$

$$\Theta(x,0) = 0 \tag{b3}$$

$$\frac{\partial\Theta}{\partial y}(x,1) + 2\Theta(x,1) = 0 \tag{b4}$$

where the Biot number, Bi = 2. Equation b1 and Equation b3 represent geometric boundary conditions, while Equation b2 and Equation b4 represent natural boundary conditions. Define $C^2[\Re]$ as the vector space of twice differentiable functions defined everywhere in the region $\Re$ of the x-y plane defined such that if (x,y) is in $\Re$, then $0 \le x \le 1$ and $0 \le x \le 1$. An element of $C^2[\Re]$ is of the form $f(x,y)$.

Let Q be the subspace of $C^2[\Re]$ spanned by the functions

$$q_1(x,y) = 1,\ q_2(x,y) = x,\ q_3(x,y) = y,\ q_4(x,y) = xy,\ q_5(x,y) = x^2,\ q_6(x,y) = y^2,$$
$$q_7(x,y) = xy^2,\ q_8(x,y) = x^2y,\ q_9(x,y) = x^3 \text{ and } q_{10}(x,y) = y^3$$

(a) Define G as a subspace of $C^2[\Re]$ such that if $f(x,y)$ is in G, then $f(0,y) = 0$ and $f(x,0) = 0$. That is, if $f(x,y)$ is in N, then f satisfies the geometric boundary conditions. Determine a basis for $Q \cap G$.

(b) Define N as a subspace of $C^2[\Re]$ such that if $f(x,y)$ is in N, then $f(0,y)=0$, $\partial f/\partial x\,(1,y)=0$, $f(x,0)=0$, and $\partial f/\partial y\,(x,1)+2f(x,1)=0$. That is, if $f(x,y)$ is in N, then f satisfies all boundary conditions. Determine a basis for $Q\cap N$.

Solution A vector in Q is of the form

$$q=C_1+C_2x+C_3y+C_4xy+C_5x^2+C_6y^2+C_7xy^2+C_8x^2y+C_9x^3+C_{10}y^3 \quad \text{(c)}$$

If q is also in G, then

$$q(0,y)=0=C_1+C_3y+C_6y^2+C_{10}y^3 \quad \text{(d)}$$

$$q(x,0)=0=C_1+C_2x+C_5x^2+C_9x^3 \quad \text{(e)}$$

Since powers of y are linearly independent, Equation d is satisfied for all y only if

$$C_1=C_3=C_6=C_{10}=0 \quad \text{(f)}$$

Similarly, since powers of x are linearly independent, Equation e is satisfied only if

$$C_2=C_5=C_9=0 \quad \text{(g)}$$

Equation c can then be reduced to

$$q(x,y)=C_4xy+C_7xy^2+C_8x^2y \quad \text{(h)}$$

Thus, one choice of a basis for $Q\cap G$ is the set $\{xy,xy^2,x^2y\}$.
(c) An element of G is also in N. Thus an element in $Q\cap G$ is of the form of Equation h, but must also satisfy

$$\frac{dq}{dx}(1,y)=0=C_4+C_7y^2+2C_8y \quad \text{(i)}$$

Since powers of y are linearly independent, Equation i is satisfied only if $C_4=C_7=C_8=0$. Hence the intersection of Q and N is the null set.

2.6 Inner products

A function of a vector is an operation that can be performed on all vectors in a vector space. The operation may involve more than one vector from the vector space. If the result of the operation is a scalar, the function is called a **scalar function.** An example of a scalar function is the dot product defined for two vectors in R^3. Properties satisfied by the dot product are summarized in section 2.2. The concept of dot product is generalized into the following definition.

Definition 2.7 Let **u** and **v** be arbitrary vectors in a vector space V. An **inner product** of **u** and **v**, written (**u**,**v**), is a scalar function which satisfies the following properties:

i. $(\mathbf{u},\mathbf{v}) = (\overline{\mathbf{v}},\overline{\mathbf{u}})$: (Commutative property)
ii. $(\mathbf{u}+\mathbf{v},\mathbf{w}) = (\mathbf{u},\mathbf{w}) + (\mathbf{v},\mathbf{w})$: (Distributive property)
iii. $(\alpha\,\mathbf{u},\mathbf{v}) = \alpha\,(\mathbf{u},\mathbf{v})$: (Associative property of scalar multiplication)
iv. $(\mathbf{u},\overline{\mathbf{u}}) \geq 0$ and $(\mathbf{u},\overline{\mathbf{u}}) = 0$ if and only if $\mathbf{u} = \mathbf{0}$: (Non-negativity property)

where α is an arbitrary scalar and an overbar denotes the complex conjugate.

Example 2.13 Which of the following constitute a valid definition of an inner product for vectors in R^3?
(a) $(\mathbf{u},\mathbf{v}) = u_1v_1 + u_2v_2 + u_3v_3$
(b) $(\mathbf{u},\mathbf{v}) = u_1v_1 + 2u_2v_2 + u_3v_3$
(c) $(\mathbf{u},\mathbf{v}) = u_1v_1 - u_2v_2 + u_3v_3$
(d) $(\mathbf{u},\mathbf{v}) = u_1v_1 + u_2v_2 + u_3v_2$
(e) $(\mathbf{u},\mathbf{v}) = u_1v_1 + u_2v_2 + u_3^2v_3^2$

Solution To determine whether each of these expressions constitutes a valid inner product on R^3, it must be determined whether or not all the properties of definition 2.7 are satisfied.

(a) Equation a is the same as the definition of the dot product of two vectors, from which the definition of inner product is a generalization, and therefore Equation a defines a valid inner product for R^3.

(b) Equation b defines a valid inner product for R^3. Its satisfaction of the properties of a valid inner product is shown below:

i. $(\mathbf{u},\mathbf{v}) = u_1v_1 + 2u_2v_2 + u_3v_3 = v_1u_1 + 2v_2u_2 + v_3u_3 = (\mathbf{v},\mathbf{u})$

ii.
$$\begin{aligned}(\mathbf{u}+\mathbf{v},\mathbf{w}) &= (u_1+v_1)w_1 + 2(u_2+v_2)w_2 + (u_3+v_3)w_3 \\ &= u_1w_1 + v_1w_1 + 2u_2w_2 + 2v_2w_2 + u_3w_3 + v_3w_3 \\ &= (u_1w_1 + 2u_2w_2 + u_3w_3) + (v_1w_1 + 2v_2w_2 + v_3w_3) \\ &= (\mathbf{u},\mathbf{w}) + (\mathbf{v},\mathbf{w})\end{aligned}$$

iii.
$$\begin{aligned}(\alpha\mathbf{u},\mathbf{v}) &= (\alpha u_1)v_1 + 2(\alpha u_2)v_2 + (\alpha u_3 v_3) \\ &= \alpha(u_1v_1 + 2u_2v_2 + u_3v_3) \\ &= \alpha(\mathbf{u},\mathbf{v})\end{aligned}$$

iv. $(\mathbf{u},\mathbf{u}) = u_1^2 + 2u_2^2 + u_3^2$. Since this is the sum of non-negative terms, it is non-negative for any **u**. In addition, it is clear that $(\mathbf{u},\mathbf{u}) \geq 0$ if and only if $\mathbf{u} = \mathbf{0}$.

(c) Equation c is not a definition of a valid inner product for R^3. Consider, for example, the vector

$$\mathbf{u} = \begin{bmatrix} 1 \\ 1 \\ 0 \end{bmatrix}$$

Using the proposed definition,

$$(\mathbf{u},\mathbf{u}) = (1)(1) - (1)(1) + (0)(0) = 0$$

Thus there exists a $\mathbf{u} \neq \mathbf{0}$ such that $(\mathbf{u},\mathbf{u}) = 0$, and property iv is violated.

(d) Equation d does not represent a valid inner product for R^3. Consider the vector

$$\mathbf{u} = \begin{bmatrix} 0 \\ 1 \\ 0 \end{bmatrix}$$

Using the proposed definition of inner product,

$$(\mathbf{u},\mathbf{u}) = (0)(0) + (1)(0) + (0)(1) = 0$$

Thus there exists a $\mathbf{u} \neq \mathbf{0}$ such that $(\mathbf{u},\mathbf{u}) = 0$, and property (iv) is violated.

(e) Equation e does not define a valid inner product for R^3. Consider property (ii):

$$(\mathbf{u}+\mathbf{v},\mathbf{w}) = (u_1 + v_1)w_1 + (u_2 + v_2)w_2 + (u_3 + v_3)^2 w_3^2$$

It is easy to find a set of vectors for which this is not equal to $(\mathbf{u},\mathbf{w}) + (\mathbf{v},\mathbf{w})$.

The definition of an inner product is not unique. There are many valid definitions of an inner product for any vector space. The inner product for R^n defined by

$$(\mathbf{u},\mathbf{v}) = \sum_{i=1}^{n} u_i v_i \tag{2.14}$$

is called the **standard inner product for R^n**. The standard inner product on C[a,b] is defined by

$$(f,g) = \int_a^b f(x)g(x)\,dx \tag{2.15}$$

Example 2.14 Let S be a subspace of $C^2[0,1]$ such that if $f(x)$ is an element of S, then $f(0)=0$ and $f(1)=0$. Show that the scalar product defined as

$$(f,g)_L = \int_0^1 \frac{d^2 f}{dx^2} g dx \quad \text{(a)}$$

is a valid inner product on S. The subscript on the definition of the scalar product is used to distinguish its definition from that of the standard inner product defined for all of $C^2[0,1]$.

Solution The scalar product defined in Equation a is a valid inner product on S if properties (i)–(iv) are true when **u** is replaced by $f(x)$ and **v** is replaced by $g(x)$ for all $f(x)$ and $g(x)$ in S. Considering property (i),

$$(f,g)_L = \int_0^1 \frac{d^2 f}{dx^2} g(x) dx \quad \text{(b)}$$

Integration by parts ($\int u dv = uv - \int v du$) is applied to the right-hand side of Equation b with $u = g(x)$ and $dv = (d^2 f/dx^2)dx$. These choices lead to $du = (dg/dx)dx$ and $v = df/dx$. Then use of integration by parts for Equation b leads to

$$(f,g)_L = g(1)\frac{df}{dx}(1) - g(0)\frac{df}{dx}(0) - \int_0^1 \frac{df}{dx}\frac{dg}{dx} dx \quad \text{(c)}$$

Application of integration by parts to Equation c with $u = dg/dx$ and $dv = (df/dx)dx$ leads to

$$(f,g)_L = g(1)\frac{df}{dx}(1) - g(0)\frac{df}{dx}(0) - \frac{dg}{dx}(1)f(1) + \frac{dg}{dx}(0)f(0) + \int_0^1 \frac{d^2 g}{dx^2} f dx \quad \text{(d)}$$

Since f and g are in S, $f(0)=0$, $f(1)=0$, $g(0)=0$, and $g(1)=0$, and Equation d reduces to

$$\begin{aligned}(f,g)_L &= \int_0^1 \frac{d^2 g}{dx^2} f dx \\ &= (g,f)_L\end{aligned} \quad \text{(e)}$$

The conclusion drawn in the last line of Equation e derives from the definition of the scalar product in Equation a.

Considering property (ii) of definition 2.7,

$$(f+g,h)_L = \int_0^1 \left(\frac{d^2f}{dx^2}+\frac{d^2g}{dx^2}\right)h(x)dx$$

$$= \int_0^1 \frac{d^2f}{dx^2}h(x)dx + \int_0^1 \frac{d^2g}{dx^2}h(x)dx \tag{f}$$

$$= (f,h)_L + (g,h)_L$$

Considering property (iii) of definition 2.7,

$$(\alpha f,g)_L = \int_0^1 \left(\alpha\frac{d^2f}{dx^2}\right)g(x)dx$$

$$= \alpha\int_0^1 \frac{d^2f}{dx^2}g(x)dx \tag{g}$$

$$= \alpha(f,g)_L$$

Considering property (iv) of definition 2.7,

$$(f,f)_L = \int_0^1 \frac{d^2f}{dx^2}f(x)dx \tag{h}$$

Integration by parts is applied to Equation h, leading to Equation c, but with $g(x)$ replaced by $f(x)$:

$$(f,f)_L = f(1)\frac{df}{dx}(1) - f(0)\frac{df}{dx}(0) - \int_0^1 \frac{df}{dx}\frac{df}{dx}dx \tag{i}$$

Since f is in S, $f(0)=0$ and $f(1)=0$, and Equation i becomes

$$(f,f)_L = \int_0^1 \left(\frac{df}{dx}\right)^2 dx \tag{j}$$

Equation j shows that $(f,f)_L \geq 0$ and $(0,0)=0$. Assume that $g(x)$ is an element of S with $(g,g)=0$. For this inner product to be zero, Equation j requires that $dg/dx=0$ for all x, $0 \leq x \leq 1$. Thus $g(x)=C$, a constant. However, since g is in S, $g(0)=0$ which implies that $C=0$, and thus $g(x)=0$. Hence property (iv) of definition 2.7 is satisfied.

Since the scalar product of Equation a satisfies the four properties of definition 2.7, Equation a defines a valid inner product on S. However, the scalar product has been shown to be a valid inner product on S only. It does not satisfy all properties of an inner product for all of $C^2[0,1]$.

Theorem 2.4 (Cauchy-Schwartz Inequality). Let $(\mathbf{u},\mathbf{v})$ represent a valid inner product defined for a real vector space V. Then

$$(\mathbf{u},\mathbf{v})^2 \leq (\mathbf{u},\mathbf{u})(\mathbf{v},\mathbf{v}) \tag{2.16}$$

Proof The proof of the Cauchy-Schwartz inequality involves the use of properties of inner products and the algebra of inner products and is thus instructive in its own right. From property (iv) of definition 2.7, for any real α and β, it should be noted that

$$(\alpha\mathbf{u}-\beta\mathbf{v},\alpha\mathbf{u}-\beta\mathbf{v}) \geq 0 \tag{2.17}$$

Use of property (iii) of definition 2.7 leads to

$$(\alpha\mathbf{u},\alpha\mathbf{u})+(\alpha\mathbf{u},-\beta\mathbf{v})+(-\beta\mathbf{v},\alpha\mathbf{u})+(-\beta\mathbf{v},-\beta\mathbf{v}) \geq 0 \tag{2.18}$$

Since V is a real vector space, property (i) of definition 2.7 becomes the commutative property. Using properties (i) and (iii) Equation (2.18) is rewritten as

$$\alpha^2(\mathbf{u},\mathbf{u})-2\alpha\beta(\mathbf{u},\mathbf{v})+\beta^2(\mathbf{v},\mathbf{v}) \geq 0 \tag{2.19}$$

Specifically define $\alpha=(\mathbf{v},\mathbf{v})^{1/2}$ and $\beta=(\mathbf{u},\mathbf{u})^{1/2}$; substitution into Equation 2.19 leads to

$$\begin{gathered}(\mathbf{v},\mathbf{v})(\mathbf{u},\mathbf{u})-2(\mathbf{u},\mathbf{u})^{\frac{1}{2}}(\mathbf{v},\mathbf{v})^{\frac{1}{2}}(\mathbf{u},\mathbf{v})+(\mathbf{u},\mathbf{u})(\mathbf{v},\mathbf{v}) \geq 0 \\ (\mathbf{u},\mathbf{u})^{\frac{1}{2}}(\mathbf{v},\mathbf{v})^{\frac{1}{2}} \geq (\mathbf{u},\mathbf{v})\end{gathered} \tag{2.20}$$

which upon squaring becomes Equation 2.16.

Definition 2.8 Two vectors **u** and **v** in a vector space V are said to be orthogonal with respect to a defined inner product if $(\mathbf{u},\mathbf{v})=0$.

The use of the term orthogonality is a generalization of the concept from R^3 in which two vectors are geometrically perpendicular or orthogonal when their dot product is zero. However note that, in general, the determination of orthogonality of two vectors is dependent upon the choice of inner product. Orthogonality of two vectors with respect to one inner product defined on a vector space V does not guarantee orthogonality with respect to any other valid inner product.

Example 2.15 Consider two vectors in R^3 defined as

$$\mathbf{u} = \begin{bmatrix} 1 \\ 3 \\ -2 \end{bmatrix} \qquad \mathbf{v} = \begin{bmatrix} -2 \\ 1 \\ 2 \end{bmatrix}$$

Determine whether these vectors are orthogonal with respect to the inner products defined in parts (a) and (b) of Example 2.13.

Solution (a) Checking orthogonality with respect to the inner product of (a) leads to

$$\begin{aligned}(\mathbf{u},\mathbf{v}) &= u_1 v_1 + u_2 v_2 + u_3 v_3 \\ &= (1)(-2) + (3)(1) + (-2)(2) \\ &= -1\end{aligned}$$

Thus the vectors are not orthogonal with respect to this inner product.

(b) Checking orthogonality with respect to the inner product of (b) leads to

$$\begin{aligned}(\mathbf{u},\mathbf{v}) &= u_1 v_1 + 2u_2 v_2 + u_3 v_3 \\ &= (1)(-2) + 2(3)(1) + (-2)(2) \\ &= 0\end{aligned}$$

Thus the vectors are orthogonal with respect to this inner product.

Example 2.16 Two polynomials in $P^2[0,1]$ are defined by $p_1(x) = a_1 x^2 + a_2 x + a_3$ and $p_2(x) = b_1 x^2 + b_2 x + b_3$. Recalling that $P^2[0,1]$ is a subspace of $C^2[0,1]$, Equation 2.14 is also a valid inner product on $P^2[0,1]$. Example 2.10 shows that a polynomial in $P^2[0,1]$ can be represented by a vector of its coefficients, an element of R^3,whose standard inner product is given by Equation 2.12. Does orthogonality of $p_1(x)$ and $p_2(x)$ with respect to the standard inner product on $C^2[0,1]$ imply orthogonality with respect to the standard inner product on R^3 and vice versa?

Solution The standard inner product on $C^2[0,1]$ of $p_1(x)$ and $p_2(x)$ is

$$(p_1,p_2)_{C^2} = \int_0^1 p_1(x)p_2(x)dx$$

$$= \int_0^1 (a_1x^2 + a_2x + a_3)(b_1x^2 + b_2x + b_3)dx$$

$$= \int_0^1 [a_1b_1x^4 + (a_1b_2 + a_2b_1)x^3 + (a_1b_3 + a_2b_2 + a_3b_1)x^2 \qquad \text{(a)}$$

$$+ (a_2b_3 + a_3b_2)x + a_3b_3]\,dx$$

$$= \frac{1}{5}a_1b_1 + \frac{1}{4}(a_1b_2 + a_2b_1) + \frac{1}{3}(a_1b_3 + a_2b_2 + a_3b_1)$$

$$+ \frac{1}{2}(a_2b_3 + a_3b_2) + a_3b_3$$

The standard inner product on R^3 of these vectors is

$$(p_1,p_2)_{R^3} = a_1b_1 + a_2b_2 + a_3b_3 \qquad \text{(b)}$$

It can be shown by counterexample, using Equation a and Equation b, that orthogonality with respect to one inner product does not imply orthogonality with respect to another inner product. Suppose $p_1(x) = 1$ and $p_2(x) = x$. Then $(p_1,p_2)_{R^{33}} = 0$, but $(p_1,p_2)_{C^2} = 1/2$. Then suppose $p_1(x) = 1$ and $p_2(x) = 2x–1$. In this case, $(p_1,p_2)_{C^2} = 0$ and $(p_1, p_2)_{R^{33}} = –2$.

2.7 *Norms*

It is often important to have a measure of the "length" of a vector. This is especially important when estimating the error in approximating one vector by another vector, such as when approximating the solution of a differential equation. A measure of the error might be the "length" of the error vector, which is defined as the difference between the exact solution and the approximate solution.

The calculation to determine the geometric length of a vector in R^3 is an operation performed on the vector which results in a scalar. Thus the length is a scalar function of the vectors that satisfies certain properties. This concept can be extended to develop a definition for the length of a vector in a general vector space.

Definition 2.9 A **norm** of a vector **u** in a vector space V, written $\|\mathbf{u}\|$, is a function of the vector whose result is a real scalar and that satisfies the following properties:

i. $\|\mathbf{u}\| \geq 0$ and $\|\mathbf{u}\| = 0$ if and only if $\mathbf{u} = \mathbf{0}$.
ii. $\|\alpha\mathbf{u}\| = |\alpha|\|\mathbf{u}\|$ for any scalar α in the associated scalar field of V.
iii. $\|\mathbf{u} + \mathbf{v}\| \leq \|\mathbf{u}\| + \|\mathbf{v}\|$, the triangle inequality.

The norm is a generalization of the concept of length of a vector. The first property requires the length of the vector to be non-negative and that only the zero vector may have a length of zero. The second property is a scaling property: if the vector is multiplied by a scalar, then the length of the resulting product is proportional to the length of the original vector, with the constant of proportionality being the absolute value of the scalar. The third property is a generalization of the triangle rule for vector addition in three-dimensional space, in which a geometric representation of the addition of two vectors is obtained by placing the head of one vector at the tail of the other, with the resultant vector being drawn from the tail of the first vector to the head of the second, the two vectors and their resultant forming a triangle. The triangle inequality is then a statement that the length of any side of a triangle must be less than the sum of the lengths of the other two sides.

Occasionally a function satisfies all requirements to be called a norm except that vectors other than the zero vector will have a norm of zero. This case is covered in the following definition.

Definition 2.10 A **semi-norm** of a vector **u** in a vector space V, **s(u)**, is a function of the vector whose result is a real scalar and that satisfies the following properties:

i. $s(\mathbf{u}) \geq 0$ and $s(\mathbf{u}) = 0$.
ii. $s(\alpha\mathbf{u}) = |\alpha| s(\mathbf{u})$ any scalar α in the associated scalar field of V.
iii. $s(\mathbf{u} + \mathbf{v}) \leq s(\mathbf{u}) + s(\mathbf{v})$, the triangle inequality.

Example 2.17 Show that the following is a valid definition of a norm on C[a,b]:

$$\|f\|_{\infty} = \max_{a \leq x \leq b} |f(x)| \tag{a}$$

Solution The proof that Equation a is a valid definition of a norm requires that properties(i)–(iii) be shown to hold for all *f(x)* in C[a,b]. Clearly, from its definition, $\|f\|_{\infty} \geq 0$ for all *f(x)* and $\|0\|_{\infty} = 0$ Since *f(x)* is a continuous function, if $f(x) \neq 0$ for some x_0, $a \leq x_0 \leq b$, then there is a finite region around x_0 where $f(x) \neq 0$. Thus it is easy to argue that any $f(x) \neq 0$ must have a maximum greater than zero.

Considering property (ii),

$$\begin{aligned}\|\alpha f\|_\infty &= \max_{a\le x\le b}|\alpha f(x)| \\ &= \max_{a\le x\le b}|\alpha||f(x)| \\ &= |\alpha|\max_{a\le x\le b}|f(x)| \\ &= |\alpha|\|f_\infty\|\end{aligned} \tag{b}$$

Finally, considering the triangle inequality,

$$\begin{aligned}\|f+g\|_\infty &= \max_{a\le x\le b}|f(x)+g(x)| \\ &\le \max_{a\le x\le b}(|f(x)|+|g(x)|) \\ &= \max_{a\le x\le b}|f(x)|+\max_{a\le x\le b}|g(x)| \\ &= \|f_\infty\|+\|g\|_\infty\end{aligned} \tag{c}$$

Hence, since the three properties of definition 2.9 are satisfied, $\|f\|_\infty$ constitutes a valid norm on C[0,1].

The definition of a norm for a vector space is not unique. One possible definition of a norm for C[a,b] is presented in Equation a of Example 2.10. Another scalar function defined for vectors in C[a,b] which satisfies the properties of definition 2.9 is

$$\|f\|_1 = \int_a^b |f(x)|\,dx$$

When comparing the closeness of two vectors, the use of $\|f\|_\infty$ determines whether the error between the two functions is small at every value of x, whereas the use of the $\|f\|_1$ norm determines whether the functions are close in an average sense.

Example 2.18 Show that the function

$$s(u) = \left[u_1^2 - 2u_1u_2 + u_2^2\right]^{\frac{1}{2}} \tag{a}$$

defined for a vector **u** in R^2 is a semi-norm for R^2, but is not a valid norm.

Solution Note that

$$s(u) = \left[(u_1-u_2)^2\right]^{\frac{1}{2}} = |u_1 - u_2|$$

and that if

$$\mathbf{u} = \begin{bmatrix} 1 \\ 1 \end{bmatrix}$$

then s(u) = (1 - 1) = 0. Thus since s(u) = 0 for $\mathbf{u} \neq \mathbf{0}$, then if s(u) is a semi-norm, it is not a norm. Clearly $s(\mathbf{u}) \geq 0$ and $s(\mathbf{0}) = 0$. Then consider

$$\begin{aligned} s(\alpha\mathbf{u}) &= \left[\left[(\alpha u_1)^1 - 2(\alpha u_1)(\alpha u_2) + \left(\alpha u_2)^2\right)\right]\right]^{\frac{1}{2}} \\ &= |\alpha| s(\mathbf{u}) \end{aligned}$$

and

$$\begin{aligned} s(\mathbf{u}+\mathbf{v}) &= |(u_1 + v_1) - (u_2 + v_2)| \\ &= \|(\mathrm{u}_1 - u_2) + (v_1 - v_2)\| \\ &\leq |(u_1 - u_2)| + |(v_1 - v_2)| \\ &= s(\mathbf{u}) + s(\mathbf{v}) \end{aligned}$$

Hence $s(\mathbf{u})$ satisfies all properties of a semi-norm.

Theorem 2.5 Let **(u,v)** be a valid inner product defined on a vector space V. Then $\|\mathbf{u}\| = (\mathbf{u},\mathbf{u})^{1/2}$ is a valid norm on V. Such a norm is called an **inner-product-generated norm.**

Proof Since **(u,v)** represents a valid inner product on V, it satisfies the properties of definition 2.7. To prove that $\|\mathbf{u}\|$ represents a valid norm, it must be shown that the properties of definition 2.9 are satisfied.

i. $\|\mathbf{u}\| = (\mathbf{u},\mathbf{u})^{1/2} \geq 0$ $\|u\| = 0$ if and only if $\mathbf{u} = \mathbf{0}$ is true because of property (iv) of definition 2.7.

ii. $\|\alpha\mathbf{u}\| = (\alpha\mathbf{u},\alpha\mathbf{u})^{1/2} = [\alpha\bar{\alpha}(\mathbf{u},\mathbf{u})]^{1/2} = |\alpha|(\mathbf{u},\mathbf{u})^{1/2} = |\alpha|\|\mathbf{u}\|$

iii. Use of the Cauchy-Schwartz inequality, Equation 2.14, leads to

$$\begin{aligned} \|\mathbf{u}+\mathbf{v}\| &\leq \left[(\mathbf{u},\mathbf{u}) + 2(\mathbf{u},\mathbf{u})^{\frac{1}{2}}(\mathbf{v},\mathbf{v})^{\frac{1}{2}} + (\mathbf{v},\mathbf{v})\right]^{\frac{1}{2}} \\ &= \left[\left[(\mathbf{u},\mathbf{u})^{\frac{1}{2}} + (\mathbf{v},\mathbf{v})^{\frac{1}{2}}\right]^2\right]^{\frac{1}{2}} \\ &= \|\mathbf{u}\| + \|\mathbf{v}\| \end{aligned}$$

The $\|f\|_\infty$ norm and the $\|f\|_1$ norm have already been defined for C[a,b]. Theorem 2.4 shows that another valid norm for C[a,b] is the inner-product-generated norm,

$$\|f\| = \left[\int_a^b [f(x)]^2 \, dx \right]^{\frac{1}{2}} \tag{2.21}$$

This norm will be used without subscript and is referred to as the standard inner-product-generated-norm for C[a,b].

Similarly, the standard inner-product-generated norm for R^n is

$$\|\mathbf{u}\| = \left[\sum_{i=1}^{n} u_i^2 \right]^{\frac{1}{2}} \tag{2.22}$$

Example 2.19 Let S be the subspace of $C^2[0,1]$ spanned by $f_1(x) = 1$, $f_2(x) = \sin(\pi x)$, and $f_3(x) = \cos(\pi x)$. Determine a set of unit vectors, with respect to the standard inner-product-generated norm for $C^2[0,1]$, which span S.

Solution A unit vector is a vector with a magnitude of one with respect to a defined norm. A set of unit vectors which span S are $f_1(x)/\|f_1(x)\| \cdot f_2(x)/\|f_2(x)\|$ and $f_3(x)/\|f_3(x)\|$ To this end,

$$\|f_1\| = \left[\int_0^1 (1)^2 dx \right]^{\frac{1}{2}} = 1, \tag{a}$$

$$\|f_2\| = \left[\int_0^1 [\sin(\pi x)]^2 \, dx \right]^{\frac{1}{2}} = \left[\frac{1}{2} \int_0^1 [1 - \cos(2\pi x)] dx \right]^{\frac{1}{2}} = \frac{1}{\sqrt{2}}, \tag{b}$$

$$\|f_3\| = \left[\int_0^1 [\cos(\pi x)]^2\,dx\right]^{\frac{1}{2}}$$

$$= \left[\frac{1}{2}\int_0^1 [1+\cos(2\pi x)]dx\right]^{\frac{1}{2}} \tag{c}$$

$$= \frac{1}{\sqrt{2}},$$

Thus a set of unit vectors which span S are $1, \sqrt{2}\sin(\pi x)$, and $\sqrt{2}\cos(\pi x)$.

2.8 Gram-Schmidt orthonormalization

The dot product defined in three-dimensional space has a geometric interpretation as the length of the projection of one vector onto another multiplied by the length of the second vector. If the two vectors are mutually orthogonal, the projection is zero, and the dot product of the vectors is zero. It is convenient to use a set of mutually orthogonal basis vectors to represent any vector in R^3 as a linear combination of these basis vectors. A common choice for an orthogonal set of basis vectors in R^3 is

$$\mathbf{i} = \begin{bmatrix}1\\0\\0\end{bmatrix} \quad \mathbf{j} = \begin{bmatrix}0\\1\\0\end{bmatrix} \quad \mathbf{k} = \begin{bmatrix}0\\0\\1\end{bmatrix} \tag{2.23}$$

Any vector $\mathbf{u}$ in R^3 can be written as

$$\mathbf{u} = u_1\mathbf{i} + u_2\mathbf{j} + u_3\mathbf{k} \tag{2.24}$$

An orthogonal basis is convenient for the representation of a vector space. It is often easy to obtain a linearly independent basis for a vector space. For example, it is clear that the vectors $p_1(x)=1$, $p_2(x)=x$, and $p_3(x)=x^2$ form a basis for $P^2[a,b]$. However, this basis is not an orthogonal basis for $P^2[a,b]$ with respect to the standard inner product. Thus it is useful to develop a scheme to determine an orthogonal basis which spans the same space as a set of basis vectors.

Definition 2.11 A set of vectors is said to be **normalized** with respect to a norm if the norm of every vector in the set is one.

Definition 2.12 A set of vectors is said to be **orthonormal** with respect to an inner product if the vectors in the set are mutually orthogonal with respect to the inner product and the set is normalized with respect to the inner-product-generated norm.

If a set of vectors $\mathbf{u}_1,\mathbf{u}_2,\ldots,\mathbf{u}_n$ is orthonormal with respect to an inner product, then

$$(\mathbf{u}_i,\mathbf{u}_j)=\delta_{ij}=\begin{cases}0 & i\neq j\\ 1 & i=j\end{cases} \tag{2.25}$$

for all $i,j=1,2,\ldots,n$.

A finite-dimensional vector space V of dimension n has n basis vectors. The following theorem provides a procedure by which another basis for V, orthonormal with respect to any valid inner product on V, may be constructed. If a basis can be found for an infinite-dimensional vector space, the Gram-Schmidt process, outlined in theorem 2.6, can be used to generate an orthonormal basis for the vector space.

Theorem 2.6 (Gram-Schmidt Orthonormalization) Let $\mathbf{u}_1,\mathbf{u}_2,\ldots$ be a finite or countably infinite set of linearly independent vectors in a vector space V with a defined inner product $(\mathbf{u},\mathbf{v})$. Let S be the subspace of V spanned by the vectors. There exists an orthonormal set of vectors $\mathbf{v}_1,\mathbf{v}_2,\mathbf{v}_3,\ldots$ whose span is also S. The members of the orthonormal basis can be calculated sequentially according to

$$\begin{aligned}
&\mathbf{w}_1=\mathbf{u}_1 && \mathbf{v}_1=\frac{\mathbf{w}_1}{\|\mathbf{w}_1\|}\\
&\mathbf{w}_2=\mathbf{u}_2-(\mathbf{u}_2,\mathbf{v}_1)\mathbf{v}_1 && \mathbf{v}_2=\frac{\mathbf{w}_2}{\|\mathbf{w}_2\|}\\
&\vdots && \vdots\\
&\mathbf{w}_n=\mathbf{u}_n-\sum_{i=1}^{n-1}(\mathbf{u}_n,\mathbf{v}_i)\mathbf{v}_i && \mathbf{v}_n=\frac{\mathbf{w}_n}{\|\mathbf{w}_n\|}\\
&\vdots && \vdots
\end{aligned} \tag{2.26}$$

Proof First consider the following lemma:

Lemma A set of vectors which are elements of a vector space V and which are mutually orthogonal with respect to any valid inner product defined on V are linearly independent.

Proof of Lemma Let $\mathbf{v}_1,\mathbf{v}_2,\ldots,\mathbf{v}_k$ be elements of a vector space V which are mutually orthogonal with respect to a valid inner product on V, $(\mathbf{u},\mathbf{v})$, Then $(\mathbf{v}_i,\mathbf{v}_j)=0$ for $i=1,2,\ldots,k$ and for $j=1,2,\ldots,k$, but $j\neq k$. Consider a linear combination of the set of vectors set equal to the zero vector,

$$0=C_1\mathbf{v}_1+C_2\mathbf{v}_2+\ldots+C_k\mathbf{v}_k \tag{a}$$

Taking the inner product of both sides of Equation a with $\mathbf{v}_j$ for an arbitrary j between 1 and k, and using properties of a valid inner product, leads to

$$(0,\mathbf{v}_j)=C_1(\mathbf{v}_1,\mathbf{v}_j)+C_2(\mathbf{v}_2,\mathbf{v}_j)+\ldots+C_1(\mathbf{v}_1,\mathbf{v}_j) \quad \text{(b)}$$

Note that $(0,\mathbf{v}_j)=0$ and that using the mutual orthogonality of the vectors in Equation b leads to $C_j=0$. Since j is arbitrary, $C_j=0$ for all j, $j=1,2,\ldots,k$. Thus the vectors are linearly independent, and the lemma is proved.

Since an orthogonal set of vectors is linearly independent, then if Equations 2.24 generate an orthonormal set, then $\mathbf{v}_1,\mathbf{v}_2,\ldots,\mathbf{v}_n,\ldots$ are linearly independent. Equations 2.24 also show that each of the vectors in the proposed orthonormal set is a linear combination of the vectors in the basis for S, and thus they are also in S. If S is a space of dimension n, then the proposed set of vectors is composed of n linearly independent vectors, and thus they span S. If S is an infinite-dimensional space, then it can be shown that the proposed set is complete in S, but that is beyond the scope of this text.

First note that

$$\|\mathbf{v}_i\|=\left\|\frac{\mathbf{w}_i}{\|\mathbf{w}_i\|}\right\|=\frac{1}{\|\mathbf{w}_i\|}\|\mathbf{w}_i\|=1$$

Hence the set is normalized. It only remains to show that $\mathbf{w}_1$, $\mathbf{w}_2,\ldots$ form an orthogonal set of vectors. This is done by induction. First, it is shown that $\mathbf{w}_2$ is orthogonal to $\mathbf{v}_1$.

$$\begin{aligned}(\mathbf{w}_2,\mathbf{v}_1)&=(\mathbf{u}_2-(\mathbf{u}_2,\mathbf{v}_1),\mathbf{v}_1)\\&=(\mathbf{u}_2,\mathbf{v}_1)-(\mathbf{u}_2,\mathbf{v}_1)(\mathbf{v}_1,\mathbf{v}_1)\end{aligned}$$

However, $\|\mathbf{v}_1\|=1=(\mathbf{v}_1,\mathbf{v}_1)^{1/2}$. Thus

$$(\mathbf{w}_2,\mathbf{v}_1)=(\mathbf{u}_2,\mathbf{v}_1)-(\mathbf{u}_2,\mathbf{v}_1)=0$$

Now assume

$$(\mathbf{v}_i,\mathbf{v}_j)=\delta_{ij} \qquad \text{for i,j}=1,2,\ldots,\text{k}-1$$

and consider

$$\begin{aligned}(\mathbf{w}_k,\mathbf{v}_i)&=(\mathbf{u}_k-\sum_{j=1}^{k-1}(\mathbf{u}_k,\mathbf{v}_j)\mathbf{v}_j,\mathbf{v}_i)\\&=(\mathbf{u}_k,\mathbf{v}_i)-\sum_{j=1}^{k-1}(\mathbf{u}_k,\mathbf{v}_j)(\mathbf{v}_j,\mathbf{v}_i)\end{aligned}$$

Since $(\mathbf{v}_i, \mathbf{v}_j) = 0$ for $i \neq j$ and for $i,j < k$, the only nonzero term in the sum occurs when $j = i$. Thus

$$(\mathbf{w}_k, \mathbf{v}_j) = (\mathbf{u}_k, \mathbf{v}_i) - (\mathbf{u}_k, \mathbf{v}_i)(\mathbf{v}_i, \mathbf{v}_i)$$
$$= (\mathbf{u}_k, \mathbf{v}_i) - (\mathbf{u}_k, \mathbf{v}_i)$$
$$= 0$$

The theorem is thus proved by induction.

If a set of vectors $\mathbf{u}_1, \mathbf{u}_2, \ldots, \mathbf{u}_{k-1}, \mathbf{u}_k, \mathbf{u}_{k+1}, \ldots$ is a complete set of linearly independent vectors in an infinite-dimensional vector space V with a defined inner product, then the Gram-Schmidt process can be used to determine a set of orthonormal vectors that is complete in V.

Example 2.20 A basis for S, the intersection of $P^6[0,1]$ with a subspace of $C^4[0,1]$, defined as those functions that satisfy the boundary conditions for the differential equation governing the vibrations of a fixed-fixed beam, is developed in Example 2.15. One basis for S is determined as

$$f_1(x) = x^6 - 4x^3 + 3x^2$$
$$f_2(x) = x^5 - 3x^3 + 2x^2 \qquad \text{(a)}$$
$$f_3(x) = x^4 - 2x^3 + x^2$$

Use the Gram-Schmidt procedure to determine a basis for S that is orthonormal with respect to the standard inner product for $C^4[0,1]$.

Solution Normalizing f_1,

$$\|f_1\| = \left[\int_0^1 (x^6 - 4x^3 + 3x^2)^2\, dx\right]^{\frac{1}{2}} = 0.171 \qquad \text{(b)}$$

Then

$$v_1(x) = \frac{f_1(x)}{\|f_1(x)\|}$$
$$v_1(x) = 5.84x^6 - 23.4x^3 + 17.5x^2 \qquad \text{(c)}$$

Calculating w_2,

$$w_2(x) = f_2(x) - (f_2(x), v_1(x))v_1(x) \qquad \text{(d)}$$

where

$$(f_2(x), v_1(x)) = \int_0^1 (x^5 - 3x^3 + 2x^2)(5.84x^6 - 23.4x^3 + 17.5x^2)dx = 0.09968 \quad \text{(e)}$$

Thus Equation d becomes

$$\begin{aligned} w_2(x) &= x^5 - 3x^3 + 2x^2 - 0.009968(5.84x^6 - 23.4x^3 + 17.5x^2) \\ &= -0.5823x^6 + x^5 - 0.6708x^3 + 0.253x^2 \end{aligned} \quad \text{(f)}$$

Normalizing,

$$\|w_2(x)\| = \left[\int_0^1 (-0.5823x^6 + x^5 - 0.6708x^3 + 0.253x^2)^2 \, dx \right]^{\frac{1}{2}} = 0.00459 \quad \text{(g)}$$

Thus

$$\begin{aligned} v_2(x) &= \frac{1}{0.00459}\left(-0.5823x^6 + x^5 - 0.6708x^3 + 0.253x^2\right) \\ &= -127.74x^6 + 219.4x^5 - 147.2x^3 + 55.5x^2 \end{aligned} \quad \text{(h)}$$

Calculating $w_3(x)$, using Equation (2.26)

$$w_3(x) = f_3(x) - (f_3(x), v_1(x))v_1(x) - (f_3(x), v_2(x))v_2(x) \quad \text{(i)}$$

where

$$(f_3(x), v_1(x)) = \int_0^1 (x^4 - 2x^3 + x^2)(5.84x^6 - 23.4x^3 + 17.5x^2)\, dx = 0.03962 \quad \text{(j)}$$

$$\begin{aligned} (f_3(x), v_2(x)) &= \int_0^1 (x^4 - 2x^3 + x^2)(-127.74x^6 + 219.4x^5 - 147.2x^3 \\ &\quad + 55.5x^2)\, dx \\ &= 0.00419 \end{aligned} \quad \text{(k)}$$

Substituting Equations j and Equation k into Equation i leads to

$$\begin{aligned} w_3(x) &= x^4 - 2x^3 + x^2 - 0.03962(5.84x^6 - 23.4x^3 + 17.5x^2) \\ &\quad - 0.00419(-127.74x^6 + 219.4x^5 - 147.2x^3 + 55.5x^2) \\ &= 0.3035x^6 - 0.9186x^5 + x^4 - 0.4580x^3 + 0.0732x^2 \end{aligned} \quad (l)$$

Finally,

$$\begin{aligned} \|w_3(x)\| &= \left[\int_0^1 \left(0.3035x^6 - 0.9186x^5 + x^4 - 0.4580x^3 + 0.0732x^2\right)^2 dx \right]^{\frac{1}{2}} \\ &= 0.0003444 \end{aligned} \quad (m)$$

which when substituted into Equation l, leads to

$$v_3(x) = 881.2x^6 - 2667.3x^5 + 2903.6x^4 - 1329.9x^3 + 212.4x^2 \quad (n)$$

Example 2.21 Use the Gram-Schmidt process to derive an orthonormal basis for $P^2[0,1]$ with respect to the standard inner product on R^7. Use the basis $f_1(x)$, $f_2(x)$, and $f_3(x)$ from Example 2.16.

Solution Polynomials in $P^6[0,1]$ can be represented as vectors in R^7, as illustrated in Example 2.8. To this end, the polynomials of Example 2.20 can be represented as

$$p_1 = \begin{bmatrix} 1 \\ 0 \\ 0 \\ -4 \\ 3 \\ 0 \\ 0 \end{bmatrix} \quad p_2 = \begin{bmatrix} 0 \\ 1 \\ 0 \\ -3 \\ 2 \\ 0 \\ 0 \end{bmatrix} \quad p_3 = \begin{bmatrix} 0 \\ 0 \\ 1 \\ -2 \\ 1 \\ 0 \\ 0 \end{bmatrix} \quad (a)$$

The standard inner-product-generated norm for p_1 is

$$\|p_1\| = \sqrt{(1)^2 + (-4)^2 + (3)^2} = \sqrt{26} \quad (b)$$

Thus p_1 can be normalized as

$$v_1 = \frac{p_1}{\|p_1\|} = \frac{1}{\sqrt{26}} \begin{bmatrix} 1 \\ 0 \\ 0 \\ -4 \\ 3 \\ 0 \\ 0 \end{bmatrix} \qquad \text{(c)}$$

Equation 2.26 is used to calculate a vector orthogonal to p_1 as

$$\begin{aligned} w_2 &= p_2 - (p_2, v_1)v_1 \\ &= \begin{bmatrix} 0 \\ 1 \\ 0 \\ -3 \\ 2 \\ 0 \\ 0 \end{bmatrix} - \frac{1}{26}[(-3)(-4) + (2)(3)] \begin{bmatrix} 1 \\ 0 \\ 0 \\ -4 \\ 3 \\ 0 \\ 0 \end{bmatrix} \\ &= \frac{1}{13} \begin{bmatrix} -9 \\ 13 \\ 0 \\ -3 \\ -1 \\ 0 \\ 0 \end{bmatrix} \end{aligned} \qquad \text{(c)}$$

The norm of w_2 is calculated as $\sqrt{260}/13$, and thus

$$v_2 = \frac{1}{\sqrt{260}} \begin{bmatrix} -9 \\ 13 \\ 0 \\ -3 \\ -1 \\ 0 \\ 0 \end{bmatrix} \qquad \text{(d)}$$

Equation 2.26 is used to construct

$$w_3 = \begin{bmatrix} 0 \\ 0 \\ 1 \\ -2 \\ 1 \\ 0 \\ 0 \end{bmatrix} - \frac{4}{13}\begin{bmatrix} 1 \\ 0 \\ 0 \\ -3 \\ 2 \\ 0 \\ 0 \end{bmatrix} - \frac{1}{52}\begin{bmatrix} -9 \\ 13 \\ 0 \\ -3 \\ -1 \\ 0 \\ 0 \end{bmatrix} = \frac{1}{52}\begin{bmatrix} -7 \\ -13 \\ 52 \\ -53 \\ 21 \\ 0 \\ 0 \end{bmatrix} \tag{e}$$

The norm of w_3 is calculated as $\sqrt{6172}/52$ and thus

$$v_3 = \frac{1}{\sqrt{6172}}\begin{bmatrix} -7 \\ -13 \\ 42 \\ -53 \\ 21 \\ 0 \\ 0 \end{bmatrix} \tag{f}$$

Converting the vectors to polynomial form, an orthonormal basis for S with respect to the standard inner product for R^7 is

$$\begin{aligned} v_1(x) &= 0.1961x^6 - 0.7845x^3 + 0.5883x^2 \\ v_2(x) &= -0.5582x^6 + 0.8062x^5 - 0.1861x^3 - 0.0620x^2 \\ v_3(x) &= -0.0819x^6 - 0.1655x^5 + 0.6619x^4 - 0.6746x^3 + 0.2673x^2 \end{aligned} \tag{g}$$

2.9 *Orthogonal expansions*

The Gram-Schmidt Theorem, theorem 2.6, shows that an orthonormal basis for any vector space can be obtained with respect to a valid inner product (**u**,**v**). Let S be a finite-dimensional vector space of dimension n, and let $\mathbf{v}_1, \mathbf{v}_2, \ldots, \mathbf{v}_n$ be a set of orthonormal vectors that span S with respect to a valid inner product on S. Then any **u** in S can be written as a linear combination of the vectors in the orthonormal basis

$$\mathbf{u} = \sum_{i=1}^{n} \alpha_i \mathbf{v}_i \tag{2.27}$$

Taking the inner product of both sides of Equation 2.27 with $\mathbf{v}_j$ for an arbitrary $j=1,2,\ldots,n$ leads to

$$(\mathbf{u},\mathbf{v}_j)=\left(\sum_{i=1}^{n}\alpha_i\mathbf{v}_i,\mathbf{v}_j\right) \tag{2.28}$$

which, using properties (ii) and (iii) of definition 2.7, becomes

$$(\mathbf{u},\mathbf{v}_j)=\sum_{i=1}^{n}\alpha_i(\mathbf{v}_i,\mathbf{v}_j) \tag{2.29}$$

Since the basis is an orthonormal basis, the only nonzero term in the summation on the right-hand side of Equation 2.29 corresponds to $i=j$. Simplification thus leads to

$$\alpha_j=(\mathbf{u},\mathbf{v}_j) \tag{2.30}$$

Substitution of Equation 2.30 into Equation 2.29 leads to

$$\mathbf{u}=\sum_{i=1}^{n}(\mathbf{u},\mathbf{v}_i)\mathbf{v}_i \tag{2.31}$$

Equation 2.31 provides an expansion for $\mathbf{u}$ in terms of the vectors in an orthonormal basis. The coefficient α_i given by Equation 2.30 is the component of $\mathbf{u}$ for the vector $\mathbf{v}_i$.

Example 2.22 The function

$$f(x)=2x^6-x^5+2x^4-9x^3+6x^2$$

is a member of Q, the vector space defined in Examples 2.11 and 2.20. Expand *f(x)* in terms of the orthonormal basis for Q determined in Example 2.20.

Solution Application of Equation 2.31 with $n=3$ leads to

$$f(x)=(f(x),v_1(x))v_1(x)+(f(x),v_2(x))v_2(x)+(f(x),v_3(x))v_3(x) \tag{a}$$

where

$$(f(x),v_1(x))=\int_0^1(2x^6-x^5+2x^4-9x^3+6x^2)(5.84x^6-23.4x^3+17.5x^2)\,dx$$

$$=0.322 \tag{b}$$

$$(f(x), v_2(x)) = \int_0^1 (2x^6 - x^5 + 2x^4 - 9x^3 + 6x^2)(-127.74x^6 + 219.4x^5 - 147.2x^3 + 55.5x^2)\, dx = 0.00382 \tag{c}$$

$$(f(x), v_3(x)) = \int_0^1 (2x^6 - x^5 + 2x^4 - 9x^3 + 6x^2)(881.2x^6 - 2667.3x^5 + 2903.6x^4 - 1329.9x^3 + 212.4x^2) dx = 0.0006888 \tag{d}$$

Thus

$$2x^6 - x^5 + 2x^4 - 9x^3 + 6x^2 = 0.322(5.84x^6 - 23.4x^3 + 17.5x^2) + 0.00382(-127.74x^6 + 219.4x^5 - 147.2x^3 + 55.5x^2) + 0.0006888(881.2x^6 - 2667.3x^5 + 2903.6x^4 - 1329.9x^3 + 212.4x^2) \tag{e}$$

The development of the expansion of a vector in an infinite-dimensional vector space is complicated by questions of completeness and convergence. A detailed discussion of these topics is important, but outside the scope of this study. However, several definitions and concepts are presented and theorems are presented without proof.

Definition 2.13 Let V be an infinite-dimensional vector space with a defined norm. A sequence of vectors $\mathbf{u}_1, \mathbf{u}_2, \mathbf{u}_3, \ldots$ is said to **converge** to a vector $\mathbf{u}$ if

$$\lim_{n\to\infty} \|\mathbf{u}_n - \mathbf{u}\| = 0 \tag{2.32}$$

Definition 2.14 Let V be an infinite-dimensional vector space with a defined norm, and let $\mathbf{u}_1, \mathbf{u}_2, \mathbf{u}_3, \ldots$ be a sequence of vectors in V. The sequence is said to be a **Cauchy sequence** if for each $\varepsilon > 0$, there exists an N such that if $m,n \geq N$, $\|\mathbf{u}_m - \mathbf{u}_n\| < \varepsilon$.

A convergent sequence is also a Cauchy sequence, but the converse is not always true. Consider, for example, the sequence of real numbers defined by $x_n = 1/n$. This sequence of real numbers is a Cauchy sequence, but it does not converge to a real number.

Definition 2.15 Let V be an infinite-dimensional vector space with a defined norm. The vector space is said to be **complete** if every Cauchy sequence in the space converges to an element of the space.

A vector space can be complete with respect to one norm, but not with respect to another. For example, the space C[a,b] is complete with respect to the norm defined $\|f\|_2 = [\int_a^b [f(x)]^2\,dx]^{1/2}$, but C[a,b] is not complete with respect to the norm $\|f\|_\infty = \max_{a \le x \le b} |f(x)|$.

Definition 2.16 An infinite-dimensional space which has an inner product defined on it and which is complete with respect to the inner-product-generated norm is called a **Hilbert space**.

Definition 2.17 An orthonormal set of vectors, S, is said to be **complete** in a Hilbert space if there is no other set of vectors of which S is a subset.

The Gram-Schmidt theorem implies that there is an orthonormal basis for every Hilbert space which is complete in the space. However, not every orthonormal set is complete. The following theorem deals with completeness of the orthonormal set.

Theorem 2.7 (Parseval's identity) Let $\mathbf{u}_1$, $\mathbf{u}_2$, $\mathbf{u}_3$,... be an orthonormal set in a Hilbert space. Then if

$$\|\mathbf{v}\| = \sum_{i=1}^{\infty} (\mathbf{v},\mathbf{u}_i)^2 \tag{2.33}$$

for every $\mathbf{v}$ in the Hilbert space, then the set is complete.

Finally, the following theorem shows how to establish an expansion of a vector in terms of a complete orthonormal set in a Hilbert space.

Theorem 2.8 Let $\mathbf{u}_1, \mathbf{u}_2, \mathbf{u}_3, \cdots$ be a complete orthonormal set in a Hilbert space V, and let $\mathbf{v}$ be an arbitrary element of V. Then

$$\mathbf{v} = \sum_{i=1}^{\infty} (\mathbf{v},\mathbf{u}_i)\mathbf{u}_i \tag{2.34}$$

Example 2.23 It is shown in chapter 5 that the set of functions defined by

$$f_i(x) = \sqrt{2}\sin(i\pi x) \tag{a}$$

is a complete orthonormal set on the Hilbert space V, which is defined as the set of all functions in C[0,1] such that if $y(x)$ is in V, then $y(0) = 0$ and $y(1) = 0$ with the standard inner product for C[0,1]. Expand $f(x) = x^2 - x$, which is a member of V, in terms of this orthonormal set.

Solution The appropriate expansion is

$$x^2 - x = \sum_{i=1}^{\infty} \alpha_i \sqrt{2}\sin(i\pi x) \tag{b}$$

where

$$\alpha_i = \int_0^1 (x^2 - x)\sqrt{2}\sin(i\pi x)\,dx$$
$$= \frac{2\sqrt{2}}{i^3\pi^3}\left[(-1)^i - 1\right] \tag{c}$$

Thus

$$x^2 - x = \frac{4}{\pi^3}\sum_{i=1}^{\infty}\frac{1}{i^3}\left[(-1)^i - 1\right]\sin(i\pi x) \tag{d}$$

The series on the right-hand side of Equation d converges to x^2-x with respect to the standard inner product generated norm defined for C[0, 1] in the sense of definition 2.13.

2.10 Linear operators

Mathematical modeling of physical systems leads to the formulation of a mathematical problem whose solution provides required information about the physical system. It is convenient to examine these equations using a consistent formulation. Let **u** represent a vector of dependent variables which is an element of a vector space $\mathcal{D}$, and let **f** represent a vector, obtained through the modeling process, which is an element of a vector space $\mathcal{R}$. The relationship between **u** and **f** can be written as

$$\mathbf{Lu} = \mathbf{f} \tag{2.35}$$

where **L** is an **operator** determined from the modeling process. A formal definition of an operator follows:

Definition 2.18 An operator **L** is a function by which an element of a vector space $\mathcal{D}$, called the domain of **L**, is mapped into an element of a vector space $\mathcal{R}$, called the range of **L**.

Equation 2.35 is the general form of an operator equation. Given a vector **f** and the definition of **L**, it is desired to find the vector or vectors for which Equation 2.35 is satisfied. The vectors **u** which satisfy Equation 2.35 are said to be **solutions** of the equation. Before considering how the solutions to Equation 2.35 are obtained, two basic questions are considered. (1) Does a solution exist? (2) If so, how many solutions exist?. The first question can be phrased as, "For a specific **f** in $\mathcal{R}$, is there at least one **u** in $\mathcal{D}$ which solves Equation 2.35?" An alternate form of the second question is, "If a solution exists, is it unique?". If for each **f** in $\mathcal{R}$, a unique solution **u**, a vector in $\mathcal{D}$, exists, then there is a one to one correspondence between the elements of $\mathcal{D}$ and $\mathcal{R}$. In this case, an inverse operator $\mathbf{L}^{-1}$ exists whose domain is $\mathcal{R}$ and whose range is $\mathcal{D}$, such that

$$\mathbf{u} = \mathbf{L}^{-1}\mathbf{f} \tag{2.36}$$

Definition 2.19 If **L** is an operator defined such that there is a one-to-one correspondence between the elements of its domain $\mathcal{D}$ and the elements of its range $\mathcal{R}$, then the inverse of **L**, denoted by $\mathbf{L}^{-1}$, exists, with domain $\mathcal{R}$ and range $\mathcal{D}$, and is defined such that if $\mathbf{Lu} = \mathbf{f}$, then $\mathbf{u} = \mathbf{L}^{-1}\mathbf{f}$.

One method of finding the solution of Equation 2.35 is to determine the inverse of **L** and apply Equation 2.36. Note that substitution for **f** from Equation 2.35 into Equation 2.36 leads to

$$\mathbf{L}^{-1}(\mathbf{Lu}) = \mathbf{u} \tag{2.37}$$

If $\mathcal{R} = \mathcal{D}$, then **u** is an element of the domain of $\mathbf{L}^{-1}$, and then Equation 2.35 and Equation 2.36 lead to

$$\mathbf{L}\left(\mathbf{L}^{-1}\mathbf{u}\right) = \mathbf{u} \tag{2.38}$$

Definition 2.20 An operator **L** is said to be a **linear operator** if for each **u** and **v** in $\mathcal{D}$ and for all scalars α and β,

$$\mathbf{L}(\alpha\mathbf{u} + \beta\mathbf{v}) = \alpha\mathbf{Lu} + \beta\mathbf{Lv} \tag{2.39}$$

An equation of the form of Equation 2.35 in which **L** is linear is said to be a linear equation. If **L** is not a linear operator, then Equation 2.35 is a nonlinear equation. The focus of this study is on linear equations.

An $n \times m$ matrix is a linear operator whose domain is R^m and whose range is a subspace of R^n. Consider a set of linear equations to solve for a set of m variables $x_1, x_2, x_3, \ldots, x_m$, as illustrated below:

$$\begin{array}{ccccccccccc} a_{1,1}x_1 & + & a_{1,2}x_2 & + & a_{1,3}x_3 & + & \ldots & + & a_{1,m}x_m & = & y_1 \\ a_{2,1}x_2 & + & a_{2,2}x_2 & + & a_{2,3}x_3 & + & \ldots & + & a_{2,m}x_m & = & y_2 \\ \vdots & & \vdots & & \vdots & & & & \vdots & & \vdots \\ a_{n,1}x_1 & + & a_{n,2}x_2 & + & a_{n,3}x_3 & + & \ldots & + & a_{n,m}x_m & = & y_n \end{array} \tag{2.40}$$

The matrix formulation of this set of equations is

$$\begin{bmatrix} a_{1,1} & a_{1,2} & a_{1,3} & \cdots & a_{1,m} \\ a_{2,1} & a_{2,2} & a_{2,3} & \cdots & a_{2,m} \\ \vdots & \vdots & \vdots & \ddots & \vdots \\ a_{n,1} & a_{n,2} & a_{n,3} & \cdots & a_{n,m} \end{bmatrix} \begin{bmatrix} x_1 \\ x_2 \\ x_3 \\ \vdots \\ x_m \end{bmatrix} = \begin{bmatrix} y_1 \\ y_2 \\ \vdots \\ y_n \end{bmatrix} \tag{2.41}$$

Equation 2.41 can be summarized as

$$\mathbf{Ax} = \mathbf{y} \tag{2.42}$$

where **A** is the matrix operator which represents the coefficients of the equations of Equation 2.40 arranged in n rows and m columns. The element in the ith row and jth column is identified as $a_{i,j}$. The solution **x** is a vector in R^m, while the input vector **y** is a vector in R^n.

When $n > m$, the number of linear equations represented by Equation 2.41 is greater than the dimension of the solution vector. If more than m equations are independent, then a solution does not exist for all **y**. That is, the range of **A** is not all of R^n When $n < m$, the number of equations is less than the dimension of the solution vector. In this case, a solution exists, but is not unique; the range is all of R^n, but for each **y** in R^n, there is more than one **x** in R^m that solves Equation 2.42.

When $n = m$ and the number of equations is equal to the dimension of the solution vector, then the domain of **A** is R^n and the range of **A** is a subspace of R^n. It can be shown that the range is all of R^n when the matrix **A** is nonsingular. That is, its determinant is not equal to zero. When the matrix is nonsingular, then $\mathbf{A}^{-1}$ exists and has the property

$$\mathbf{A}^{-1}(\mathbf{A}\mathbf{x}) = \mathbf{x} \tag{2.43}$$

An associative property can be used on Equation 2.43, giving $(\mathbf{A}^{-1}\mathbf{A})\mathbf{x} = \mathbf{x}$. The operator in parentheses defines the $n \times n$ identity matrix (a matrix with ones along the diagonal and zeros for all off-diagonal elements). A formal procedure exists to determine the inverse of a nonsingular square matrix.

When the determinant of a square matrix is zero, the matrix is said to be singular. In this case, the range of **A** is only a subset of R^n. That is, a solution does not exist for all **y** in R^n. When a solution does exist for a system with a singular matrix, the solution is not unique.

Example 2.24 The equation of a plane in three-dimensional space is of the form

$$ax + by + cz = d \tag{a}$$

where a, b, c, and d are constants. If $\mathbf{n} = n_x\mathbf{i} + n_y\mathbf{j} + n_z\mathbf{k}$ is a unit vector normal to the plane, Equation a can be rewritten as

$$n_x x + n_y y + n_z z = \hat{d} \tag{b}$$

where $\hat{d} = d/\sqrt{a^2 + b^2 + c^2}$. Defining $\mathbf{r} = x\mathbf{i} + y\mathbf{j} + z\mathbf{k}$ as the position vector from the origin to the point (x,y,z), Equation b can be written in the form

$$\mathbf{r} \cdot \mathbf{n} = \hat{d} \tag{c}$$

Equation c can be interpreted as a statement that the length of the projection of a position vector from the origin to any point on the plane is the constant $\hat{d}$.

The equations of three separate planes may be written as

$$a_1x + b_1y + c_1z = d_1 \tag{d}$$

$$a_2x + b_2y + c_2z = d_2 \tag{e}$$

$$a_3x + b_3y + c_3z = d_3 \tag{f}$$

Values of x, y, and z which satisfy Equation d, Equation e, and Equation f are the coordinates of a point that lies on all three planes, a point on the intersection of the planes. Equation d, Equation e, and Equation f may be summarized in matrix form as

$$\begin{bmatrix} a_1 & b_1 & c_1 \\ a_2 & b_2 & c_2 \\ a_3 & b_3 & c_3 \end{bmatrix} \begin{bmatrix} x \\ y \\ z \end{bmatrix} = \begin{bmatrix} d_1 \\ d_2 \\ d_3 \end{bmatrix} \tag{g}$$

Use the matrix formulation given in Equation g to discuss the existence and uniqueness of a point of intersection of three planes.

Solution Three planes may (a) intersect at a single point, (b) intersect along a line, or (c) have no common point of intersection. If the planes intersect at a single point, then Equation g has a unique solution. In this case, the determinant of the matrix on the left hand side of Equation g is nonzero. Noting the geometric interpretation of Equation c, this implies that the normal vectors to the planes are linearly independent.

If the vectors are linearly dependent, then without loss of generality it can be assumed that there exist constants α and β such that

$$\mathbf{n}_1 = \alpha\mathbf{n}_2 + \beta\mathbf{n}_3 \tag{h}$$

where $\mathbf{n}_1$, $\mathbf{n}_2$ and $\mathbf{n}_3$ are unit vectors normal to the planes, and that a position vector, if it exists, satisfies equations of the form of Equation c for each plane. These equations are used in the following derivation:

$$\begin{aligned} &\mathbf{r}\cdot\mathbf{n}_1 = \hat{d}_1 \\ &\mathbf{r}\cdot(\alpha\mathbf{n}_2 + \beta\mathbf{n}_3) = \hat{d}_1 \\ &\alpha\mathbf{r}\cdot\mathbf{n}_2 + \beta\mathbf{r}\cdot\mathbf{n}_3 = \hat{d}_1 \\ &\alpha\hat{d}_2 + \beta\hat{d}_1 = \hat{d}_3 \end{aligned} \tag{i}$$

Equation i must be satisfied for a solution of Equation g to exist when the normal vectors to the planes are linearly dependent. A condition of the form of Equation i is called a solvability condition. When the solvability condition is satisfied, the three planes intersect in a line, and the solution is not unique.

If the solvability condition of Equation i is not satisfied, then there is no intersection between the three planes.

Example 2.25 The state of stress at a point in a material is uniquely defined by the stress tensor at the point,

$$\sigma = \begin{pmatrix} \sigma_{xx} & \tau_{xy} & \tau_{xz} \\ \tau_{yx} & \sigma_{yy} & \tau_{yz} \\ \tau_{zx} & \tau_{zy} & \sigma_{zz} \end{pmatrix} \tag{a}$$

where σ_{xx} is the stress acting normal to a plane whose normal is in the $+x$ direction, τ_{xy} and τ_{xz} are the components of the shear stress acting in this plane with τ_{xy} the component in the $+y$ direction, and τ_{xz} is the component in the $+z$ direction. The angular momentum equation can be used to show that the stress tensor is symmetric, $\tau_{yx} = \tau_{xy}$, $\tau_{xz} = \tau_{zx}$ and $\tau_{yz} = \tau_{zy}$, in the absence of unusual phenomena such as body moments or intrinsic angular momentum.

Let $\mathbf{n} = n_x\mathbf{i} + n_y\mathbf{j} + n_z\mathbf{k}$ be a unit vector normal to a plane through the point for which the stress tensor is known. The stress vector acting on the plane is calculated by

$$\begin{bmatrix} \sigma_x \\ \sigma_y \\ \sigma_z \end{bmatrix} = \begin{bmatrix} \sigma_{xx} & \tau_{xy} & \tau_{xz} \\ \tau_{yx} & \sigma_{yy} & \tau_{yz} \\ \tau_{zx} & \tau_{zy} & \sigma_{zz} \end{bmatrix} \begin{bmatrix} n_1 \\ n_2 \\ n_3 \end{bmatrix} \tag{b}$$

The stress tensor is a matrix operator. In general, its domain and range are R^3. However, in application, its domain is restricted to the set of all unit vectors in R^3. The resulting range is thus also restricted. In operator notation,

$$\mathbf{A}\mathbf{n} = \sigma \tag{c}$$

The stress tensor at a point in a solid is determined to be

$$\sigma = \begin{pmatrix} 900 & -600 & 300 \\ -600 & 1200 & -600 \\ 300 & -600 & -1500 \end{pmatrix} \text{ Pa} \tag{d}$$

Determine (a) the stress vector, (b) the normal stress, and (c) the shear stress acting on a plane whose normal is $\mathbf{n} = 2/3\mathbf{i} + 1/3\mathbf{j} - 2/3\mathbf{k}$.

Solution The stress vector acting on this plane can be calculated using Equation c:

$$\begin{bmatrix}\sigma_x\\ \sigma_y\\ \sigma_z\end{bmatrix}=\begin{bmatrix}900 & -600 & 300\\ -600 & 1200 & -600\\ 300 & -600 & -1500\end{bmatrix}\begin{bmatrix}\frac{2}{3}\\ \frac{1}{3}\\ -\frac{2}{3}\end{bmatrix} \tag{e}$$

$$=\begin{bmatrix}200\\ 400\\ 1000\end{bmatrix}\text{ Pa}$$

The normal stress is the projection of the stress vector onto the unit vector normal to the plane, or

$$\sigma_n=\sigma\cdot\mathbf{n} \tag{f}$$

Substitution of Equation c into Equation f leads to

$$\sigma_n=(\mathbf{An},\mathbf{n}) \tag{g}$$

where the inner product of Equation g is the standard inner product for R^3. Thus

$$\sigma_n=(200)\left(\frac{2}{3}\right)+(400)\left(\frac{1}{3}\right)+(1000)\left(-\frac{2}{3}\right) \tag{h}$$

$$=-400\text{ Pa}$$

(c) The shear stress is the component of the stress vector perpendicular to the unit normal. Thus if τ is the shear stress vector, then $\tau\cdot\mathbf{n}=0$. Also $\tau=\sigma-\sigma_n$ $\mathbf{n}$. Hence, for this problem,

$$\tau=\begin{bmatrix}200\\ 400\\ 1000\end{bmatrix}-(-400)\begin{bmatrix}\frac{2}{3}\\ \frac{1}{3}\\ -\frac{2}{3}\end{bmatrix} \tag{i}$$

$$=\frac{1}{3}\begin{bmatrix}1400\\ 1600\\ 2400\end{bmatrix}\text{ Pa}$$

The magnitude of the shear stress is calculated as

$$\tau = \left|\tau \cdot \tau\right|^{1/2} = 1.07 \times 10^3 \text{ Pa} \tag{j}$$

Example 2.26 Consider an n-degree-of-freedom linear mechanical system as illustrated in Figure 2.6. The displacements of the particles as functions of time are defined as $x_1, x_2, \ldots, x_n$ and are referred to as generalized coordinates. The potential energy of the system, V, at any instant is a function of the generalized coordinates, $V = V(x_1, x_2, \ldots, x_n)$. The stiffness matrix for a linear system is the matrix $\mathbf{K}$ whose element in the ith row and jth column, $k_{i,j}$, is calculated by

$$k_{i,j} = \frac{\partial^2 V}{\partial x_i \partial x_j} \tag{a}$$

Since the order of differentiation is interchangeable, $k_{i,j} = k_{j,i}$, and therefore the stiffness matrix is symmetric.

The potential-energy functional for a linear system can be written in a quadratic form as

$$V = \frac{1}{2}\sum_{i=1}^{n}\sum_{j=1}^{n} k_{ij} x_i x_j \tag{b}$$

(a) Determine the quadratic form of potential energy for the system shown in Figure 2.7 and use it to determine the stiffness matrix for the system.

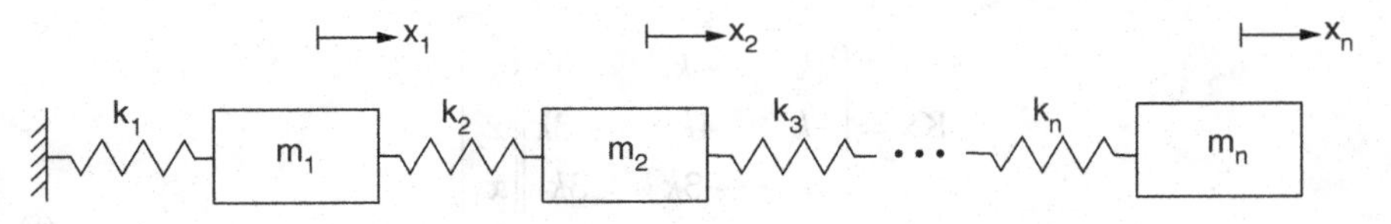

Figure 2.6 The stiffness matrix for the system of Example 2.26 is obtained using the potential-energy functional written at an arbitrary instant.

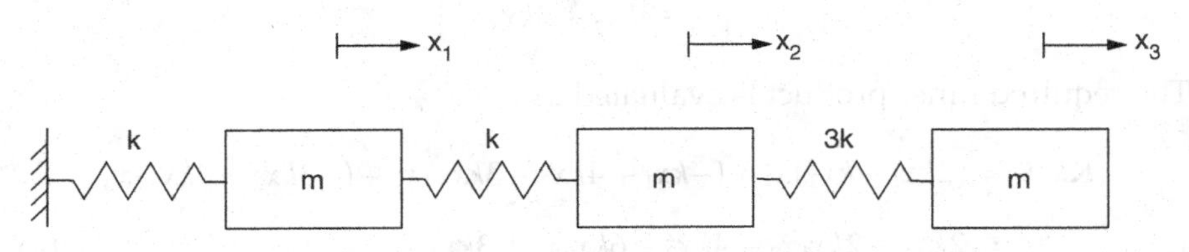

Figure 2.7 The stiffness matrix for this three-degree-of-freedom mechanical system is a 3×3 matrix.

(b) Show that, for this system, if $\mathbf{x}$ is the vector of generalized coordinates at an arbitrary instant,

$$(\mathbf{Kx},\mathbf{x}) = 2V \tag{c}$$

where the inner product is the standard inner product for R^n.

Solution (a) Note that the potential energy developed in a spring of stiffness k when the spring is subject to a force which leads to a change in length of x, measured from the spring's unstretched length, is $V = 1/2(kx^2)$. The potential energy for the system of Figure 2.7 at an arbitrary instant is

$$\begin{aligned} V &= \frac{1}{2}kx_1^2 + \frac{1}{2}k(x_2 - x_1)^2 + \frac{1}{2}3k(x_3 - x_2)^2 \\ &= \frac{1}{2}\left[2kx_1^2 - 2kx_1x_2 + 4kx_2^2 - 6kx_2x_3 + 3kx_3^2\right] \end{aligned} \tag{d}$$

The stiffness matrix is obtained through comparison of Equation d with the general quadratic form of Equation c as

$$\mathbf{K} = \begin{bmatrix} 2k & -k & 0 \\ -k & 4k & -3k \\ 0 & -3k & 3k \end{bmatrix} \tag{e}$$

To obtain Equation d, the symmetry of the stiffness matrix is used, so that two terms within the double summation in Equation b are combined to yield $k_{i,j}x_ix_j + k_{j,i}x_jx_i = 2k_{i,j}x_ix_j$.

Equation e is used to calculate

$$\begin{aligned} \mathbf{Kx} &= \begin{bmatrix} 2k & -k & 0 \\ -k & 4k & -3k \\ 0 & -3k & 3k \end{bmatrix}\begin{bmatrix} x_1 \\ x_2 \\ x_3 \end{bmatrix} \\ &= \begin{bmatrix} 2kx_1 - kx_2 \\ -kx_1 + 4kx_2 - 3kx_3 \\ -3kx_2 + 3kx_3 \end{bmatrix} \end{aligned} \tag{f}$$

The required inner product is evaluated as

$$\begin{aligned} (\mathbf{Kx},\mathbf{x}) &= (2kx_1 - kx_2)x_1 + (-kx_1 + 4kx_2 - 3kx_3)x_2 + (-3kx_2 + 3kx_3)x_3 \\ &= 2kx_1^2 - 2kx_1x_2 + 4kx_2^2 - 6kx_2x_3 + 3kx_3^2 \\ &= 2V \end{aligned} \tag{g}$$

There are many examples of linear operators defined for infinite-dimensional vector spaces. The proof of existence and uniqueness of solutions of equations of the form of Equation 2.35 when the domain of **L** is an infinite-dimensional vector space is beyond the scope of this study. However, the problems considered are formulated from the viewpoint of the mathematical modeling of a physical system. The physics of the system often dictate that a solution must exist and it must be unique. For example, the temperature distribution in a solid must be continuous and single-valued, requiring a unique solution. If the mathematical problem correctly models the physics, then unique solutions should exist.

Example 2.27 The nondimensional differential equation for the steady-state temperature distribution in a thin rod subject to an internal heat generation is

$$\frac{d^2\Theta}{dx^2} - Bi\Theta = \alpha + \beta\sin(\pi x) \tag{a}$$

where α and β are nondimensional constants and $Bi = hp/kA$ is the Biot number. The left end of the rod is maintained at a constant temperature, while the right end is insulated. The temperature distribution satisfies the boundary conditions

$$\Theta(0) = 0 \tag{b}$$

$$\frac{d\Theta}{dx}(1) = 0 \tag{c}$$

Write this problem in the operator form of Equation 2.35 and define $\mathcal{D}$ and $\mathcal{R}$.

Show, by obtaining the solution of the differential equation, the existence and uniqueness of the solution.

Solution (a) Equation a can be written in the form $L\Theta = f$, where $L\Theta = d^2\Theta/$ and $dx^2 - Bi\Theta$ and $f(x) = \alpha + \beta\sin(\pi x)$ The domain $\mathcal{D}$ is the subspace of $C^2[0,1]$ of all functions $g(x)$ such that $g(0) = 0$ and $dg/dx(1) = 0$. The range $\mathcal{R}$ is the set of all elements of $PC[0,1]$, the space of all piecewise continuous functions defined on $[0,1]$.

The homogeneous solution of Equation a is of the form

$$\Theta_h(x) = C_1\cosh\left(\sqrt{Bi}x\right) + C_2\sinh\left(\sqrt{Bi}x\right) \tag{d}$$

where C_1 and C_2 are arbitrary constants of integration. The particular solution of Equation a is

$$\Theta_p(x) = -\frac{\alpha}{Bi} - \frac{\beta}{\pi^2 + Bi}\sin(\pi x) \tag{e}$$

The general solution of Equation a is

$$\Theta_p(x) = C_1 \cosh\left(\sqrt{Bi}x\right) + C_2 \sinh\left(\sqrt{Bi}x\right) - \frac{\alpha}{Bi} - \frac{\beta}{\pi^2 + Bi}\sin(\pi x) \tag{f}$$

Application of the boundary condition of Equation b to Equation f leads to

$$0 = C_1 - \frac{\alpha}{Bi} \Rightarrow C_1 = \frac{\alpha}{Bi} \tag{g}$$

Application of the boundary condition of Equation c to Equation f used Equation g leads to

$$\begin{aligned} 0 &= \frac{\alpha}{\sqrt{Bi}}\cosh\left(\sqrt{Bi}\right) + C_2\sqrt{Bi}\sinh\left(\sqrt{Bi}\right) + \frac{\beta\pi}{\pi^2 + Bi} \\ C_2 &= -\frac{\beta\pi}{\sqrt{Bi}\left(\pi^2 + Bi\right)\sinh\left(\sqrt{Bi}\right)} - \frac{\alpha}{Bi}\coth\left(\sqrt{Bi}\right) \end{aligned} \tag{h}$$

Substitution of Equation g and Equation h into Equation f leads to

$$\begin{aligned} \Theta(x) = {} & \frac{\alpha}{Bi}\cosh\left(\sqrt{Bi}x\right) + \left[\frac{\beta\pi}{\sqrt{Bi}\left(\pi^2 + Bi\right)\sinh\left(\sqrt{Bi}\right)} - \frac{\alpha}{Bi}\coth\left(\sqrt{Bi}\right)\right] \\ & \sinh\left(\sqrt{Bi}x\right) - \frac{\alpha}{Bi} - \frac{\beta}{\pi^2 + \beta}\sin(\pi x) \end{aligned} \tag{i}$$

Equation i satisfies the differential equation, Equation a, as well as the boundary conditions of Equation b and Equation c. Thus a solution exists. If a second solution $\hat{\Theta}(x)$ exists such that $\hat{\Theta}(x) \neq \Theta(x)$ for some values of x, then the temperature is multi-valued at those values, a physical impossibility. Therefore, if Equation a, Equation b, and Equation c are a true mathematical model of the physical system, the solution must be unique.

Example 2.28 The nondimensional differential equations for the transverse displacements $w_1(x)$ and $w_2(x)$ of two elastically coupled, statically loaded beams with an axial load P are

$$\frac{d^4 w_1}{dx^4} - \varepsilon\frac{d^2 w_1}{dx^2} + \eta(w_1 - w_2) = 0 \tag{a}$$

$$\mu\frac{d^4 w_1}{dx^4} - \varepsilon\frac{d^2 w_1}{dx^2} + \eta(w_2 - w_1) = \Lambda f(x) \tag{b}$$

where $\varepsilon = PL^2/E_1I_1$, $\eta = kL^4/E_1I_1$ and $\mu = E_2I_2/E_1I_1$. The load per unit length, $F(x)$, is nondimensionalized according to $f(x) = F(x)/F_{max}$, and then $\Lambda = F_{max}L^3/E_1I_1$. The boundary conditions for pinned-pinned beams are:

$$w_1(0) = 0 \qquad w_1(1) = 0$$
$$\frac{d^2w_1}{dx^2}(0) = 0 \qquad \frac{d^2w_1}{dx^2}(1) = 0 \tag{c}$$

$$w_2(0) = 0 \qquad w_2(1) = 0$$
$$\frac{d^2w_2}{dx^2}(0) = 0 \qquad \frac{d^2w_2}{dx^2}(1) = 0 \tag{d}$$

(a) Write this problem in the operator form of Equation 2.33 and define $\mathcal{D}$ and $\mathcal{R}$.

(b) Show, by obtaining the solution of the differential equation, the existence and uniqueness of the solution. Use $\varepsilon = 2$, $\eta = 1$, $\mu = 1$ and $\Lambda = 1$ with $f(x) = \sin(\pi x)$.

Solution (a) The problem defined by Equation a, Equation b, Equation c, and Equation d can be written in the form of Equation 2.35 as $\mathbf{Lw} = \mathbf{f}$ with $\mathbf{w} = [w_1(x) \quad w_2(x)]^T$,

$$\mathbf{Lw} = \begin{bmatrix} \frac{d^4}{dx^4} - \varepsilon\frac{d^2}{dx^2} + \eta & -\eta \\ -\eta & \mu\frac{d^4}{dx^4} - \varepsilon\frac{d^2}{dx^2} + \eta \end{bmatrix} \begin{bmatrix} w_1(x) \\ w_2(x) \end{bmatrix} \tag{e}$$

and

$$\mathbf{f} = \begin{bmatrix} 0 \\ \Lambda f(x) \end{bmatrix} \tag{f}$$

Let Q be the vector space defined as $Q = C^4[0,1] \times R^2$. An element of Q is a two-dimensional vector of functions with each element a member of $C^4[0,1]$. That is, if $\mathbf{g}$ is in Q, then $\mathbf{g} = \begin{bmatrix} g_1(x) \\ g_2(x) \end{bmatrix}$, where $g_1(x)$ and $g_2(x)$ each an element of $C^4[0,1]$. Define S as the subspace of Q defined such that if $\mathbf{g}$ is in Q, then

$$g_1(0) = 0, \frac{d^2g_1}{dx^2}(0) = 0, g_1(1) = 0, \frac{d^2g_1}{dx^2}(1)$$
$$= 0, g_2(0) = 0, \frac{d^2g_2}{dx^2}(0) = 0, g_2(1) = 0 \text{ and } \frac{d^2g_2}{dx^2}(1) = 0$$

Then $\mathcal{D} = S$. The range of $\mathbf{L}$ is $\mathcal{R} = C^4[0,1] \times R^2$.

(b) Substitution of the given parameters into Equation a and Equation b leads to

$$\frac{d^4w_1}{dx^4} - 2\frac{d^2w_1}{dx^2} + (w_1 - w_2) = 0 \tag{g}$$

$$\frac{d^4w_2}{dx^4} - 2\frac{d^2w_2}{dx^2} + (w_2 - w_1) = \sin(\pi x) \tag{h}$$

The solutions of Equation g and Equation h are of the form

$$\mathbf{w} = \mathbf{w}_h + \mathbf{w}_p \tag{i}$$

where $\mathbf{w}_h$ is the homogeneous solution, the solution obtained if $\mathbf{f} = \mathbf{0}$, and $\mathbf{w}_p$ is the particular solution, the solution particular to the specific form of **f**.

The homogeneous solution is assumed to be of the form

$$\mathbf{w}_h = \begin{bmatrix} a \\ b \end{bmatrix} e^{\alpha x} \tag{j}$$

Substitution of Equation j into Equation g and Equation h with $\mathbf{f} = \mathbf{0}$ leads to

$$(\alpha^4 - 2\alpha^2 + 1)a - b = 0 \tag{k}$$

$$-a + (\alpha^4 - 2\alpha^2 + 1)b = 0 \tag{l}$$

A nontrivial solution to Equation k and Equation i exists only if

$$\begin{vmatrix} \alpha^4 - 2\alpha^2 + 1 & -1 \\ -1 & \alpha^4 - 2\alpha^2 + 1 \end{vmatrix} = 0 \tag{m}$$

Evaluation of the determinant in Equation k leads to the following eighth-order polynomial equation:

$$\alpha^8 - 4\alpha^6 + 6\alpha^4 - 4\alpha^2 = 0 \tag{n}$$

The solutions of Equation l are

$$\alpha = 0, 0, 1.414, -1.414, 1.0987 \pm 0.4551i, -1.0987 \pm 0.4551i \tag{o}$$

Equation k implies that for an appropriate value of $\propto$, α, $b = (\alpha^4 - 2\alpha^2 + 1)$. The homogeneous solution is a linear combination of all possible solutions of the form of Equation j, where the appropriate values of α are given in Equation o.

Arbitrarily choosing $a = 1$ in each case and using trigonometric functions to replace complex exponentials leads to a homogeneous solution of

$$\begin{bmatrix} w_1(t) \\ w_2(t) \end{bmatrix} = C_1\begin{bmatrix} 1 \\ 1 \end{bmatrix} + C_2\begin{bmatrix} 1 \\ 1 \end{bmatrix}t + C_3\begin{bmatrix} 1 \\ 1 \end{bmatrix}e^{1.414t} + C_4\begin{bmatrix} 1 \\ 1 \end{bmatrix}e^{-1.414t}$$

$$+C_5\begin{bmatrix} 1 \\ -1 \end{bmatrix}\cosh(1.0987t)\cos(0.4551t)$$

$$+C_6\begin{bmatrix} 1 \\ -1 \end{bmatrix}\cosh(1.0987t)\sin(0.4551t)$$

$$+C_7\begin{bmatrix} 1 \\ -1 \end{bmatrix}\sinh(1.0987t)\cos(0.4551t)$$

$$+C_8\begin{bmatrix} 1 \\ -1 \end{bmatrix}\sinh(1.0987t)\sin(0.4551t) \tag{p}$$

The particular solution is obtained using the method of undetermined coefficients, discussed in Chapter 3. A particular solution is assumed as

$$\begin{bmatrix} w_1(x) \\ w_2(x) \end{bmatrix} = \begin{bmatrix} W_1 \\ W_2 \end{bmatrix}\sin(\pi x) \tag{q}$$

Substitution of Equation q into Equation g and Equation h, using each to develop a relation between W_1 and W_2 and solving simultaneously leads to

$$\begin{bmatrix} w_1(x) \\ w_2(x) \end{bmatrix} = \begin{bmatrix} \dfrac{1}{\left(\pi^2+1\right)^4+1} \\ \dfrac{\left(\pi^2+1\right)^2}{\left(\pi^2+1\right)^4+1} \end{bmatrix}\sin(\pi x) \tag{r}$$

The general solution is the sum of the homogeneous solution, Equation q and the particular solution Equation r. Application of the boundary conditions, Equation c and Equation d leads to a set of eight simultaneous equations to solve for the constants of integration for which a unique solution is attained.

2.11 Adjoint operators

Each linear operator has a related operator, called its adjoint, which has important properties.

Definition 2.21 Let V be a vector space with an inner product $(\mathbf{u}, \mathbf{v})$ defined for all $\mathbf{u}$ and $\mathbf{v}$ in V. Let $\mathbf{L}$ be a linear operator whose domain is D, a subspace of V, and whose range is R, also a subspace of V. The adjoint of **L** with respect to the defined inner product, written **L***, is an operator whose domain is R and whose range is D, such that

$$(\mathbf{Lu},\mathbf{v}) = (\mathbf{u},\mathbf{L}^*\mathbf{v}) \tag{2.44}$$

Definition 2.22 If D = R and **L*** = **L**, then **L** is said to be self-adjoint, that is,

$$(\mathbf{Lu},\mathbf{v}) = (\mathbf{u},\mathbf{Lv}) \tag{2.45}$$

for all **u** and **v** in D.

Example 2.29 Let **A** be a $n \times n$ matrix of the form

$$\mathbf{A} = \begin{bmatrix} a_{1,1} & a_{1,2} & a_{1,3} & \cdots & a_{1,n} \\ a_{2,1} & a_{2,2} & a_{2,3} & \cdots & a_{2,n} \\ a_{3,1} & a_{3,2} & a_{3,3} & \cdots & a_{3,n} \\ \vdots & \vdots & \vdots & \ddots & \vdots \\ a_{n,1} & a_{n,2} & a_{n,3} & \cdots & a_{n,n} \end{bmatrix} \tag{a}$$

(a) Determine **A*** with respect to the standard inner product on R^n.
(b) Under what conditions is **A** self-adjoint with respect to the standard inner product on R^n?
(c) Determine **A*** with respect to an inner product on R^n defined by

$$(\mathbf{u},\mathbf{v}) = 2u_1v_1 + u_2v_2 + u_3v_3 + \ldots + u_nv_n \tag{b}$$

Solution (a) Let **u** and **v** be arbitrary vectors in R^n. Then by definition of the standard inner product for R^n,

$$\begin{aligned} (\mathbf{Au},\mathbf{v}) &= \sum_{i=1}^{n} (\mathbf{Au})_i \mathbf{v}_i \\ &= \sum_{i=1}^{n}\sum_{j=1}^{n} a_{i,j}u_jv_i \end{aligned} \tag{c}$$

Since **A*** is an operator whose domain and range are R^n, it has a matrix representation of the form

$$\mathbf{A}^* = \begin{bmatrix} a_{1,1}^* & a_{1,2}^* & a_{1,3}^* & \cdots & a_{1,n}^* \\ a_{2,1}^* & a_{2,2}^* & a_{2,3}^* & \cdots & a_{2,n}^* \\ a_{3,1}^* & a_{3,2}^* & a_{3,3}^* & \cdots & a_{3,n}^* \\ \vdots & \vdots & \vdots & \ddots & \vdots \\ a_{n,1}^* & a_{n,2}^* & a_{n,3}^* & \cdots & a_{n,n}^* \end{bmatrix} \tag{d}$$

Then

$$\begin{aligned}(\mathbf{u},\mathbf{A}^{*}\mathbf{v}) &= \sum_{i=1}^{n} u_i(\mathbf{A}^{*}\mathbf{v})_i \\ &= \sum_{i=1}^{n}\sum_{j=1}^{n} u_i a_{i,j}^{\ *} v_j\end{aligned} \tag{e}$$

Interchanging the names of the indices in Equation e leads to

$$(\mathbf{u},\mathbf{A}^{*}\mathbf{v}) = \sum_{j=1}^{n}\sum_{i=1}^{n} a_{j,i}^{\ *} u_j v_j \tag{f}$$

For the expression in Equation c to be equal to the expression in Equation f for all possible **u** and **v**, it is required that

$$a_{i,j}^{\ *} = a_{j,i} \quad \begin{cases} i = 1,2,\ldots,n \\ j = 1,2,\ldots,n \end{cases} \tag{g}$$

Thus

$$\mathbf{A}^{*} = \begin{bmatrix} a_{1,1} & a_{2,1} & a_{3,1} & \cdots & a_{n,1} \\ a_{1,2} & a_{2,2} & a_{3,3} & \cdots & a_{n,2} \\ a_{11,3} & a_{2,3} & a_{3,3} & \cdots & a_{n,3} \\ \vdots & \vdots & \vdots & \ddots & \vdots \\ a_{1,,n} & a_{2,n} & a_{3,n,,} & \cdots & a_{n,n} \end{bmatrix} \tag{h}$$

Thus the adjoint of **A** with respect to the standard inner product on R^n is the matrix obtained by interchanging the rows and columns of **A**. Such a matrix is called the transpose matrix, $\mathbf{A}^{\mathrm{T}}$. Thus $\mathbf{A}^{*} = \mathbf{A}^{\mathrm{T}}$.

(b) From Equation g, it is clear that **A** is self-adjoint with respect to the standard inner product if

$$a_{i,j} = a_{j,i} \quad \begin{cases} i = 1,2,\ldots,n \\ j = 1,2,\ldots,n \end{cases} \tag{i}$$

Such a matrix, whose columns can be interchanged with its rows without changing the matrix, is called a symmetric matrix. Thus an $n \times n$ matrix **A** is self-adjoint with respect to the standard inner product for R^n if and only if **A** is a symmetric matrix.

Following along the same lines as part (a), and using the inner product defined in Equation b, leads to

$$
\begin{aligned}
(\mathbf{Au},\mathbf{v}) &= 2(\mathbf{Au})_1 v_1 + \sum_{i=2}^{n} (\mathbf{Au})_i \mathbf{v}_i \\
&= 2\sum_{j=1}^{n} a_{1,j}u_j v_1 + \sum_{i=2}^{n}\sum_{j=1}^{n} a_{i,j}u_j v_i \\
&= \sum_{j=1}^{n} 2a_{1,j}u_j v_1 + \sum_{i=2}^{n} a_{i,1}u_1 v_i + \sum_{i=2}^{n}\sum_{j=2}^{n} a_{i,j}u_j v_i
\end{aligned} \tag{j}
$$

Assuming $\mathbf{A}^*$ in the form of Equation d leads to

$$
\begin{aligned}
(\mathbf{u},\mathbf{A}^*\mathbf{v}) &= 2u_1(\mathbf{A}^*\mathbf{v})_1 + \sum_{i=2}^{n} u_i(\mathbf{A}^*\mathbf{v})_i \\
&= 2u_1\sum_{j=1}^{n} a_{1,j}{}^* v_j + \sum_{i=2}^{n}\sum_{j=1}^{n} u_i a_{i,j}{}^* v_j \\
&= \sum_{j=1}^{n} 2a_{1,j}{}^* u_1 v_j + \sum_{i=2}^{n} u_i a_{i,1}{}^* v_1 + \sum_{i=2}^{n}\sum_{j=2}^{n} u_i a_{i,j}{}^* v_j
\end{aligned} \tag{k}
$$

Renaming the indices in Equation k and rearranging leads to

$$
(\mathbf{u},\mathbf{A}^*\mathbf{v}) = \sum_{i=1}^{n} 2a_{1,i}{}^* v_i u_1 + \sum_{j=2}^{n} u_j a_{j,1}{}^* v_1 + \sum_{i=2}^{n}\sum_{j=1}^{n} u_i a_{j,i}{}^* v_j \tag{l}
$$

Equation j and Equation l are identical for all possible $\mathbf{u}$ and $\mathbf{v}$ if and only if

$$
\begin{aligned}
a_{1,1}{}^* &= a_{1,1} \\
a_{1,j}{}^* &= \frac{1}{2} a_{j,1} \quad j = 2,3,\ldots,n \\
a_{j,1}{}^* &= 2a_{1,j} \quad j = 2,3,\ldots,n \\
a_{j,i}{}^* &= a_{i,j} \quad i,j = 2,3,\ldots,n
\end{aligned} \tag{l}
$$

For example, a pair of adjoints with respect to the inner product of Equation b is

$$\mathbf{A} = \begin{bmatrix} 4 & 2 & -1 & 3 \\ -2 & 4 & 2 & 4 \\ -4 & 2 & 1 & 0 \\ 2 & 3 & 1 & 5 \end{bmatrix} \quad \mathbf{A}^* = \begin{bmatrix} 4 & -1 & -2 & 1 \\ 4 & 4 & 2 & 3 \\ -2 & 2 & 1 & 1 \\ 6 & 4 & 0 & 5 \end{bmatrix} \tag{m}$$

Part (a) of Example 2.29 illustrates that a symmetric $n \times n$ matrix is self-adjoint with respect to the standard inner product for R^n. Thus the matrices developed from the stress tensor in Example 2.25 and the stiffness matrix of Example 2.26 are self-adjoint with respect to the standard inner product.

Example 2.30 The nondimensional differential equation governing the deflection, $w(x)$, of a non-uniform beam due to a distributed load per unit length, $f(x)$, is

$$\frac{d^2}{dx^2}\left(\alpha(x)\frac{d^2w}{dx^2}\right) = f(x) \tag{a}$$

where $\alpha(x)$ is a function describing the variation of material and geometric parameters across the span of the beam. The boundary conditions are dependent on the end constraints. For a beam fixed at $x=0$ and free at $x=1$, the appropriate boundary conditions are

$$w(0) = 0 \tag{b1}$$

$$\frac{dw}{dx}(0) = 0 \tag{b2}$$

$$\frac{d^2w}{dx^2}(1) = 0 \tag{b3}$$

$$\frac{d}{dx}\left(\alpha(x)\frac{d^2w}{dx^2}\right)(1) = 0 \tag{b4}$$

Equation a may be written in the form of Equation 2.35 with $Lw = d^2/dx^2(\alpha(x)\, d^2w/dx^2)$. Define V as $C^4[0,1]$. The domain of L is the subspace of V such that if $g(x)$ is in D, then $g(x)$ satisfies the boundary conditions of Equation b. Consider the range of V to be the same as its domain. The standard inner product on V is

$$(f(x), g(x)) = \int_0^1 f(x)g(x)dx \tag{c}$$

Show that L is self-adjoint on D with respect to the inner product of Equation C.

Solution If **L** is self-adjoint on D, then (**Lf, g**) = (**g, Lf**) for all $f(x)$ and $g(x)$ in D. That is,

$$\int_0^1 \frac{d^2}{dx^2}\left(\alpha(x)\frac{d^2 f}{dx^2}\right)g(x)dx = \int_0^1 f(x)\frac{d^2}{dx^2}\left(\alpha(x)\frac{d^2 g}{dx^2}\right)dx \tag{d}$$

Recall the integration by parts formula, $\int u dv = uv - \int v du$. Using integration by parts on the integral on the left-hand side of Equation d with $u = g(x)$ and $dv = \left[d^2(\alpha d^2 f/dx^2)/dx^2\right]dx$ gives

$$\int_0^1 \frac{d^2}{dx^2}\left(\alpha(x)\frac{d^2 f}{dx^2}\right)g(x)dx = g(x)\frac{d}{dx}\left(\alpha\frac{d^2 f}{dx^2}\right)_{x=0}^{x=1} - \int_0^1 \frac{d}{dx}\left(\alpha\frac{d^2 f}{dx^2}\right)\frac{dg}{dx}dx \tag{e}$$

Using integration by parts on the remaining integral of Equation e with $u = dg/dx$ and $dv = \left[d(\alpha d^2 f/dx^2)/dx^2\right]dx$ leads to

$$\begin{aligned}\int_0^1 \frac{d^2}{dx^2}\left(\alpha(x)\frac{d^2 f}{dx^2}\right)g(x)dx &= g(x)\frac{d}{dx}\left(\alpha\frac{d^2 f}{dx^2}\right)_{x=0}^{x=1} - \left[\frac{dg}{dx}\,\alpha(x)\frac{d^2 f}{dx^2}\right]_{x=0}^{x=1} \\ &\quad + \int_0^1 \alpha(x)\frac{d^2 f}{dx^2}\frac{d^2 g}{dx^2}dx \\ &= g(1)\frac{d}{dx}\left(\alpha\frac{d^2 f}{dx^2}\right)(1) - g(0)\frac{d}{dx}\left(\alpha\frac{d^2 f}{dx^2}\right)(0) \\ &\quad - \frac{dg}{dx}(1)\alpha(1)\frac{d^2 f}{dx^2}(1) - \frac{dg}{dx}(0)\alpha(0)\frac{d^2 f}{dx^2}(0) \\ &\quad + \int_0^1 \alpha(x)\frac{d^2 f}{dx^2}\frac{d^2 g}{dx^2}dx\end{aligned} \tag{f}$$

Since $f(x)$ and $g(x)$ are both in D, $d/dx\,(\alpha(d^2f/dx^2))(1) = 0$ $g(0) = 0$, $d^2f/dx^2\,(1) = 0$, and $dg/dx\,(0) = 0$. Thus Equation f reduces to

$$\int_0^1 \frac{d^2}{dx^2}\left(\alpha(x)\frac{d^2 f}{dx^2}\right)g(x)dx = \int_0^1 \alpha(x)\frac{d^2 f}{dx^2}\frac{d^2 g}{dx^2}dx \tag{g}$$

Using similar steps, it can be shown that for all $f(x)$ and $g(x)$ in D,

$$\int_0^1 f(x)\frac{d^2}{dx^2}\left(\alpha(x)\frac{d^2 g}{dx^2}\right)dx = \int_0^1 \alpha(x)\frac{d^2 f}{dx^2}\frac{d^2 g}{dx^2}dx \tag{h}$$

Thus Equation a is proved, and L is self-adjoint on D with respect to the standard inner product.

Example 2.31 The nondimensional partial differential equation for the steady-state temperature distribution, $\Theta(\mathbf{r})$ in a bounded three-dimensional region V is

$$\nabla^2\Theta = -f(\mathbf{r}) \tag{a}$$

where **r** is a position vector from the origin of the coordinate system to a point in the body and ∇^2 is the Laplacian operator where Bi is the Biot number. The surface of the region is described by $S(\mathbf{r}) = 0$. The surface of the body is open, and heat transfer occurs though convection, leading to a boundary condition of the form

$$\nabla\Theta\cdot\mathbf{n} = -Bi\Theta \quad \text{on S} \tag{b}$$

Let Q be the space of all functions defined in V. The Laplacian is a linear operator. Define $\mathcal{D}$ as the subspace of V consisting of all functions $g(\mathbf{r})$ that satisfy the boundary condition of Equation b. The standard inner product on V is

$$(f(\mathbf{r}), g(\mathbf{r})) = \int_V f(\mathbf{r})g(\mathbf{r})dV \tag{c}$$

Show that $L = \nabla^2$ is a self-adjoint operator on $\mathcal{D}$ with respect to the standard inner product for Q.

Solution If **L** is self-adjoint, then for any vectors $f(\mathbf{r})$ and $g(\mathbf{r})$ both in D,

$$\int_V (\nabla^2 f)g dV = \int_V f(\nabla^2 g)dV \tag{d}$$

Recall the vector identity,

$$\nabla\cdot(g\nabla f) = \nabla f\cdot\nabla g + g\nabla^2 f \tag{e}$$

Using the identity of Equation e in the integral on the left-hand side of Equation d leads to

$$\int_V (\nabla^2 f) g dV = \int_V [-\nabla f \cdot \nabla g + \nabla \cdot (g \nabla f)] dV \qquad \text{(f)}$$

The divergence theorem implies

$$\int_V \nabla \cdot (g \nabla f) dV = \int_S (g \nabla f) \cdot n dS \qquad \text{(g)}$$

where **n** is a unit r(a) vector normal to S. Equation g leads to

$$\int_V (\nabla^2 f) g dV = -V \int \nabla f \cdot \nabla g dV - \int_S g \nabla f \cdot n dS \qquad \text{(h)}$$

Since both f and g are in D, $\nabla f_g \cdot \mathbf{n} = -Bif$ on S and $\nabla g \cdot \mathbf{n} = -Big$ on S. Thus Equation h can be written as

$$\int_V (\nabla^2 f) g dV = -\int_V \nabla f \cdot \nabla g dV - \mathrm{Bi} \int_S g f dS \qquad \text{(i)}$$

In a similar fashion it can be shown that

$$\int_V f (\nabla^2 g) dV = -\int_V \nabla f \cdot \nabla g dV - \mathrm{Bi} \int_S g f dS \qquad \text{(j)}$$

The equality of the right-hand sides of Equations i and Equation j proves that **L** is self-adjoint.

Example 2.32 A Fredholm integral equation is of the form

$$\int_0^x f(x) K(x,y) dy = g(x) \qquad \text{(a)}$$

Equation a can be formulated as **Lf** = g, where $\mathbf{L}f = \int_0^x f(x) K(x,y) dy = g(x)$. The domain and range of **L** are $C[0,a]$. Assuming that the adjoint operator is of the from $\mathbf{L}^* f = \int_0^x f(x) K^*(x,y) dy$, determine the form of $K^*(x,y)$ using the standard inner product

$$(f,g) = \int_0^a f(x) g(x) dx$$

Solution If L^* is the adjoint of **L** with respect to the standard inner product, then

$$\int_0^a \left(\int_0^x f(x)K(x,y)dy \right) g(x)dx = \int_0^a f(x) \left(\int_0^x K^*(x,y)g(y)dy \right) dx \qquad \text{(b)}$$

Working with the left-hand side of Equation b by interchanging the order of integration leads to

$$\int_0^a \left(\int_0^x f(y)K(x,y)dy \right) g(x)dx = \int_0^a \int_0^y f(x)K(x,y)g(y)dxdy \qquad \text{(c)}$$

Renaming the variables of integration, replacing x by λ and y by τ in Equation c, leads to

$$(Lf,g) = \int_0^a \int_0^\tau f(\lambda)g(\tau)K(\lambda,\tau)d\lambda d\tau \qquad \text{(d)}$$

Performing similar steps on the right-hand side of Equation a leads to

$$\begin{aligned}(f,L^*g) &= \int_0^a \int_0^x f(y)g(x)K^*(x,y)dydx \\ &= \int_0^\lambda \int_0^\tau f(\lambda)g(\tau)K^*(\tau,\lambda)d\lambda d\tau\end{aligned} \qquad \text{(e)}$$

It is clear that Equation d and Equation e are equivalent for all f and g if and only if $K^*(x,y) = K(y,x)$ for all x and y such that $0 \le x \le a$ and $0 \le y \le a$.

2.12 *Positive definite operators*

Positive definiteness is a property of many linear operators that, if proven for the operator, has many ramifications regarding approximate and exact solutions of an equation involving the operator.

Definition 2.23 Let **L** be a linear operator whose domain D is a subspace of a vector space V, on which an inner product **(u,v)** is defined. **L** is positive definite with respect to the inner product if $(\mathbf{Lu},\mathbf{u}) \ge 0$ for all **u** in D, and $(\mathbf{Lu},\mathbf{u}) = 0$ if and only if $\mathbf{u} = \mathbf{0}$.

Some operators studied will not satisfy the "only if" clause which is required for definition 2.23 to apply. For such cases, definition 2.23 can be modified as:

Definition 2.24 Let **L** be a linear operator whose domain D is a subspace of a vector space V, on which an inner product (**u**,**v**) is defined. **L** is positive semi-definite with respect to the inner product if $(\mathbf{Lu},\mathbf{u})\geq 0$ for all **u** in D, and $(\mathbf{Lu},\mathbf{u})=0$ when $\mathbf{u}=0$, but there are vectors, $\mathbf{u}\neq 0$, such that $(\mathbf{Lu},\mathbf{u})=0$.

Example 2.33 The stiffness matrix for a two-degree-of-freedom system is

$$\mathbf{K}=\begin{bmatrix} 4 & -2 \\ -2 & 3 \end{bmatrix} \tag{a}$$

Determine whether or not **K** is positive definite with respect to the standard inner product for R^2.

Solution Let **u** be an arbitrary vector in R^2. Then

$$\mathbf{Ku}=\begin{bmatrix} 4 & -2 \\ -2 & 3 \end{bmatrix}\begin{bmatrix} u_1 \\ u_2 \end{bmatrix}=\begin{bmatrix} 4u_1-2u_2 \\ -2u_1+3u_2 \end{bmatrix} \tag{b}$$

and

$$\begin{aligned}(\mathbf{Ku},\mathbf{u})&=(4u_1-2u_2)u_1+(-2u_1+3u_2)u_2\\ &=4u_1^2-4u_1u_2+3u_2^2\\ &=2u_1^2+2\left(u_1^2-2u_1u_2+u_2^2\right)+u_2^2\\ &=2u_1^2+2(u_1-u_2)^2+u_2^2\end{aligned} \tag{c}$$

Thus from Equation c, since (**Ku**, **u**) is the sum of three non-negative terms, $(\mathbf{Ku},\mathbf{u})\geq 0$ for all **u**. Clearly the right-hand side of Equation c is zero if $\mathbf{u}=\mathbf{0}$, and if either component of **u** is not zero, then (**Ku**, **u**) is greater than zero. Thus **K** is positive definite according to definition 2.23.

Example 2.34 Consider the operator defined in Example 2.30 which is used to solve for the deflection of a beam. Show that the operator is positive definite on the domain specified in Example 2.30 with respect to the standard inner product for $C^4[0,1]$. Note that $\alpha(x)>0$ for all x, $0\leq x\leq 1$.

Solution Algebraic manipulation of (Lf,g) for arbitrary f and g both in D led to Equation g of Example 2.30, which is repeated following:

$$(Lf,g)=\int_0^1 \frac{d^2}{dx^2}\left(\alpha(x)\frac{d^2 f}{dx^2}\right)g(x)dx=\int_0^1 \alpha(x)\frac{d^2 f}{dx^2}\frac{d^2 g}{dx^2}dx \tag{a}$$

Using Equation a with $g=f$ leads to

$$(Lf,f)=\int_0^1 \alpha(x)\left(\frac{d^2 f}{dx^2}\right)^2 dx \tag{b}$$

The integrand of the integral on the right-hand side of Equation b is non-negative for all x within the range of integration. Thus $(Lf, f) \geq 0$ for all f. If $f(x) = 0$, $d^2 f/dx^2 = 0$, and the definite integral in Equation b is zero. The integral is zero for any f such that $d^2 f/dx^2 = 0$ for all x, $0 \leq x \leq 1$. The only continuous function that has a second derivative identically equal to zero is a linear function of the form

$$f(x) = c_1 + c_2 x \tag{c}$$

However, f must be in D and thus satisfy the boundary conditions of the fixed-free beam given in Equation b of Example 2.30. Applying $f(0) = 0$ leads to $c_1 = 0$. Then application of $df/dx(0) = 0$ requires $c_2 = 0$. Thus the only linear function in D is $f(x) = 0$. Thus the only $f(x)$ in D such that $(Lf, f) = 0$ is $f(x) = 0$.

From the above and definition 2.23, it is clear that L is positive definite.

Example 2.35 Reconsider the heat transfer problem of Example 2.31. Define $L\Theta = -\nabla^2\Theta$. Determine whether or not L is positive definite with respect to the standard inner product for C^2 (V) on the domain D defined such that if $f(\mathbf{r})$ is in D, then (a) $f(\mathbf{r}) = 0$ everywhere on S, the surface of V(this condition corresponds to the surface at a prescribed temperature), (b) $\nabla f \cdot \mathbf{n} = 0$ everywhere on S (this corresponds to a completely insulated surface), and (c) $\nabla f \cdot \mathbf{n} = -Bi(f)$ everywhere on S (this corresponds to a condition of heat transfer by convection from the ambient to the surface).

Solution Using the same algebraic manipulations as in Example 2.31, an equation similar to that of Equation i of Example 2.31 can be obtained as

$$(Lf, g) = \int_V \left(-\nabla^2 f\right) g dV = \int_V \nabla f \cdot \nabla g dV - \int_S g \nabla f \cdot \mathbf{n} dS \tag{a}$$

Applying Equation a with $g = f$ leads to

$$\begin{aligned} (Lf, f) &= \int_V \nabla f \cdot \nabla f dV - \int_S f \nabla f \cdot \mathbf{n} dS \\ &= \int_V |\nabla f|^2 dV - \int_S f \nabla f \cdot \mathbf{n} dS \end{aligned} \tag{b}$$

The integrand of the volume integral of Equation a is non-negative everywhere in V, thus the integral is clearly non-negative. It is also equal to zero when $f = 0$. However, it could also be zero when ∇f is zero everywhere in V. This may occur if $f(\mathbf{r}) = C$, a nonzero constant.

(a) If $f(\mathbf{r}) = 0$ everywhere on S, then the integrand of the surface integral of Equation a is zero everywhere on S and the integral is zero. Since f is prescribed to be zero on S, if $f(\mathbf{r}) = C$, then C must be zero. Thus the only f in D such that $(L, f, f) = 0$ is $f = 0$. Hence L is positive definite on D.

If $\nabla f \cdot \mathbf{n} = 0$ everywhere on S, then the surface integral of Equation a is identically zero. However, the surface condition is satisfied by $f(\mathbf{r}) = C$ for any value of C. Thus there exists $f(\mathbf{r}) \neq 0$ such that $(L,f,f) = 0$. Hence L is positive semi-definite on D.

If $\nabla f \cdot n = -Bi(f)$ everywhere on S, then Equation a can be rewritten as

$$(Lf, f) = \int_V |\nabla f|^2 dV + Bi \int_S f^2 dS \tag{b}$$

Clearly the right-hand side of Equation b is non-negative. From the boundary condition, if $\nabla f = 0$ everywhere in V, which implies that $f(\mathbf{r}) = C$, then $f = 0$ everywhere on S, which implies $C = 0$, and thus the right-hand side of Equation b is zero only when $f = 0$ and L is positive definite on D.

2.13 *Energy inner products*

Theorem 2.10 If L is a positive definite and self-adjoint linear operator with respect to a defined inner product on a domain D, then an inner product, called an energy inner product, can be defined by

$$(\mathbf{u}, \mathbf{v})_L = (\mathbf{Lu}, \mathbf{v}) \tag{2.46}$$

Proof It is required to show that properties (i)–(iv) of definition 2.7 are valid for all **u**, **v**, and **w** in D when Equation 2.46 is used as an inner-product definition. To this end,

Property (i)	$(\mathbf{u}, \mathbf{v})_L = (\mathbf{Lu}, \mathbf{v})$	Definition of proposed inner product
	$= (\mathbf{u}, \mathbf{Lv})$	By hypothesis, **L** is self-adjoint
	$= (\mathbf{v}, \mathbf{Lu})$	By hypothesis, **(u,v)** is a valid inner product, thus property (i) of definition 2.7 applies
	$= (\mathbf{v}, \mathbf{u})_L$	Definition of proposed inner product
Property (ii)	$(\mathbf{u} + \mathbf{v}, \mathbf{w})_L = (\mathbf{L}(\mathbf{u} + \mathbf{v}), \mathbf{w})$	Definition of proposed inner product
	$= (\mathbf{Lu} + \mathbf{Lv}, \mathbf{w})$	Linearity of **L**
	$= (\mathbf{Lu}, \mathbf{w}) + (\mathbf{Lv}, \mathbf{w})$	Property (ii) of Definition 2.7
	$= (\mathbf{u}, \mathbf{w})_L + (\mathbf{v}, \mathbf{w})_L$	Definition of proposed inner product

Property (iii)	$(\alpha\mathbf{u},\mathbf{v})_L = (\mathbf{L}(\alpha\mathbf{u}),\mathbf{v})$	Definition of proposed inner product
	$= (\alpha\mathbf{L}\mathbf{u},\mathbf{v})$	Linearity of **L**
	$= \alpha(\mathbf{L}\mathbf{u},\mathbf{v})$	Property (iii) of definition 2.7
	$= \alpha(\mathbf{u},\mathbf{v})$	Definition of proposed inner product
Property (iv)	$(\mathbf{u},\mathbf{u})_L = (\mathbf{L}\mathbf{u},\mathbf{u})$	Definition of proposed inner product
	$\begin{cases} \geq 0 & \textbf{for all u in } D \\ = 0 & \textbf{if and only if u} = \mathbf{0} \end{cases}$	**L** is positive definite.

Thus since Properties (i)–(iv) of definition 2.7 are satisfied by the proposed definition of Equation 2.46, the energy inner product is a valid inner product on D.

Corollary If L is a positive-definite and self-adjoint linear operator with respect to an inner product (u,v) defined for its domain D, then an energy norm defined by

$$\|\mathbf{u}\|_L = [(\mathbf{u},\mathbf{u})_L]^{\frac{1}{2}} \tag{2.47}$$

is a valid norm on D

Proof Since L is positive definite and self adjoint with respect to (u,v), then by theorem 2.10, the energy inner product is a valid inner product on D. Then by theorem 2.5, any inner product can generate a norm of the form of Equation 2.47.

The term "energy" inner product is applied to inner products of the form of Equation 2.46, because positive definite and self-adjoint operators that arise from the mathematical modeling of physical systems can often be derived using energy methods. In such cases, **(Lu,u)** is related to some form of energy.

Example 2.36 The stiffness matrix of Example 2.33 is the stiffness matrix obtained in the modeling of the two-degree-of-freedom system shown Figure 2.8. Since the matrix is symmetric, it is self-adjoint with respect to the standard inner product on R^2. It is shown in Example 2.18 that the matrix is positive definite. (a) Determine the form of the energy inner product and confirm the commutativity property of the energy inner product. (b) If u_1 and u_2 represent the displacements of the two blocks, measured from equilibrium, show how the energy norm relates to the potential energy of the system.

Solution (a) Using the stiffness matrix of Example 2.21,

$$
\begin{aligned}
(\mathbf{u},\mathbf{v})_K &= (\mathbf{K}\mathbf{u},\mathbf{v}) \\
&= (4u_1 - 2u_2)v_1 + (-2u_1 + 3u_2)v_2 \qquad \text{(a)} \\
&= 4u_1v_1 - 2u_2v_1 - 2u_1v_2 + 3u_2v_2
\end{aligned}
$$

The commutativity of the energy inner product is confirmed by

$$\begin{aligned}(\mathbf{v},\mathbf{u})_K &= (\mathbf{u},\mathbf{K}\mathbf{v}) \\ &= u_1(4v_1 - 2v_2) + u_2(-2v_1 + 3v_2) \\ &= 4u_1v_1 - 2u_1v_2 + -2u_2v_1 + 3u_2v_2 \\ &= (\mathbf{u},\mathbf{v})_K\end{aligned} \tag{b}$$

(b) The potential energy of the system when the blocks have displacements u_1 and u_2, respectively is

$$\begin{aligned}V &= \frac{1}{2}2u_1^2 + \frac{1}{2}2(u_2 - u_1)^2 + \frac{1}{2}u_2^2 \\ &= \frac{1}{2}\left[2u_1^2 + 2(u_2 - u_1)^2 + u_2^2\right]\end{aligned} \tag{c}$$

Comparing Equation c with Equation C of Example 2.33 leads to

$$\begin{aligned}V &= \frac{1}{2}(\mathbf{K}\mathbf{u},\mathbf{u}) \\ &= \frac{1}{2}(\mathbf{u},\mathbf{u})_K \\ &= \frac{1}{2}\|\mathbf{u}\|_K^2\end{aligned} \tag{d}$$

Example 2.37 The differential equation for the transverse displacement of a beam due to transverse loading is

$$EI\frac{d^4w}{dx^4} = F(x) \tag{a}$$

The end conditions for a beam fixed at $x = 0$ and with a spring of stiffness k attached at $x = L$ are

$$w(0) = 0 \tag{b1}$$

$$\frac{dw}{dx}(0) = 0 \tag{b2}$$

$$\frac{d^2w}{dx^2}(L) = 0 \tag{b3}$$

$$EI\frac{d^3w}{dx^3}(L) - kw(L) = 0 \tag{b4}$$

Nondimensionalization of Equation a and Equations b as in Example 2.28 leads to

$$\frac{d^4w}{dx^4} = \Lambda f(x) \tag{c}$$

$$w(0) = 0 \tag{d1}$$

$$\frac{dw}{dx}(0) = 0 \tag{d2}$$

$$\frac{d^2w}{dx^2}(1) = 0 \tag{d3}$$

$$\frac{d^3w}{dx^3}(1) - \eta w(1) = 0 \tag{d4}$$

where $\eta = kL^3/EI$.

Define the operator L by $Lw = d^4w/dx^4$. The domain of L is S, the subspace of $C^4[0,1]$ defined such that if $g(x)$ is in S, then $g(0) = 0, dg/dx(0) = 0, d^2g/dx^2(1) = 0$, and $d^3g/dx^3(1) - \eta g(1) = 0$. It is easily shown that L is self-adjoint with respect to the standard inner product for $C^4[0,1]$.

(a) Show that L is positive definite on S with respect to the standard inner product for $C^4[0,1]$ and that thus an energy inner product is defined on S $(f,g,)_E = (Lf,g)$.

(b) Consider the cross-section shown in Figure 2.8. The normal stress along a line at distance z from the neutral axis (centroidal axis) is

$$\sigma = \frac{Mz}{I} \tag{e}$$

where the bending moment M in the cross-section is

$$M = EI\frac{d^2w}{dx^2} \tag{f}$$

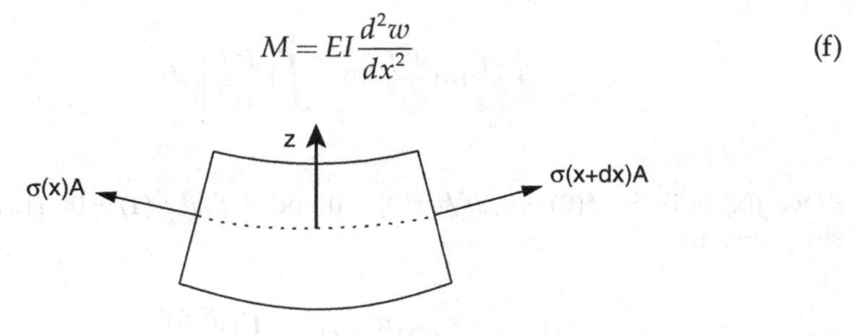

Figure 2.8 The normal stress in an Euler-Bernoulli beam varies across the span of the beam.

and $I = \int z^2 dA$ is the area moment of inertia about the neutral axis.

The strain energy per unit volume at a point in the beam, assuming elastic behavior throughout, is

$$e = \frac{\sigma^2}{2E} \tag{g}$$

If $w(x)$ is the transverse displacement of the beam, show how $(w, w)_L$ is related to the total potential energy of the system (total strain energy plus potential energy developed in the spring).

(c) Discuss the physical meaning of (w, f).

Solution (a) For any $f(x)$ in S,

$$(Lf, f) = \int_0^1 \frac{d^4 f}{dx^4} f dx \tag{h}$$

Application of integration by parts to the right-hand side of Equation e with $u = f$ and $dv = (d^4 f / dx^4)dx$ leads to

$$(Lf, f) = f(x)\frac{d^3 f}{dx^3}\bigg|_{x=0}^{x=1} - \int_0^1 \frac{df}{dx}\frac{d^3 f}{dx^3} dx \tag{i}$$

Application of integration by parts to the right-hand side of Equation i with $u = df/dx$ and $dv = d^3f/dx^3 dx$ gives

$$\begin{aligned}(Lf, f) &= f(x)\frac{d^3 f}{dx^3}\bigg|_{x=0}^{x=1} - \frac{df}{dx}\frac{d^2 f}{dx^2}\bigg|_{x=1}^{x=1} + \int_0^1 \left(\frac{d^2 f}{dx^2}\right)^2 dx \\ &= f(1)\frac{d^3 f}{dx^3}(1) - f(0)\frac{d^3 f}{dx^3}(0) - \frac{df}{dx}(1)\frac{d^2 f}{dx^2}(1) \\ &\quad + \frac{df}{dx}(0)\frac{d^2 f}{dx^2}(0) + \int_0^1 \left(\frac{d^2 f}{dx^2}\right)^2 dx\end{aligned} \tag{j}$$

Since $f(x)$ is in S, $f(0) = 0, df/dx(0) = 0$ and $d^2 f/dx^2(1) = 0$. Thus Equation j simplifies to

$$(Lf, f) = f(1)\frac{d^3 f}{dx^3}(1) + \int_0^1 \left(\frac{d^2 f}{dx^2}\right)^2 dx \tag{k}$$

Also, since $f(x)$ is in S, $d^3f/dx^3(1) = \eta f(1)$ which when substituted into Equation k gives

$$(Lf, f) = \eta[f(1)]^2 + \int_0^1 \left(\frac{d^2 f}{dx^2}\right)^2 dx \tag{l}$$

Equation l clearly shows that since $\eta > 0$, $(Lf, f) \geq 0$ for all f in S and that $(L0,0) = 0$. Furthermore, $(Lf, f) = 0$ implies that $d^2f/dx^2 = 0$ for all x, $0 \leq x \leq 1$. Thus $f(x) = ax + b$. However, if f is in S, then $f(0) = 0$, implying $b = 0$ and $df/dx(1) = 0$, implying a = 0, and L is positive definite on S with respect to the standard inner product for $C^4[0,1]$.

(a) Since L is self-adjoint and positive definite on S with respect to the standard inner product for $C^4[0,1]$, an energy inner product of the form $(f, g)_L = (Lf, g)$ can be defined on S.

(b) Let $w(x)$ be the dimensional transverse displacement of the beam. The potential energy developed in the spring is

$$V_s = \frac{1}{2} k[w(L)]^2 \tag{m}$$

Combining Equation e, Equation f, and Equation g, the strain energy per volume is

$$e = \frac{1}{2} E z^2 \left(\frac{d^2 w}{dx^2}\right)^2 \tag{n}$$

The total strain energy in the beam is

$$\begin{aligned} V_b &= \int e dV \\ &= \int_0^L \int_A \left[\frac{1}{2} E z^2 \left(\frac{d^2 w}{dx^2}\right)^2\right] dA dx \\ &= \frac{E}{2} \int_0^L \left[\left(\frac{d^2 w}{dx^2}\right)^2 \int_A z^2 dA\right] \end{aligned} \tag{o}$$

Recalling that $I = \int_A z^2 dA$, where I is the area moment of inertia of the cross sectional area about its centroidal axis Equation o becomes

$$V_b = \frac{EI}{2} \int_0^L \left(\frac{d^2 w}{dx^2}\right)^2 dx \tag{p}$$

The total potential energy is

$$V = V_s + V_b = \frac{1}{2}k[w(L)]^2 + \frac{1}{2}EI\int_0^L \left[\frac{d^2w}{dx^2}\right]^2 dx \tag{q}$$

Introduction of nondimensional variables $(x^* = x/L, w^* = w/L)$ into Equation q leads to

$$V = \frac{1}{2}kL^2[w^*(1)]^2 + \frac{1}{2}EI\int_0^1 \left(\frac{d^2w^*}{dx^{*2}}\right)^2 Ldx^* \tag{r}$$

Dividing Equation r by EIL and dropping the *s from the nondimensional variables leads to

$$\frac{V}{EIL} = \frac{1}{2}\left\{\eta[w(1)]^2 + \int_0^1 \left(\frac{d^2w}{dx^2}\right)^2 dx\right\} \tag{s}$$

Thus, from Equation l,

$$\frac{V}{EIL} = \frac{1}{2}(w,w)_{\mathrm{L}} \tag{t}$$

The energy inner product of the transverse displacement with itself is proportional to the total potential energy of the system. The quantity *V/(EIL)* is a nondimensional potential energy; the energy inner product of the transverse displacement with itself is twice the nondimensional potential energy. This inner product is often called the potential-energy inner product.

(c) Consider a differential element of the beam of length *dx*. Let *w* be the dimensional displacement of the beam and *F(x)* the transverse load per length. The work done during application of the load is

$$dU = \frac{1}{2}wF(x)dx \tag{u}$$

The total work done by the external load over the length of the beam is

$$U = \int dU = \int_0^L \frac{1}{2}w(x)F(x)dx$$

$$= \frac{1}{2}\int_0^1 Lw^*(x^*)F_{\max}f(x^*)Ldx^* \tag{v}$$

$$= \frac{L^2F_{mas}}{2}\int_0^1 w(x)f(x)dx$$

By defining a nondimensional work variable as $U^* = U/(EIL)$, Equation v can be rewritten as

$$\frac{U^*EIL}{L^2F_{max}} = \frac{1}{2}\int_0^1 w(x)f(x)dx$$

$$U^* = \frac{1}{2}\Lambda\int_0^1 w(x)f(x)dx \tag{w}$$

$$= \frac{1}{2}\Lambda(w,f)$$

Problems

2.1. The triangle inequality for all vectors in R^3 states that if $\mathbf{A}=\mathbf{B}+\mathbf{C}$, **then** $\|\mathbf{A}\| \le \|\mathbf{B}\|+\|\mathbf{C}\|$. Prove that the triangle inequality holds for all vectors **A, B,** and **C** in R^3.

2.2. Prove that the triangle inequality, as stated in problem 2.1, holds for all vectors in R^n for any finite value of n where the norm is the standard inner-product-generated norm.

2.3. Let $\mathbf{A}=2\mathbf{i}+3\mathbf{j}-\mathbf{k}$ and $\mathbf{B}=3\mathbf{j}+2\mathbf{k}$. Let V be the set of vectors in the span of **A** and **B**.
   a. Show that V is a subspace of R^3.
   b. Determine an orthonormal basis for V with respect to the standard inner product for R^3.

2.4. Let V be the set of vectors in R^3 such that if $\mathbf{u}=[u_1 \;\; u_2 \;\; u_3]$ is in V, then $u_1+2u_2-u_3=0$ Prove or disprove that V is a subspace of R^3. If V is a subspace of R^3, what is its dimension?

2.5. The boundary conditions satisfied by a nondimensional temperature distribution, $\theta(x)$, over an extended surface subject to a constant temperature at $x=0$ and insulated at $x=1$ are $\theta(0)=0$ and $d\theta/dx(1)=0$. Define S as the set of all functions in $C^2[0,1]$ which satisfy these boundary conditions. Show that S is a subspace of $C^2[0,1]$.

2.6. Let V be the space of functions which are in both $P^4[0,1]$, the space of all polynomials of degree four or less, and the space S as defined in problem 2.5. Determine a basis for V.

2.7. Determine an orthonormal basis with respect to the standard inner product on $C^2[0,1]$ for the vector space V defined in problem 2.6.

2.8. Expand $f(x)=x^2(1-x)^2$ in terms of (a) the basis vectors obtained in the solution of problem 2.6 and (b) the orthonormal basis obtained in problem 2.7.

2.9. Let $p(x)=a_nx^n+a_{n-1}x^{n-1}+\ldots+a_1x+a_0$ and $q(x)=b_nx^n+b_{n-1}x^{n-1}+\ldots+b_1x+b_0$ be elements of $P^n[0,1]$. Show that $(p,q)=\Sigma_{i=0}^n a_ib_i$ is a valid inner product defined for $P^n[0,1]$.

2.10. Determine an orthonormal basis with respect to the inner product as defined in problem 2.9 for the vector space V as defined in problem 2.6.

2.11. Expand $f(x)=x^2(1-x)^2$ in terms of the basis vectors obtained in problem 2.10.

2.12. Determine an orthonormal basis for the vector space V as defined in problem 2.6 with respect to the inner product $(p,q)=2a_0b_0+\Sigma_{i=1}^{n} a_i b_i$ defined on $P^n[0,1]$.

2.13. Let V be the space of all continuously differentiable functions of two independent variables x and y over the region $0 \le x \le 1$ and $0 \le y \le 1$. Show that $(f,g)=\int_0^1\int_0^1 f(x,y)g(x,y)dydx$ is a valid inner product on V.

2.14. Let S be the subspace of the vector space V of problem 2.13 spanned by the functions $1,x,y,x^2,xy,y^2,x^3,x^2y,xy^2,y^3$ Determine an orthonormal basis for V with respect to the inner product as defined in problem 2.11.

2.15. The deflection $w(x, y)$ of a square membrane fixed along its edge is subject to the boundary conditions $w(0,y)=0$, $w(1,y)=0$, $w(x,0)=0, w(x,1)=0$. Let Q be the subspace of the vector space S defined in problem 2.14 such that if $f(x, y)$ is in Q, then $f(0,y)=0, f(1,y)=0, f(x,0)=0, f(x,1)=0$. Determine a basis for Q.

2.16. Determine an orthonormal basis with respect to the inner product defined in problem 2.13 for the vector space Q defined in Problem 2.15.

2.17. The boundary conditions for the deflection, $w(x)$, of a fixed-free Euler-Bernoulli beam with an axial load are $w(0)=0, dw/dx(0)=0, d^2w/dx^2(1)=0, d^3w/dx^3(1)+\varepsilon dwdx(1)=0$. Let Q be the subspace of $P^6[0,1]$ such that if $f(x)$ is in Q, $f(0)=0, df/dx(0)=0, d^2f/dx^2(1)=0, d^3f/dx^3(1)+2df/dx(1)=0$. Determine a basis for Q.

2.18. Determine an orthonormal basis for the vector space Q as defined in Problem 2.15 with respect to the standard inner product on $C^4[0,1]$

2.19. For what value(s) of c_3 are the vectors **a**, **b**, and **c** linearly independent?

$$\mathbf{a}=\begin{bmatrix}2\\-2\\-1\end{bmatrix} \qquad \mathbf{b}=\begin{bmatrix}4\\-1\\3\end{bmatrix} \qquad \mathbf{c}=\begin{bmatrix}-2\\-1\\c_3\end{bmatrix}$$

2.20. For what value(s) of c_3 are the vectors **b** and **c** of problem 2.19 orthogonal with respect to the standard inner product for R^3?

2.21. For what value(s) of c_3 are the vectors **b** and **c** of problem 2.19 orthogonal with respect to the inner product defined by $(\mathbf{a},\mathbf{b})=a_1b_1+3a_2b_2+2a_3b_3$?

2.22. What is the adjoint of the matrix $\begin{bmatrix}1&3&2\\2&2&1\\4&2&1\end{bmatrix}$ with respect to (a) the standard inner product for R^3 and (b) the inner $(\mathbf{x},\mathbf{y})=2x_1y_1+2x_2y_2+3x_3y_3$?

2.23. Determine sufficient conditions for a 2×2 matrix to be self adjoint with respect to the standard inner product on R^2.

2.24. Determine sufficient conditions for a 2×2 matrix to be self adjoint with respect to the inner $(\mathbf{x},\mathbf{y}) = 2x_1y_1 + x_2y_2$.

2.25. For what values of ε is the operator $\mathbf{L}f = -d^2f/dx^2$ positive definite with respect to the standard inner product for $C^2[0,1]$ if the domain of L is specified such that $f(0) = 0$ and $df/dx(1) + \varepsilon f(1) = 0$?

2.26. Show that the biharmonic operator $\nabla^4\phi = \nabla^2(\nabla^2\phi)$ is self adjoint and positive definite with respect to the inner product $(f(\mathbf{r}), g(\mathbf{r})) = \int_R f(x,y) g(x,y)dV$ when $\phi = 0$ and $\partial\phi/\partial n = 0$ everywhere on the boundary of R.

For problems 2.27–2.31, consider the boundary value problem defined by

$$\frac{d^2y}{dx^2} + 2\frac{dy}{dx} - 10y = f(x) \tag{a}$$

$$y(0) = 0 \tag{b}$$

$$\frac{dy}{dx}(1) + 2y(1) = 0 \tag{c}$$

2.27. Show that Equation a can be rewritten as

$$-\frac{1}{e^{2x}}\frac{d}{dx}\left(e^{2x}\frac{dy}{dx}\right) + 10y = f(x) \tag{d}$$

2.28. Define the operator **L** whose domain is the subspace of $C^2[0,1]$ in which all elements satisfy the boundary conditions of Equation b and Equation c by $\mathbf{L}y = -\frac{1}{e^{2x}}\frac{d}{dx}\left(e^{2x}\frac{dy}{dx}\right) + 10y$. Show that **L** is self-adjoint with respect to the inner product defined by

$$(f,g) = \int_0^1 f(x)g(x)e^{2x}dx \tag{e}$$

2.29. Show that **L**, as defined in Problem 2.28 is positive definite with respect to the inner product defined by Equation e.

2.30. Since **L** is self-adjoint and positive definite with respect to the inner product of Equation e, **L** may be used to define an energy inner product $(f,g)_{\mathbf{L}}$. Calculate

$$\left(x^2 - \frac{3}{4}x, x^3 - \frac{3}{5}x\right)_{\mathbf{L}} \tag{f}$$

2.31. Define the vector space Q as the intersection of $P^4[0,1]$ with the domain of **L**. Determine a basis of **L**, which is orthonormal with respect to the energy inner product of Problem 2.30.

2.32. Consider the operator **L** defined as

$$\mathbf{L}f(x) = \int_0^1 f(y)\sin(\pi x)\sin(\pi y)dy \tag{g}$$

(a) Determine the conditions under which **L** is self-adjoint with respect to the standard inner product on C[0,1].

(b) Determine the condition under which **L** is positive definite with respect to the standard inner product on C[0,1].

2.33. Demonstrate the definition of an energy inner product $(f,g)_L$ derived from the operator of Problem 2.32 by calculating $(x,x^2)_L$.

Chapter 3

Ordinary differential equations

3.1 Linear differential equations

A differential equation is an operator equation of the form

$$\mathbf{Lu}=\mathbf{f} \tag{3.1}$$

where $\mathbf{u}=u(x)$ is an element of $S_{\mathcal{D}}$, a subspace of $C^n[a,b]$, and $f=f(x)$ is an element of $S_{\mathcal{R}}$, a subspace of $PC[a,b]$, the space of piecewise continuous functions on [a,b]. $\mathbf{L}$ is a differential operator with domain $S_{\mathcal{D}}$ and range $S_{\mathcal{R}}$. The order of the differential equation is the order of the highest-order derivative in $\mathbf{L}$. The differential equation is linear if $\mathbf{L}$ is a linear operator; otherwise, it is nonlinear. The differential equation has constant coefficients if for arbitrary values of x_1 and x_2 such that $a \leq x_1, x_2 \leq b$, the operator definition of $\mathbf{L}$ is the same for x_1 and x_2.

The general form of an nth-order linear differential equation is

$$a_n(x)\frac{d^n u}{dx^n}+a_{n-1}(x)\frac{d^{n-1}u}{dx^{n-1}}+\cdots+a_1(x)\frac{du}{dx}+a_0(x)u=f(x) \tag{3.2}$$

The differential equation has constant coefficients if $a_n, a_{n-1}, \ldots, a_n, a_0$ are all independent of x. The differential equation is called homogeneous if $f(x)=0$; otherwise, it is nonhomogeneous.

A homogeneous solution of Equation 3.2 is a solution of $\mathbf{Lu}=\mathbf{0}$. It should be noted that if $u_1(x)$ and $u_2(x)$ are homogeneous solutions, then because of the linearity of $\mathbf{L}$,

$$\mathbf{L}(\alpha u_1(x)+\beta u_2(x))=\alpha\mathbf{L}u_1(x)+\beta\mathbf{L}u_2(x) \tag{3.3}$$

As a result of the hypothesis that $u_1(x)$ and $u_2(x)$ are homogeneous solutions of Equation 3.2, the right-hand side of Equation 3.3 is zero. This shows that the homogeneous solution is not unique and that any linear combination of homogeneous solutions is also a homogeneous solution. The homogeneous solution space for a linear differential operator $\mathbf{L}$ is the space of functions which satisfy $\mathbf{Lu}=0$. This space is independent of $S_{\mathcal{D}}$ and $S_{\mathcal{R}}$. The dimension of the homogeneous solution space for an nth-order differential operator is n. This is proved in Section 3.2 for $n=2$. The proof can be generalized for any n.

Thus, for an nth-order linear differential equation, the operator's homogeneous solution space has a basis of n functions, say $u_1(x),u_2(x),\ldots,u_n(x)$, such that the general homogeneous solution is a linear combination of the basis elements and thus is of the form

$$u_h(x) = C_1u_1(x) + C_2u_2(x) + \ldots C_nu_n(x) \tag{3.4}$$

where $C_1, C_2, \ldots, C_n$ are arbitrary constants, often called constants of integration.

Let $u_p(x)$ be a particular solution of Equation 3.4, a solution specific to the form of $f(x)$. A particular solution satisfies $\mathbf{L}u_p = f$. Define

$$u(x) = u_h(x) + u_p(x) \tag{3.5}$$

Then $\mathbf{L}u = \mathbf{L}(u_h + u_p) = \mathbf{L}u_h + \mathbf{L}u_p = 0 + f = f$. Thus a function of the form of Equation 3.5 satisfies Equation 3.2. This solution is called the general solution. The particular solution is linearly independent of the homogenous solution, because otherwise it could be expressed as a linear combination of the basis vectors of the homogeneous solution space.

The general solution of an nth-order linear ordinary differential equation is of the form

$$u(x) = C_1u_1(x) + C_2u_2(x) + \cdots + C_nu_n(x) + u_p(x) \tag{3.6}$$

where $u_1(x),u_2(x),\ldots,u_n(x)$ is a basis for the homogeneous solution space, $C_1, C_2, \ldots, C_n$ are arbitrary constants, and $u_p(x)$ is the particular solution. The constants of integration are determined through application of constraints specified by the physical problem. The constraints are conditions which must be satisfied at specific values of x. A unique solution can be attained only if n independent constraints are specified. If n constraints are specified, they are applied to the solution of Equation 3.6, leading to a set of n simultaneous linear algebraic equation to solve for the constants of integration. Unless the determinant of the resulting coefficient matrix is zero, a unique solution exists for the constants, and a unique solution exists for the differential problem.

An initial value problem is one in which the constraints are developed at a single value of the independent variable x, perhaps $x=0$. The range over which the solution is defined is $0 \le x < \infty$. The initial conditions are of the form $u(0) = u_0$, $du/dx(0) = u_0', \ldots, d^{(n-1)}u/dx^{n-1}(0) = u_0^{(n-1)}$. Application of initial conditions to Equation 3.6 leads to

$$\begin{bmatrix} u_1(0) & u_2(0) & u_3(0) & \ldots & u_n(0) \\ u_1'(0) & u_2'(0) & u_3'(0) & \ldots & u_n'(0) \\ u_1''(0) & u_2''(0) & u_3''(0) & \ldots & u_n''(0) \\ \vdots & \vdots & \vdots & \ddots & \vdots \\ u_1^{(n-1)}(0) & u_2^{(n-1)}(0) & u_3^{(n-1)}(0) & \ldots & u_n^{(n-1)}(0) \end{bmatrix} \begin{bmatrix} C_1 \\ C_2 \\ C_3 \\ \vdots \\ C_n \end{bmatrix} = \begin{bmatrix} u_0 - u_p(0) \\ u_0' - u_p'(0) \\ u_0'' - u_p''(0) \\ \vdots \\ u_0^{(n-1)} - u_p^{(n-1)}(0) \end{bmatrix} \tag{3.7}$$

The determinant of the coefficient matrix in Equation 3.7 is the Wronskian evaluated at $x=0$ of the basis functions for the homogeneous solution space. It can be shown that $W(0) \neq 0$ and a unique solution of Equation 3.7 exists.

Boundary-value problems are those for which the constraints are specified at two values of x, say $x=a$ and $x=b$. Boundary conditions may specify that a linear combination of the solution and some of its derivatives have specific values at a or b. The boundary condition is homogeneous if that specific value is zero. The differential operators for most boundary-value problems are of even order. Half the n boundary conditions are applied at $x=a$, and half are applied at $x=b$. If all boundary conditions are homogeneous, then $S_{\mathfrak{D}}$ is defined as the subspace of $C^n[a,b]$ consisting of all functions satisfying all boundary conditions. The boundary conditions are applied to Equation 3.6, leading to a set of simultaneous equations to solve for the constants of integration. If the determinant of the coefficient matrix is zero, then $f(x)$ must satisfy a solvability condition (Section 5.11) for a solution to exist.

If any boundary condition is nonhomogeneous, then $S_{\mathfrak{D}}$ is not a vector space in its own right, because the nonhomogeneous boundary condition prevents closure under addition, closure under scalar multiplication, and the existence of the zero vector in $S_{\mathfrak{D}}$. In this case, the nonhomogeneity can be transferred to the differential equation through a redefinition of dependent variables.

The nondimensional differential equation governing the temperature distribution over an extended surface is $d^2\theta/dx^2 - Bi\theta = 0$. The end at $x=0$ is subject to a fixed temperature, while the end at $x=1$ is subject to a heat flux for which the appropriate boundary conditions are $\theta(0)=1$ and $d\theta/dx\,(1)=q$. The boundary conditions are nonhomogeneous. A new dependent variable is defined as $\phi(x)=\theta(x)-1-qx$. The differential equation governing $\phi(x)$ is $d^2\phi/dx^2 - Bi\phi = Bi(1+qx)$. The boundary conditions become $\phi(0)=0$ and $d\phi/dx\,(1)=0$. Thus the solution for $\phi(x)$ is in $S_{\mathfrak{D}}$, a subspace of $C^2[0,1]$ defined such that if $f(x)$ is in $S_{\mathfrak{D}}$, then $f(0)=0$ and $df/dx\,(1)=0$.

Methods for determining particular solutions are presented in Section 3.3. Superposition of particular solutions can be used for linear problems. Consider a linear differential equation of the form

$$\mathbf{Lu} = f_1(x) + f_2(x) \tag{3.8}$$

Let $u_{p,1}(x)$ be the particular solution for $\mathbf{Lu}=f_1(x)$ and let $u_{p,2}(x)$ be the particular solution for $\mathbf{Lu}=f_2(x)$. The linearity of $\mathbf{L}$ implies that

$$\begin{aligned} \mathbf{L}(u_{p,1}+u_{p,2}) &= \mathbf{L}u_{p,1} + \mathbf{L}u_{p,2} \\ &= f_1(x) + f_2(x) \end{aligned} \tag{3.9}$$

Equation 3.9 implies that $u_{p,1}+u_{p,2}$ is the particular solution of Equation 3.8. The general statement of the principle of linear superposition is that if the nonhomogeneous term in a linear ordinary differential equation is the sum of a finite number of terms, then the particular solution is the sum of the particular solutions corresponding to each term in the sum.

3.2 General theory for second-order differential equations

In this section, a general theory for second-order differential equations is developed and then generalized to higher-order equations. The general form of a linear second-order ordinary differential equation is

$$\frac{d^2y}{dx^2} + a_1(x)\frac{dy}{dx} + a_0(x)y = f(x) \tag{3.10}$$

Equation 3.10 is supplemented by two initial conditions, $y(0)=y_0$ and $dy/dx(0)=y'_0$, or by two boundary conditions expressed as linear combinations of y and dy/dx. If the interval over which the differential equation is to be solved is $a \leq x \leq b$, one boundary condition is developed at $x=a$ and one at $x=b$. The following discussion is directed toward initial value problems. However the results can be extended to boundary-value problems, which are the main focus of this book.

The following theorem is fundamental, but is presented without proof because the proof is lengthy and noninstructive toward the goals of this work.

Theorem 3.1 Existence and uniqueness of solutions. If $a_1(x)$, $a_0(x)$ and $f(x)$ are all continuous for all x, then, there exists a unique solution for the differential equation, Equation 3.1, subject to the initial conditions $y(0)=y_0$ and $dy/dx(0)=y'_0$.

The homogeneous solution space is the subspace V of $C^2[0,\infty]$ consisting of functions which satisfy

$$\frac{d^2y}{dx^2} + a_1(x)\frac{dy}{dx} + a_0(x)y = 0 \tag{3.11}$$

as well as two initial conditions of the form specified in theorem 3.1. The space of functions is not dependent on the values of the initial conditions, but only on the existence of such conditions. It is desired to determine the dimension of V and find a basis for V.

Theorem 3.2 Abel's formula. Let $y_1(x)$ and $y_2(x)$ be two solutions of Equation 3.11. The Wronskian of these solutions defined by

$$W(x) = \begin{vmatrix} y_1(x) & y_2(x) \\ \dfrac{dy_1}{dx} & \dfrac{dy_2}{dx} \end{vmatrix} = y_1(x)\frac{dy_2}{dx} - y_2(x)\frac{dy_1}{dx} \tag{a}$$

satisfies

$$W(x) = Ce^{-\int a_1(x)dx} \tag{b}$$

Proof Since $y_1(x)$ and $y_2(x)$ solve Equation 3.11,

$$\frac{d^2y_1}{dx^2} + a_1(x)\frac{dy_1}{dx} + a_0(x)y_1 = 0 \tag{c}$$

$$\frac{d^2y_2}{dx^2} + a_1(x)\frac{dy_2}{dx} + a_0(x)y_2 = 0 \tag{d}$$

Multiplying Equation c by $y_2(x)$ and Equation d by $y_1(x)$ and then subtracting the resulting equations leads to

$$y_2\frac{d^2y_1}{dx^2} - y_1\frac{d^2y_2}{dx^2} + a_1(x)\left(y_2\frac{dy_1}{dx} - y_1\frac{dy_2}{dx}\right) = 0 \tag{e}$$

Noting that $dW/dx = y_2(d^2y_1/dx^2) - y_1(d^2y_2/dx^2)$, Equation e becomes

$$\frac{dW}{dx} + a_1(x)W = 0 \tag{f}$$

Equation b is a solution of Equation f.

The Wronskian of two linearly independent functions is nonzero for some value of x, while the Wronskian of two linearly dependent functions is zero for all values of x. Let $y_1(x)$ and $y_2(x)$ be two linearly independent functions which solve Equation 3.11. A linear combination of two solutions is also a solution of the differential equation.

Suppose that the dimension of V is 1, and let $y_1(x)$ be a solution of Equation 3.11. The general form of a solution is $y(x) = cy_1(x)$. However, all elements in V must satisfy two initial conditions, which is impossible if the dimension of V is one. Suppose that the dimension of V is two. Let $y_1(x)$ and $y_2(x)$ be two linearly independent solutions of Equation 3.11. Then a linear combination is of the form $y(x) = C_1y_1(x) + C_2y_2(x)$. Satisfaction of the initial conditions requires that $C_1y_1(0) + C_2y_2(0) = y_0$ and $C_1(dy_1/dx)(0) + C_2(dy_2/dx)(0) = y_0'$. A unique solution of these equations exists as long as $W(0) \neq 0$. Therefore, the dimension of V is at least 2. If there are only two linearly independent solutions of Equation 3.11, then the dimension of V is two.

Theorem 3.3 Let V be the vector space of solutions of Equation 3.11 which satisfy two arbitrary initial conditions. The dimension of V is two.

Proof In view of the previous discussion, it is necessary only to show that Equation 3.11 has only two linearly independent solutions, say $y_1(x)$ and $y_2(x)$.

Suppose that $y_3(x)$ is another solution of Equation 3.11. Then from Abel's formula, the Wronskian for the solution combination $y_1(x)$ and $y_3(x)$ can be obtained using Abel's formula as

$$y_1(x)\frac{dy_3}{dx} - y_3(x)\frac{dy_1}{dx} = C_{1,3}e^{-\int a_1(x)dx} \tag{a}$$

and the Wronskian for the solution combination $y_2(x)$ and $y_3(x)$ is

$$y_2(x)\frac{dy_3}{dx} - y_3(x)\frac{dy_2}{dx} = C_{2,3}e^{-\int a_1(x)dx} \tag{b}$$

Equation a and Equation b can be rearranged to yield

$$\begin{bmatrix} y_1(x) & \dfrac{dy_1}{dx} \\ y_2(x) & \dfrac{dy_2}{dx} \end{bmatrix} \begin{bmatrix} \dfrac{dy_3}{dx} \\ y_3(x) \end{bmatrix} = \begin{bmatrix} C_{1,3} \\ C_{2,3} \end{bmatrix} e^{-\int a_1(x)dx} \tag{c}$$

Equation c can be solved, perhaps using Cramer's rule, for $y_3(x)$, leading to

$$y_3(x) = \frac{[C_{2,3}y_1(x) - C_{1,3}y_2(x)]e^{-\int a_1(x)dx}}{y_1(x)\dfrac{dy_2}{dx} - y_2(x)\dfrac{dy_1}{dx}} \tag{d}$$

Note that the denominator of Equation d is the Wronskian of the solutions $y_1(x)$ and $y_2(x)$, which can be replaced using Abel's formula by $C_{1,2}e^{-\int a_1(x)dx}$. Thus Equation d reduces to

$$y_3(x) = \frac{C_{2,3}}{C_{1.2}}y_1(x) - \frac{C_{1,3}}{C_{1.2}}y_2(x) \tag{e}$$

Equation e implies that $y_3(x)$ is not linearly independent of $y_1(x)$ and $y_2(x)$. Thus Equation 3.11 has two linearly independent homogeneous solutions, from which it follows that the dimension of V is 2.

Since the dimension of V is 2, any set of two linearly independent solutions of Equation 3.11 is a basis for V.

Abel's formula provides a method for finding a second linearly independent solution if one is known. Suppose that $y_1(x)$ is known, and it is desired to determine a second linearly independent solution. Abel's formula implies that the second solution, $y_2(x)$, satisfies

$$y_1(x)\frac{dy_2}{dx} - y_2(x)\frac{dy_1}{dx} = e^{-\int a_1(x)dx} \tag{3.12}$$

where the constant C is chosen to be 1. If a different value of C is chosen, say 43.1, the solution for $y_2(x)$ is 43.1 times the solution for C = 1. Dividing Equation 3.12 by $y_1(x)$ leads to

$$\frac{dy_2}{dx} - \frac{\frac{dy_1}{dx}}{y_1} y_2 = \frac{1}{y_1} e^{-\int a_1(x)dx} \tag{3.13}$$

Equation 3.13 is a first-order linear differential equation whose solution is $y_2(x)$. Equation 3.13 can be solved using the integrating factor technique where the integrating factor is $1/y_1(x)$. Multiplying by the integrating factor, Equation 3.13 can be written as

$$\frac{d}{dx}\left(\frac{y_2}{y_1}\right) = \frac{1}{y_1^2} e^{-\int a_1(x)dx} \tag{3.14}$$

Integration of Equation 3.14 leads to

$$y_2(x) = y_1(x) \int \frac{e^{-\int a_1(x)}}{y_1^2} dx + Cy_1(x) \tag{3.15}$$

Since $y_1(x)$ has already been identified as a homogeneous solution, the constant of integration can be taken to be zero, leading to

$$y_2(x) = y_1(x) \int \frac{e^{-\int a_1(x)}}{y_1^2} dx \tag{3.16}$$

The general solution of Equation 3.10 is a linear combination of the basis functions of V plus a particular solution.

3.3 *Differential equations with constant coefficients*

The general form of an nth-order linear differential operator with constant coefficients is

$$\mathbf{L}y = a_n \frac{d^n y}{dx^n} + a_{n-1} \frac{d^{n-1} y}{dx^{n-1}} + \ldots a_1 \frac{dy}{dx} + a_0 y \tag{3.17}$$

where $a_n, a_{n-1}, \ldots, a_1$ and a_0 are called the coefficients of the differential operator.

A homogeneous solution for a linear differential equation with constant coefficients is assumed to be

$$y_h(x) = Ce^{\alpha x} \tag{3.18}$$

where C is an arbitrary constant and α is a constant to be determined. Substitution of Equation 3.18 into Equation 3.17 leads to

$$\left(a_n\alpha^n + a_{n-1}\alpha^{n-1} + \ldots a_1\alpha + a_0\right)Ce^{\alpha x} = 0 \tag{3.19}$$

Since $e^{\alpha x} \neq 0$ for all x for any value of α, and since $C=0$ implies $y_h(x)=0$ for all x, Equation 3.19 leads to a nontrivial solution if and only if

$$a_n\alpha^n + a_{n-1}\alpha^{n-1} + \ldots a_1\alpha + a_0 = 0 \tag{3.20}$$

Equation 3.20 is an nth-order polynomial equation whose roots are the values of α such that Equation 3.18 is a nontrivial solution of Equation 3.1 when **L** is defined by Equation 3.5. The polynomial in Equation 3.20 has n roots $\alpha_1, \alpha_2, \ldots, \alpha_n$, each leading to a solution of the form of Equation 3.18. If all values of α are distinct, then n linearly independent solutions have been found, and thus a basis for the homogeneous solution space is $e^{\alpha_1 x}, e^{\alpha_2 x}, \ldots, e^{\alpha_n x}$ and the general homogeneous solution is

$$y_h(x) = \sum_{i=1}^{n} C_i e^{\alpha_i x} \tag{3.21}$$

The coefficients of the polynomial of Equation 3.20 are all real, and therefore complex roots occur in conjugate pairs. Let $\sigma \pm i\tau$ be complex conjugate solutions of Equation 3.20. Their contribution to the homogeneous solution of Equation 3.21 through application of Euler's identity ($e^{i\theta} = \cos\theta + i\sin\theta$) is

$$\begin{aligned}
D_1e^{(\sigma+i\tau)x} + D_2e^{(\sigma-i\tau)x} &= e^{\sigma x}\left(D_1e^{i\tau x} + D_2e^{-i\tau x}\right) \\
&= e^{\sigma x}[D_1(\cos\tau x + i\sin\tau x) + D_2(\cos\tau x - i\sin\tau x)] \\
&= e^{\sigma x}[(D_1 + D_2)\cos\tau x + i(D_1 - D_2)\sin\tau x] \\
&= e^{\sigma x}(C_1\cos\tau x + C_2\sin\tau x)
\end{aligned} \tag{3.22}$$

Let α be a repeated root of Equation 3.20. The second-order differential equation satisfied by this repeated root is $d^2y/dx^2 + 2\alpha\, dy/dx + y = 0$. One solution is $y_1(x) = e^{\alpha x}$. The second solution, which is linearly independent with $y_1(x) = e^{\alpha x}$, can be obtained using Equation 3.16 as

$$y_2(x) = e^{\alpha x}\int \frac{e^{-\int 2\alpha dx}}{\left(e^{\alpha x}\right)^2}dx = e^{\alpha x}\int dx = xe^{\alpha x} \tag{3.23}$$

The results of Equation 3.23 can be generalized as follows. If α is a root of Equation 3.20 of multiplicity m, the linearly independent solutions corresponding to this root are $e^{\alpha x}, xe^{\alpha x}, \ldots, x^{m-1}e^{\alpha x}$. The result can be summarized as

Theorem 3.4 A basis for the homogeneous solution space for an nth-order linear differential equation with constant coefficients whose operator is of the form of Equation 3.17 is the set of functions $e^{\alpha_1 x}, e^{\alpha_2 x}, \ldots, e^{\alpha_n x}$, where α_i, $i=1,2,\ldots,n$ are the roots of Equation 3.20. If Equation 3.20 has a root of multiplicity m, then m linearly independent basis vectors are $e^{\alpha x}, xe^{\alpha x}, \ldots, x^{m-2}e^{\alpha x}, x^{m-1}e^{\alpha x}$.

Example 3.1 The free response of the single-degree-of-freedom system shown in Figure 3.1 is governed by the differential equation

$$m\frac{d^2x}{dt^2}+c\frac{dx}{dt}+kx=0 \tag{a}$$

The block has a mass of 1 kg, and the spring has a stiffness of 100 N/m. Determine the free response of the block when it is displaced 1 mm from equilibrium and released if (a) $c=4$ N·s/m, (b) $c=30$ N·s/m, and (c) $c=20$ N·s/m.

Solution The initial conditions for the problem are $x\,(0)=0.001$ m and $dx/dt\,(0)=0$ m/s. Substitution of a solution of the form of Equation 3.18 into Equation a leads to the polynomial equation,

$$\alpha^2+c\alpha+100=0 \tag{b}$$

(a) For $c=4$ N·s/m, the roots of Equation b are $\alpha_{1,2}=-2\pm\sqrt{-99}=-2\pm 9.950i$. The roots are complex conjugates, and therefore the general solution is obtained using Equation 3.22 as:

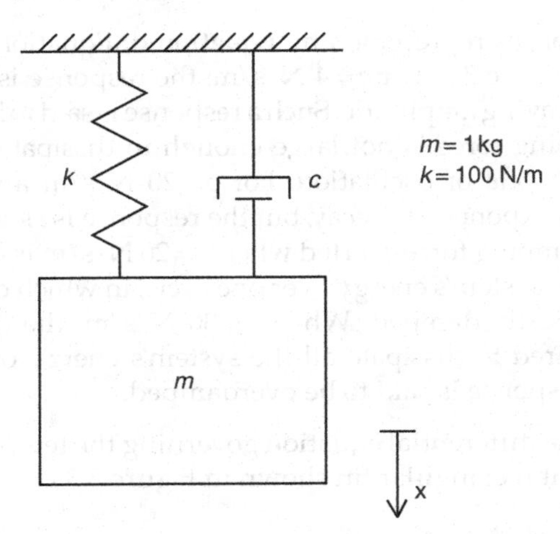

Figure 3.1 The free response of the mass-spring-viscous-damper system of Example 3.1 is governed by a second-order differential equation.

$$x(t) = e^{-2t}[C_1 \cos(9.950t) + C_2 \sin(9.950t)] \qquad \text{(c)}$$

Setting $x(0) = 0.001$ leads to $C_1 = 0.001$, and setting $dx/dt\,(0) = 0$ leads to $C_2 = 2.01 \times 10^{-4}$. The time-dependent response is

$$x(t) = e^{-2t}[\cos(9.950t) + 0.201 \sin(9.950t)] \quad \text{mm} \qquad \text{(d)}$$

(b) For $c = 30$ N·s/m, the roots of Equation b are $\alpha_{1,2} = -15 \pm \sqrt{125}$ or $\alpha_1 = -26.18$ and $\alpha_2 = -3.819$. The general solution of Equation a is

$$x(t) = C_1 e^{-3.819t} + C_2 e^{-26.18t} \qquad \text{(e)}$$

Setting $x(0) = 0.001$ leads to $C_1 + C_2 = 0.001$, while setting $dx/dt\,(0) = 0$ leads to $-3.819C_1 - 26.18C_2 = 0$. Simultaneous solution of these equations gives $C_1 = 1.117 \times 10^{-3}$ and $C_2 = -1.708 \times 10^{-4}$, which when substituted into Equation e give

$$x(t) = 1.117e^{-3.819t} - 0.171C_2 e^{-26.18t} \quad \text{mm} \qquad \text{(f)}$$

(c) For $c = 20$ N·s/m, Equation b has a double root of $\alpha = -10$. The appropriate form of the general solution is

$$x(t) = C_1 e^{-10t} + C_2 t e^{-10t} \qquad \text{(g)}$$

Setting $x(0) = 0.001$ in Equation g gives $C_1 = 0.001$, while setting $dx/dt\,(0) = 0$ leads to $C_2 = 0.01$. The system response for $c = 20$ N·s/m is

$$x(t) = e^{-10t} + 10te^{-10t} \quad \text{mm} \qquad \text{(h)}$$

The system responses represented by Equation d, Equation f, and Equation h are plotted in Figure 3.2. For $c = 4$ N·s/m, the response is oscillatory with exponentially decaying amplitude. Such a response is said to be underdamped in that the damping force is not large enough to dissipate all the system's energy over one cycle of oscillation. For $c = 20$ N·s/m and $c = 30$ N·s/m, responses exhibit exponential decay, but the response is slower for the larger value of c. The damping force exerted when $c = 20$ N·s/m is just large enough to dissipate all the system's energy over one cycle, in which case the response is said to be critically damped. When $c = 30$ N·s/m, the damping force is larger than required to dissipate all the system's energy over one cycle, in which case the response is said to be overdamped.

Example 3.2 The differential equation governing the temperature distribution in the straight rectangular fin shown in Figure 3.3 is

$$\frac{d^2\theta}{dx^2} - m^2\theta = 0 \qquad \text{(a)}$$

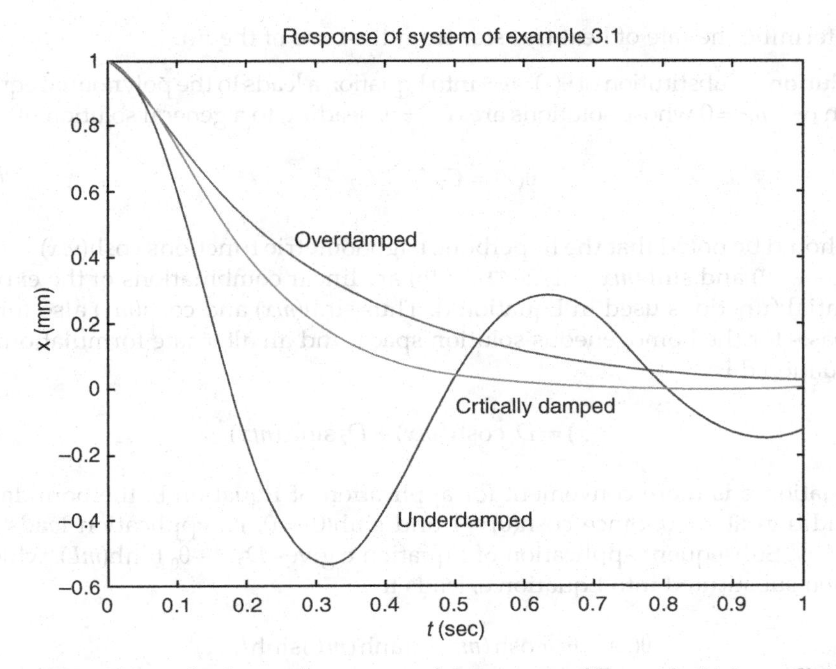

Figure 3.2 The underdamped response of the system of Example 3.1 is oscillatory with exponentially decaying amplitude. The critically damped and overdamped responses are not oscillatory and decay exponentially.

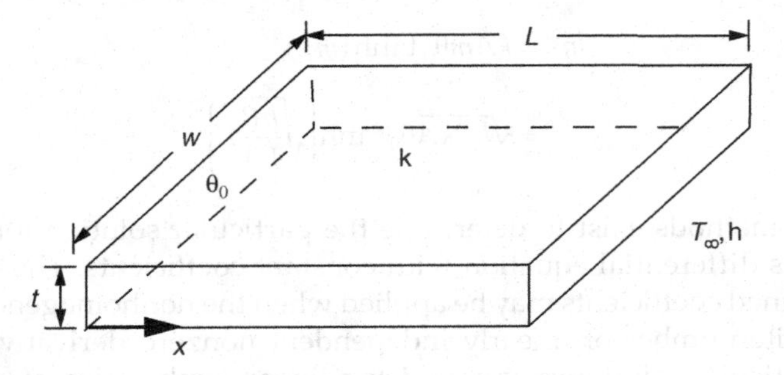

Figure 3.3 The temperature distribution in the extended surface of Example 3.2 is obtained from the solution of a second-order boundary value problem.

where $\theta = T - T_\infty$ and $m = \sqrt{hP/hA}$ with $P = 2(w + t)$ and $A = wt$. The fin is maintained at a constant temperature at its base and is insulated at its tip, leading to the boundary conditions

$$\theta(0) = \theta_0 \tag{b}$$

$$\frac{d\theta}{dx}(L) = 0 \tag{c}$$

Determine the rate of heat transfer from the base of the fin.

Solution Substitution of $\theta(x)=e^{\alpha x}$ into Equation a leads to the polynomial equation $\alpha^2-m^2=0$ whose solutions are $\alpha=\pm m$, leading to a general solution of

$$\theta(x)=C_1e^{mx}+C_2e^{-mx} \tag{d}$$

It should be noted that the hyperbolic trigonometric functions $\cosh(mx)=1/2(e^{mx}+e^{-mx})$ and $\sinh(mx)=1/2(e^{mx}-e^{-mx})$ are linear combinations of the exponential functions used in Equation d. Thus $\sinh(mx)$ and $\cosh(mx)$ also form a basis for the homogeneous solution space, and an alternate formulation of Equation d is

$$\theta(x)=D_1\cosh(mx)+D_2\sinh(mx) \tag{e}$$

Equation e is more convenient for application of Equation b, the boundary condition at $x=0$. Since $\cosh(0)=1$ and $\sinh(0)=0$, its application leads to $D_1=\theta_0$. Subsequent application of Equation c gives $D_2=-\theta_0\tanh(mL)$, which when substituted into Equation e, leads to

$$\theta(x)=\theta_0[\cosh(mx)-\tanh(mL)\sinh(mx)] \tag{f}$$

The rate of heat transfer from the base of the fin is $q=-kA\,(d\theta/dx)(0)$, which, using Equation f, becomes

$$\begin{aligned} q &= kAm\theta_0\tanh(mL) \\ &= \sqrt{hPkA}\,\theta_0\tanh\left(\sqrt{\frac{hP}{kA}}L\right) \end{aligned} \tag{g}$$

Several methods exist to determine the particular solution for a nonhomogeneous differential equation with constant coefficients. The method of undetermined coefficients may be applied when the nonhomogeneous terms have a finite number of linearly independent nonzero derivatives. In this case, a particular solution is assumed as a linear combination of all nonzero derivatives with the coefficients unspecified. The coefficients are determined by substituting the assumed particular solution into the differential equation and equating coefficients of like terms. Consider a linear ordinary differential equation with constant coefficients of the form $\mathbf{L}y=f$, where f has a finite number of linearly independent derivatives $f'(x), f''(x),\ldots, f^{(k)}(x)$. A trial solution for the particular solution of differential equation is assumed to be

$$y_p(x)=A_0f(x)+A_1f'(x)+\cdots+A_kf^{(k)}(x) \tag{3.24}$$

where $A_0, A_1, \ldots, A_k$ are coefficients which are as yet undetermined. Equation 3.24 is then substituted into the differential equation. The resulting equation is rearranged such that each side expresses a linear combination of linearly independent functions. Coefficients of functions on the left-hand side are equated with coefficients of like functions on the right-hand side. This leads to a set of $k+1$ simultaneous equations to solve for the coefficients.

The method of undetermined coefficients is applied to equations in which the nonhomogeneous terms have a finite number of linearly independent derivatives. These include polynomials, exponential functions, trigonometric functions, and products of these functions. Table 3.1 lists functions for which the method of undetermined coefficients leads to successful solutions, as well as the form of the trial solutions.

A special case occurs when a nonhomogeneous term is a solution of the corresponding homogeneous equation. In this case, a trial solution of the form of Equation 3.24 leads to $\mathrm{L}y_p = 0$, indicating that the trial solution does not lead to a particular solution. Noting that $d^k[xf(x)]/dx^k = x d^k f/dx^k + k d^{k-1} f/dx^{k-1}$, substitution into the operation defined by Equation 3.17 leads to

$$
\begin{aligned}
\mathbf{L}(xf(x)) &= a_n\left[x\frac{d^n f}{dx^n} + n\frac{d^{n-1} f}{dx^{n-1}}\right] + a_{n-1}\left[x\frac{d^{n-1} f}{dx^{n-1}} + (n-1)\frac{d^{n-2} f}{dx^{n-2}}\right] \\
&\quad + \cdots + a_1\left[x\frac{df}{dx} + f\right] + a_0 f \\
&= x\left(a_n\frac{d^n f}{dx^n} + a_{n-1}\frac{d^{n-1} f}{dx^{n-2}} + \cdots + a_1\frac{df}{dx} + a_0\right) \\
&\quad + na_n\frac{d^{n-1} f}{dx^{n-1}} + (n-1)a_{n-1}\frac{d^{n-2} f}{dx^{n-1}} + \cdots + a_1 f \\
&= x\mathbf{L}f(x) + \mathbf{K}f
\end{aligned} \tag{3.25}
$$

Table 3.1 Trial Solutions for Use in Method of Undetermined Coefficients

Nonhomogeneous term	Trial solution
x^n where n is an integer	$C_n x^n + C_{n-1}x^{n-1} + \cdots + C_1 x + C_0$
e^{ax}	Ce^{ax}
$\sin(ax)$ or $\cos(ax)$	$C_1\sin(ax) + C_2\cos(ax)$
$x^n e^{ax}$	$(C_n x^n + C_{n-1}x^{n-1} + \cdots + C_1 x + C_0)e^{ax}$
$x^n\sin(ax)$ or $x^n\cos(ax)$	$(C_n x^n + C_{n-1}x^{n-1} + \cdots + C_1 x + C_0)\sin(ax)$ $+ (D_n x^n + D_{n-1}x^{n-1} + \cdots + D_1 x + D_0)\cos(ax)$
$e^{bx}\sin(ax)$ *or* $e^{bx}\cos(ax)$	$[C_1\sin(ax) + C_2\cos(ax)]e^{bx}$

Since $f(x)$ is a homogeneous solution, $\mathbf{L}f=0$, and Equation 3.25 reduces to

$$\mathbf{L}(xf(x)) = \mathbf{K}(f) \tag{3.26}$$

where $\mathbf{K}f$ is a linear combination of f and its derivatives. This suggests that if $f(x)$ is a homogeneous solution, then an appropriate trial solution is of the form

$$y_p(x) = x\left[A_0 f(x) + A_1 f'(x) + \ldots + A_k f^{(k)}(x)\right] \tag{3.27}$$

If $xf(x)$ is also a homogeneous solution, then the trial solution can be again multiplied by x to obtain the solution using the method of undetermined coefficients.

Example 3.3 An extended surface is subject to resistive heating which is modeled as an internal heat source within the surface. The differential equation governing the temperature distribution over the surface is

$$\frac{d^2\theta}{dx^2} - m^2\theta = -\Lambda u(x) \tag{a}$$

where $\Lambda = u_{\max} L^2/k$ and m is as defined in Example 3.2. The surface at $x=0$ is maintained at a constant temperature, while the surface at $x=\mathrm{L}$ is insulated. Determine the heat transfer across the surface of the fin if $u = x - x^2/L$.

Solution The general homogeneous solution of Equation a as determined in Example 3.2 is

$$\theta_h(x) = D_1 \cosh(mx) + D_2 \sinh(mx) \tag{b}$$

The assumed form of the particular solution is

$$\theta_p(x) = A + Bx + Cx^2 \tag{c}$$

Substitution of Equation c into Equation a leads to

$$2C - m^2(A + Bx + Cx^2) = \Lambda\left(x - \frac{x^2}{L}\right) \tag{d}$$

Collecting coefficients of like powers of x leads to

$$-m^2 C = -\frac{\Lambda}{L} \tag{e}$$

$$-m^2 B = \Lambda \tag{f}$$

$$2C - m^2 A = 0 \tag{g}$$

Solving Equation e, Equation f, and Equation g for A, B, and C and substituting into Equation c gives

$$\theta_p(x) = \frac{2\Lambda}{m^4 L} - \frac{\Lambda}{m^2}x + \frac{\Lambda}{m^2 L}x^2 \tag{h}$$

The general solution is

$$\begin{aligned}\theta &= \theta_h(x) + \theta_p(x) \\ &= D_1 \cosh(mx) + D_2 \sinh(mx) + \frac{2\Lambda}{m^4 L} - \frac{\Lambda}{m^2}x + \frac{\Lambda}{m^2 L}x^2\end{aligned} \tag{i}$$

Application of $\theta(0) = \theta_0$ to Equation i leads to

$$\theta_0 = D_1 + \frac{2\Lambda}{m^4 L} \tag{j}$$

Application of $d\theta/dx\ (L) = 0$ to Equation i, using Equation k, gives

$$\begin{aligned}0 &= m\left(\theta_0 - \frac{2\Lambda}{m^4 L}\right)\sinh(mL) + D_2 m \cosh(mL) - \frac{\Lambda}{m^2} + \frac{2\Lambda}{m^2} \\ D_2 &= -\left(\theta_0 - \frac{2\Lambda}{m^4 L}\right)\tanh(mL) - \frac{\Lambda}{m^3\cosh(mL)}\end{aligned} \tag{k}$$

The solution of Equation a subject to the boundary conditions of Equation b and Equation c is

$$\begin{aligned}\theta(x) &= \left(\theta_0 - \frac{2\Lambda}{m^4 L}\right)[\cosh(mx) - \tanh(mL)\sinh(mx)] \\ &\quad - \frac{\Lambda}{m^3\cosh(mL)}\sinh(mx) + \frac{\Lambda}{m^4 L}(2 - m^2 Lx + m^2 x^2)\end{aligned} \tag{l}$$

The total rate of heat transfer across the surface of the surface is

$$\begin{aligned}q &= \int_0^L h\theta(x)dx \\ &= \frac{h}{m}\left(\theta_0 - \frac{2\Lambda}{m^4 L}\right)[\sinh(mx) - \tanh(mL)\cosh(mL)]_{x=0}^{x=L} \\ &\quad - \frac{h\Lambda}{m^4\cosh(mL)}\cosh(mx)\Big|_{x=0}^{x=L} + \frac{h\Lambda}{m^4 L}\left(2x - \frac{m^2 L x^2}{2} + \frac{m^2 x^3}{3}\right)\Bigg|_{x=0}^{x=L} \\ &= \frac{h}{m}\left[\theta_0 + \frac{\Lambda}{m^3}\left(1 - \frac{2}{mL} + \mathrm{sec}h(mL) - \frac{L^2}{m^3}\right)\right]\end{aligned} \tag{m}$$

Example 3.4 The nondimensional differential equation for the transverse deflection of an axially loaded beam with a transverse load per unit length is

$$\frac{d^4w}{dx^4} - \varepsilon\frac{d^2w}{dx^2} = \Lambda f(x) \tag{a}$$

where $\varepsilon = PL^2/EI$ and $\Lambda = F_0L^3/EI$. The boundary conditions for a fixed-pinned beam are

$$w(0) = 0 \tag{b}$$

$$\frac{dw}{dx}(0) = 0 \tag{c}$$

$$w(1) = 0 \tag{d}$$

$$\frac{d^2w}{dx^2}(1) = 0 \tag{e}$$

Let $\Lambda = 1$ and

$$f(x) = \sin(\pi x) \tag{f}$$

(a) Determine and plot the transverse deflection for $\varepsilon = -4$,0 and 4; (b) Determine and plot the transverse deflection when $\varepsilon = -\pi^2$.

Solution The homogeneous solution is obtained by assuming $w(x) = e^{\alpha x}$, which when substituted into Equation a, leads to $\alpha^4 - \varepsilon\alpha^2$, whose solutions are $\alpha = 0, 0, \pm\sqrt{\varepsilon}$. If $\varepsilon < 0$, the load is compressive, and the general form of the homogeneous solution is

$$w_h(x) = C_1\cos\left(\sqrt{-\varepsilon}x\right) + C_2\sin\left(\sqrt{-\varepsilon}x\right) + C_3 + C_4x \tag{g}$$

If $\varepsilon = 0$, the beam is not subject to an axial load, and the general form of the homogeneous solution is

$$w_h(x) = C_1 + C_2x + C_3x^2 + C_4x^3 \tag{h}$$

If $\varepsilon > 0$, the load is tensile, and the general form of the homogeneous solution is

$$w_h(x) = C_1\cosh\left(\sqrt{\varepsilon}x\right) + C_2\sinh\left(\sqrt{\varepsilon}x\right) + C_3 + C_4x \tag{i}$$

Using the method of undetermined coefficients, the particular solution of Equation a when the beam is subject to the load per unit length given by Equation f can be assumed to be

$$w_p(x) = A\sin(\pi x) + B\cos(\pi x) \tag{j}$$

Substitution of Equation l into Equation a leads to

$$\pi^4 A\sin(\pi x) + \pi^4 B\cos(\pi x) + \varepsilon\pi^2 A\sin(\pi x) + \varepsilon\pi^2 B\cos(\pi x) = \Lambda\sin(\pi x) \tag{k}$$

Equating coefficients of sin (πx) on both sides of Equation k gives $A = \Lambda/\pi^2(\pi^2 + \varepsilon)$, while equating coefficients of $\cos(\pi x)$ on both sides of Equation k leads to $B = 0$. It should be noted that $\varepsilon = -\pi^2$ is a special case. The particular solution of Equation a for $\varepsilon < 0$, $\varepsilon \neq -\pi^2$ is

$$w_p(x) = \frac{\Lambda}{\pi^2\left(\pi^2 + \varepsilon\right)}\sin(\pi x) \tag{l}$$

The general solution for $\varepsilon < 0$ is

$$w(x) = C_1\cos\left(\sqrt{-\varepsilon}x\right) + C_2\sin\left(\sqrt{-\varepsilon}x\right) + C_3 + C_4 x + \frac{\Lambda}{\pi^2\left(\pi^2 + \varepsilon\right)}\sin(\pi x) \tag{m}$$

Application of Equation b, Equation c, Equation d, and Equation e to Equation m leads to

$$w(0) = 0 \Rightarrow C_1 + C_3 = 0 \tag{n}$$

$$\frac{dw}{dx}(0) = 0 \Rightarrow \sqrt{-\varepsilon}C_2 + C_4 = -\frac{\Lambda}{\pi\left(\pi^2 - \varepsilon\right)} \tag{o}$$

$$w(1) = 0 \Rightarrow \cos\left(\sqrt{-\varepsilon}\right)C_1 + \sin\left(\sqrt{-\varepsilon}\right)C_2 + C_3 + C_4 = 0 \tag{p}$$

$$\frac{d^2w}{dx^2}(1) = 0 \Rightarrow \varepsilon\cos\left(\sqrt{-\varepsilon}\right)C_1 + \varepsilon\sin\left(\sqrt{-\varepsilon}\right)C_2 = 0 \tag{q}$$

Equation n, Equation o, Equation p, and Equation q can be solved simultaneously, leading to

$$C_1 = -\frac{\Lambda\sin\left(\sqrt{-\varepsilon}\right)}{\pi\left(\pi^2 + \varepsilon\right)\left[\sqrt{-\varepsilon}\cos\left(\sqrt{-\varepsilon}\right) - \sin\left(\sqrt{-\varepsilon}\right)\right]} \tag{r}$$

$$C_2 = \frac{\Lambda\cos\left(\sqrt{-\varepsilon}\right)}{\pi\left(\pi^2 + \varepsilon\right)\left[\sqrt{-\varepsilon}\cos\left(\sqrt{-\varepsilon}\right) - \sin\left(\sqrt{-\varepsilon}\right)\right]} \tag{s}$$

$$C_3 = \frac{\Lambda \sin\left(\sqrt{-\varepsilon}\right)}{\pi(\pi^2+\varepsilon)\left[\sqrt{-\varepsilon}\cos\left(\sqrt{-\varepsilon}\right)-\sin\left(\sqrt{-\varepsilon}\right)\right]} \tag{t}$$

$$C_4 = -\frac{\Lambda \sin\left(\sqrt{-\varepsilon}\right)}{\pi(\pi^2+\varepsilon)\left[\sqrt{-\varepsilon}\cos\left(\sqrt{-\varepsilon}\right)-\sin\left(\sqrt{-\varepsilon}\right)\right]} \tag{u}$$

The resulting general solution for $\varepsilon < 0$ is

$$w(x) = \frac{\Lambda \sin\left(\sqrt{-\varepsilon}\right)}{\pi(\pi^2+\varepsilon)\left[\sqrt{-\varepsilon}\cos\left(\sqrt{-\varepsilon}\right)-\sin\left(\sqrt{-\varepsilon}\right)\right]} \left[\cos\left(\sqrt{-\varepsilon}x\right) - \frac{\cos\left(\sqrt{-\varepsilon}\right)}{\sin\left(\sqrt{-\varepsilon}\right)}\sin\left(\sqrt{-\varepsilon}x\right) - 1 + x\right] + \frac{\Lambda}{\pi^2(\pi^2+\varepsilon)}\sin(\pi x) \tag{v}$$

The general solution for $\varepsilon = 0$ is

$$w(x) = C_1 + C_2 x + C_3 x^2 + C_4 x^3 + \frac{\Lambda}{\pi^4}\sin(\pi x) \tag{w}$$

Application of the boundary conditions, Equation b, Equation c, Equation d, and Equation e, to Equation w results in

$$w(x) = \frac{\Lambda}{\pi^4}\left[\sin(\pi x) - \pi x + \frac{3\pi}{2}x^2 - \frac{\pi}{2}x^3\right] \tag{x}$$

The general solution for $\varepsilon > 0$ is

$$w_h(x) = C_1 \cosh\left(\sqrt{\varepsilon}x\right) + C_2 \sinh\left(\sqrt{\varepsilon}x\right) + C_3 + C_4 x + \frac{\Lambda}{\pi^2\left(\pi^2+\varepsilon\right)}\sin(\pi x) \tag{y}$$

Application of Equation b, Equation c, Equation d, and Equation e to Equation y leads to

$$w(x) = \frac{\Lambda \sinh\left(\sqrt{\varepsilon}\right)}{\pi(\pi^2+\varepsilon)\left[\sqrt{\varepsilon}\cosh\left(\sqrt{\varepsilon}\right)-\sinh\left(\sqrt{\varepsilon}\right)\right]} \left[\cosh\left(\sqrt{\varepsilon}x\right) - \frac{\cosh\left(\sqrt{\varepsilon}\right)}{\sinh\left(\sqrt{\varepsilon}\right)}\sinh\left(\sqrt{\varepsilon}x\right) - x\right] + \frac{\Lambda}{\pi^2\left(\pi^2+\varepsilon\right)}\sin(\pi x) \tag{z}$$

Equation u, Equation x, and Equation z are plotted as functions of x for $\varepsilon = -4$, $\varepsilon = 0$, and $\varepsilon = 4$, respectively in Figure 3.4.

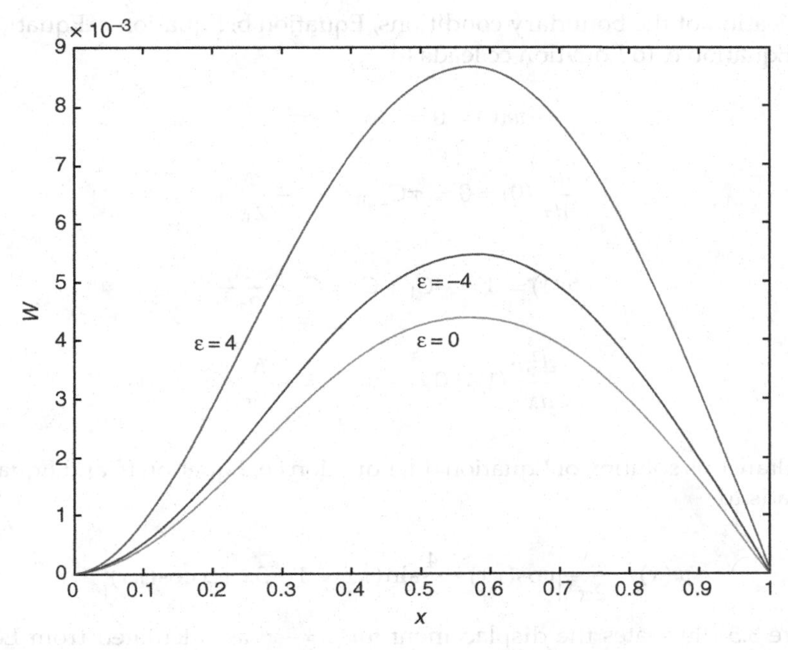

Figure 3.4 The form of the homogeneous solution for the deflection of an axially loaded beam of Example 3.4 is dependent on the value of ε. The load is compressive if $\varepsilon < 1$. tensile for $\varepsilon > 1$, and does not exist for $\varepsilon = 0$.

If $\varepsilon = -\pi^2$, the nonhomogeneous term is also a homogeneous solution. The appropriate form of the particular solution for this special case is

$$w_p(x) = Ax\sin(\pi x) + Bx\cos(\pi x) \tag{aa}$$

Substitution of Equation aa into Equation a for $\varepsilon = -\pi^2$ leads to

$$\begin{aligned}
&-4A\pi^3\cos(\pi x) + Ax\pi^4\sin(\pi x) + 4B\pi^3\sin(\pi x) + Bx\pi^4\cos(\pi x) \\
&+\pi^2\left[2A\pi\cos(\pi x) - Ax\pi^2\sin(\pi x) - 2B\pi\sin(\pi x) - Bx\pi^2\cos(\pi x)\right] \\
&= \Lambda\sin(\pi x)
\end{aligned}$$

$$-2\pi^3 A\cos(\pi x) + 2B\pi^3\sin(\pi x) = \Lambda\sin(\pi x) \tag{bb}$$

Equating coefficients of $\sin(\pi x)$ on both sides of Equation bb leads to $B = \Lambda/2\pi^3$, while equating coefficients of $\cos(\pi x)$ on both sides of Equation bb leads to $A = 0$, leading to the general solution,

$$w(x) = C_1\cos(\pi x) + C_2\sin(\pi x) + C_3 + C_4 x + \frac{\Lambda}{2\pi^3}x\cos(\pi x) \tag{cc}$$

Application of the boundary conditions, Equation b, Equation c, Equation d, and Equation e, to Equation cc leads to

$$w(0)=0 \Rightarrow C_1+C_3=0 \tag{dd}$$

$$\frac{dw}{dx}(0)=0 \Rightarrow \pi C_2+C_4=-\frac{\Lambda}{2\pi^3} \tag{ee}$$

$$w(1)=0 \Rightarrow -C_1+C_3+C_4=\frac{\Lambda}{2\pi^3} \tag{ff}$$

$$\frac{d^2w}{dx^2}(1)=0 \Rightarrow -\pi^2 C_1=-\frac{\Lambda}{2\pi} \tag{gg}$$

Simultaneous solution of Equation dd, Equation ee, Equation ff, and Equation gg leads to

$$w(x)=\frac{\Lambda}{2\pi^3}\left[\cos(\pi x)-\frac{4}{\pi}\sin(\pi x)-1+3x+x\cos(\pi x)\right] \tag{hh}$$

Figure 3.5 illustrates the displacement for $\varepsilon=-\pi^2$ as calculated from Equation hh.

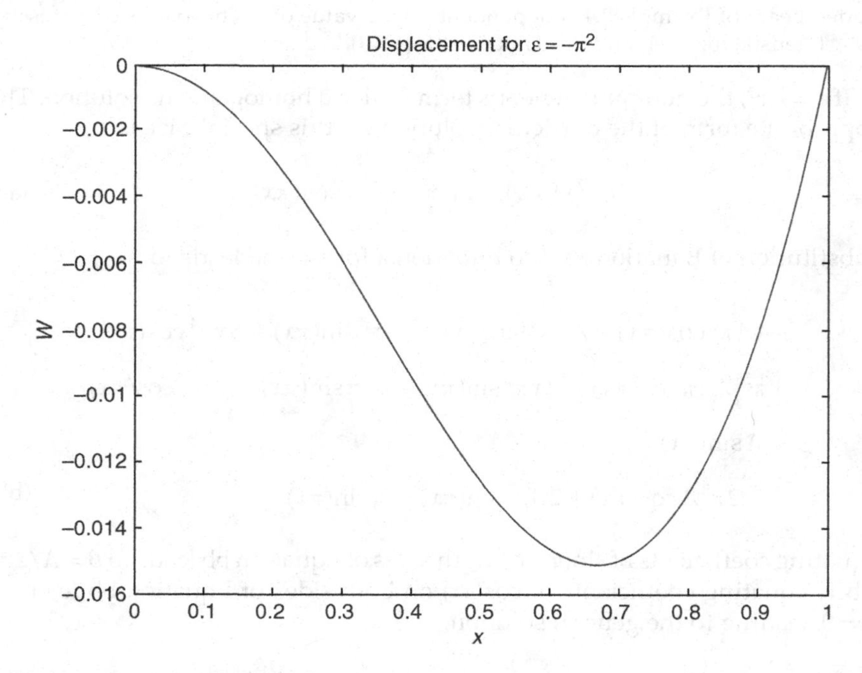

Figure 3.5 When $\varepsilon=-\pi^2$, a special case occurs for the beam of Example 3.4; the nonhomogeneous term is a homogeneous solution.

The method of variation of parameters is another technique used to derive a particular solution of a nonhomogeneous differential equation. The method can be applied to linear ordinary differential equations with constant or variable coefficients. As an illustration, consider a linear second-order differential equation of the form

$$a_2(x)\frac{d^2y}{dx^2}+a_1(x)\frac{dy}{dx}+a_0(x)y=f(x) \tag{3.28}$$

Let $y_1(x)$ and $y_2(x)$ be linearly independent solutions of the corresponding homogeneous equation. A solution of Equation 3.28 is assumed in the form of

$$y(x)=C_1(x)y_1(x)+C_2(x)y_2(x) \tag{3.29}$$

where $C_1(x)$ and $C_2(x)$ are functions to be determined by substitution into Equation 3.28. Differentiation of Equation 3.29 with respect to x leads to

$$\frac{dy}{dx}=C_1'(x)y_1(x)+C_1(x)y_1'(x)+C_2'(x)y_2(x)+C_2(x)y_2'(x) \tag{3.30}$$

If the coefficients are constants, then $dy/dx=C_1y_1'(x)+C_2y_2'(x)$. The same form of dy/dx is achieved if

$$C_1'(x)y_1(x)+C_2'(x)y_2(x)=0 \tag{3.31}$$

Differentiation of Equation 3.30 when Equation 3.31 is enforced leads to

$$\frac{d^2y}{dx^2}=C_1'(x)y_1'(x)+C_1(x)y_1''(x)+C_2'(x)y_2'(x)+C_2(x)y_2''(x) \tag{3.32}$$

Substitution of Equation 3.29, Equation 3.30, Equation 3.31, and Equation 3.32 into Equation 3.28 leads to

$$\begin{aligned}a_2(x)[C_1'(x)y_1'(x)+C_1(x)y_1''(x)+C_2'(x)y_2'(x)+C_2(x)y_2''(x)]\\+a_1(x)[C_1(x)y_1'(x)+C_2(x)y_2'(x)]\\+a_0(x)[C_1(x)y_1(x)+C_2(x)y_2(x)]=f(x)\end{aligned} \tag{3.33}$$

Equation 3.33 can be rearranged to yield

$$\begin{aligned}C_1(x)\left[a_2(x)y_1''(x)+a_1(x)y_1'(x)+a_0(X)y_1(x)\right]\\+C_2(x)\left[a_2(x)y_2''(x)+a_1(x)y_2'(x)+a_0(x)y_{21}(x)\right]\\+a_2(x)\left[C_1'(x)y_1'(x)+C_2'(x)y_2'(x)\right]=f(x)\end{aligned} \tag{3.34}$$

Recalling that $y_1(x)$ and $y_2(x)$ are homogeneous solutions, Equation 3.34 simplifies to

$$y_1'C_1' + y_2'C_2' = \frac{f(x)}{a_2(x)} \tag{3.35}$$

Equation 3.35 can be summarized in matrix form as

$$\begin{bmatrix} y_1(x) & y_2(x) \\ y_1'(x) & y_2'(x) \end{bmatrix} \begin{bmatrix} C_1'(x) \\ C_2'(x) \end{bmatrix} = \begin{bmatrix} 0 \\ \dfrac{f(x)}{a_2(x)} \end{bmatrix} \tag{3.36}$$

Simultaneous solution of Equation 3.36 leads to

$$C_1'(x) = -\frac{1}{W(x)} \frac{f(x)y_2(x)}{a_2(x)} \tag{3.37}$$

$$C_2'(x) = \frac{1}{W(x)} \frac{f(x)y_1(x)}{a_2(x)} \tag{3.38}$$

where $W(x)$ is the Wronskian of the homogeneous solutions,

$$W(x) = y_1(x)y_2'(x) - y_2(x)y_1'(x) \tag{3.39}$$

Equation 3.37 and Equation 3.38 are then integrated, leading to $C_1(x)$ and $C_2(x)$, which are used in Equation 3.29.

The method of variation of parameters is easily generalized to determine the particular solution of higher-order differential equations. Let $y_1, y_2, \ldots, y_n$ be n linearly independent homogeneous solutions for an nth-order linear differential equation. A particular solution can be formed as

$$y(x) = C_1(x)y_1(x) + C_2(x)y_2(x) + \cdots + C_n(x)y_n(x) \tag{3.40}$$

A system of n simultaneous equations is then developed to solve for $C_k'(x)$ $k = 1, 2, .., n$ by requiring that $d^jy/dx^j = C_1y_1^{(j)}(x) + C_2y_2^{(j)}(x) + \cdots + C_ny_n^{(j)}(x)$ for $j = 1, 2, \ldots, n-1$. Successive enforcement of these conditions leads to $C_1'(x)y_1^{(j-1)}(x) + C_2'(x)y_2^{(j-1)}(x) + \cdots + C_n'(x)y_n^{(j-1)}(x)$ for $j = 1, 2, \ldots, n-1$. The nth equation results from eventual substitution into the differential equation which gives $C_1'y_1^{(n-1)}(x) + C_2'y_2^{(n-1)}(x) + \cdots + C_n'y_n^{(n-1)}(x) = f(x)/a_n(x)$. The resulting simultaneous equations can be written in matrix form as

$$\begin{bmatrix} y_1 & y_2 & \cdots & y_n \\ y_1' & y_2' & \cdots & y_n' \\ \vdots & \vdots & \ddots & \vdots \\ y_1^{(n-1)} & y_2^{(n-1)} & \cdots & y_n^{(n-1)} \end{bmatrix} \begin{bmatrix} C_1' \\ C_2' \\ C_3' \\ C_4' \end{bmatrix} = \begin{bmatrix} 0 \\ 0 \\ \vdots \\ f(x)/a_n(x) \end{bmatrix} \tag{3.41}$$

Equation 3.41 can be solved simultaneously, perhaps using Cramer's rule. The determinant of the $n \times n$ matrix on the right-hand side of Equation 3.41 is the Wronskian of the homogeneous solutions.

Example 3.5 Use the method of variation of parameters to derive a solution for the transverse deflection of the beam of Example 3.4 when $\varepsilon = -1$ for an arbitrary distributed load per length $f(x)$.

Solution When $\varepsilon = -1$, the homogeneous solutions of Equation a of Example 3.4 are $w_1(x) = \cos(x)$, $w_2(x) = \sin(x)$, $w_3(x) = 1$, and $w_4(x) = x$. The general solution is assumed to be of the form

$$w(x) = C_1(x)\cos(x) + C_2(x)\sin(x) + C_3(x) + C_4(x)x \tag{a}$$

The derivatives of the coefficients are obtained by solving a system of the form $\mathbf{WC}' = \mathbf{f}$, where $\mathbf{W}$ is the matrix from which the Wronskian of the homogeneous solutions is formed, $\mathbf{C}'$ and the vector of derivatives of the coefficients, is $\mathbf{f} = [0 \;\; 0 \;\; 0 \;\; f(x)]^T$. This system can be expressed as

$$\begin{bmatrix} \cos(x) & \sin(x) & 1 & x \\ -\sin(x) & \cos(x) & 0 & 1 \\ -\cos(x) & -\sin(x) & 0 & 0 \\ \sin(x) & -\cos(x) & 0 & 0 \end{bmatrix} \begin{bmatrix} C_1' \\ C_2' \\ C_3' \\ C_4' \end{bmatrix} = \begin{bmatrix} 0 \\ 0 \\ 0 \\ f(x) \end{bmatrix} \tag{b}$$

The Wronskian is computed as

$$W(x) = \begin{vmatrix} \cos(x) & \sin(x) & 1 & x \\ -\sin(x) & \cos(x) & 0 & 1 \\ -\cos(x) & -\sin(x) & 0 & 0 \\ \sin(x) & -\cos(x) & 0 & 0 \end{vmatrix} = 1 \tag{c}$$

Cramer's rule is used to solve Equation (b) as

$$C_1' = \begin{vmatrix} 0 & \sin(x) & 1 & x \\ 0 & \cos(x) & 0 & 1 \\ 0 & -\sin(x) & 0 & 0 \\ f(x) & -\cos(x) & 0 & 0 \end{vmatrix} = f(x)\sin(x) \tag{d}$$

$$C_2' = \begin{vmatrix} \cos(x) & 0 & 1 & x \\ -\sin(x) & 0 & 0 & 1 \\ -\cos(x) & 0 & 0 & 0 \\ \sin(x) & f(x) & 0 & 0 \end{vmatrix} = -f(x)\cos(x) \tag{e}$$

$$C_3' = \begin{vmatrix} \cos(x) & \sin(x) & 0 & x \\ -\sin(x) & \cos(x) & 0 & 1 \\ -\cos(x) & -\sin(x) & 0 & 0 \\ \sin(x) & -\cos(x) & f(x) & 0 \end{vmatrix} = -f(x)x \tag{f}$$

$$C_4' = \begin{vmatrix} \cos(x) & \sin(x) & 1 & 0 \\ -\sin(x) & \cos(x) & 0 & 0 \\ -\cos(x) & -\sin(x) & 0 & 0 \\ \sin(x) & -\cos(x) & 0 & f(x) \end{vmatrix} = f(x) \tag{g}$$

The variation-of-parameters solution is

$$\begin{aligned} w(x) = {} & \cos(x)\int_0^x f(\xi)\sin(\xi)d\xi - \sin(x)\int_0^x f(\xi)\cos(\xi)d\xi \\ & -\int_0^x f(\xi)\xi d\xi + x\int_0^x f(\xi)d\xi + c_1\cos(x) + c_2\sin(x) + c_3 + c_4 x \end{aligned} \tag{h}$$

Equation h can be rearranged as

$$w(x) = \int_0^x [\sin(\xi - x) + x - \xi] f(\xi)d\xi + c_1\cos(x) + c_2\sin(x) + c_3 + c_4 x \tag{i}$$

Application of the boundary conditions requires calculations of the first and second derivatives of $w(x)$. Leibnitz's rule for differentiation of a definite integral whose limits depend on the variable of differentiation is

$$\frac{d}{dx}\left[\int_{a(x)}^{b(x)} g(x,\xi)d\xi\right] = \int_{a(x)}^{b(x)} \frac{\partial g}{\partial x}d\xi + g(x,a(x))\frac{da}{dx} - g(x,b(x))\frac{db}{dx} \tag{j}$$

Application of Leibnitz's rule to Equation i leads to

$$\frac{dw}{dx} = \int_0^x [1 - \cos(\xi - x)] f(\xi)d\xi \tag{k}$$

$$\frac{d^2w}{dx^2} = -\int_0^x \sin(x - \xi) f(\xi)d\xi \tag{l}$$

Application of the boundary conditions to Equation j, using Equation k and Equation l, leads to

$$w(0)=0 \Rightarrow c_1+c_3=0 \tag{m}$$

$$\frac{dw}{dx}(0)=0 \Rightarrow c_2+c_4=0 \tag{n}$$

$$w(1)=0 \Rightarrow \int_0^1 [\sin(\xi-1)+1-\xi] f(\xi) d\xi + c_1 \cos(1)$$

$$+c_2 \sin(1)+c_3+c_4=0 \tag{o}$$

$$\frac{d^2w}{dx^2}=0 \Rightarrow -\int_0^1 \sin(\xi-1) f(\xi) d\xi - c_1 \cos(1) - c_2 \sin(1) = 0 \tag{p}$$

Equation m, Equation n, Equation o, and Equation p are solved simultaneously for the constants of integration. Subsequent substitution into Equation i leads to the solution of the system for any $f(x)$. The integrals can be evaluated for any $f(x)$.

If a solution is desired for only one form of $f(x)$, say $f(x)=\sin(\pi x)$ as in Example 3.5, then substitution into Equation i leads to

$$w(x)=\cos(x)\int_0^x \sin(\pi x)\sin(x)dx - \sin(x)\int_0^x \sin(\pi x)\cos(x)dx - \int_0^x x\sin(\pi x)dx$$

$$+x\int_0^x \sin(\pi x)dx + C_1\cos(x)+C_2\sin(x)+C_3+C_4x \tag{q}$$

Evaluation of the integrals in Equation q results in

$$w(x)=\frac{1}{\pi^2(\pi^2+1)}\sin(\pi x)x\int_0^x \sin(\pi x)dx + C_1\cos(x)+C_2\sin(x)+C_3+C_4x \tag{r}$$

Equation r is identical to Equation m of Example 3.5 with $\varepsilon=-1$. Boundary conditions are applied to Equation r to obtain Equation v of Example 3.5.

3.4 Differential equations with variable coefficients

Let $b(x)$ be a function of an independent variable x Let x_0 be a specific value of x. The function $b(x)$ is said to be analytic at x_0 if there exists a power series expansion of $b(x)$, written as

$$b(x)=\sum_{n=0}^{\infty} c_n(x-x_0)^n \tag{3.42}$$

such that the expansion converges pointwise to $b(x)$ within an interval $x_0 - R < x < x_0 + R$. The value R is called the radius of convergence. The constants c_n $n = 0, 1, 2, \ldots$ are called the coefficients of the expansion.

The function $(1-x)^{-1}$ is analytic at $x=0$ with a radius of convergence of 1 because it can be expressed using a binomial expansion as

$$(1-x)^{-1} = 1 + x + x^2 + \cdots + x^{k-1} + x^k + x^{k+1} + \cdots \tag{3.43}$$

The series expressed by the right-hand side of Equation 3.43 is a geometric series of ratio x. The series converges if $|x| < 1$ and diverges if $|x| \geq 1$.

Taylor's theorem provides conditions under which a function has a convergent expansion about point x_0, a method for calculating the coefficients, and an estimate of the remainder after n terms. A form of the theorem presented below is found in most elementary calculus books.

Theorem 3.5 Taylor's Theorem. Let $f(x)$ be a function defined such that $f(x)$ and its first $n+1$ derivatives are continuous within a finite interval of a real value x_0. For an x in this interval,

$$f(x) = f(x_0) + (x - x_0)\frac{df}{dx}(x_0) + \frac{(x-x_0)^2}{2!}\frac{d^2 f}{dx^2}(x_0) + \frac{(x-x_0)^3}{3!}\frac{d^3 f}{dx^3}(x_0) +$$

$$+ \ldots + \frac{(x-x_0)^n}{n!}\frac{d^n f}{dx^n}(x_0) + R_n(x_0, x) \tag{3.44}$$

where the remainder is

$$R_n(x_0, x) = \frac{(x-x_0)^{n+1}}{(n+1)!}\frac{d^{n+1} f}{dx^{n+1}}(\xi) \tag{3.45}$$

for some ξ between x and x_0.

Taylor's theorem implies that a function is analytic at x_0 if the function and all its higher-order derivatives exist at x_0.

The general form of a linear nth-order differential equation with variable coefficients can be written as

$$\frac{d^n y}{dx^n} + a_{n-1}(x)\frac{d^{n-1} y}{dx^{n-1}} + \cdots + a_1(x)\frac{dy}{dx} + a_0(x)y = f(x) \tag{3.46}$$

The point $x = x_0$ is said to be an ordinary point of the differential equation if $a_{n-1}(x), a_{n-2}(x), \ldots, a_1(x), a_0(x)$ and $f(x)$ are all analytic at x_0. The following theorem, presented without proof, shows that if x_0 is an ordinary point of the differential equation, then $y(x)$ is analytic at x_0.

Theorem 3.6 Let $a_{n-1}(x), a_{n-2}(x), \ldots, a_1(x), a_0(x)$ and $f(x)$ be coefficients in a differential equation of the form of Equation 3.46. If all coefficients are analytic at x_0, then $x=x_0$ is an ordinary point of the differential equation and all solutions are analytic at x_0. Furthermore, if R_m is the minimum of the radii of convergence of the coefficients of Equation 3.46, then the radius of convergence of $y(x)$ is at least R_m.

Theorem 3.6 implies that if x_0 is an ordinary point of the differential equation, then $y(x)$ has a Taylor series expansion about x_0. This suggests that a Taylor series expansion may be assumed for a solution and the method of undetermined coefficients applied to determine the coefficients in the expansion. Such a method is called a power series method.

The power series method is usually applied to determine homogeneous solutions. If the differential equation is nonhomogenous, once the homogeneous solutions are determined, the method of variation of parameters may be used to determine the particular solution.

Example 3.6 The differential equation governing the temperature distribution over a nonuniform extended surface is

$$\frac{d}{dx}\left[\alpha(x)\frac{d\theta}{dx}\right]-m^2\theta=0 \tag{a}$$

Use a series solution to determine the homogeneous solutions to Equation a if $\alpha(x)=1$ where μ is a constant.

Solution It is noted that $x_0=0$ is an ordinary point of Equation a as long as $\alpha(0)\neq 0$. (a) A series solution is assumed to be of the form

$$\theta(x)=\sum_{n=0}^{\infty}c_n x^n \tag{b}$$

Theorem 3.6 implies that Equation b converges pointwise over a finite interval about $x=0$. Thus it can be differentiated term by term, leading to

$$\frac{d\theta}{dx}=\sum_{n=1}^{\infty}nc_n x^{n-1} \tag{c}$$

$$\frac{d^2\theta}{dx^2}=\sum_{n=2}^{\infty}n(n-1)c_n x^{n-2} \tag{d}$$

Substitution of Equation b, Equation c, and Equation d into Equation a with $\alpha(x)=1$ leads to

$$\sum_{n=2}^{\infty}n(n-1)c_n x^{n-2}-m^2\sum_{n=0}^{\infty}c_n x^n=0 \tag{e}$$

The name of an index of summation is arbitrary and does not affect the value of the sum. Thus

$$\begin{aligned}\sum_{n=2}^{\infty} n(n-1)c_n x^{n-2} &= \sum_{k=0}^{\infty}(k+2)(k+1)c_{k+2}x^k \\ &= \sum_{n=0}^{\infty}(n+2)(n+1)c_{n+2}x^n\end{aligned} \tag{f}$$

Substituting Equation f into Equation e leads to

$$\begin{aligned}&\sum_{n=0}^{\infty}(n+2)(n+1)c_{n+2}x^n - m^2\sum_{n=0}^{\infty}c_n x^n = 0 \\ &\sum_{n=0}^{\infty}\left[(n+2)(n+1)c_{n+2} - m^2 c_n\right]x^n = 0\end{aligned} \tag{g}$$

Powers of x are linearly independent, and therefore for Equation g to hold for all x, coefficients of all powers of x in the summation in Equation g must be zero:

$$(n+2)(n+1)c_{n+2} - m^2 c_n = 0 \quad n = 0, 1, 2, \ldots \tag{h}$$

Equation h is a recurrence relation which is satisfied by the coefficients. It is convenient to consider even and odd values of n separately. Writing Equation h for the first four even values of n gives

$$2c_2 - m^2 c_0 = 0 \Rightarrow c_2 = \frac{m^2}{2}c_0 \tag{i}$$

$$(3)(4)c_4 - m^2 c_2 = 0 \Rightarrow c_4 = \frac{m^2}{(3)(4)}c_2 = \frac{m^4}{(2)(3)(4)}c_0 = \frac{m^4}{4!}c_0 \tag{j}$$

$$(5)(6)c_6 - m^2 c_4 = 0 \Rightarrow c_6 = \frac{m^2}{(5)(6)}c_4 = \frac{m^4}{(2)(3)(4)(5)(6)} = \frac{m^6}{6!}c_0 \tag{k}$$

Mathematical induction with Equation i, Equation j, and Equation k can be used to prove that

$$c_n = \frac{m^n}{n!}c_0 \quad n = 2, 4, 6, \ldots \tag{l}$$

In a similar fashion, Equation h can be used to show that

$$c_n = \frac{m^{n-1}}{n!} c_1 \quad n = 3, 5, 7, \ldots \tag{m}$$

The coefficients c_0 and c_1 are arbitrary. They cannot be specified from the differential equation. The solution can be written in terms of these two arbitrary constants as

$$y(x) = c_0 \sum_{n=0,2,4,}^{\infty} \frac{(mx)^n}{n!} + \frac{c_1}{m} \sum_{n=1,3,5,}^{\infty} \frac{(mx)^n}{n!} \tag{n}$$

Equation n can then be rewritten as

$$y(x) = c_0 \sum_{n=0}^{\infty} \frac{(mx)^{2n}}{(2n)!} + \frac{c_1}{m} \sum_{n=1}^{\infty} \frac{(mx)^{2n+1}}{(2n+1)!} \tag{o}$$

The functions $y_1(x) = \Sigma_{n=0}^{\infty} (mx)^{2n}/(2n)!$ and $y_2(x) = \Sigma_{n=1}^{\infty} (mx)^{2n+1}/(2n+1)!$ are linearly independent and form a basis for the homogeneous solution space for the differential equation, Equation a.

It should be noted that $y_1(x)$ is the Taylor series expansion for $\cosh(mx)$ about $x = 0$ and that $y_2(x)$ is the Taylor series expansion for $\sinh(mx)$ about $x = 0$.

Example 3.7 The natural frequencies and mode shapes for the longitudinal motion of a bar of constant area, but of varying density, are obtained from the solution of the non-dimensional differential equation

$$\frac{d^2u}{dx^2} + \beta(x)\omega^2 u = 0 \tag{a}$$

Determine power series representations of the basis functions for the homogeneous solution space of Equation a if $\beta(x) = 1 + \mu x$.

Solution: A power series solution of Equation a is assumed as

$$u(x) = \sum_{k=0}^{\infty} c_k x^k \tag{b}$$

Substitution of Equation b into Equation a leads to

$$\sum_{k=2}^{\infty} c_k(k)(k-1)x^{k-2} + \omega^2 \sum_{k=0}^{\infty} c_k\left(x^k + \mu x^{k+1}\right) = 0 \tag{c}$$

Redefinition of indices in the summations of Equation c leads to

$$2c_2 + \omega^2 c_0 + \sum_{k=1}^{\infty} \left[(k+1)(k+2)c_{k+2} + \omega^2 c_k + \mu\omega^2 c_{k-1} \right] = 0 \tag{d}$$

Setting coefficients of powers of x to zero independently leads to

$$2c_2 + \omega^2 c_0 = 0 \tag{e}$$

$$(k+1)(k+2)c_{k+2} + \omega^2 c_k + \mu\omega^2 c_{k-1} \qquad \mathrm{k} = 1,2,... \tag{f}$$

Using Equation e and Equation f to solve for c_2, c_3, c_4, c_5 and c_6 in terms of c_0 and c_1 leads to

$$c_2 = -\frac{\omega^2}{2} c_0 \tag{g}$$

$$c_3 = -\frac{\omega^2}{(2)(3)} c_1 - \frac{\mu\omega^2}{(2)(3)} c_0 \tag{h}$$

$$\begin{aligned} c_4 &= -\frac{\omega^2}{(3)(4)} c_2 - \frac{\mu\omega^2}{(3)(4)} c_1 \\ &= \frac{\omega^4}{(2)(3)(4)} c_0 - \frac{\mu\omega^2}{(3)(4)} c_1 \end{aligned} \tag{i}$$

$$\begin{aligned} c_5 &= -\frac{\omega^2}{(4)(5)} c_3 - \frac{\mu\omega^2}{(4)(5)} c_2 \\ &= \frac{\omega^4}{(2)(3)(4)(5)} c_1 + \frac{\mu\omega^4}{(2)(3)(3)(4)} c_0 + \frac{\mu\omega^4}{(2)(4)(5)} c_0 \\ &= \frac{\omega^4}{(2)(3)(4)(5)} c_1 + \frac{\mu\omega^4 [5 + (3)(3)]}{(2)(3)(3)(4)(5)} c_0 \end{aligned} \tag{j}$$

$$\begin{aligned} c_6 &= -\frac{\omega^2}{(5)(6)} c_4 - \frac{\mu\omega^2}{(5)(6)} c_3 \\ &= -\frac{\omega^6}{(2)(3)(4)(5)(6)} c_0 + \frac{\mu\omega^4}{(3)(4)(5)(6)} c_1 + \frac{\mu\omega^4}{(2)(3)(5)(6)} c_1 + \frac{\mu^2\omega^4}{(2)(3)(5)(6)} c_0 \end{aligned} \tag{k}$$

It is clear from Equation g, Equation h, Equation i, Equation j and Equation k that

$$c_k = \alpha_k c_0 + \beta_k c_1 \tag{l}$$

where the values of α_k and β_k are obtained recursively from Equation f. The general solution of Equation a is

$$u(x) = c_0 \sum_{k=0}^{\infty} \alpha_k x^k + c_1 \sum_{k=1}^{\infty} \beta_k x^k \tag{m}$$

3.5 Singular points of second-order equations

A singular point x_0 of a second-order differential equation is called a regular singular point if the differential equation can be written as

$$(x - x_0)^2 \frac{d^2y}{dx^2} + (x - x_0)q(x)\frac{dy}{dx} + r(x)y = 0 \tag{3.47}$$

where $q(x)$ and $r(x)$ are analytic at x_0.

An example of a second-order differential equation with a regular singular point is the Cauchy-Euler differential equation whose general form is

$$a_n x^n \frac{d^n y}{dx^n} + a_{n-1}x^{n-1}\frac{d^{n-1}y}{dx^{n-1}} + \cdots + a_1 x \frac{dy}{dx} + a_0 y = 0 \tag{3.48}$$

Consider a change of independent variables specified by

$$u = \ln(x) \tag{3.49}$$

The chain rule is applied to change derivatives with respect to x into derivatives with respect to u. The results are

$$\begin{aligned}
\frac{dy}{dx} &= \frac{dy}{du}\frac{du}{dx} = e^{-u}\frac{dy}{du} \\
\frac{d^2y}{dx^2} &= \frac{d}{du}\left(e^{-u}\frac{dy}{du}\right)e^{-u} = e^{-2u}\left(\frac{d^2y}{du^2} - \frac{dy}{du}\right) \\
&\vdots \\
\frac{d^n y}{dx^n} &= e^{-nu}L_n y
\end{aligned} \tag{3.50}$$

where L_n is a linear differential operator of derivatives with respect to u with constant coefficients. Noting that $x^n = e^{nu}$, substitution of Equation 3.49 into Equation 3.48 leads to

$$a_n L_n y + a_{n-1}L_{n-1}y + \cdots + a_1 L_1 y + a_0 y = 0 \tag{3.51}$$

Equation 3.51 is a linear differential equation with constant coefficients, and therefore its homogeneous solutions are of the form $y_h(u) = e^{\alpha u}$. Using Equation 3.50, the homogeneous solutions written in terms of x are of the form

$$y_h(x) = x^{\alpha} \tag{3.52}$$

If α is a root of multiplicity m of the resulting polynomial, the resulting linearly independent solutions are

$$x^{\alpha}, x^{\alpha}\ln(x), x^{\alpha}[\ln(x)]^2, \ldots, x^{\alpha}[\ln(x)]^{k-1} \tag{3.53}$$

Example 3.8 Solve Example 3.6 if $\alpha(x) = (1 + \mu x)^2$ and the boundary conditions are $\theta(0) = 1$ and $d\theta/dx\ (1) = 0$.

Solution Equation a of Example 3.6 becomes

$$\frac{d}{dx}\left[(1+\mu x)^2 \frac{d\theta}{dx}\right] - m^2\theta = 0 \tag{a}$$

Define a new independent variable by

$$z = 1 + \mu x \tag{b}$$

Equation a can be rewritten, with z as the independent variable, as

$$\frac{d}{dx}\left(z^2 \frac{d\theta}{dz}\right) - \left(\frac{m}{\mu}\right)^2 \theta = 0 \tag{c}$$

Expansion of the derivative in Equation c leads to

$$z^2 \frac{d^2\theta}{dz^2} + 2z\frac{d\theta}{dz} - \left(\frac{m}{\mu}\right)^2 \theta = 0 \tag{d}$$

Equation d is of the form of a Cauchy-Euler differential equation. A solution is assumed of the form $\theta = z^{\alpha}$ which when substituted in Equation d, leads to

$$\alpha^2 + \alpha - \left(\frac{m}{\mu}\right)^2 = 0 \tag{e}$$

Application of the quadratic formula to Equation e leads to

$$\alpha_{1,2} = \frac{1}{2}\left[-1 \pm \sqrt{1 + 4\left(\frac{m}{\mu}\right)^2}\right] \tag{f}$$

The general form solution to Equation a is

$$\theta(z) = C_1 z^{\alpha_1} + C_2 z^{\alpha_2} \tag{g}$$

which can be written in terms of x as

$$\theta(x) = C_1(1+\mu x)^{\alpha_1} + C_2(1+\mu x)^{\alpha_2} \tag{h}$$

The constants of integration are determined through application of the boundary conditions, leading to

$$\theta(x) = \frac{2\left(\frac{m}{\mu}\right)^2}{1-\sqrt{1+4\left(\frac{m}{\mu}\right)^2}}(1+\mu x)^{\frac{1}{2}\left[-1+\sqrt{1+4\left(\frac{m}{\mu}\right)^2}\right]} - \frac{1-\sqrt{1+4\left(\frac{m}{\mu}\right)^2}+2\left(\frac{m}{\mu}\right)^2}{1-\sqrt{1+4\left(\frac{m}{\mu}\right)^2}}(1+\mu x)^{\frac{1}{2}\left[-1-\sqrt{1+4\left(\frac{m}{\mu}\right)^2}\right]} \tag{i}$$

The general form of a second-order Cauchy-Euler equation is

$$x^2\frac{d^2y}{dx^2} + ax\frac{dy}{dx} + by = 0 \tag{3.54}$$

The general solution of Equation 3.54 is obtained by assuming $y = x^v$, which leads to

$$\nu(\nu-1) + a\nu + b = 0 \tag{3.55}$$

The form of the solution depends on the values of a and b. If $(a-1)^2 > 4b$, Equation 3.55 has two real solutions, v_1 and v_2, such that

$$y_1(x) = x^{\nu_1} \quad y_2(x) = x^{\nu_2} \tag{3.56}$$

If in addition $a > 1$, then at least one root is negative, and the corresponding solution is not analytic at $x = 0$.

If $(a-1)^2 = 4b$ then Equation 3.55 has a single repeated root ν, and the linearly independent solutions of Equation 3.54 are

$$y_1(x) = x^{\nu} \quad y_2(x) = x^{\nu}\ln|x| \tag{3.57}$$

If $(a-1)^2 < 4b$, then Equation 3.55 has a pair of complex conjugate roots, $v = v_r \pm iv_i$ and linearly independent solutions of Equation 3.54 are

$$y_1(x) = x^{\nu_r} \cos\left(x^{\nu_i} \ln|x|\right) \quad y_2(x) = x^{\nu_r} \sin\left(x^{\nu_i} \ln|x|\right) \tag{3.58}$$

Equation 3.59 illustrates that solutions of a second-order equation expanded about a regular singular point are not of the form of Equation b of Example 3.7. Indeed, such solutions need not be analytic. Solutions of the Cauchy-Euler equation suggest that solutions about a regular singular point involve noninteger powers of x and perhaps logarithmic terms.

The technique used to obtain series solutions of second-order equations about a regular singular point is called the method of Frobenius and is described by the following theorem, presented without proof.

Theorem 3.7 (Method of Frobenius). Let x_0 be a regular singular point of a second-order differential equation, Equation 3.47. A solution to this equation can be expressed as

$$y(x) = (x - x_0)^{\nu} \sum_{k=0}^{\infty} c_k (x - x_0)^k \tag{3.59}$$

where v is the largest solution of the indicial equation,

$$\nu(\nu - 1) + \nu q(x_0) + r(x_0) = 0 \tag{3.60}$$

and $c_0 \neq 0$. Let μ be the second solution of Equation 3.60. A second linearly independent solution can be obtained dependent on the value of μ:

Case (1) If $\mu - v$ is not an integer, then the second solution is

$$y_2(x) = (x - x_0)^{\mu} \sum_{k=0}^{\infty} d_k (x - x_0)^k \tag{3.61}$$

Case (2) If $\mu = v$, then the second solution is

$$y_2(x) = (x - x_0)^{\mu} \sum_{k=0}^{\infty} d_k (x - x_0)^k + y_1(x) \ln|x - x_0| \tag{3.62}$$

Case (3) If $\mu - v$ is an integer, then

$$y_2(x) = (x - x_0)^{\mu} \sum_{k=0}^{\infty} d_k (x - x_0)^k + c y_1(x) \ln|x - x_0| \tag{3.63}$$

where c is a constant.

Example 3.9 Apply the method of Frobenius to obtain solutions of Bessel's equation of order n,

$$x^2\frac{d^2y}{dx^2}+x\frac{dy}{dx}+\left(x^2-n^2\right)y=0 \tag{a}$$

Solution Note that $x=0$ is a regular singular point of Bessel's equation with $q(x)=1$ and $r(x)=x^2-n^2$. A solution to Equation a is assumed to be

$$y(x)=x^{\nu}\sum_{k=0}^{\infty}c_kx^k \tag{b}$$

Noting that

$$\frac{dy}{dx}=\sum_{k=0}^{\infty}(k+\nu)c_kx^{k+\nu-1} \tag{c}$$

$$\frac{d^2y}{dx^2}=\sum_{k=0}^{\infty}(k+\nu)(k+\nu-1)c_kx^{k+\nu-2} \tag{d}$$

substitution of Equation b into Equation a leads to

$$\sum_{k=0}^{\infty}\left[(k+\nu)(k+\nu-1)+(k+\nu)-n^2\right]c_kx^{k+\nu}+\sum_{k=0}^{\infty}c_kx^{k+\nu+2}=0 \tag{e}$$

Noting that

$$\sum_{k=0}^{\infty}c_kx^{k+\nu+2}=\sum_{m=2}^{\infty}c_{m-2}x^{m+\nu}=\sum_{k=2}^{\infty}c_{k-2}x^{k+\nu} \tag{f}$$

Equation e can be rewritten as

$$\sum_{k=2}^{\infty}\left\{\left[(k+\nu)^2-n^2\right]c_k+c_{k-2}\right\}x^k+c_0\left(\nu^2-n^2\right)+c_1\left[(1+\nu)^2-n^2\right]x=0 \tag{g}$$

Setting coefficients of each power of x to zero independently leads to

$$\left(\nu^2-n^2\right)c_0=0 \tag{h}$$

$$\left[(1+\nu)^2-n^2\right]c_1=0 \tag{i}$$

$$\left[(k+\nu)^2-n^2\right]c_k+c_{k-2}\quad k=2,3,\ldots \tag{j}$$

Equation h is the indicial equation. Either $c_0 = 0$, or $v = \pm n$. However, the method of Frobenius requires that $c_0 \neq 0$. The largest solution is $v = n$.

If $c_0 \neq 0$, then Equation h and Equation i are satisfied by $c_1 \neq 0$ unless $v = -1/2$. However, since it is assumed that $v > \mu$, this can never occur, although this phenomenon does provide an alert that $n = 1/2$ is a special case in that case (3) of theorem 3.7 applies to determine the second solution. The first solution is determined with $v = n$, $c_0 \neq 0$, $c_1 = 0$.

The recurrence relation, Equation j, can be rearranged as

$$c_k = -\frac{c_{k-2}}{k(2n+k)} \tag{k}$$

Repeated application of Equation k shows that because $c_1 = 0$, $c_k = 0$ for all odd k. Application of Equation k for $k = 2,4$, and 6 leads to

$$c_2 = -\frac{c_0}{2(2+2n)} \tag{l}$$

$$c_4 = -\frac{c_2}{4(2n+4)} = \frac{c_0}{(2)(4)(2n+2)(2n+4)} \tag{m}$$

$$c_6 = -\frac{c_4}{6(2n+6)} = -\frac{c_0}{(2)(4)(6)(2n+2)(2n+4)(2n+6)} \tag{n}$$

Use of Equation l, Equation m, and Equation n and mathematical induction lead to

$$\begin{aligned} c_{2k} &= \frac{(-1)^k c_0}{[(2)(4)...(2k)][(2n+2)(2n+4)...(2n+2k)]} \\ &= \frac{(-1)^k c_0}{2^{2k} k![(n+1)(n+2)...(n+k)]} \end{aligned} \tag{o}$$

Substitution of Equation o into Equation a leads to

$$y(x) = c_0 \sum_{k=0}^{\infty} \frac{(-1)^k x^{2k+n}}{2^{2k} k![(n+1)(n+2)...(n+k)]} \tag{p}$$

Equation p is valid for any value of n, integer or noninteger. The form of the second linearly independent solution depends on the value of n. If n is neither an integer nor the midpoint between two integers, then the second solution is obtained from case (1) of theorem 3.7:

$$y_2(x) = x^{-n}\sum_{k=1}^{\infty} d_k x^k \tag{q}$$

A similar procedure is followed with the recurrence relation reduced to

$$d_k = \frac{d_{k-2}}{k(k-2n)} \tag{r}$$

Mathematical induction is used to obtain

$$d_{2k} = \frac{(-1)^k d_0}{2^{2k} k![(1-n)(2-n)...(k-n)]} \tag{s}$$

and the second solution becomes

$$y_2(x) = d_0 \sum_{k=0}^{\infty} \frac{(-1)^k x^{2k-n}}{2^{2k} k![(1-n)(2-n)...(k-n)]} \tag{t}$$

It is noted that $y_2(x)$ could be obtained by replacing n by $-n$ in the expression for $y_1(x)$. It is clear from Equation t that such a solution cannot be obtained when n is an integer because the denominator will be zero for $k=n$.

The roots of the indicial equation are equal when $n=0$ ($\nu=\mu=0$). The second solution is obtained using case (2) of theorem 3.7:

$$y_2(x) = \sum_{k=0}^{\infty} d_k x^k + y_1(x)\ln|x| \tag{u}$$

Note that $d/dx(y_1 \ln|x|) = \ln|x|(dy_1/dx) + (1/x)y_1$ and $d^2/dx^2(y_1 \ln|x|) = \ln|x|$ $(d^2y_1/dx^2) + 2/x\,(dy_1/dx) - (1/x^2)y_1$. Thus,

$$x^2\frac{d^2}{dx^2}(y_1\ln|x|) + x\frac{d}{dx}(y_1\ln|x|) + x^2 y_1\ln|x| = \ln|x|\left[x^2\frac{d^2y_1}{dx^2} + x\frac{dy_1}{dx} + x^2 y_1\right]$$

$$+2x\frac{dy_1}{dx} = 2x\frac{dy_1}{dx} \tag{v}$$

Thus, substitution of Equation u into Equation a with $n=0$ leads to

$$\sum_{k=2}^{\infty} k(k-1)d_k x^k + \sum_{k=1}^{\infty} k d_k x^k + \sum_{k=0}^{\infty} d_k x^{k+2} + 2\sum_{k=1}^{\infty} k c_k x^k = 0 \tag{w}$$

where $y_1(x)$ is used in the form of Equation b. Equation w can be rearranged to yield

$$\sum_{k=2}^{\infty}\left[k^2 d_k + 2kc_k + d_{k-2}\right]x^k + (d_1 + 2c_1)x = 0 \tag{x}$$

Recall that c_1 is determined to be zero; thus $d_1 = 0$. The recurrence relation obtained from Equation x is

$$d_k = -\frac{d_{k-2}}{k^2} - \frac{2}{k}c_k \tag{y}$$

Without loss of generality, $c_0 = 1$ and $d_0 = 1$. Noting that for $n = 0$, $c_k = 0$ for odd k and $c_k = 1/2^k[(k/2)!]^2$ for even k. Mathematical induction and Equation y are used to determine an explicit expression for d_k

$$d_k = \frac{(-1)^{\frac{k}{2}} d_0}{(k!)^2} + 2\sum_{c=0}^{k-2}\frac{(-1)^{k+i}c_{2n}\left[(2i)!\right]^2}{2i(k!)^2} \tag{z}$$

3.6 Bessel functions

The differential equation whose solutions were obtained in Example 3.8 is called Bessel's equation:

$$x^2\frac{d^2y}{dx^2} + x\frac{dy}{dx} + (x^2 - n^2)y = 0 \tag{3.64}$$

The first solution of Equation 3.64 valid for any n is given by Equation p of Example 3.8 as

$$y_1(x) = \sum_{k=1}^{\infty}\frac{(-1)^k x^{2k+n}}{2^{2k}k![(1+n)(2+n)\ldots(k+n)]} \tag{3.65}$$

If n is an integer, an alternate representation of Equation 3.65 is

$$y_1(x) = 2^n n!\sum_{k=0}^{\infty}\frac{(-1)^k\left(\frac{x}{2}\right)^{2k+n}}{k!(n+k)!} \tag{3.66}$$

It should be noted that $2^n n!$ is a constant and can be incorporated into a constant of integration.

The Bessel function of the first kind of order n is defined for integer n as

$$J_n(x) = \sum_{k=0}^{\infty}\frac{(-1)^k\left(\frac{x}{2}\right)^{2k+n}}{k!(n+k)!} \tag{3.67}$$

$J_n(x)$ is the first solution of Bessel's equation for integer n.

The Bessel function of the first kind for a noninteger value of n is defined in terms of the gamma function,

$$\Gamma(p) = \int_0^{\infty} e^{-t} t^{p-1} dt \tag{3.68}$$

Integration by parts is applied to derive an important property of the gamma function,

$$\Gamma(p+1) = p\Gamma(p) \tag{3.69}$$

Noting that

$$\Gamma(1) = \int_0^{\infty} e^{-t} dt = 1 \tag{3.70}$$

Equation 3.69 shows that $\Gamma(2)=1, \Gamma(3)=2, \Gamma(4)=6, \ldots$. Induction can be used to show that for integer n,

$$\Gamma(n) = (n-1)! \tag{3.71}$$

Thus, the gamma function can be thought of as a generalization of the factorial to noninteger values.

The properties of the gamma function can be used to rewrite Equation 3.67 as

$$y_1(x) = 2^n \Gamma(n+1) \sum_{k=0}^{\infty} \frac{(-1)^k \left(\frac{x}{2}\right)^{2k+n}}{\Gamma(k+1)\Gamma(n+k+1)} \tag{3.72}$$

Equation 3.72 leads to the definition of the Bessel function of the first kind of order n as

$$J_n(x) = \sum_{k=0}^{\infty} \frac{(-1)^k \left(\frac{x}{2}\right)^{2k+n}}{\Gamma(k+1)\Gamma(n+k+1)} \tag{3.73}$$

When n is not an integer or a half integer, the second solution of Bessel's equation is obtained by replacing n with $-n$ in the first solution. In this case, the Bessel function of the second kind is defined by

$$Y_n(x) = J_{-n}(x) = \sum_{k=0}^{\infty} \frac{(-1)^k \left(\frac{x}{2}\right)^{2k-n}}{\Gamma(k+1)\Gamma(k+1-n)} \tag{3.74}$$

An alternate and commonly used definition for the Bessel function of the second kind is a linear combination of Equation 3.73 and Equation 3.74, defined by

$$Y_n(x) = \frac{\cos(n\pi)J_n(x) - J_{-n}(x)}{\sin(n\pi)} \tag{3.75}$$

For an integer value of n, Equation 3.75 can be replaced by

$$Y_n(x) = \lim_{\nu \to n} \frac{\cos(\nu\pi)J_\nu(x) - J_{-\nu}(x)}{\sin(\nu\pi)} \tag{3.76}$$

The series form of Equation 3.76 is

$$Y_n(x) = \frac{2}{\pi} J_n(x)\left(\ln\left|\frac{x}{2}\right| + \gamma\right) - \sum_{k=0}^{n-1} \frac{\Gamma(n-k)}{\Gamma(k+1)}\left(\frac{x}{2}\right)^{2k-n}$$
$$-\frac{1}{\pi}\sum_{k=0}^{\infty}\left\{\left[\frac{(-1)^k}{\Gamma(k+1)\Gamma(n+k+1)}\left(\frac{x}{2}\right)^{2k+n}\right]\left[\sum_{\ell=1}^{k}\frac{1}{\ell} + \sum_{\ell=1}^{n+k}\frac{1}{\ell}\right]\right\} \tag{3.77}$$

where Euler's constant is defined as

$$\gamma = \lim_{m \to \infty}\left(\sum_{\ell=1}^{m}\frac{1}{\ell} - \ln(m)\right) = 0.5772... \tag{3.78}$$

The general solution of Bessel's equation of order n is

$$y(x) = C_1 J_n(x) + C_2 Y_n(x) \tag{3.79}$$

Bessel functions of the first and second kind of integer orders are plotted in Figure 3.6 and Figure 3.7, respectively. Bessel functions of the first order are finite for all values of x, whereas Bessel functions of the second kind are infinite at $x=0$ and finite everywhere else. Both Bessel functions have an infinite, but countable, number of zeroes. However, they are not periodic.

Consider the equation

$$x^2\frac{d^2y}{dx^2} + x\frac{dy}{dx} + (b^2x^2 - n^2)y = 0 \tag{3.80}$$

Let $z=bx$. Rewriting Equation 3.80 with z as the independent variable leads to

$$z^2\frac{d^2y}{dz^2} + z\frac{dy}{dz} + (z^2 - b^2)y = 0 \tag{3.81}$$

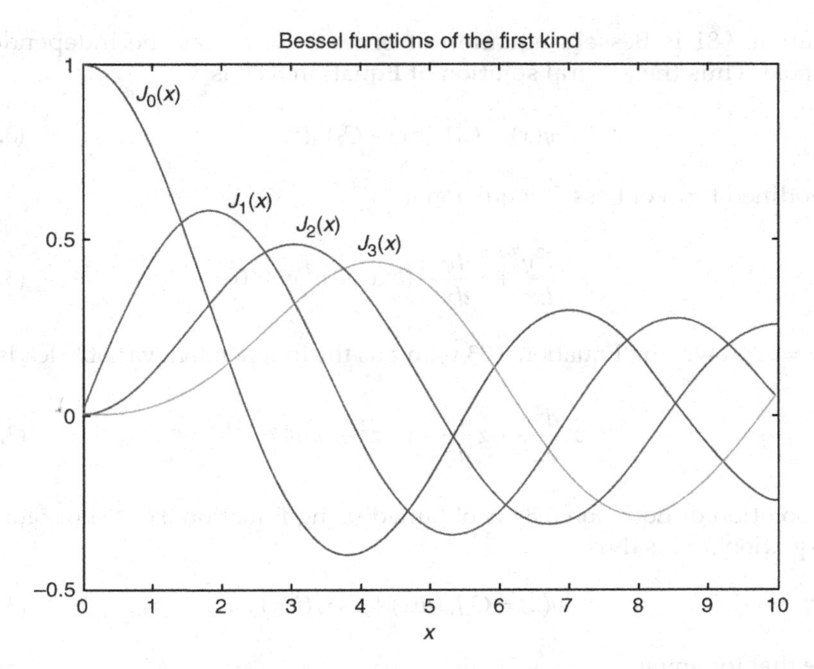

Figure 3.6 Bessel functions of the first kind are solutions of Bessel's equation which are finite for all values of x.

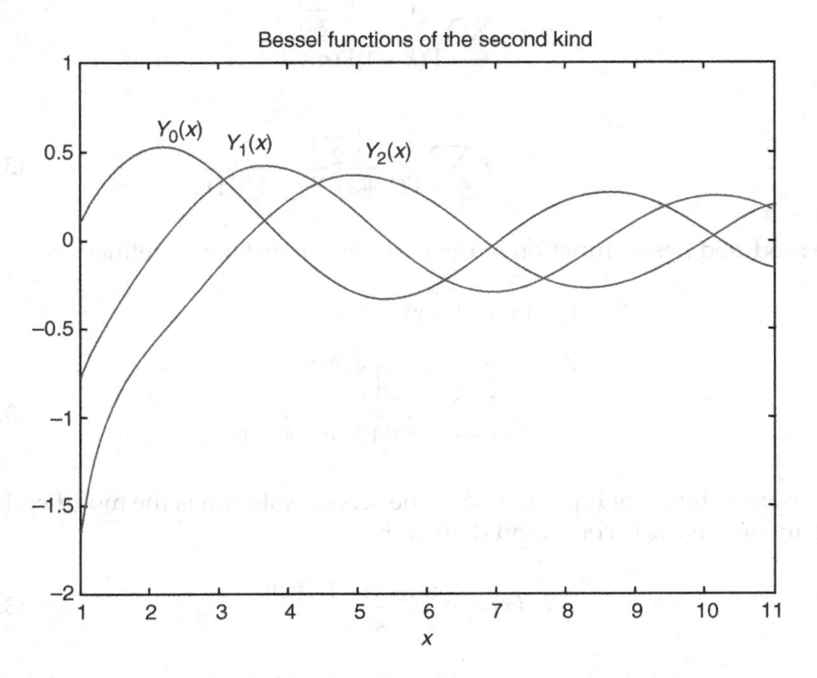

Figure 3.7 Bessel functions of the second kind do not exist at $x = 0$.

Equation 3.81 is Bessel's equation of order n with z as the independent variable. Thus the general solution of Equation 3.80 is

$$y(x) = C_1 J_n(bx) + C_2 Y_n(bx) \tag{3.82}$$

A modified form of Bessel's equation is

$$x^2 \frac{d^2y}{dx^2} + x\frac{dy}{dx} - (b^2x^2 + n^2)y = 0 \tag{3.83}$$

Let $z = ix$. Rewriting Equation 3.83 with z as the independent variable leads to

$$z^2 \frac{d^2y}{dz^2} + z\frac{dy}{dz} + (b^2z^2 - n^2)y = 0 \tag{3.84}$$

The solution of Equation 3.84 is obtained using Equation 3.64. The solution of Equation 3.83 is thus

$$y(z) = C_1 J_n(ibx) + C_2 Y_n(ibx) \tag{3.85}$$

Note that for any n,

$$J_n(ibx) = \sum_{k=0}^{\infty} \frac{(-1)^k \left(\frac{ibx}{2}\right)^{2k+n}}{\Gamma(k+1)\Gamma(n+k+1)}$$

$$= i^n \sum_{k=0}^{\infty} \frac{\left(\frac{bx}{2}\right)^{2k+n}}{\Gamma(k+1)\Gamma(n+k+1)} \tag{3.86}$$

The modified Bessel function of the first kind of order n is defined as

$$I_n(x) = i^{-n} J_n(ix)$$

$$= \sum_{k=0}^{\infty} \frac{\left(\frac{x}{2}\right)^{2k+n}}{\Gamma(k+1)\Gamma(n+k+1)} \tag{3.87}$$

and is one solution of Equation 3.83. The second solution is the modified Bessel function of the second kind defined by

$$K_n(x) = -\frac{\pi}{2} \frac{I_n(x) - I_{-n}(x)}{\sin(n\pi)} \tag{3.88}$$

A limiting process similar to that used to define $Y_n(x)$ when n is an integer, Equation 3.88, is used to define $K_n(x)$ for n an integer.

Modified Bessel functions of the first and second kind are plotted in Figure 3.8 and Figure 3.9, respectively. The modified Bessel functions of the first kind are finite for all values of their argument, while the modified Bessel functions of the second kind are infinite when evaluated at zero, but finite for all other values of their argument. Contrary to the regular Bessel functions, the modified Bessel functions do not have zeroes.

Example 3.10 The nondimensional differential equation governing the temperature distribution in the straight triangular fin shown in Figure 3.10 is

$$\frac{d}{dx}\left(x\frac{d\theta}{dx}\right) - m^2\theta = 0 \tag{a}$$

where x is nondimensionalized with respect to L, $\theta = (T - T_\infty)/(T_0 - T_\infty)$, and $m = \sqrt{2hL^2/kb}$. The boundary condition is

$$\theta(1) = 1 \tag{b}$$

The temperature must be finite everywhere over the interval, $0 \leq x \leq L$. Determine the rate of heat transfer at the base of the fin.

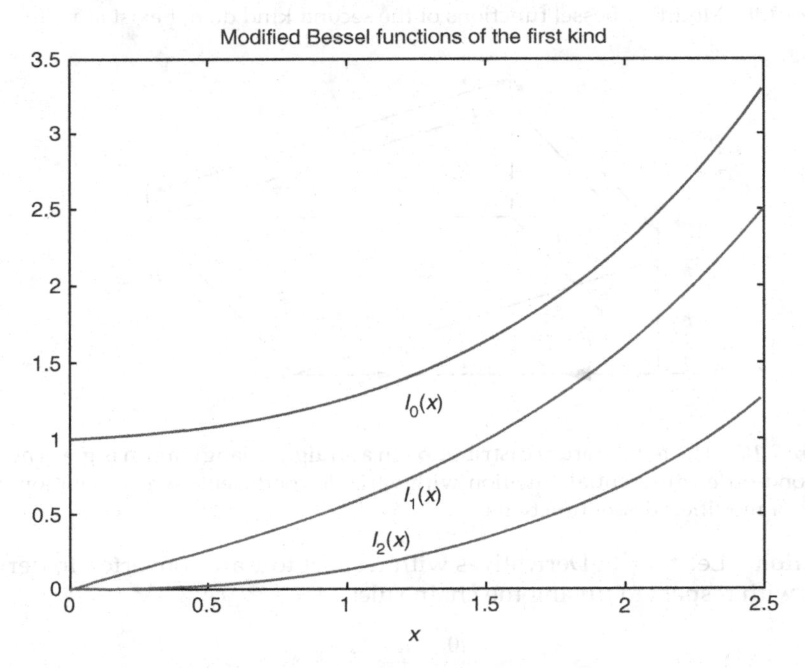

Figure 3.8 Modified Bessel functions of the first kind are finite at all finite x.

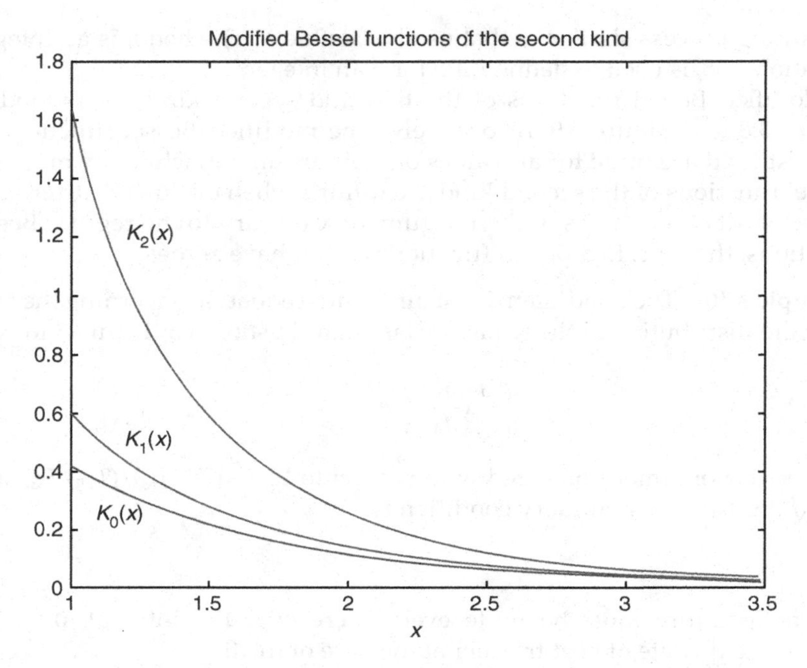

Figure 3.9 Modified Bessel functions of the second kind do not exist at $x=0$.

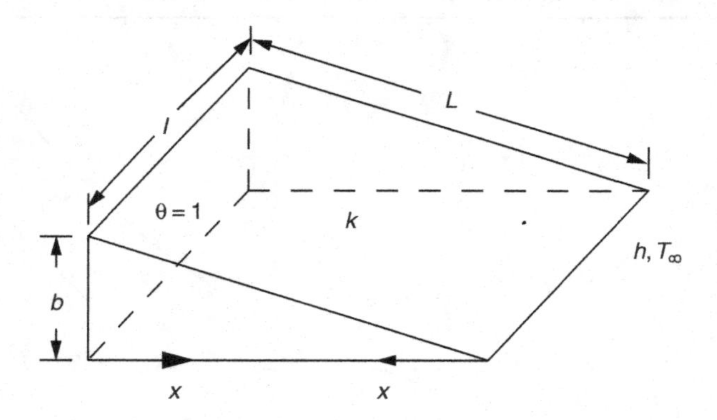

Figure 3.10 The temperature distribution in a straight triangular fin is governed by a second-order differential equation with variable coefficients whose solution is in terms of modified Bessel functions.

Solution Let $z=x^{1/2}$. Derivatives with respect to x are converted to derivatives with respect to z using the chain rule:

$$\frac{d\theta}{dx}=\frac{d\theta}{dz}\frac{dz}{dx}$$

$$=\frac{1}{2z}\frac{d\theta}{dz} \tag{c}$$

$$\frac{d}{dx}\left(x\frac{d\theta}{dx}\right)=\frac{1}{2z}\frac{d}{dz}\left(\frac{z}{2}\frac{d\theta}{dz}\right)$$

$$=\frac{1}{4}\frac{d^2\theta}{dz^2}+\frac{1}{4z}\frac{d\theta}{dz} \tag{d}$$

Substitution of Equation d into Equation a leads to

$$\frac{1}{4}\frac{d^2\theta}{dz^2}+\frac{1}{4z}\frac{d\theta}{dz}-m^2\theta=0 \tag{e}$$

Multiplying Equation e by $4z^2$ gives

$$z^2\frac{d^2\theta}{dz^2}+z\frac{d\theta}{dz}-4m^2z^2\theta=0 \tag{f}$$

Equation f is the same as the modified Bessel's equation, Equation 3.83, with $n=0$ and $b=2m$. Thus the general solution of Equation f is

$$\theta(z)=C_1I_0(2mz)+C_2K_0(2mz) \tag{g}$$

Substituting for z in term of x leads to

$$\theta(x)=C_1I_0\left(2mx^{\frac{1}{2}}\right)+C_2K_0\left(2mx^{\frac{1}{2}}\right) \tag{h}$$

It should be noted $I_0(0)=0$, but $\lim_{x\to 0}K_0(x)=\infty$. Thus, the temperature is finite at $x=0$ if and only if $C_2=0$. Application of Equation b to Equation h with $C_2=0$ leads to

$$C_1=\frac{1}{I_0(2m)} \text{ and}$$

$$\theta(x)=\frac{I_0\left(2mx^{\frac{1}{2}}\right)}{I_0(2m)} \tag{i}$$

The rate of heat transfer from the base is

$$q=\frac{kb\ell(T_0-T_\infty)}{L}\frac{d\theta}{dx}(1) \tag{j}$$

where

$$\frac{d\theta}{dx}=\frac{(2m)\left(\frac{1}{2}x^{-\frac{1}{2}}\right)}{I_0(2m)}I_0'\left(2mx^{\frac{1}{2}}\right) \tag{k}$$

Note that in Equation k, the prime represents differentiation with respect to the argument of the function. Substitution of Equation k into Equation j gives

$$q=\frac{kb\ell m}{I_0(2m)}I_0'(2m) \tag{l}$$

Term-by-term differentiation of Equation 3.87 for $n=0$ leads to

$$\begin{aligned}
\frac{d}{dx}[I_0(x)]&=\frac{d}{dx}\left[\sum_{k=0}^{\infty}\frac{\left(\frac{x}{2}\right)^{2k}}{(k!)^2}\right]\\
&=\sum_{k=1}^{\infty}\frac{k\left(\frac{x}{2}\right)^{2k-1}}{(k!)^2}\\
&=\sum_{p=0}^{\infty}\frac{(p+1)\left(\frac{x}{2}\right)^{2(p+1)-1}}{[(p+1)!]^2}\\
&=\sum_{p=1}^{\infty}\frac{\left(\frac{x}{2}\right)^{2p+1}}{(p+1)!(p+2)!}\\
&=I_1(x)
\end{aligned} \tag{m}$$

The rate of heat transfer from the base of the fin is

$$\begin{aligned}
q&=\frac{kb\ell m}{I_0(2m)}I_1(2m)\\
&=\frac{\ell L\sqrt{2hkb}}{I_0\left(2L\sqrt{\frac{2h}{kb}}\right)}I_1\left(2L\sqrt{\frac{2h}{kb}}\right)
\end{aligned} \tag{n}$$

It is shown in Equation m of Example 3.9 that $I_0'(x) = I_1(x)$. This is an example of a identity involved Bessel functions which can be derived using the series definitions of Bessel functions. These identities are useful in applications where differentiation and integration of Bessel functions are required. Some useful identities are stated below.

1. Derivatives of Bessel functions of order zero

$$J_0'(x) = -J_1(x) \tag{3.89}$$

$$Y_0'(x) = -Y_1(x) \tag{3.90}$$

$$I_0'(x) = I_1(x) \tag{3.91}$$

$$K_0'(x) = -K_1(x) \tag{3.92}$$

2. Derivatives of Bessel functions

$$\frac{d}{dx}\left[x^n J_n(x)\right] = x^n J_{n-1}(x) \tag{3.93}$$

$$\frac{d}{dx}\left[x^n Y_n(x)\right] = x^n Y_{n-1}(x) \tag{3.94}$$

$$\frac{d}{dx}\left[x^n I_n(x)\right] = x^n I_{n-1}(x) \tag{3.95}$$

$$\frac{d}{dx}\left[x^n K_n(x)\right] = -x^n K_{n-1}(x) \tag{3.96}$$

$$\frac{d}{dx}\left[x^{-n} J_n(x)\right] = -x^{-n} J_{n+1}(x) \tag{3.97}$$

$$\frac{d}{dx}\left[x^{-n} Y_n(x)\right] = -x^{-n} Y_{n+1}(x) \tag{3.98}$$

$$\frac{d}{dx}\left[x^{-n} I_n(x)\right] = x^{-n} I_{n+1}(x) \tag{3.99}$$

$$\frac{d}{dx}\left[x^{-n} K_n(x)\right] = -x^{-n} K_{n+1}(x) \tag{3.100}$$

3. Recurrence Relations

$$J_{n+1}(x) + J_{n-1}(x) = \frac{2n}{x} J_n(x) \tag{3.101}$$

$$Y_{n+1}(x)+Y_{n-1}(x)=\frac{2n}{x}Y_n(x) \tag{3.102}$$

$$J_n'(x)=J_{n-1}(x)-\frac{n}{x}J_n(x) \tag{3.103}$$

$$J_n'(x)=\frac{n}{x}J_n(x)-J_{n+1}(x) \tag{3.104}$$

Example 3.11 An integral that is useful in applications is $\int_0^1 x[J_n(\alpha x)]^2 dx$. During the evaluation of the integral, it is convenient to use the identity

$$\frac{d}{dx}\left[x^2 J_{n-1}(x)J_{n+1}(x)\right]=2x^2 J_n(x)\frac{d}{dx}[J_n(x)] \tag{a}$$

(a) Use the properties of Bessel functions, Equations 3.93–3.104, to derive Equation a, (b) use Equation a in the evaluation of $\int_0^1 x[J_n(\alpha x)]^2\,dx$, (c) simplify the result if $J_n(\alpha)=0$, and (d) simplify the result if $J_n'(\alpha)=0$.

Solution (a) The product rule for differentiation is used to expand the derivative on the left-hand side of Equation a as

$$\frac{d}{dx}\left[x^2 J_{n+1}(x)J_{n-1}(x)\right]=2xJ_{n+1}(x)J_{n-1}(x)+x^2 J_{n+1}'(x)J_{n-1}(x)+x^2 J_{n+1}(x)J_{n-1}'(x) \tag{b}$$

Using Equation 3.103 to substitute for $J_{n+1}'(x)$ and Equation 3.104 to substitute for $J_{n-1}'(x)$, Equation b can be rewritten as

$$\begin{aligned}\frac{d}{dx}\left[x^2 J_{n+1}(x)J_{n-1}(x)\right] &= 2xJ_{n+1}(x)J_{n-1}(x)+x^2 J_{n-1}(x)\left[J_n(x)-\frac{n+1}{x}J_{n+1}(x)\right]\\ &\quad +x^2 J_{n+1}(x)\left[\frac{n-1}{x}J_{n-1}(x)-J_n(x)\right]\\ &=[2x-x(n+1)+x(n-1)]J_{n+1}(x)J_{n-1}(x)\\ &\quad +x^2 J_n(x)[J_{n-1}(x)-J_{n+1}(x)]\\ &=x^2 J_n(x)[J_{n-1}(x)-J_{n+1}(x)]\end{aligned} \tag{c}$$

Addition of Equation 3.103 and Equation 3.104 leads to

$$J_n'(x)=\frac{1}{2}[J_{n-1}(x)-J_{n+1}(x)] \tag{d}$$

Substitution of Equation d on the right-hand side of Equation c leads to Equation a.

(b) First consider the indefinite integral

$$I = \int x[J_n(\alpha x)]^2\,dx \tag{e}$$

The variable of integration is changed from x to z using $z = \alpha x$, leading to

$$I = \frac{1}{\alpha^2}\int z[J_n(z)]^2\,dz \tag{f}$$

Integration by parts is used to evaluate the integral in Equation f by defining $u = [J_n(z)]^2$ and $dv = zdz$ such that $du = 2J_n(x)J_n'(z)dz$ and $v = z^2/2$. Equation f becomes

$$I = \frac{1}{\alpha^2}\left\{\frac{z^2}{2}[J_n(z)]^2 - \int \frac{z^2}{2}[2J_n(z)J_n'(z)]dz\right\} \tag{g}$$

Equation a is used to simplify the integral on the right-hand side of Equation g:

$$\begin{aligned} I &= \frac{1}{\alpha^2}\left\{\frac{z^2}{2}[J_n(z)]^2 - \frac{1}{2}\int \frac{d}{dz}[z^2 J_{n+1}(z)J_{n-1}(z)]dz\right\} \\ &= \frac{z^2}{2\alpha^2}\{[J_n(z)]^2 - [J_{n+1}(z)J_{n-1}(z)]\} \end{aligned} \tag{h}$$

Recalling that $z = \alpha x$, Equation h becomes

$$I = \frac{x^2}{2}\{[J_n(\alpha x)]^2 - [J_{n+1}(\alpha x)J_{n-1}(\alpha x)]\} \tag{i}$$

Noting that $J_n(\alpha) = 0$, the definite integral becomes

$$\begin{aligned} \int_0^1 x[J_n(\alpha x)]^2\,dx &= \left[\frac{x^2}{2}\{[J_n(\alpha x)]^2 - [J_{n+1}(\alpha x)J_{n-1}(\alpha x)]\}\right]_{x=0}^{x=1} \\ &= \frac{1}{2}\{[J_n(\alpha)]^2 - [J_{n+1}(\alpha)J_{n-1}(\alpha)]\} \end{aligned} \tag{j}$$

(c) If $J_n(\alpha) = 0$, Equation j becomes

$$\int_0^1 x[J_n(\alpha x)]^2\,dx = -\frac{1}{2}J_{n+1}(\alpha)J_{n-1}(\alpha) \tag{k}$$

and from Equation 3.103 and Equation 3.104, $J_{n+1}(\alpha) = -J_n'(\alpha)$ and $J_{n-1}(\alpha) = J_n'(\alpha)$. Thus Equation k becomes

$$\int_0^1 x[J_n(\alpha x)]^2\,dx = \frac{1}{2}[J_n'(\alpha)]^2$$

$$= \frac{1}{2}[J_{n+1}(\alpha)]^2 \tag{l}$$

Equation l is the form of the integral used in applications when $J_n(\alpha) = 0$.
(d) If $J_n'(\alpha) = 0$, Equation 3.103 and Equation 3.104 lead to $J_{n+1}(\alpha) = (n/\alpha)\,J_n(\alpha)$ and $J_{n-1}(\alpha) = (n/\alpha)\,J_n(\alpha)$. In this case, Equation j becomes

$$\int_0^1 x[J_n(\alpha x)]^2\,dx = \frac{1}{2}\left(1 - \frac{n^2}{\alpha^2}\right)[J_n(\alpha)]^2 \tag{m}$$

Example 3.12 Evaluate $\int x^5 J_2(x)dx$.

Solution Integrals of the form of $\int x^m J_n(x)dx$ when m and n are integers can be evaluated using integration by parts in conjunction with Equation 3.93 and Equation 3.94. If $m+n$ is odd, repeated integration by parts will lead to a closed-form evaluation of the integral. If $m+n$ is even, repeated application of integration by parts eventually leads to $\int J_0(x)dx$, which does not have a closed-form evaluation.

The initial strategy in evaluating these integrals is, when using integration by parts, to choose dv such that v can be obtained using either Equation 3.93 or Equation 3.94. For the integral at hand, if $dv = x^3 J_2(x)$, then, from Equation 3.93, $v = x^3 J_3(x)$. For this choice of v, $u = x^2$ and $du = 2xdx$. Thus application of integration by parts leads to

$$\int x^5 J_2(x)dx = x^2\left[x^3 J_3(x)\right] - \int \left[x^3 J_3(x)\right] 2xdx$$

$$= x^5 J_3(x) - 2\int x^4 J_3(x)dx \tag{a}$$

The remaining integral can be evaluated directly from Equation 3.93, leading to

$$\int x^5 J_2(x)dx = x^5 J_3(x) - 2x^4 J_4(x) \tag{b}$$

3.7 Differential equations whose solutions are expressible in terms of Bessel functions

Example 3.9 illustrates that some differential equations may, through a change of independent variable, be transformed into Bessel's equation. There are several transformations which enable differential equations of certain forms to be rewritten as Bessel's equation.

Consider the differential equation,

$$x^2\frac{d^2y}{dx^2}+ax\frac{dy}{dx}+b^2x^2y=0 \tag{3.105}$$

Equation 3.105 is similar to Bessel's equation of order zero which has a coefficient of 1 multiplying the $x(dy/dx)$ term. A change of dependent variable $y=x^{(1-a)/2}z$ leads to

$$x^2\frac{d^2z}{dx^2}+x\frac{dz}{dx}+\left[b^2x^2-\left(\frac{1-a}{2}\right)^2\right]z=0 \tag{3.106}$$

Equation 3.106 is Bessel's equation of order $(1-a)/2$, and thus the general solution of Equation 3.106 is

$$y(x)=C_1x^{\frac{1-a}{2}}J_{\frac{1-a}{2}}(bx)+C_1x^{\frac{1-a}{2}}Y_{\frac{1-a}{2}}(bx) \tag{3.107}$$

If $b^2<0$, then modified Bessel functions are used in Equation 3.106 with argument $|b|x$.

Next consider

$$\frac{d}{dx}\left(x^r\frac{dy}{dx}\right)+b^2x^sy=0 \tag{3.108}$$

where r and s are nonnegative. A change of independent variable $x=u^{2/(s-r+2)}$ leads to the differential equation

$$u^2\frac{d^2y}{du^2}+\left[1+\frac{2(r-1)}{s-r+2}\right]u\frac{dy}{du}+b^2\left(\frac{2}{s-r+2}\right)^2y=0 \tag{3.109}$$

Equation 3.109 is of the form of Equation 3.106 with $a=1+2(r-1)/s-r+2$. Noting that $(1-a)/2=(1-r)/s-r+2$, the general solution of Equation 3.109 is

$$y(u)=C_1u^{\frac{1-r}{s-r+2}}J_{\frac{1-r}{s-r+2}}\left(\frac{2b}{s-r+2}u\right)+C_2u^{\frac{1-r}{s-r+2}}Y_{\frac{1-r}{s-r+2}}\left(\frac{2b}{s-r+2}u\right) \tag{3.110}$$

The solution of Equation 3.110 in terms of the original independent variable x is

$$y(x) = C_1 x^{\frac{1-r}{2}} J_{\frac{1-r}{s-r+2}}\left(\frac{2b}{s-r+2} x^{\frac{s-r+2}{2}}\right) + C_2 x^{\frac{1-r}{2}} J_{\frac{1-r}{s-r+2}}\left(\frac{2b}{s-r+2} x^{\frac{s-r+2}{2}}\right) \tag{3.111}$$

The differential equation

$$\frac{d}{dx}\left(x^r \frac{dy}{dx}\right) + (b^2 x^s + c x^{r-2}) y = 0 \tag{3.112}$$

has a general solution of

$$y(x) = C_1 x^{\frac{1-r}{2}} J_\nu\left(\frac{2b}{s-r+2} x^{\frac{s-r+2}{2}}\right) + C_2 x^{\frac{1-r}{2}} J_\nu\left(\frac{2b}{s-r+2} x^{\frac{s-r+2}{2}}\right) \tag{3.113}$$

where

$$\nu = \frac{\sqrt{(1-r)^2 - 4c}}{s-r+2} \tag{3.114}$$

Example 3.13 The differential equation governing the temperature distribution in an annular fin of hyperbolic profile is

$$\frac{d}{dr}\left[\frac{4\pi}{R_0^{\frac{1}{2}}} r^{\frac{3}{2}} \frac{dT}{dr}\right] - \frac{4\pi h}{k} r (T - T_\infty) = 0 \tag{a}$$

The boundary conditions are

$$T(R_0) = T_0 \tag{b}$$

$$\frac{dT}{dr}(R_1) = 0 \tag{c}$$

Nondimensional variables are introduced as $r^* = r/R_0$ and $\theta = (T - T_\infty)/(T_0 - T_\infty)$. The nondimensional formulation of the problem is

$$\frac{d}{dr}\left(r^{\frac{3}{2}} \frac{d\theta}{dr}\right) - (Bi) r\theta = 0 \tag{d}$$

$$\theta(1) = 1 \tag{e}$$

$$\frac{d\theta}{dr}\left(\frac{R_1}{R_0}\right) = 0 \tag{f}$$

where the Biot number is $Bi = hR_0/K$. Determine the temperature distribution in the fin.

Solution Equation d is in the form of Equation 3.108 with $r = (3/2)$, $s = 1$, and $b^2 = -Bi$. Thus the solution is in the form of Equation 3.111, except that the Bessel functions J and Y are replaced by the modified Bessel functions I and K and b is replaced by $\sqrt{Bi}$. The general solution is

$$\theta(r) = C_1 r^{-\frac{1}{4}} I_{-\frac{1}{3}}\left(\frac{4}{3}\sqrt{Bi}\, r^{\frac{3}{4}}\right) + C_2 r^{-\frac{1}{4}} K_{-\frac{1}{3}}\left(\frac{4}{3}\sqrt{Bi}\, r^{\frac{3}{4}}\right) \tag{g}$$

Application of the boundary condition, Equation f, leads to

$$C_2 = -\frac{\sqrt{Bi}\left(\frac{R_1}{R_0}\right)^{\frac{3}{4}} I'_{-\frac{1}{3}}\left[\frac{4}{3}\sqrt{Bi}\left(\frac{R_1}{R_0}\right)^{\frac{3}{4}}\right] - \frac{1}{4} I_{-\frac{1}{3}}\left[\frac{4}{3}\sqrt{Bi}\left(\frac{R_1}{R_0}\right)^{\frac{3}{4}}\right]}{\sqrt{Bi}\left(\frac{R_1}{R_0}\right)^{\frac{3}{4}} K'_{-\frac{1}{3}}\left[\frac{4}{3}\sqrt{Bi}\left(\frac{R_1}{R_0}\right)^{\frac{3}{4}}\right] - \frac{1}{4} K_{-\frac{1}{3}}\left[\frac{4}{3}\sqrt{Bi}\left(\frac{R_1}{R_0}\right)^{\frac{3}{4}}\right]} C_1$$

$$= \mu C_1 \tag{h}$$

Application of Equation e to Equation g, using Equation h, gives

$$C_1 = \frac{1}{I_{-\frac{1}{3}}\left(\frac{4}{3}\sqrt{Bi}\right) - \mu K_{-\frac{1}{3}}\left(\frac{4}{3}\sqrt{Bi}\right)} \tag{i}$$

The total heat transfer at the base of the fin is

$$Q = -k\left(4\pi R_0^{\frac{3}{2}}\right)\frac{dT}{dr}(R_0) \tag{j}$$

Equation j can be rewritten in terms of nondimensional variables as

$$Q = -4\pi R_0^{\frac{1}{2}} \frac{d\theta}{dr}(1) \tag{k}$$

Substitution of Equation g, Equation h, and Equation i into Equation k leads to

$$Q = -4\pi k R_0^{\frac{1}{2}} C_1 \left[-\frac{1}{4} I_{-\frac{1}{3}}\left(\frac{4}{3}\sqrt{Bi}\right) + \sqrt{Bi}\, I'_{-\frac{1}{3}}\left(\frac{4}{3}\sqrt{Bi}\right) - \frac{\mu}{4} K_{-\frac{1}{3}}\left(\frac{4}{3}\sqrt{Bi}\right)\right.$$

$$\left. + \mu\sqrt{Bi}\, K'_{-\frac{1}{3}}\left(\frac{4}{3}\sqrt{Bi}\right)\right] \tag{l}$$

A differential equation that often occurs in applications is

$$x^2\frac{d^2y}{dx^2}+2x\frac{dy}{dx}+[x^2-m(m+1)]y=0 \tag{3.115}$$

where m is an integer. Equation 3.115 is in the form of Equation 3.112 with $r=2$, $s=2$, $b=1$, and $c=-m(m+1)$. The general solution of Equation 3.115 is obtained using Equation 3.113 as

$$y(x)=C_1x^{-(1/2)}J_{m+(1/2)}(x)+C_2x^{-(1/2)}Y_{m+(1/2)}(x) \tag{3.116}$$

The Bessel functions used in Equation 3.116 are of fractional order which is an odd integer times one half.

Spherical Bessel functions of the first kind and second kind are defined as

$$j_m(x)=\left(\frac{\pi}{2x}\right)^{(1/2)}J_{m+(1/2)}(x) \tag{3.117}$$

$$y_m(x)=\left(\frac{\pi}{2x}\right)^{(1/2)}Y_{m+(1/2)}(x) \tag{3.118}$$

Thus, the general solution of Equation 3.115 is written using spherical Bessel functions as

$$y(x)=C_1j_m(x)+C_2y_m(x) \tag{3.119}$$

Graphs of spherical Bessel functions are presented in Figure 3.11 and Figure 3.12.

It is noted that $j_0(0)=1$ while $j_m(0)=0$ for $m>0$ and $\lim_{x\to 0}y_m(x)=-\infty$. All spherical Bessel functions have an infinite number of zeroes.

Spherical Bessel functions satisfy recurrence relations of the form

$$j_{m+1}(x)-\frac{2m+1}{x}j_m(x)+j_{m-1}(x)=0 \tag{3.120}$$

$$y_{m+1}(x)-\frac{2m+1}{x}y_m(x)+y_{m-1}(x)=0 \tag{3.121}$$

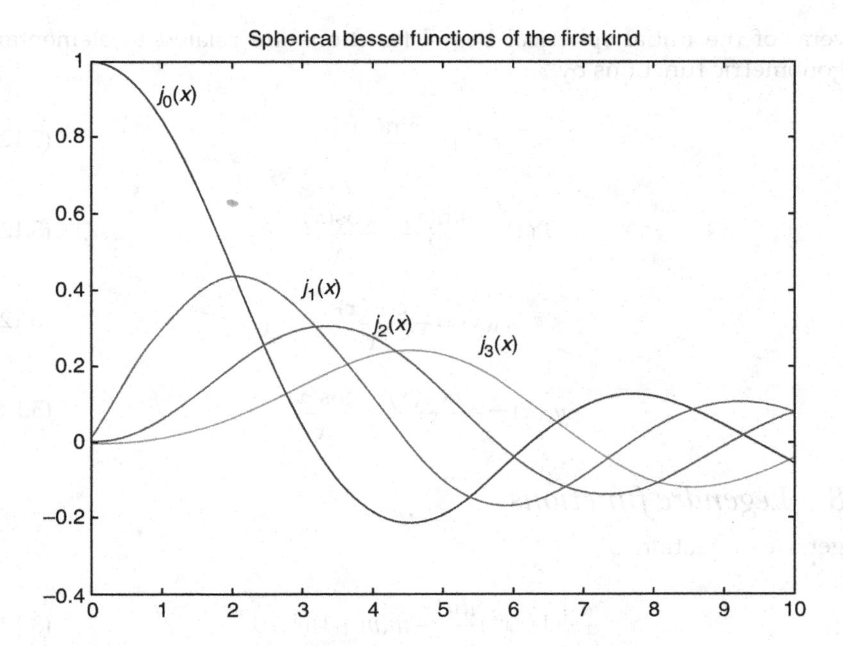

Figure 3.11 Spherical Bessel functions of the first kind.

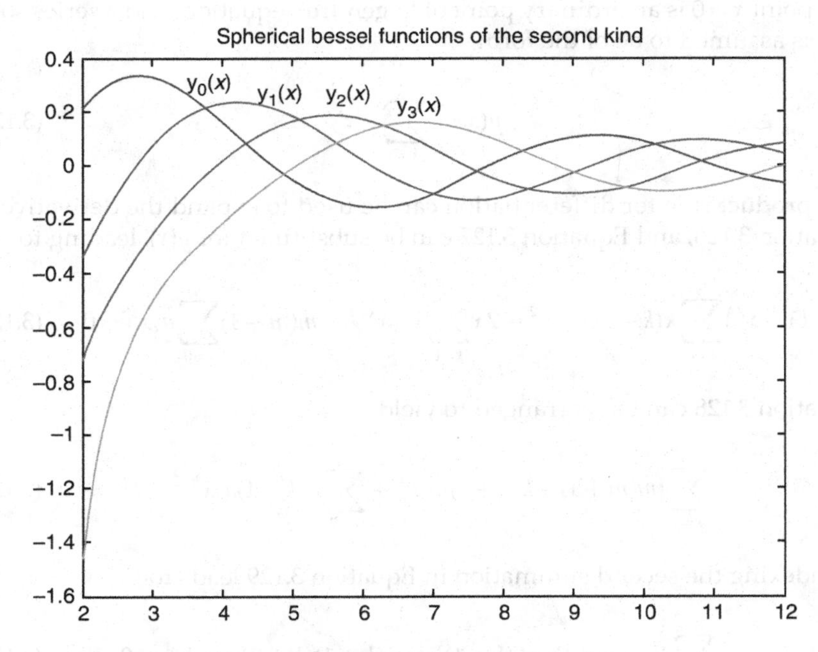

Figure 3.12 Spherical Bessel functions of the second kind.

Several of the initial spherical Bessel functions are related to elementary trigonometric functions by

$$j_0(x) = \frac{\sin(x)}{x} \tag{3.122}$$

$$j_1(x) = \frac{\sin(x)}{x^2} - \frac{\cos(x)}{x} \tag{3.123}$$

$$y_0(x) = -\frac{\cos(x)}{x} \tag{3.124}$$

$$y_1(x) = -\frac{\cos(x)}{x^2} - \frac{\cos(x)}{x} \tag{3.125}$$

3.8 *Legendre functions*

Legendre's equation is

$$\frac{d}{dx}\left[(1-x^2)\frac{dy}{dx}\right] + m(m+1)y = 0 \tag{3.126}$$

The point $x=0$ is an ordinary point of Legendre's equation, and a series solution is assumed to be of the form

$$y(x) = \sum_{k=0}^{\infty} a_k x^k \tag{3.127}$$

The product rule for differentiation can be used to expand the derivative in Equation 3.126, and Equation 3.127 can be substituted for $y(x)$, leading to

$$\left(1-x^2\right)\sum_{k=0}^{\infty} k(k-1)a_k x^{k-2} - 2x\sum_{k=1}^{\infty} k a_k x^{k-1} + m(m+1)\sum_{k=0}^{\infty} a_k x^k = 0 \tag{3.128}$$

Equation 3.128 can be rearranged to yield

$$\sum_{k=0}^{\infty}[m(m+1) - k(k+1)]a_k x^k + \sum_{k=2}^{\infty} k(k-1)a_k x^{k-2} = 0 \tag{3.129}$$

Reindexing the second summation in Equation 3.129 leads to

$$\sum_{k=0}^{\infty}\{[m(m+1) - k(k+1)]a_k + (k+1)(k+2)a_{k+2}\}x^k = 0 \tag{3.130}$$

Setting coefficients of powers of x to zero gives

$$a_{k+2} = \frac{k(k+1) - m(m+1)}{(k+1)(k+2)} a_k \tag{3.131}$$

Equation 3.120 is a recursion relation between the coefficients. Its application for even values of k leads to definition of coefficients with even subscripts in terms a_0, while its application for odd values of k leads to definition of coefficients with odd subscripts in terms a_1. Leaving these initial coefficients arbitrary, the general solution of Equation 3.131 is of the form

$$y(x) = a_0 y_0(x) + a_1 y_1(x) \tag{3.132}$$

where the expansion for $y_0(x)$ contains only even powers of x and the expansion for $y_1(x)$ contains only odd powers of x.

First consider integer values of m. Equation 3.131 shows that $a_{m+2} = 0$. If m is even, then $y_0(x)$ is a polynomial of order m, while $y_1(x)$ has an infinite series expansion. If m is odd, then $y_1(x)$ is a polynomial of order m, while $y_0(x)$ has an infinite series expansion. It thus becomes convenient to rename the functions such that for an integer m,

$$y_m(x) = c_1 z_{1,m}(x) + c_2 z_{2,m}(x) \tag{3.133}$$

where $z_{1,m}(x)$ is a polynomial of order n and $z_{2,m}(x)$ has an infinite series expansion.

The Legendre polynomial of the first kind of order m is defined by

$$P_m(x) = \frac{z_1(x)}{z_1(1)} \tag{3.134}$$

Equation 3.133 and Equation 3.134 are used to evaluate the first four Legendre polynomials as

$$P_0(x) = 1 \tag{3.135a}$$

$$P_1(x) = x \tag{3.135b}$$

$$P_2(x) = \frac{1}{3}(3x^2 - 1) \tag{3.135c}$$

$$P_3(x) = \frac{1}{2}(5x^3 - 3x) \tag{3.135d}$$

The Legendre function of the second kind of order m is defined by

$$Q_m(x) = \begin{cases} -z_1(1)z_2(x) & \text{m} = 0, 2, 4, \ldots \\ z_1(1)z_2(x) & \text{m} = 1, 3, 5, \ldots \end{cases} \tag{3.136}$$

It can be shown that the infinite series representations of the Legendre functions of the second kind for an integer order converge within the interval $-1 < x < 1$ to known functions. For example,

$$Q_0(x) = \tanh^{-1}(x) \tag{3.137a}$$

$$Q_1(x) = \frac{x}{2}\ln\frac{1+x}{1-x} - 1 \tag{3.137b}$$

Legendre's equation, Equation 3.126, can be used to derive recurrence relations between the Legendre polynomials and Legendre functions as

$$mP_m(x) - (2m-1)xP_{m-1}(x) - (m-1)P_{m-2}(x) \tag{3.138a}$$

$$mQ_m(x) - (2m-1)xQ_{m-1}(x) - (m-1)Q_{m-2}(x) \tag{3.138b}$$

If m is not an integer, a change of variables $u = 1 - x$ is used in Equation 3.126. The resulting equation has a regular singular point at $u = 0$ $(x = 1)$, with one root of its indicial equation as zero. The second solution is obtained using the method of Frobenius and has a logarithmic singularity at $u = 0$. The solution which remains finite at $x = 1$, but is undefined at $x = -1$, is called the Legendre function of the first kind of order m, $P_m(x)$. The solution which is undefined at $x = \pm 1$ is called the Legendre function of the second kind of order m, $Q_m(x)$.

The associated Legendre's equation is

$$\frac{d}{dx}\left[(1-x^2)\frac{dy}{dx}\right] + \left[m(m+1) - \frac{n^2}{1-x^2}\right]y = 0 \tag{3.139}$$

The associated Legendre functions of the first kind, $P_m^n(x)$, and of the second kind, $Q_m^n(x)$, of order m and index n satisfy Equation 3.139 and are defined as

$$P_m^n(x) = (1-x^2)^{\frac{n}{2}}\frac{d^n P_m(x)}{dx^n} \tag{3.140}$$

$$Q_m^n(x) = (1-x^2)^{\frac{n}{2}}\frac{d^n Q_m(x)}{dx^n} \tag{3.141}$$

Note that by definition, if n and m are integers and $n>m$, then $P_m^n(x)=0$.

The associated Legendre function of the first kind of order m is finite at $x=\pm 1$ only if m is an integer.

Problems

3.1. Prove that the homogeneous solution space of a linear differential operator **L** is a vector space.

3.2. Consider the set of solutions of Equation (3.10) which satisfy the boundary conditions specified in theorem 3.1 for arbitrary values of y_0 and $\dot{y}_0$. That is, the functions in the set satisfy boundary conditions of this form, but not for specific values of y_0 and $\dot{y}_0$. Prove that this set of functions is a vector space.

3.3.–3.13. Solve the following differential equations subject to the given boundary conditions:

3.3. $(d^2y/dx^2)-4y=0$ with $y(0)=2$ and $y(1)=0$.

3.4. $(d^2y/dx^2)+16y=0$ with $y(0)=2$ and $dy/dx(1)+3y(1)=0$.

3.5. $(d^2y/dx^2)+4(dy/dx)+4y=0$ with $y(0)=0$ and $dy/dx(1)=1$.

3.6. $(d^2y/dx^2)+5(dy/dx)-14y=0$ with $dy/dx(0)=0$ and $y(1)=1$.

3.7. $(d^2y/dx^2)+5(dy/dx)+14y=0$ with $dy/dx(0)=0$ and $y(1)=1$.

3.8. $(d^2y/dx^2)+2(dy/dx)+10y=0$ with $dy/dx(0)=0$ and $(dy/dx)(1)=1$.

3.9. $(d^2y/dx^2)-y=3e^{-2x}$ with $y(0)=0$ and $y(1)=0$.

3.10. $(d^2y/dx^2)+3y=4e^{-3x}+5$ with $y(0)=0$ and $dy/dx(1)=0$.

3.11. $(d^2y/dx^2)+9y=0.1\sin(4x)$ with $y(0)=1$ and $y(1)=0$.

3.12. $(d^2y/dx^2)-3y=2e^{-3x}\sin(2x)$ with $y(0)=1$ and $y(1)=0$.

3.13. $(d^2y/dx^2)+3(dy/dx)=3x+2e^{-3x}$ with $y(0)=0$ and $dy/dx(1)=0$.

3.14. The nondimensional equation for the temperature distribution over the extended surface shown in Figure P3.14 is

$$\frac{d^2\theta}{dx^2}-Bi\theta=0 \tag{a}$$

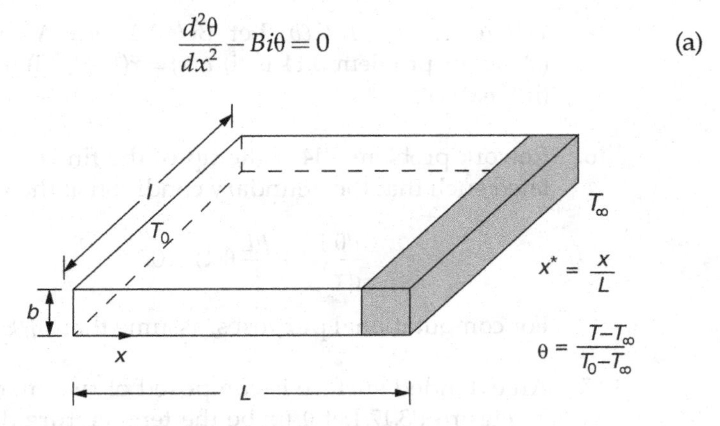

Figure P3.14 System of problems 3.14–3.16.

Where $Bi = 2hL^2(\ell + b)/k\ell b$. The face at $x = 0$ is maintained at a constant nondimensional temperature of 1, $\theta(0) = 1$ and is insulated at its right end, $d\theta/dx(1) = 0$.

a. Determine $\theta(x)$.
b. Determine the rate at which heat is transferred from the base of the surface,

$$q_0 = \frac{-k\ell b(T_0 - T_\infty)}{L}\frac{d\theta}{dx}(0).$$

c. Determine the rate at which heat is transferred over the surface by convection,

$$q_c = \int_0^2 2h(\ell + b)(T_0 - T_\infty)\theta(x)dx\,.$$

d. Compare the answers to (b) and (c) and explain.
e. The efficiency of the extended surface is defined as the ratio of total heat transfer from the surface of the fin as calculated in part (c) to the heat transfer from the surface if were held at the base temperature. Determine the efficiency for the extended surface.

3.15. Rework problem 3.14 if the surface is subject to an internal heat generation per unit volume, *u(x)*, such that the nondimensional equation for the temperature distribution is

$$\frac{d^2\theta}{dx^2} - Bi\theta = -\Lambda u(x),$$

Where $\Lambda = u_{\max}L^2/k\ell b$. Let $\mathrm{Bi} = 0.2$ and $\Lambda = 0.5$. Solve parts (a)–(c) of problem 3.14 if (i) $u(x) = x(1-x)$, (ii) $u(x) = \sin(\pi x)$, and (iii) $u(x) = e^{-x}$.

3.16. Rework problem 3.14 if the tip of the fin is not perfectly insulated such that the boundary condition at the right end is

$$\frac{d\theta}{dx}(1) + \frac{hL}{k}\theta(1) = 0.$$

For computational purposes, assume that $hl/k = 0.5$ and $\mathrm{Bi} = 2$.

3.17. An extended surface is composed of two materials as shown in Figure P3.17. Let $\theta_1(x)$ be the temperature distribution from the base to $x=\lambda$ where the material changes, and let $\theta_2(x)$ be the temperature distribution in the fin from $x=\lambda$ to $x=1$, the tip of the extended surface. The governing differential equations are

$$\frac{d^2\theta_1}{dx^2}-Bi_1\theta_1=0 \quad 0\leq x\leq\lambda \tag{a}$$

$$\frac{d^2\theta_2}{dx^2}-Bi_2\theta_2=0 \quad \lambda\leq x\leq 1 \tag{b}$$

where $Bi_1=2hL^2(\ell+b)/k_1\ell b$ and $Bi=2hL^2(\ell+b)/k_2\ell b$. The appropriate boundary conditions are $\theta_1(0)=1$ and $(d\theta_2/dx)(1)=0$. The temperature must be continuous across the surface, requiring that $\theta_1(\lambda)=\theta_2(\lambda)$. The rate of heat transfer must also be continuous across the surface, requiring that $d\theta_1/dx(\lambda)=k_2/k_1(d\theta_2/dx)(\lambda)$. Rework problem 3.14 for this extended surface. For the computations, assume $k_2/k_1=1.2$ and $\lambda=0.42$.

3.18. The transverse deflection of the uniform beam shown in Figure P3.18 is governed by the differential equation

$$EI\frac{d^4w}{dx^4}=f(x) \tag{a}$$

where $f(x)$ is the distributed load per unit length. Determine the transverse deflection of a pinned-pinned beam which is subject to the boundary conditions $w(0)=0, (d^2w/dx^2)(0)=0$, $w(L)=0$ and $(d^2w/dx^2)(L)=0$ for (a) $f(x)=F_0x(L-x)$ and (b) $f(x)=F_m\sin(3\pi x/L)$.

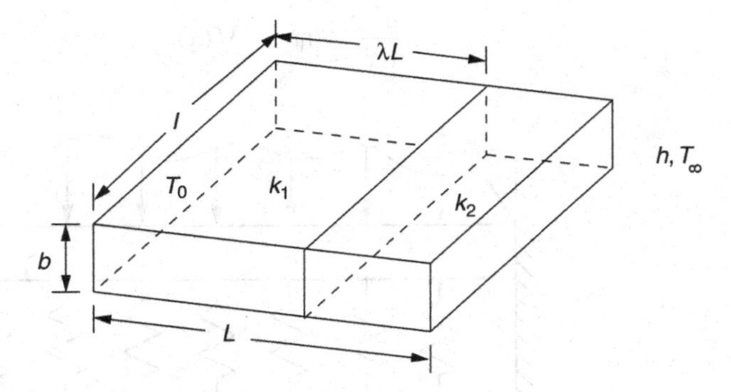

Figure P3.17 System of problem 3.17.

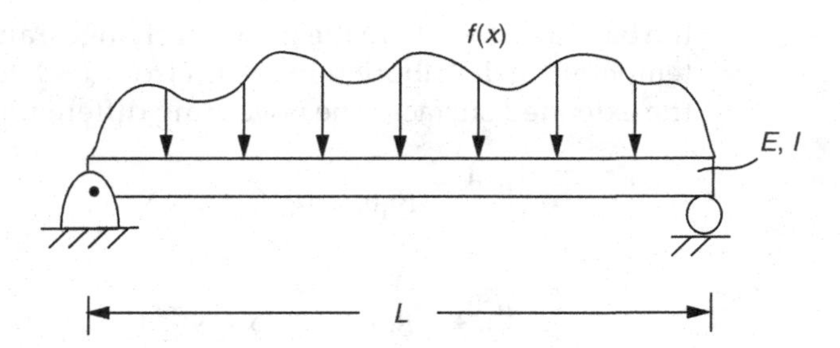

Figure P3.18 System for problems 3.18–3.21.

3.19. Repeat problem 3.18 for a fixed-pinned beam which has boundary a $w(0) = 0$, $(dw/dx)(0) = 0$, $w(L) = 0$ and $(d^2w/dx^2)(L) = 0$.

3.20. Repeat problem 3.19 for a fixed-pinned beam with a concentrated load of magnitude F_0 applied at the midspan. One approach to solve this problem is to define $w_1(x)$ as the transverse deflection between $x = 0$ and $x = L/2$ and $w_2(x)$ as the deflection between $x = L/2$ and $x = L$. The boundary conditions of problem 3.18 apply appropriately to $w_1(0)$ and $w_2(L)$. Matching conditions at $x = L/2$ are that the deflection is continuous, $w_1(L/2) = w_2(L/2)$, the slope of the elastic curve is continuous, $(dw_1/dx)(L/2) = (dw_2/dx)(L/2)$, the bending moment is continuous, $(d^2w_1/dx^2)(L/2) = (d^2w_2/dx^2)(L/2)$, and the shear force is discontinuous at $x = L/2$ due to the presence of the concentrated load, $EI(d^3w_1/dx^3)(L/2) - EI(d^3w_2/dx^3)(L/2) = F_0$.

3.21. Repeat problem 3.18 for a beam pinned at $x = 0$, but with a linear spring of stiffness k attached at $x = L$. The appropriate boundary conditions are $w(0) = 0$, $(d^2w/dx^3)(0) = 0$, $(d^2w/dx^2)(L) = 0$, and $EI(d^3w/dx^3)(L) + kw(L) = 0$.

3.22. The transverse deflection of a beam on an elastic foundation (a Winkler foundation), as shown in Figure P3.22, is governed by the nondimensional differential equation

$$\frac{d^4w}{dx^4} + \eta w = \Lambda f(x) \tag{a}$$

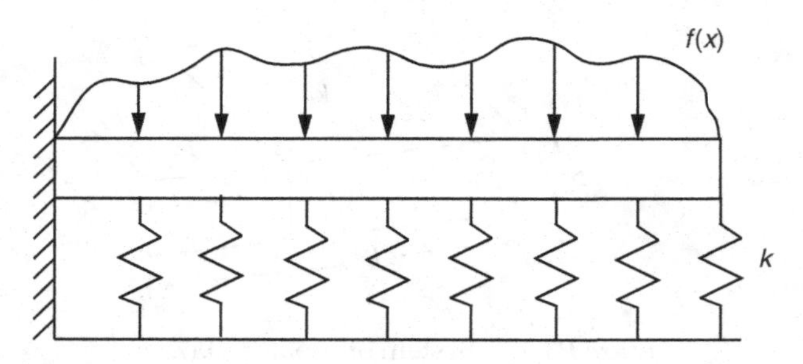

Figure P3.22 System for problems 3.22 and 3.23.

Determine $w(x)$ for a fixed-free beam when (a) $f(x)=1-x$ and (b) $f(x)=\sin(\pi x)$. The appropriate boundary conditions are $w(0)=x$, $(dw/dx)(0)=0$, $(d^2w/dx^2)(1)=0$, and $(d^3w/dx^3)(1)=0$. Use $\eta=4$ and $\Lambda=2$.

3.23. Determine the static deflection of a fixed-free beam on a Winkler foundation when a force is applied at the free end of the beam such that its deflection is δ. The problem is $(d^4w/dx^4)+\eta w=0$ subject to $w(0)=x$, $(dw/dx)(0)=0$, $(d^2w/dx^2)(1)=0$, and $w(1)=\delta$. Use $\eta=4$. What is the value of the required force which is equal to $(d^3w/dx^3)(1)$?

3.24. Solve

$$\frac{d}{dr}\left(r^2\frac{dw}{dr}\right)+\frac{1}{2}w=0$$

subject to $w(1)=4$.

3.25. Solve

$$\frac{d}{dr}\left(r^2\frac{dw}{dr}\right)-\frac{1}{2}w=0$$

subject to $w(0)=2$ and $w(1)=4$.

3.26. Solve

$$\frac{d}{dr}\left(r^2\frac{dw}{dr}\right)+w=0$$

subject to $w(1)=2$ and $w(2)=0$.

3.27. Solve

$$\frac{d}{dr}\left(r^2\frac{dw}{dr}\right)+3w=0$$

subject to $w(1)=2$ and $w(2)=-2$.

3.28. Solve

$$r^2\frac{d^2w}{dr^2}-6r\frac{dw}{dr}+2w=0$$

subject to $w(1)=2$ and $w(2)=1$.

3.29. Solve

$$r^2\frac{d^2w}{dr^2}-6r\frac{dw}{dr}+9w=0$$

subject to $w(1)=-1$ and $w(2)=0$.

3.30. Solve

$$r^2\frac{d^2w}{dr^2}-6r\frac{dw}{dr}+10w=0$$

subject to $w(0)=3$ and $w(1)=1$.

3.31. Solve

$$r^2\frac{d^2w}{dr^2}-6r\frac{dw}{dr}+10w=2r+3$$

subject to $w(1)=3$ and

$w(2)=1$.

3.32. Use the power series method to determine two linearly independent solutions to Airy's equation, $(d^2y/dx^2)+xy=0$.

3.33. Use the power series method to determine two linearly independent solutions to $(d^2y/dx^2)+4y=0$.

3.34. Chebyshev functions are solutions of the differential equation

$$(1-x^2)\frac{d^2y}{dx^2}-\frac{dy}{dx}+n^2y=0 \qquad \text{(a)}$$

a. When n is an integer, one solution of Equation a is a polynomial of order n. These are called Chebyshev polynomials. Determine the polynomial form of these solutions and show that the first four Chebyshev polynomials can be defined by

$$T_0(x)=1 \quad T_1(x)=x \quad T_2(x)=2x^2-1 \quad T_3(x)=4x^3-3x \qquad \text{(b)}$$

b. Show that the Chebyshev polynomials can also be represented as

$$T_n(x)=\cos\left(n\cos^{-1}x\right) \qquad \text{(c)}$$

c. Determine the second solution of Equation a for $n=0$.

d. Determine the second solution of Equation a for $n=3$.

3.35. Hermite's equation is

$$\frac{d^2y}{dx^2} - 2x\frac{dy}{dx} + 2ny = 0 \qquad \text{(a)}$$

a. When n is an integer, one solution of Equation a is a polynomial of order n. These are called the Hermite polynomial, $H_n(x)$. Determine the first four Hermite polynomials.
b. Determine the second solution of Equation a for $n=0$.
c. Determine the second solution of Equation a for $n=3$.

3.36. Use a series solution to obtain two linearly independent solutions $d^2y/dx^2 - 4y = 0$.

3.37. Use the method of Frobenius to determine solutions of

$$x\frac{d^2y}{dx^2} + 3\frac{dy}{dx} + xy = 0$$

3.38. Use the method of Frobenius to determine solutions of

$$x^2\frac{d^2y}{dx^2} - 2x\frac{dy}{dx} + x^2y = 0$$

3.39.–3.41. Determine the solution of each of the following in terms of Bessel functions:

3.39. $\dfrac{d}{dx}\left(x^3\dfrac{dy}{dx}\right) + 4x^2y = 0$

3.40. $\dfrac{d}{dx}\left(x\dfrac{dy}{dx}\right) - 16xy = 0$

3.41. $\dfrac{d^2y}{dx^2} + 2x^2y = 0$

3.42. Evaluate the integral,

$$\int_0^1 \left[J_{6n}\left(\sqrt{\lambda}r\right)\right]^2 r\,dr$$

3.43. Evaluate the integral,

$$\int_0^1 rJ_0(ar)\,dr$$

Chapter 4

Variational methods

4.1 Introduction

The general form of an equation involving a linear operator L whose domain is D and whose range is *R* is

$$\mathbf{L}\mathbf{u} = \mathbf{f} \tag{4.1}$$

where **f** is a vector in *R* and the solution **u** is in D. As noted in Chapter 3, one method of solving Equation 4.1 for an operator where D = *R* and there is a one-to-one correspondence between the elements of the domain and the elements of the range is to obtain the inverse of **L** and solve $\mathbf{u} = \mathbf{L}^{-1}\mathbf{f}$. Unfortunately, the inverse is not often readily attainable. Other methods of obtaining an exact solution to Equation 4.1 are often not successful because of the complexity of **L**. For example, it is difficult to obtain an exact solution of a linear differential equation with variable coefficients. Even when an exact solution is available in such cases, it is usually in terms of special functions with which subsequent numerical computations are difficult.

Because of this, approximate solutions of Equation 4.1 are often desired. There are three basic categories of approximation methods:

- The operator **L** is approximated by an operator $\hat{\mathbf{L}}$ whose domain and range are the same as **L**, and an exact solution $\hat{\mathbf{u}}$ is obtained for the approximate equation

$$\hat{\mathbf{L}}\hat{\mathbf{u}} = \mathbf{f} \tag{4.2}$$

  For example, the differential equation for the vibrations of a beam with a slowly varying cross-section might be approximated by the differential equation for a beam with a uniform cross-section. Obviously when the operator is approximated, $\hat{\mathbf{u}}$ not the exact solution. Improvements to the approximation are often obtained using asymptotic expansions. If, when nondimensionalized, the problem formulation involves a small nondimensional parameter, often denoted by ε, the solution may be expanded into an asymptotic expansion in terms of linearly independent functions of ε.

- A numerical solution is obtained when **L** is approximated by an operator $\mathbf{L}_n$, whose domain is R^n, and the solution is approximated at discrete values of independent variables. If **L** is a differential operator with D=C[a,b], a numerical method is used to provide an approximation to **u** at a finite number of points, $a \leq x_1 < x_2 < \cdots < x_{n-1} < x_n \leq b$. The range of $\mathbf{L}_n$ is R^n.

- Methods which use the operator **L** but provide approximations for **u** at every value of x are called variational methods. Such methods are based on minimizing the distance between the exact solution and the approximate solution.

Variational methods are the topics of this chapter. The general theory behind variational methods is that of calculus of variations. A varied path is any alternate path that the solution can take from the exact solution while satisfying the same beginning and ending conditions. The true path is obtained by minimizing the energy between the true path and all alternate paths.

The variational methods discussed in this chapter are approximate methods in which the choice of alternate paths is limited to those that are in the span of a finite-dimensional subspace of the vector space in which the true path resides. The best approximation for this subspace is the one in which the distance between the true solution and the approximate solution is minimized, the distance being measured by an inner-product-generated norm.

The method of least squares uses the standard inner product defined for the vector space. If the operator of Equation 4.1 is self-adjoint and positive definite with respect to the standard inner product, then an energy inner product can be defined. The Rayleigh-Ritz method uses an energy-inner-product-generated norm to minimize the distance between the true solution and other approximate solutions.

It can be shown that when using the energy inner product, it is possible to ease the requirements on the vector space where the possible approximate solutions reside. Only geometric boundary conditions need to be directly satisfied by all elements in the vector space of approximate solutions. The continuity requirements on approximate solutions can be eased. These changes lead to approximate solutions being drawn from vector spaces of piecewise continuous functions and the development of a method called the finite-element method. This method is well known and extensively used in modeling engineering problems. In keeping with the objectives of this text, only the basic variational theory behind the finite-element method is presented, along with several examples.

4.2 The general variational problem

Let **L** be a linear operator defined on a domain D which is a subspace of a vector space V. Let **(u,v)** represent a valid inner product on D. Let Q be an n-dimensional subspace of D that is spanned by a set of vectors $\mathbf{v}_1, \mathbf{v}_2, \ldots, \mathbf{v}_n$. The Gram-Schmidt process can always be used to generate an orthonormal basis for Q using $\mathbf{v}_1, \mathbf{v}_2, \ldots, \mathbf{v}_n$. Thus, without loss of generality, it can be assumed that the basis is an orthonormal basis for Q.

It is desired to find an approximation to **u**, the solution of Equation 4.1, from the vectors in Q. This concept is illustrated in Figure 4.1. The quality of the approximation must be assessed by the application of some standard. One possibility is to measure the length of the difference between the exact solution **u** and the approximate solution. An appropriate norm to use for measuring the length of this difference is an inner-product-generated norm. When the norm of the difference is smaller, the approximation is better. The best approximation to **u** from the vectors in Q is the vector **q** in Q such that $||u - q||$ is a minimum. Defining

$$\lambda(\mathbf{q}) = \|u - q\| \tag{4.3}$$

the best approximation is the vector that minimizes λ.

Since **q** is in Q, it can be represented by a linear combination of its basis elements,

$$\mathbf{q} = \sum_{i=1}^{n} \alpha_i \mathbf{v}_i \tag{4.4}$$

Substituting Equation 4.4 into Equation 4.3 and using the properties of the inner product assuming it only has real value

$$\begin{aligned}
\lambda^2 &= \|\mathbf{u} - \mathbf{q}\|^2 = (\mathbf{u} - \mathbf{q}, \mathbf{u} - \mathbf{q}) \\
&= (\mathbf{u} - \sum_{i=1}^{n} \alpha_i \mathbf{v}_i, \mathbf{u} - \sum_{i=1}^{n} \alpha_i \mathbf{v}_i) \\
&= (\mathbf{u}, \mathbf{u}) - 2\sum \alpha_i (\mathbf{u}, \mathbf{v}_i) + \sum_{i=1}^{n} \sum_{j=1}^{n} \alpha_i \alpha_j (\mathbf{v}_i, \mathbf{v}_j)
\end{aligned} \tag{4.5}$$

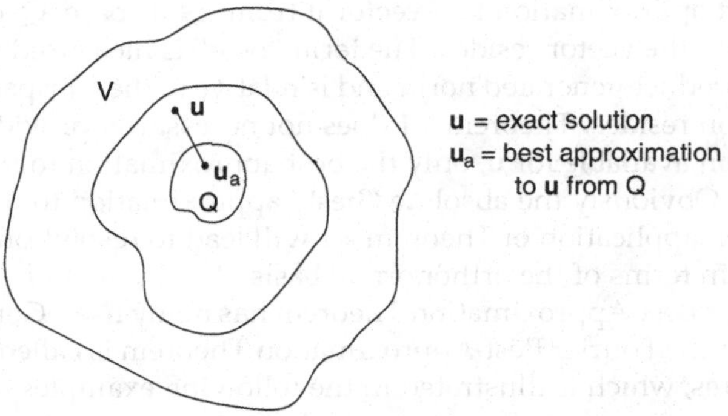

Figure 4.1 Variational methods determine the best approximation to a vector **u**, an element of a vector space V, from Q, a finite-dimensional subspace of V.

Since the basis is assumed to be orthonormal, $(\mathbf{v}_i,\mathbf{v}_j) = \delta_{ij}$, and the inner sum of the double sum in Equation 4.5 collapses to a single sum, yielding

$$\lambda^2 = (\mathbf{u},\mathbf{u}) - 2\sum_{i=1}^{n} \alpha_i(\mathbf{u},\mathbf{v}_i) + \sum_{i=1}^{n} \alpha_i^2 \tag{4.6}$$

The right-hand side of Equation 4.6 is a function of the coefficients $\alpha_1, \alpha_2, \ldots, \alpha_n$. Thus, $\lambda\ (\alpha_1, \alpha_2, \ldots, \alpha_n)$ is stationary when $d\lambda = 0$ or

$$\frac{\partial \lambda^2}{\partial \alpha_1^2} = \frac{\partial \lambda^2}{\partial \alpha_2^2} = \cdots = \frac{\partial \lambda^2}{\partial \alpha_n^2} = 0 \tag{4.7}$$

Applying Equation 4.7 to Equation 4.6 leads to

$$\frac{\partial \lambda^2}{\partial \alpha_k} = 2(\mathbf{u},\mathbf{v}_k) + 2\alpha_k$$
$$\alpha_k = (\mathbf{u},\mathbf{v}_k) \tag{4.8}$$

Thus, the following theorem is proved.

Theorem 4.1 Fourier Best-Approximation Theorem: Let **u** be an element of a vector space V with a defined inner product **(u,v)**. Let $\mathbf{v}_1, \mathbf{v}_2, \ldots, \mathbf{v}_n$ be an orthonormal basis for Q, a subspace of V. The best approximation to **u** from among all elements of Q, measured with respect to the inner-product-generated norm, is the vector

$$\mathbf{q} = \sum_{i=}^{n} (\mathbf{u},\mathbf{v}_i)\mathbf{v}_i \tag{4.10}$$

The Fourier Best-Approximation Theorem provides an algorithm to determine the best approximation to a vector **u** from a subspace, Q, of the vector space in which the vector resides. The term "best" is measured with respect to an inner-product-generated norm and is relative to the subspace where the approximation resides. Theorem 4.1 does not necessarily provide the "best" approximation available for **u**, only the best approximation to **u** from all the vectors in Q. Obviously the absolute "best" approximation to **u** is **u** itself. If **u** is in Q, then application of Theorem 4.1 will lead to resolution of **u** into its components in terms of the orthonormal basis.

The Fourier Best-Approximation Theorem has many uses. Generically, the application of the Fourier Best-Approximation Theorem is called the method of least squares, which is illustrated in the following examples.

Example 4.1 Find the best approximation to $f(x) = x\ (1-e^{1-x})$ on the interval [0,1] with respect to the inner product on C[0,1] from the following subspaces, S, of C[0,1], (a) S is spanned by $u_1(x) = 1$, $u_2(x) = \sin(2\pi x)$, $u_3(x) = \cos(2\pi x)$. Improve the approximation by including additional basis functions of $\sin(2\pi ix)$ and $\cos(2\pi ix)$ for i = 2,3,...,50. Develop an expression for the error the approximation. (b) S is spanned by $u_i(x)=\sin(i\pi x)$ for $i=1,2,\ldots,n$. Develop the expressions for the expansion coefficients, (c) S is spanned by $u_1(x)=1$, $u_2(x)=x$, $u_3(x)=x^2$, $u_4(x)=x^3$. Use an orthonormal basis for S to develop the approximation.

Solution MATHCAD is used to perform the calculations and develop the comparison graphs.

(a) The basis functions are orthogonal with respect to the standard inner product on C[0,1]. The normalized set of basis functions is $v_1(x)=1$, $v_2(x)=\sqrt{2}\sin(2\pi x)$, $v_3(x)=\sqrt{2}\cos(2\pi x)$. The coefficients in the approximation can be calculated using Equation 4.8 as:

$$\alpha_1 = \int_0^1 x(1-e^{1-x})dx = -0.218$$

$$\alpha_2 = \int_0^1 x(1-e^{1-x})\sqrt{2}\sin(2\pi x)dx = -0.024 \tag{a}$$

$$\alpha_3 = \int_0^1 x(1-e^{1-x})\sqrt{2}\cos(2\pi x)dx = 0.092$$

The least-squares approximation for $f(x)$ using this orthonormal basis is

$$q(x) = -0.218 - 0.024\sqrt{2}\sin(2\pi x) + 0.092\sqrt{2}\cos(2\pi x) \tag{b}$$

A comparison of $f(x)$ and $q(x)$ is illustrated in Figure 4.2 The norm of the error of the approximation is $\|e\| = \sqrt{\int_0^1 (f(x)-q(x))^2 dx} = 0.028$. Much of the error occurs near the ends of the interval. Only one basis function satisfies the end conditions of $f(x)$, $f(0)=0$ and $f(1)=0$.

(b) The basis functions are mutually orthogonal with respect to this inner product. The orthonormal basis functions are $v_i(x)=\sqrt{2}\sin(i\pi x)$. The least-squares approximation is of the form of Equation 4.4 with

$$\begin{aligned}\alpha_i &= \int_0^1 f(x)v_i(x)dx \\ &= \int_0^1 x(1-e^{1-x})\sqrt{2}\sin(i\pi x)dx \\ &= \frac{\pi i\sqrt{2}}{(\pi^2 i^2+1)^2}\left[(\pi^2 i^2+3)(-1)^i - e(\pi^2 i^2+2)\right]\end{aligned} \tag{c}$$

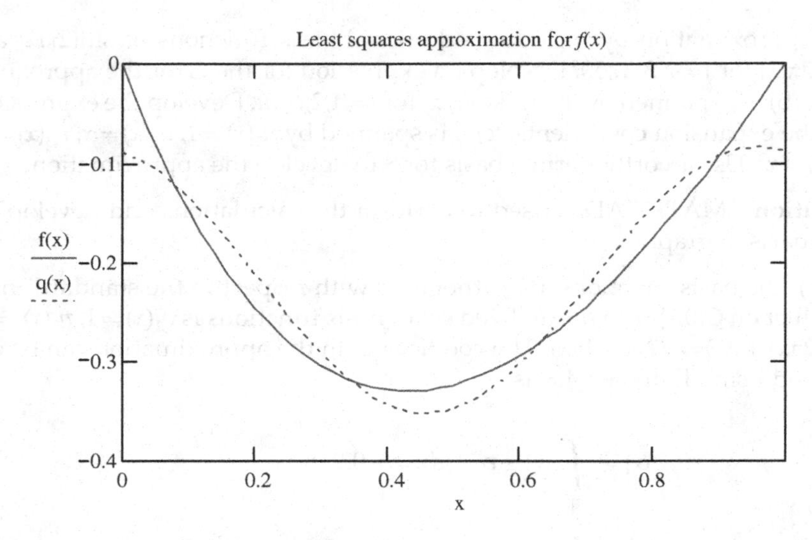

Figure 4.2 The least-squares approximation to $f(x)=x(1-e^{1-x})$ from a subspace spanned by $u_1(x)=1$, $u_2(x)=\sin(2\pi x)$, $u_3(x)=\cos(2\pi x)$ on [0,1] is illustrated. The plot was generated using MATHCAD.

The least-squares approximation for $f(x)$ using n basis functions is

$$q_n(x)=\sum_{i=1}^{n}\alpha_i\sqrt{2}\sin(i\pi x) \tag{d}$$

A comparison of the least-squares approximations using $n=3$, 5, and 10 is shown below. The norm of the error is calculated has $\|e_n\|=\sqrt{\int_0^1 (f(x)-q_n(x))^2 dx}$.

Example 4.1

Part (a)

$f(x):=x\cdot(1-e^{1-x})$

$u1(x):=1$

$u2(x):=\sin(2\cdot\pi x)$ Definition of basis functions

$u3(x):=\cos(2\cdot\pi\cdot x)$

$$\int_0^1 u1(x)\cdot u2(x)dx=0$$

$$\int_0^1 u1(x)\cdot u3(x)dx=0$$

Demonstrating orthogonality with respect to standard inner product

$$\int_0^1 u2(x)\cdot u3(x)dx=0$$

$$c1 := \sqrt{\int_0^1 u1(x)\cdot u1(x)dx} \quad c1 = 1$$

$$c2 := \sqrt{\int_0^1 u2(x)\cdot u2(x)dx} \quad c2 = 0.707$$ Calculating inner product generated norms

$$c3 := \sqrt{\int_0^1 u3(x)\cdot u3(x)dx} \quad c3 = 0.707$$

Developing the orthonormal basis

$$v1(x) := \frac{u1(x)}{c1} \qquad v2(x) := \frac{u2(x)}{c2} \qquad v3(x) := \frac{u3(x)}{c3}$$

Inner product evaluation

$$\alpha 1 := \int_0^1 f(x)\cdot v1(x)dx \quad \alpha 1 = -0.218$$

$$\alpha 2 := \int_0^1 f(x)\cdot v2(x)dx \quad \alpha 2 = -0.24$$

$$\alpha 3 := \int_0^1 f(x)\cdot v3(x)dx \quad \alpha 3 = 0.092$$

Least squares Approximation

$$q(x) := \alpha 1\cdot v1(x) + \alpha 2\cdot v2(x) + \alpha 3\cdot v3(x)$$

$$x := 0, .02 .. 1$$

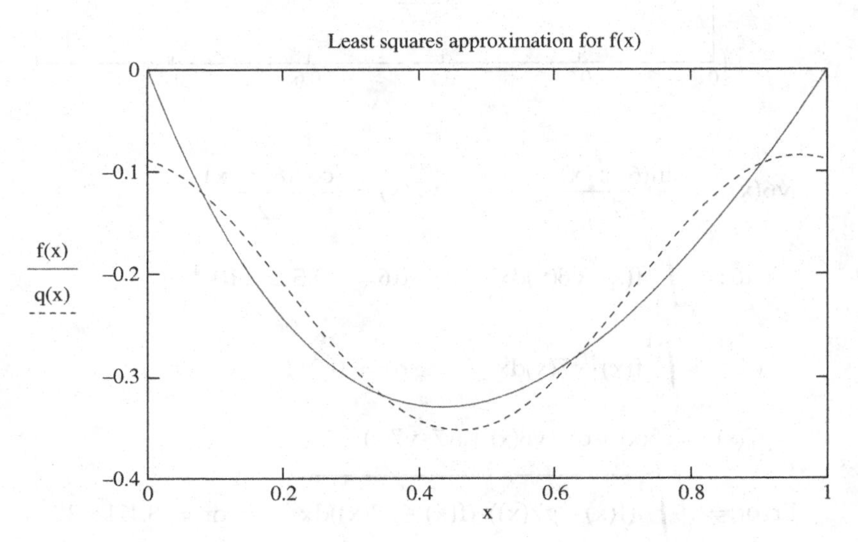

Error computation

$$\text{Error} := \sqrt{\int_0^1 (f(x)-q(x))\cdot(f(x)-q(x))dx} \qquad \text{Error} = 0.028$$

$$v4(x) := \frac{\sin(4\cdot\pi\cdot x)}{c2} \qquad v5(x) := \frac{\cos(4\cdot\pi\cdot x)}{c3}$$

$$\alpha 4 := \int_0^1 f(x)\cdot v4(x)dx \qquad \alpha 4 = -3.127\times 10^{-3}$$

$$\alpha 5 := \int_0^1 f(x)\cdot v5(x)dx \qquad \alpha 5 = 0.024$$

$$g5(x) := q(x) + \alpha 4\cdot v4(x) + \alpha 5\cdot v5(x)$$

$$\text{Error} := \sqrt{\int_0^1 (f(x)-g5(x))\cdot(f(x)-g5(x))dx} \qquad \text{Error} = 0.014$$

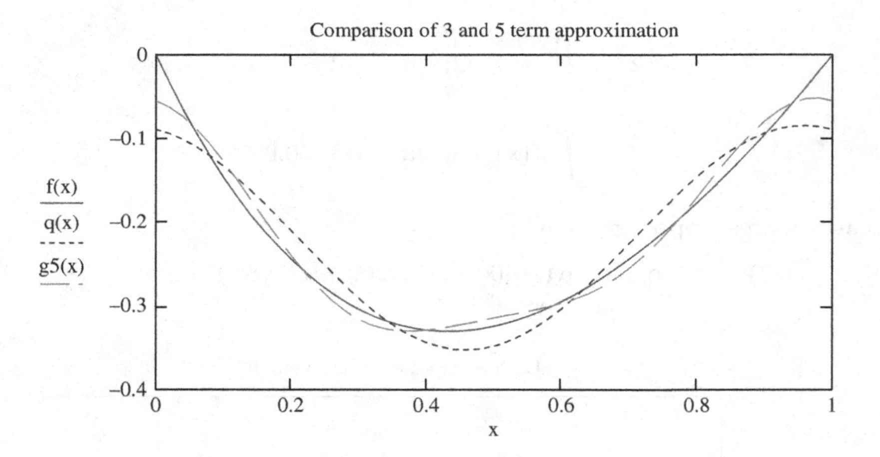

$$v6(x) := \frac{\sin(6\cdot\pi\cdot x)}{c2} \qquad v7(x) := \frac{\cos(6\cdot\pi\cdot x)}{c2}$$

$$\alpha 6 := \int_0^1 f(x)\cdot v6(x)dx \qquad \alpha 6 = -9.322\times 10^{-4}$$

$$\alpha 7 := \int_0^1 f(x)\cdot v7(x)dx \qquad \alpha 6 = 0.011$$

$$g7(x) := g5(x) + \alpha 6\cdot v6(x) + \alpha 7\cdot v7(x)$$

$$\text{Error} := \sqrt{\int_0^1 (f(x)-g7(x))\cdot(f(x)-g7(x))dx} \qquad \text{Error} = 8.411\times 10^{-3}$$

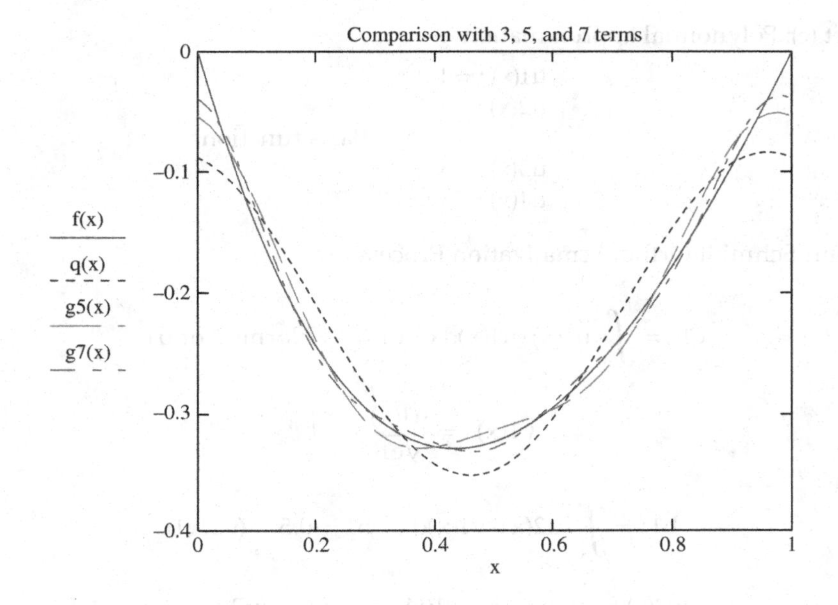

$$gh(x) := \sum_{n=1}^{50}\left(\int_0^1 f(x)\cdot\sqrt{2}\cdot\sin(2\cdot n\cdot\pi\cdot x)dx\cdot\sqrt{2}\cdot\sin(\pi 2\cdot n\cdot x)\right)$$

$$+\sum_{n=1}^{50}\left(\int_0^1 f(x)\cdot\sqrt{2}\cdot\cos(2\cdot n\cdot\pi\cdot x)dx\cdot\sqrt{2}\cdot\cos(\pi 2\cdot n\cdot x)\right)+\alpha 1$$

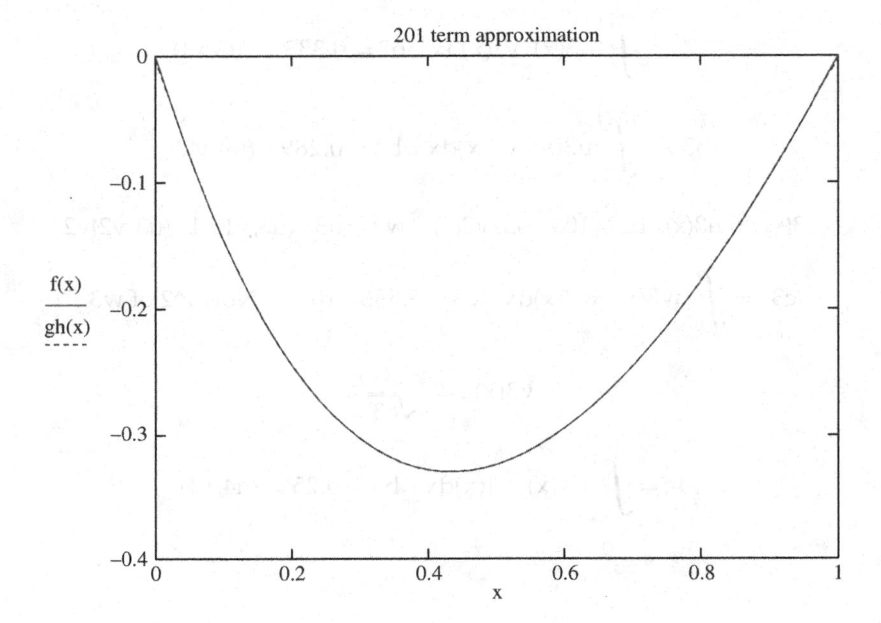

Part (c): Polynomial approximation

$$u1(x) := 1$$
$$u2(x) := x$$
Basis functions
$$u3(x) := x^2$$
$$u4(x) := x^3$$

Gram-Schmidt Orthonormalization Process

$$c1 := \int_0^1 u1(x) \cdot u1(x)dx \quad c1 = 1 \quad \text{Norm^2 of u1}$$

$$v1(x) := \frac{u1(x)}{\sqrt{c1}} \quad v1$$

$$b1 := \int_0^1 u2(x) \cdot v1(x)dx \quad b1 = 0.5 \quad (u2, v1)$$

$$w2(x) := u2(x) - b1 \cdot v1(x) \quad w2 = u2 - (u2, v1)v1$$

$$c2 := \int_0^1 w2(x) \cdot w2(x)dx \quad c2 = 0.083 \quad \text{Norm^2 of w2}$$

$$v2(x) := \frac{w2(x)}{\sqrt{c2}} \quad v2$$

$$b2 := \int_0^1 u3(x) \cdot v1(x)dx \quad b2 = 0.333 \quad (u3, v1)$$

$$b3 := \int_0^1 u3(x) \cdot v2(x)dx \quad b3 = 0.289 \quad (u3, v2)$$

$$w3(x) := u3(x) - b2 \cdot v1(x) - b3 \cdot v2(x) \quad w3 = u3 - (u3, v1)v1 - (u3, v2)v2$$

$$c3 := \int_0^1 w3(x) \cdot w3(x)dx \quad c3 = 5.556 \times 10^{-3} \quad \text{Norm^2 of w3}$$

$$v3(x) := \frac{w3(x)}{\sqrt{c3}}$$

$$b4 := \int_0^1 u4(x) \cdot v1(x)dx \quad b4 = 0.25 \quad (u4, v1)$$

$$b5 := \int_0^1 u4(x)\cdot v2(x)dx \quad b5 = 0.026 \quad (u4, v2)$$

$$b6 := \int_0^1 u4(x)\cdot v3(x)dx \quad b6 = 0.112 \quad (u4, u3)$$

$$w4(x) := u4(x) - b4\cdot v1(x) - b5\cdot v2(x) - b6\cdot v3(x) \qquad w4 = u4 - (u4, v1)v1 - (u4, u2)v3 - (u4, u3)v3$$

$$c4 := \int_0^1 w4(x)\cdot w4(x)dx \quad \text{Norm^2 of w4}$$

$$v4(x) := \frac{w4(x)}{\sqrt{c4}} \quad v4$$

v1, v2, v3 and v4 form an orthonormal basis.

Check orthogonality

$$\int_0^1 v4(x)\cdot v1(x)dx = 3.089\times 10^{-15}$$ Small non-zero value is round-off error

$$\int_0^1 v3(x)\cdot v2(x)dx = 0$$

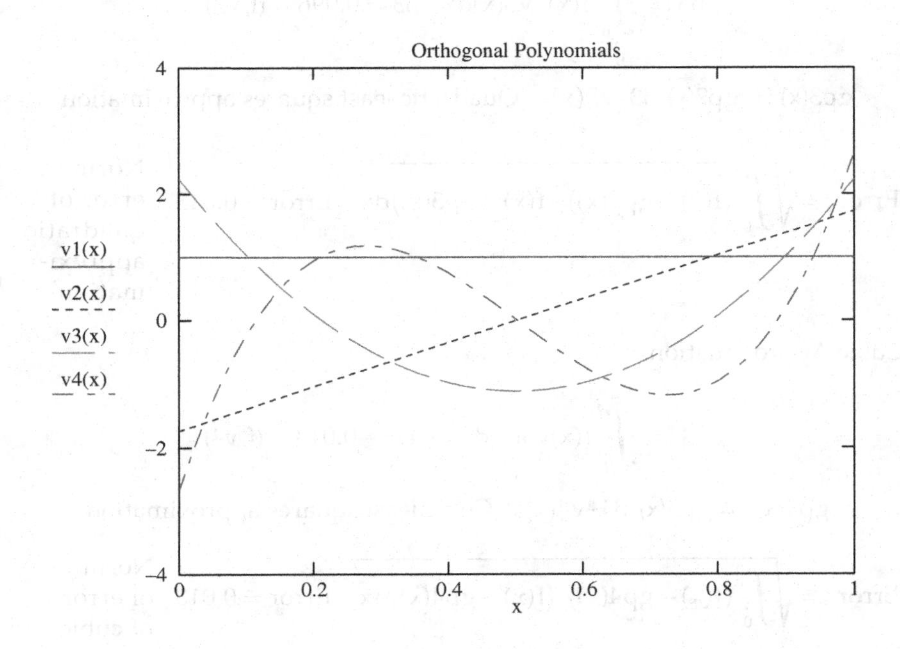

Least squares approximation

$$f(x) := x \cdot (1 - e^{1-x}) \quad \text{Given function}$$

Linear Approximation

$$\beta 1 := \int_0^1 f(x) \cdot v1(x)dx \quad \beta 1 = -0.218 \quad (f, v1)$$

$$\beta 2 := \int_0^1 f(x) \cdot v2(x)dx \quad \beta 2 = 0.02 \quad (f, v2)$$

$$gp2(x) := \beta 1 \cdot v1(x) + \beta 2 \cdot v2(x) \quad \text{Linear least squares approximation}$$

$$\text{Error} := \sqrt{\int_0^1 (f(x) - gp2(x)) \cdot (f(x) - gp2(x))dx} \quad \text{Error} = 0.097$$ Norm of error of linear approximation

Quadratic Approximation

$$\beta 3 := \int_0^1 f(x) \cdot v3(x)dx \quad \beta 3 = 0.096 \quad (f, v3)$$

$$gp3(x) := gp2(x) \;\; \beta 3 \cdot v3(x) \quad \text{Quadratic least squares approximation}$$

$$\text{Error} := \sqrt{\int_0^1 (f(x) - gp3(x)) \cdot (f(x) - gp3(x))dx} \quad \text{Error} = 0.013$$ Norm of error of quadratic approximation

Cubic Approximation

$$\beta 4 := \int_0^1 f(x)v4(x)dx \quad \beta 4 = -0.013 \quad (f, v4)$$

$$gp4(x) := gp3(x) \;\; \beta 4 \cdot v4(x) \quad \text{Cubic least squares approximation}$$

$$\text{Error} := \sqrt{\int_0^1 (f(x) - gp4(x)) \cdot (f(x) - gp4(x))dx} \quad \text{Error} = 0.013$$ Norm of error of cubic approximation

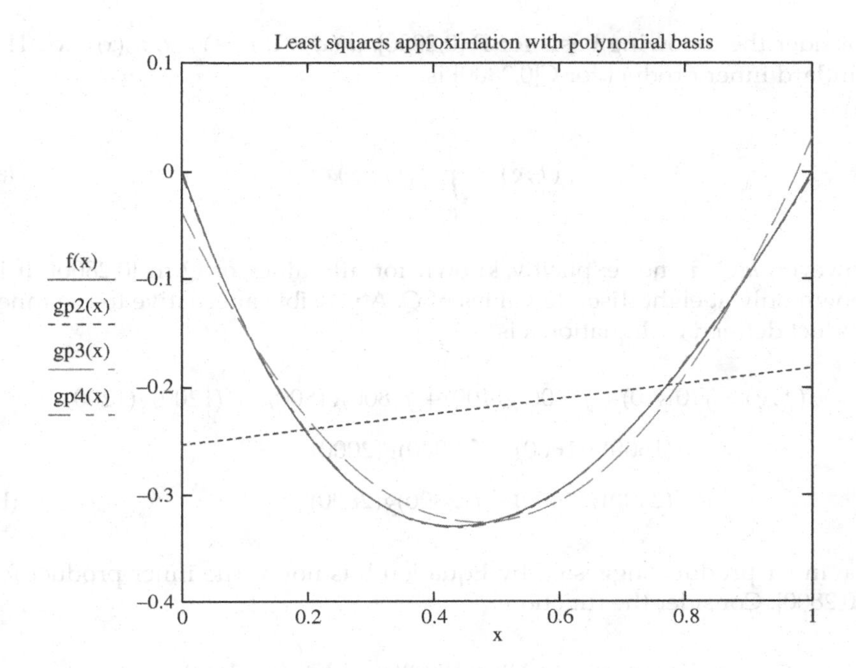

(c) The basis functions in part (b) are not orthogonal, and therefore the Gram-Schmidt procedure is used first to determine an orthogonal basis. The MATHCAD file shown above is set up so that the effect of adding each additional term is evident. The linear approximation is very poor whereas the cubic approximation is excellent.

Example 4.2 An engineer runs a test on a centrifugal pump to determine the head developed at a constant speed, but for varying flow rate. The data in Table 4.1 are obtained at 2000 rpm.

The engineer wants to use these data to generate a performance curve by fitting the "best" parabola through the data. Show how the least-squares method can be used to do this.

Solution: It is desired to fit the best parabola through eight data points. In terms of the previous discussion, it is desired to find the best approximation to an assumed continuous function, *h(Q)*, on C[0,2800] from $P^2[0,2800]$.

Table 4.1 Head of a Centrifugal Pump Measured for Varying Flow Rate

Q (gpm)	h (ft)	Q (gpm)	h (ft)
0	115	1600	112
400	122	2000	105
800	122	2400	96
1200	118	2800	87

Consider the following basis for $P^2[0,2800]$: $u_1(x)=1$, $u_2(x)=x$, $u_3(x)=x^2$. The standard inner product for C[0,2800] is

$$(f,g)=\int_0^{2800} f(x)g(x)dx \tag{a}$$

However, $h(Q)$ is not explicitly known for all values of Q in [0,2800]. It is known only at eight discrete values of Q. A possible alternative to the inner product defined by Equation a is

$$\begin{aligned}(f,g)=&f(0)g(0)+f(400)g(400)+f(800)g(800)+f(1200)g(1200)\\&+f(1600)g(1600)+f(2000)g(2000)\\&+f(2400)g(2400)+f(2800)g(2800)\end{aligned} \tag{b}$$

The inner product suggested by Equation b is not a true inner product for C[0,2800]. Consider the function

$$\begin{aligned}f(x)=&x(x-400)(x-800)(x-1200)(x-1600)\\&(x-2000)(x-2400)(x-2800)\end{aligned} \tag{c}$$

The function defined by Equation c is in C[0,2800]. Using the inner-product definition of Equation b, $(f,f)=0$. Thus, there exists $f\neq 0$ such that $(f,f)=0$. The inner product proposed by Equation b violates property (iv) of the inner-product definition. Interestingly, the inner product of Equation b is a valid inner-product definition for $P^2[0,2800]$, because the only quadratic polynomial which can be zero for $x=0$, 400, 1200,…, 2800 is the zero polynomial. Since $h(Q)$ is not known exactly, it can be assumed that $h(Q)$ is actually a member of a space of functions that cannot have eight zeros on [0,2800] without the function itself being the zero function.

An alternate interpretation is that Equation b could be obtained from Equation a through application of some quadrature formula. In this case, it would be possible to think of $h(Q)$ as a member of R^8 and Equation b as the standard inner product on R^8.

The first step is to apply the Gram-Schmidt procedure to determine an orthonormal basis for $P^2[0,2800]$ with respect to the inner product defined in Equation b. To this end,

$$\|u_1\|=(1,1)^{\frac{1}{2}}=\sqrt{8}$$

$$v_1(x)=\frac{u_1}{\|u_1\|}=\frac{1}{\sqrt{8}} \tag{d}$$

$$w_2(x) = u_2(x) - (u_2, v_1)v_1$$
$$= x - (x, 8^{-\frac{1}{2}})8^{-\frac{1}{2}}$$
$$= x - \frac{1}{8}[(0)(0) + 400(1) + (800(1) + (1200)(1) + (1600(1)$$
$$+(2000)(1) + (2400)(1) + (2800)(1)](1)$$
$$= x - 1400 \quad \text{(e)}$$

$$\|w_2(x)\| = (x - 1400, x - 1400)^{\frac{1}{2}}$$
$$= [(-1400)(-1400) + (-1000)(-1000) + (-600)(-600)$$
$$+ (-200)(-200) + (200)(200) + (600)(600)$$
$$+ (1000)(1000) + (1400)(1400)]^{\frac{1}{2}}$$
$$= 2592.3 \quad \text{(f)}$$

$$v_2(x) = \frac{w_2(x)}{\|w_2(x)\|} = \frac{x - 1400}{2592.3} = 3.86 \times 10^{-4} x - 5.41 \times 10^{-1} \quad \text{(g)}$$

$$w_3(x) = u_3(x)(u_3, v_1)v_1(u_3, v_2)v_2$$
$$= x^2(x^2, 3.86 \times 10^{-4} x - 5.41 \times 10^{-1})$$
$$(3.86 \times 10^{-4} x - 5.41 \times 10^{-1}) - (x^2, \frac{1}{\sqrt{8}})\frac{1}{\sqrt{8}}$$

$$v_3(x) = \frac{w_3(x)}{\|w_3(x)\|} = 4.82 \times 10^{-7} x^2 - 1.35 \times 10^{-3} x + 0.54 \quad \text{(h)}$$

The least-squares approximation is

$$q(x) = (h, v_1)v_1 + (h, v_2)v_2 + (h, v_3)v_3$$
$$= -7.55 \times 10^{-6} x^2 + 1.03 \times 10^{-2} x + 1.16 \times 10^2 \quad \text{(i)}$$

Equation i is illustrated in Figure 4.3 along with the data. The norm of the error is calculated as

$$\|e(x)\| = \{[q(0) - h(0)]^2 + [q(400) - h(400)]^2 + [q(800) - h(800)]^2$$
$$+ [q(1200) - h(1200)]^2 + [q(1600) - h(1600)]^2 + [q(2000) - h(2000)]^2$$
$$+ [q(2400) - h(2400)]^2 + [q(2800) - h(2800)]^2\}^{\frac{1}{2}}$$
$$= 4.91$$

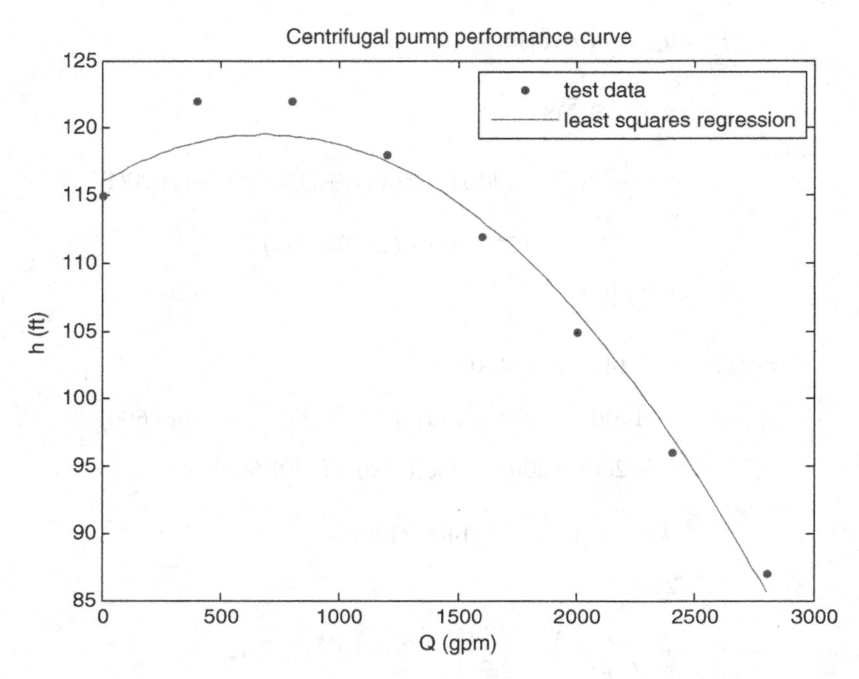

Figure 4.3 A least-squares regression on the data in Table 4.1 is performed using quadratic polynomials.

4.3 Variational solutions of operator equations

The Fourier Best-Approximation Theorem was applied in the previous section to approximate known vectors. These approximations required evaluation of inner products of the exact vector using members of a predetermined basis. However, the approximation of the solution **u** of an equation of the form **Lu** = **f**, where **L** is a linear operator, is more difficult because **u** is not known and the inner products of the exact solution with the basis vectors cannot be easily determined.

Two methods of approximating solutions of **Lu** = **f** are presented in this section. The first method, the method of least squares, provides an approximation to **f**, while the second, the Rayleigh-Ritz method, is applied to approximate solutions to equations in which the operator is positive definite and self-adjoint.

4.3.1 Method of least squares

As usual, let **L** be a linear operator whose domain is a vector space D and whose range is R. An inner product (**u**,**v**) is defined for all vectors in R. Let Q be a finite-dimensional subspace of D with a (not necessarily orthonormal) basis $q_1, q_2, \ldots, q_n$. An arbitrary element of Q is of the form

$$q = \sum_{i=1}^{n} \beta_i q_i \tag{4.11}$$

The least-squares approximation provides the approximation to q such that $\boldsymbol{Lq}$ best approximates f. Thus, it is necessary to minimize $\mu(q) = \|\boldsymbol{Lq} - f\|^2$. Using Equation 4.10 and the properties of inner products,

$$\begin{aligned}\mu(q) &= (Lq - f, Lq - f) \\ &= (Lq, Lq) - (f, Lq) - (Lq, f) + (f, f) \\ &= \left(L\left[\sum_{i=1}^{n} \beta_i q_i\right], L\left[\sum_{j=1}^{n} \beta_j q_j\right]\right) - 2\left(L\left[\sum_{i=1}^{n} \beta_i q_i\right], f\right) + (f, f)\end{aligned} \tag{4.12}$$

Using the linearity of L and further properties of inner products, Equation 4.12 reduces to

$$\mu(q) = \sum_{i=1}^{n}\sum_{j=1}^{n} \beta_i \beta_j (Lq_i, Lq_j) - 2\sum_{i=1}^{n} \beta_i (Lq_i, f) + (f, f) \tag{4.13}$$

Minimization of $\mu(q)$ is achieved by requiring that $\partial\mu/\partial\beta_k = 0$ $k = 1, 2, \ldots, n$. To this end,

$$\begin{aligned}0 &= \frac{\partial}{\partial\beta_k}\left[\sum_{i=1}^{n}\sum_{j=1}^{n} \beta_i \beta_j (Lq_i, Lq_j)\right] - 2\frac{\partial}{\partial\beta_k}\left[\sum_{i=1}^{n} \beta_i (Lq_i, f)\right] \\ &= \sum_{i=1}^{n}\sum_{j=1}^{n} (Lq_i, Lq_j)\frac{\partial}{\partial\beta_k}(\beta_i\beta_j) - 2\sum_{j=1}^{n} (Lq_i, f)\frac{\partial\beta_i}{\partial\beta_k} \\ &= \sum_{i=1}^{n}\sum_{j=1}^{n} (Lq_i, Lq_j)(\beta_j\delta_{ik} + \beta_i\delta_{jk}) - 2\sum_{i=1}^{n} (Lq_i, f)\delta_{ik} \\ &= \sum_{j=1}^{n} (Lq_j, Lq_k)\beta_j + \sum_{i=1}^{n} (Lq_i, Lq_k)\beta_i - 2(Lq_k, f) \\ &= 2\sum_{i=1}^{n} (Lq_i, Lq_k)\beta_i - 2(Lq_k, f)\end{aligned} \tag{4.14}$$

Equation 4.14 can be rearranged to yield

$$\sum_{i=1}^{n} (Lq_i, Lq_k)\beta_i = (Lq_k, f) \tag{4.15}$$

Equation 4.15 represents a set of simultaneous linear algebraic equations whose solutions are the coefficients in the linear combination. The least-squares method finds the best approximation to **u**, interpreted as the vector in Q which minimizes the norm of the difference between **Lq** and **f**.

4.3.2 *Rayleigh-Ritz method*

Recall that if L is a positive definite and self-adjoint operator with a defined inner product (**u**,**v**), then an energy inner product and a corresponding energy-inner-product-generated norm are defined by

$$(u,v)_L = (Lu,v) \tag{4.16}$$

$$\|u\|_L = (u,u)_L^{1/2} \tag{4.17}$$

The Rayleigh-Ritz method finds the best approximation to **u**, using the energy-inner-product-generated norm to measure the distance between two vectors. That is, $v(q) = \|u - q\|_L{}^2$ is the minimization functional. It should be noted that:

$$\begin{aligned} \nu(q) &= (u-q,u-q)_L \\ &= (L(u-q),u-q) \\ &= (Lu-Lq,u-q) \\ &= (Lu,u)-(Lu,q)-(Lq,u)+(Lq,q) \\ &= (Lu,u)-2(Lu,q)+(Lq,q) \end{aligned} \tag{4.18}$$

Noting that for the exact **u**, **Lu** $= f$, substituting for q from Equation 4.11, and applying properties of the inner product leads to

$$\begin{aligned} \nu(q) &= (f,u) - 2\sum_{i=1}^{n} \beta_i (f,q_i) + \sum_{i=1}^{n}\sum_{j=1}^{n} \beta_i\beta_j (Lq_i,q_j) \\ &= (f,u) - 2\sum_{i=1}^{n} \beta_i (f,q_i) + \sum_{i=1}^{n}\sum_{j=1}^{n} \beta_i\beta_{uj} (q_i,q_j)_L \end{aligned} \tag{4.19}$$

Minimization of $\nu(q)$ requires that $\partial\nu/\partial\beta_k = 0$ for all $k = 1, 2, \ldots, n$. The minimization procedure is similar to that performed in Equation 4.14 and leads to a set of simultaneous linear algebraic equations to solve for the coefficients in the linear combination that leads to the best Rayleigh-Ritz approximation for **u**. The resulting equations can be summarized by

$$\sum_{i=1}^{n} (q_i,q_k)_L \beta_i = (f,q_k) \quad k = 1,2,\ldots,n \tag{4.20}$$

If the basis vectors are chosen to be an orthonormal set with respect to the energy inner product, then $(q_i, q_k)_L = \delta_{ik}$, and Equation 4.20 simplifies to

$$\beta_k = (f, q_k) \quad k = 1,2,\ldots,n \tag{4.21}$$

Example 4.3 The differential equation for the transverse displacement, $w(x)$, of a uniform beam on an elastic foundation with vertical loading is

$$EI\frac{d^4w}{dx^4} + kw = f(x) \tag{a}$$

where E is the elastic modulus of the material from which the beam is made, I is the moment of inertia of the cross-section about the beam's neutral axis, k is the stiffness per unit length of the elastic foundation, and $f(x)$ is the load per unit length. Equation a is nondimensionalized through introduction of $x^* = x/L$, $w^* = w/L$ and definition of $f^*(x^*) = f(Lx)/F_{max}$, where F_{max} is the maximum value of the vertical load. The resulting nondimensional equation is

$$\frac{d^4w}{dx^4} + \phi w = \lambda f(x) \tag{b}$$

where, as is customary, the *s have been dropped from nondimensional variables and it is understood that all variables and functions in Equation a are nondimensional. The nondimensional parameters are defined as

$$\phi = \frac{kL^3}{EI} \tag{c}$$

$$\lambda = \frac{F_{max}L^3}{EI} \tag{d}$$

Consider a beam that is fixed at both ends and is loaded such that

$$f(x) = x - x^2 \tag{e}$$

Equation b is written in the form of Equation 4.1 with $Lw = d^4w/dx^4 + \phi w$. The domain of L, D, is the subspace of $C^4[0,1]$ such that all elements of D satisfy the boundary conditions of a fixed-fixed beam. It can be shown that L is self-adjoint and positive definite on D with respect to the standard inner product for $C^4[0,1]$.

Let $P^6[0,1]$ be the space of all polynomials of degree six or less defined on [0,1], and let Q be the intersection of D with $P^6[0,1]$. A basis for Q was determined in Example 3.7 as

$$u_1(x) = x^6 - 4x^3 + 3x^2 \tag{f}$$

$$u_2(x) = x^5 - 3x^3 + 2x^2 \quad \text{(g)}$$

$$u_3(x) = x^4 - 2x^3 + x^2 \quad \text{(h)}$$

a. Determine the exact solution for $w(x)$ for $\phi = 0$ and $\phi = 1$. Assume that $\lambda = 1$.
b. Determine the best least-squares approximation from Q for $w(x)$ for $\phi = 1$. Use $\lambda = 1$ and compare with the exact solutions.
c. Determine the best Rayleigh-Ritz approximation for $w(x)$ from Q, using the basis defined in Equation f, Equation g, and Equation h, for $\phi = 1$. Use $\lambda = 1$ and compare with the exact solutions.
d. Determine the best Rayleigh-Ritz approximation from Q for $w(x)$ using an orthonormal basis for Q.

Solution: (a) For $\phi = 0$, Equation b reduces to

$$\frac{d^4 w}{dx^4} = x - x^2 \quad \text{(i)}$$

Equation i is easily integrated four times with respect to x, leading to

$$w(x) = C_1 + C_2 x + C_3 x^2 + C_4 x^3 + \frac{x^5}{120} - \frac{x^6}{360} \quad \text{(j)}$$

The constants of integration are obtained by application of the boundary conditions

$$w(0) = 0 \Rightarrow C_1 = 0 \quad \text{(k)}$$

$$\frac{dw}{dx}(0) = 0 \Rightarrow C_2 = 0 \quad \text{(l)}$$

$$w(1) = 0 \Rightarrow C_3 + C_4 = -\frac{1}{60} \quad \text{(m)}$$

$$\frac{dw}{dx}(1) = 0 \Rightarrow 2C_3 + 3C_4 = -\frac{1}{40} \quad \text{(n)}$$

Equation m and Equation n are solved simultaneously, leading to $C_3 = -1/40$ and $C_4 = 1/120$. Thus,

$$w(x) = -\frac{x^2}{40} + \frac{x^3}{120} + \frac{x^5}{120} \frac{x^6}{360} \quad \text{(o)}$$

For $\phi=1$, Equation b becomes

$$\frac{d^4w}{dx^4}+w=x-x^2 \tag{p}$$

The homogeneous solution of Equation p is obtained by assuming a solution of the form $w_h(x)=e^{\alpha x}$, which leads to $\alpha^4+1=0$, whose solutions are $\sqrt{2}/2\pm\sqrt{2}/2(i)$ and $-\sqrt{2}/2\pm\sqrt{2}/2(i)$. The particular solution is readily obtained as $w_p(x)=x-x^2$. Hence the general solution of Equation p is

$$\begin{aligned} w(x) = {} & C_1\cos\mu x\cosh\mu x+C_2\cos\mu x\sinh\mu x \\ & +C_3\sin\mu x\cosh\mu x+C_4\sin\mu x\sinh\mu x+x-x^2 \end{aligned} \tag{q}$$

where $\mu=\sqrt{2}/2$. Application of the initial conditions leads to

$$w(0)=0\Rightarrow C_1=0 \tag{r}$$

$$\frac{dw}{dx}(0)=0\Rightarrow C_2+C_3=-\frac{1}{\mu} \tag{s}$$

$$w(1)=0\Rightarrow C_2\cos\mu\sinh\mu+C_3\sin\mu\cosh\mu+C_4\sin\mu\sinh\mu=0 \tag{t}$$

$$\begin{aligned} \frac{dw}{dx}(1)=0\Rightarrow {} & C_2(\cos\mu\cosh\mu-\sin\mu\sinh\mu) \\ & +C_3(\sin\mu\sinh\mu+\cos\mu\cosh\mu) \\ & +C_4(\sin\mu\cosh\mu+\cos\mu\sinh\mu)=\frac{1}{\mu} \end{aligned} \tag{u}$$

Equation s, Equation t, and Equation u are solved simultaneously yielding $C_2=-0.648$, $C_3=-0.766$, and $C_4=2.017$.

MATHCAD is used to provide solutions for parts (b)–(d). Since the exact solution of Equation b is an element of $P^6[0,1]$ and D for $\phi=0$, it is an element of Q. Thus, the best approximation to **u** from the elements of Q is **u** itself. Both the least-squares approximation and the Rayleigh-Ritz approximation are identical to **u** for $\phi=0$.

The results for $\phi=1$ are presented in the MATHCAD file for Example 4.3. These results indicate excellent agreement among the least-squares approximation, the Rayleigh-Ritz approximation, and the exact solution for $\phi=1$.

Example 4.3 Least squares and Rayleigh-Ritz approximations for the displacement of a fixed-fixed beam on an elastic foundation

Beam properties

$f(x) := x - x^2$ Non-dimensional load per length

$\phi := 1$ Ratio of elastic stress to bending stress

$\lambda := 1$ Ratio of stress due to applied load to bending stress

Basis functions

$u1(x) := x^6 - 4x^3 + 3x^2$

$u2(x) := x^5 - 3x^3 + 2x^2$ Basis functions

$u3(x) := x^4 - 2x^3 + x^2$

(a) **Least squares method**

$$Lu1(x) := \frac{d^4}{dx^4}u1(x) + \phi \cdot u1(x)$$

$$Lu2(x) := \frac{d^4}{dx^4}u2(x) + \phi \cdot u2(x) \quad \text{Operator acting on basis functions}$$

$$Lu3(x) := \frac{d^4}{dx^4}u3(x) + \phi \cdot u3(x)$$

Evaluation of inner products to calculate coefficient matrix

$$A(i,j) = (Lui,Luj)$$

$$A_{1,1} := \int_0^1 Lu1(x) \cdot Lu1(x)dx \quad A_{1,2} := \int_0^1 Lu1(x) \cdot Lu2(x)dx$$

$$A_{1,3} := \int_0^1 Lu1(x) \cdot Lu3(x)dx \quad A_{2,2} := \int_0^1 Lu2(x) \cdot Lu2(x)dx$$

$$A_{2,3} := \int_0^1 Lu2(x) \cdot Lu3(x)dx \quad A_{3,3} := \int_0^1 Lu3(x) \cdot Lu3(x)dx$$

Symmetry is used to calculate remaining elements

$$A_{2,1} := A_{1,2} \quad A_{3,1} := A_{1,3} \quad A_{3,2} := A_{2,3}$$

Coefficient matrix of Eq. (4.15)

$$A = \begin{pmatrix} 2.592 \times 10^4 & 1.08 \times 10^4 & 2.88 \times 10^3 \\ 1.08 \times 10^4 & 4.8 \times 10^3 & 1.44 \times 10^3 \\ 2.88 \times 10^3 & 1.44 \times 10^3 & 576 \end{pmatrix}$$

Calculation of right-hand side vector of Eq. (4.15)

$$b(i) = (f, Lui)$$

$$b_1 := \int_0^1 f(x) \cdot Lu1(x)dx \quad b_2 := \int_0^1 f(x) \cdot Lu2(x)dx \quad b_3 := \int_0^1 Lu3(x) \cdot f(x)dx$$

$$b = \begin{pmatrix} 18 \\ 10 \\ 4 \end{pmatrix}$$

Solution of simultaneous equations to determine coefficients of least squares approximation

Solution is simply obtained by multiplying the inverse of A by b leading to

$$\beta := A^{-1} \cdot b$$

$$\beta = \begin{pmatrix} -2.778 \times 10^{-3} \\ 8.333 \times 10^{-3} \\ -1.341 \times 10^{-10} \end{pmatrix}$$

The resulting least squares approximation is

$$w(x) := \beta_1 \cdot u1(x) + \beta_2 \cdot u2(x) + \beta_3 \cdot u3(x)$$

$$x := 0, .02 .. 1$$

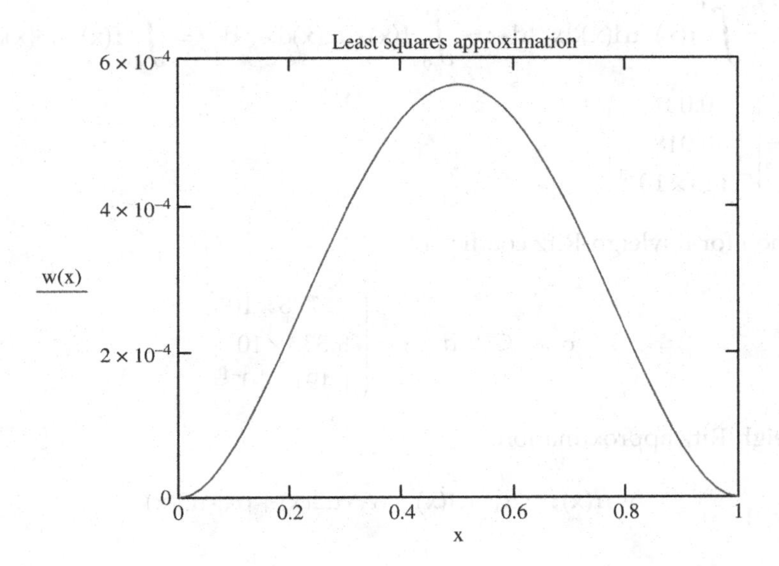

(b) **Rayleigh-Ritz method**

Energy inner product evaluations

$$C(i,j) = (ui,uj)L$$

$$C_{1,1} := \int_0^1 u1(x) \cdot Lu1(x)dx \quad C_{1,2} := \int_0^1 Lu1(x) \cdot u2(x)dx$$

$$C_{1,3} := \int_0^1 Lu1(x) \cdot u3(x)dx \quad C_{2,2} := \int_0^1 Lu2(x) \cdot u2(x)dx$$

$$C_{2,3} := \int_0^1 Lu2(x) \cdot u3(x)dx \quad C_{3,3} := \int_0^1 Lu3(x) \cdot u3(x)dx$$

$$C_{2,1} := \int_0^1 Lu2(x) \cdot u1(x)dx \quad C_{3,1} := \int_0^1 Lu3(x) \cdot u1(x)dx$$

$$C_{3,2} := \int_0^1 Lu3(x) \cdot u2(x)dx$$

$$C = \begin{pmatrix} 16 & 9 & 3.429 \\ 9 & 5.143 & 2 \\ 3.429 & 2 & 0.8 \end{pmatrix}$$ Symmetry of C is due to self-adjointness of L

Evaluation of right-hand side vector of Eq. (4.20)

$d(i) = (f,ui)$

$$d_1 := \int_0^1 f(x) \cdot u1(x)dx \quad d_2 := \int_0^1 f(x) \cdot u2(x)dx \quad d_3 := \int_0^1 f(x) \cdot u3(x)dx$$

$$d = \begin{pmatrix} 0.031 \\ 0.018 \\ 7.143 \times 10^{-3} \end{pmatrix}$$

Solution for Rayleigh-Ritz coefficients

$$\alpha := C^{-1} \cdot d \quad \alpha = \begin{pmatrix} -2.778 \times 10^{-3} \\ 8.333 \times 10^{-3} \\ 1.491 \times 10^{-10} \end{pmatrix}$$

Rayleigh-Ritz approximation

$$wrr(x) := \alpha_1 \cdot u1(x) + \alpha_2 \cdot u2(x) + \alpha_3 \cdot u3(x)$$

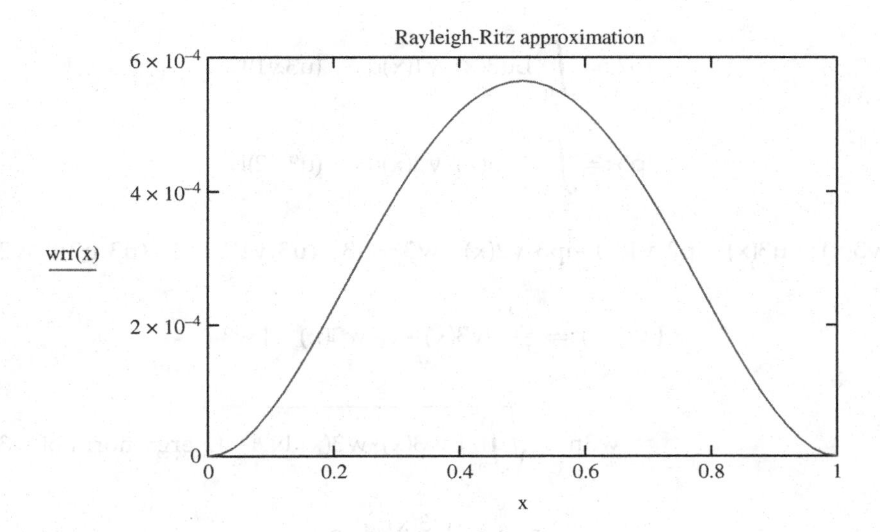

(c) **Rayleigh-Ritz method using on orthonormal basis with respect to energy inner product**

The Gram Schmidt process is used to develop an orthonormal basis with respect to the energy inner product which spans the same space as u1, u2, and u3.

$$w1(x) := u1(x)$$

$$w1n := \sqrt{\int_0^1 Lu1(x) \cdot u1(x)dx} \qquad \text{Energy norm of w1}$$

$$v1(x) := \frac{w1(x)}{w1n} \qquad v1$$

$$ip1 := \int_0^1 Lu2(x) \cdot v1(x)dx \qquad (u2,v1)L$$

$$w2(x) := u2(x) - ip1 \cdot v1(x) \qquad w2(x){=}u2\text{-}{*}u2,v1)L{*}v1$$

$$Lw2(x) := \frac{d^4}{dx^4}(w2(x)) + \phi \cdot w2(x) \qquad L(w2)$$

$$w2n(x) := \sqrt{\int_0^1 Lw2(x) \cdot w2(x)dx} \qquad \text{Energy norm of w2}$$

$$v2(x) := \frac{w2(x)}{w2n} \qquad v2$$

$$ip2 := \int_0^1 Lu3(x) \cdot v1(x)dx \quad (u3,v1)L$$

$$ip3 := \int_0^1 Lu3(x) \cdot v2(x)dx \quad (u3,v2)L$$

$$w3(x) := u3(x) - ip2 \cdot v1(x) - ip3 \cdot v2(x) \quad w3 = u3 - (u3,v1)L * v1 - (u3,u2)L * v2$$

$$Lw3(x) := \frac{d^4}{dx^4} w3(x) + \phi \cdot w3(x) \quad Lw3$$

$$w3n := \sqrt{\int_0^1 Lw3(x) \cdot w3(x)dx} \quad \text{Energy norm of w3}$$

$$v3(x) := \frac{w3(x)}{w3n} \quad v3$$

Inner product evaluation for coefficients

$$\delta1 := \int_0^1 f(x) \cdot v1(x)dx \quad (f,v1)$$

$$\delta2 := \int_0^1 f(x) \cdot v2(x)dx \quad (f,v2)$$

$$\delta3 := \int_0^1 f(x) \cdot v3(x)dx \quad (f,v3)$$

Rayleigh-Ritz approximation

$$wgs(x) := \delta1 \cdot v1(x) + \delta2 \cdot v2(x) + \delta3 \cdot v3(x)$$

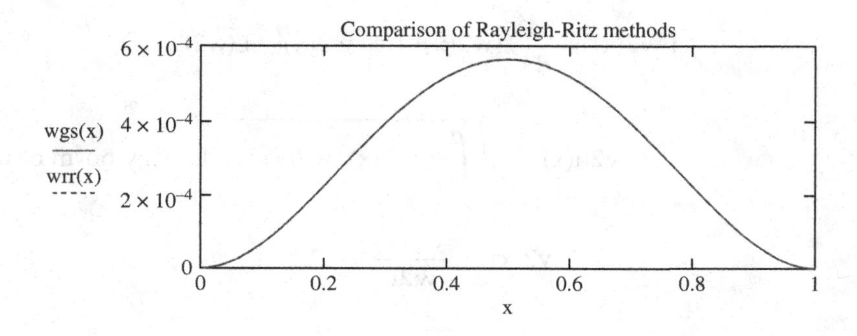

Exact Solution ($\phi = 1$)

$$\mu := \frac{\sqrt{2}}{2} \quad \mu = 0.707$$

$ct := \cos(\mu) \quad st := \sin(\mu)$ Definitions made for convenience

$ch := \cosh(\mu) \quad sh := \sinh(\mu)$

$$E := \begin{pmatrix} 1 & 1 & 0 \\ ct \cdot sh & st \cdot ch & st \cdot sh \\ ct \cdot ch - st \cdot sh & st \cdot sh + ct \cdot ch & st \cdot ch + ct \cdot sh \end{pmatrix}$$

$$E = \begin{pmatrix} 1 & 1 & 0 \\ 0.584 & 0.819 & 0.499 \\ 0.46 & 1.457 & 1.402 \end{pmatrix}$$

$$e := \begin{pmatrix} -1 \\ 0 \\ 1 \end{pmatrix} \cdot \frac{1}{\mu}$$

$$CEX := E^{-1} \cdot e \quad CEX = \begin{pmatrix} -0.648 \\ -0.766 \\ 2.017 \end{pmatrix}$$

$$w1(x) := CEX_1 \cdot \cos(\mu \cdot x) \cdot \sinh(\mu \cdot x) + CEX_2 \cdot \sin(\mu \cdot x) \cdot \cosh(\mu \cdot x)$$

$$wexact(x) := w1(x) + CEX_3 \cdot \sin(\mu \cdot x) \cdot \sinh(\mu \cdot x) + x - x^2$$

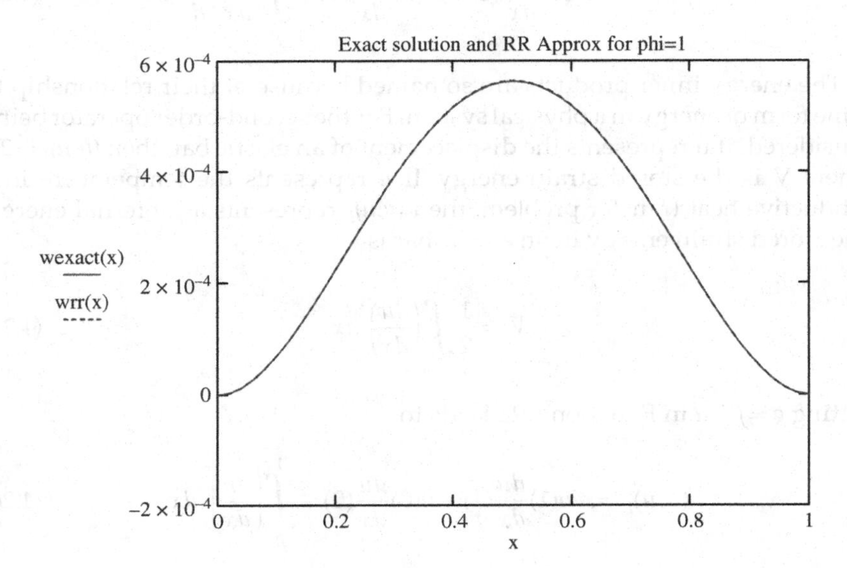

The least-squares approximation and the Rayleigh-Ritz approximation are often close to one another, as illustrated in Example 4.3, when the approximations are drawn from the same vector space. However, as shown below, the Rayleigh-Ritz method is more flexible regarding the vector spaces which can be used to generate an approximation.

Consider the second-order differential operator defined by $Lf=-d^2f/dx^2$, where f is an arbitrary element of D, the domain of the operator. Consider the standard inner product on D defined by $(f,g)=\int_0^1 f(x)g(x)dx$. Assuming that D is defined such that L f L is self-adjoint and positive definite with respect to the standard inner product, an energy inner product is defined by $(f,g)_L=(Lf,g)=\int_0^1(-d^2f/dx^2)gdx$. Integration by parts on the energy inner product leads to

$$(Lf,g)_L=-g(1)\frac{df}{dx}(1)+g(0)\frac{df}{dx}(0)+\int_0^1\frac{df}{dx}\frac{dg}{dx}dx \tag{4.22}$$

Let $q_1,q_2,\ldots,q_n$ be a set of basis functions chosen for a Rayleigh-Ritz approximation for the solution $Lu=f$. The approximation requires evaluation of energy inner products of the form $(q_i,q_j)_L$, and since L is self-adjoint, $(q_i,q_j)_L=(q_i,q_j)_L$. To this end, using Equation 4.22,

$$(q_i,q_j)_L=-q_j(1)\frac{dq_i}{dx}(1)+q_j(0)\frac{dq_j}{dx}(0)+\int_0^1\frac{dq_i}{dx}\frac{dq_j}{dx}dx \tag{4.23}$$

$$(q_j,q_i)_L=-q_i(1)\frac{dq_j}{dx}(1)+q_i(0)\frac{dq_j}{dx}(0)+\int_0^1\frac{dq_j}{dx}\frac{dq_i}{dx}dx \tag{4.24}$$

The energy inner products are so named because of their relationship to some form of energy in a physical system. For the second-order operator being considered, if u represents the displacement of an elastic bar, then $(u,u)_L=2V$, where V is the stored strain energy. If u represents the temperature in a conductive heat transfer problem, then $(u,u)_L$ represents an internal energy. The stored strain energy in an elastic bar is

$$V=\frac{1}{2}\int_0^1\left(\frac{du}{dx}\right)^2dx \tag{4.25}$$

Setting $g=f=u$ in Equation 4.22 leads to

$$(u,u)_L=-u(1)\frac{du}{dx}(1)+u(0)\frac{du}{dx}(0)+\int_0^1\left(\frac{du}{dx}\right)^2dx \tag{4.26}$$

Comparison of Equation 4.25 and Equation 4.26 shows that

$$(u,u)_L = 2V \quad \text{if} - u(1)\frac{du}{dx}(1) + u(0)\frac{du}{dx}(0) = 0$$

Consider a bar with a discrete spring of nondimensional stiffness k attached at $x=1$. When the end of the bar is displaced, a potential energy develops in the spring. In this case, the total potential energy of the system is the sum of the strain energy and the potential energy in the spring,

$$V = \frac{1}{2}\int_0^1 \left(\frac{du}{dx}\right)^2 dx + \frac{1}{2}ku(1)^2 \tag{4.27}$$

Noting that the boundary condition for the displacement of the bar with a discrete spring at its end is $du/dx\,(1) = -ku(1)$ Equation 4.26 becomes

$$(u,u)_L = u(0)\frac{du}{dx}(0) + ku(1)^2 + \int_0^1 \left(\frac{du}{dx}\right)^2 dx \tag{4.28}$$

and comparison of Equation 4.27 and Equation 4.28 leads to $u(0)$ or $du/dx\,(0) = 0$.

The boundary condition applied to the end of the bar at $x=1$, whether the bar is free or has an attached spring, is a natural boundary condition. The boundary condition applied to the fixed end of the bar is a geometric boundary condition.

The above discussion leads to the consideration of several points regarding the application of the Rayleigh-Ritz method. The Rayleigh-Ritz method may be interpreted as a method which minimizes the difference between the system's energy for the exact solution and the energy for the approximation. Interpreted in this light, using Equations 4.22–4.28, it appears that the restrictions on the choices of the basis functions can be relaxed. Evaluation of energy inner products using Equation 4.22 only requires that the basis functions be differentiable, not twice differentiable as required by the exact solution of the differential equation. Furthermore, realizing that, for the cases discussed, the potential energy of the system is independent of the value of $du/dx(1)$, it is not necessary that the basis functions satisfy the boundary condition at $x=1$ as long as the potential energy is still represented by Equation 4.27. Under these conditions, an alternate representation of the energy inner product is

$$(f,g)_L = kf(1)g(1) + \int_0^1 \frac{df}{dx}\frac{dg}{dx}dx \tag{4.29}$$

The above suggests that the requirements for the basis functions used in a Rayleigh-Ritz approximation can be relaxed from requiring that all basis functions be in D, the domain of L. A basis chosen from D is called a set of comparison functions. The functions used in Example 4.3 are a set of comparison functions. Use of Equation 4.29 for the energy inner product requires only that the basis functions be differentiable, without having to satisfy the natural boundary condition at $x=1$. The chosen basis functions must still satisfy the geometric boundary condition. Functions that satisfy such conditions are called admissible functions.

The general requirements for admissible functions are that (1) admissible functions satisfy all geometric boundary conditions, and (2) admissible functions must have a level of differentiability required by the energy formulation. Admissible functions need not satisfy natural boundary conditions, but when using admissible functions that do not satisfy natural boundary conditions, the energy inner product must include, if appropriate, terms to account for nonzero energy from the natural boundary conditions.

If A is the space of admissible functions, then Equation 4.29 is a valid definition of an inner product for elements of A. That is, the inner product defined in Equation 4.29 satisfies the four properties required for inner products specified in Definition 3.7. The inner product of Equation 4.29 is then used for determination of the elements of the coefficient matrix, $(q_i, q_j)_L$, used in the Rayleigh-Ritz approximation.

Example 4.4 The differential equation governing the displacement, $w(x)$, of a beam of variable cross-sectional moment of inertia $I(x)$ due to a uniform distributed load per unit length, $f(x)$, is

$$\frac{d^2}{dx^2}\left(EI\frac{d^2w}{dx^2}\right)=f(x) \tag{a}$$

Consider a beam fixed at $x=0$ with a spring of stiffness k attached at $x=L$. (a) Determine the appropriate formulation of the energy inner product to use for a Rayleigh-Ritz approximation. (b) Find the best approximation to $w(x)$ from the space of polynomials of degree four or less that satisfy only the geometric boundary conditions. Use

$$f(x)=F_{\max}\left[x/L-(x/L)^2\right],$$

$$I(x)=I_0\left(1+0.1(x/L)^2\right) \text{ with } F_{\max}=1000\text{ N},\ I_0=1\times10^{-5}\text{ m}^4,$$

$$E=200\times10^9\text{ N/m}^2,\ k=1\times10^6\text{ N/m and } L=2\text{m}$$

Solution The boundary conditions at $x=0$ are geometric boundary conditions:

$$w(0)=0 \tag{b}$$

$$\frac{dw}{dx}(0)=0 \tag{c}$$

The boundary conditions at $x=L$ are natural boundary conditions given by

$$EI(L)\frac{d^2w}{dx^2}(L)=0 \tag{d}$$

$$\frac{d}{dx}\left(EI\frac{d^2w}{dx^2}\right)(L)=kw(L) \tag{e}$$

The energy inner product is defined by

$$(f,g)_L=\int_0^L \frac{d^2}{dx^2}\left(EI\frac{d^2f}{dx^2}\right)g\,dx \tag{f}$$

Using integration by parts twice on Equation f leads to

$$\begin{aligned}(f,g)_L={}&g(L)\frac{d}{dx}\left(EI\frac{d^2f}{dx^2}\right)(L)-g(0)\frac{d}{dx}\left(EI\frac{d^2f}{dx^2}\right)(0)\\&-\frac{dg}{dx}(L)EI(L)\frac{d^2w}{dx^2}(L)+\frac{dg}{dx}(0)EI(0)\frac{d^2w}{dx^2}(0)\\&+\int_0^1 EI\frac{d^2f}{dx^2}\frac{d^2g}{dx^2}dx\end{aligned} \tag{g}$$

Requiring $f(x)$ and $g(x)$ to satisfy the geometric boundary conditions reduces Equation g to

$$\begin{aligned}(f,g)_L={}&g(L)\frac{d}{dx}\left(EI\frac{d^2f}{dx^2}\right)(L)-\frac{dg}{dx}(L)EI(L)\frac{d^2f}{dx^2}(L)\\&+\int_0^1 EI\frac{d^2f}{dx^2}\frac{d^2g}{dx^2}dx\end{aligned} \tag{h}$$

Satisfaction of Equation e for all f and g in D leads to

$$(f,g)_L=kf(L)g(L)-\frac{dg}{dx}(L)EI(L)\frac{d^2f}{dx^2}(L)+\int_0^1 EI\frac{d^2f}{dx^2}\frac{d^2g}{dx^2}dx \tag{i}$$

If only comparison functions are used in the Rayleigh-Ritz approximation, then Equation i reduces to

$$(f,g)_L = kf(L)g(L) + \int_0^1 EI \frac{d^2 f}{dx^2}\frac{d^2 g}{dx^2}dx \qquad \text{(j)}$$

Equation j is the form of the inner product that should be used when admissible functions are used. With admissible functions that do not satisfy the natural boundary conditions, Equation d and Equation e, are used, then Equation j is used for inner-product evaluation, because it is clear that this definition is a valid definition of the inner product on the space of admissible functions.

(b) The subspace of the space of admissible functions from which a Rayleigh-Ritz approximation is sought is the intersection of $P^4[0,L]$ with the space of functions that satisfy the geometric boundary conditions $f(0)=0$ and $df/dx\ (0)=0$. It is not hard to show that $u_1(x)=x^2$, $u_2(x)=x^3$ and $u_3(x)=x^4$ form a basis for this space. Hence the form of the Rayleigh-Ritz approximation is

$$q(x) = \alpha_1 x^2 + \alpha_2 x^3 + \alpha_3 x^4 \qquad \text{(g)}$$

A MATHCAD file showing the details of the solution is presented below.

Example 4.4
Parameters

$$E := 200 \cdot 10^9$$

$$L := 2$$

$$I0 := 1 \cdot 10^{-5}$$

$$k := 1 \cdot 10^7$$

$$Fmax := 100$$

Known functions

$$f(x) := Fmax \cdot \left(\frac{x}{L} - \frac{x^2}{L^2}\right)$$

$$I(x) := I0 \cdot \left(1 + 0.1 \cdot \frac{x^2}{L^2}\right)$$

Basis functions

$$u1(x) := x^2$$
$$u2(x) := x^3$$
$$u3(x) := x^4$$

$$u(x) := \begin{pmatrix} u1(x) \\ u2(x) \\ u3(x) \end{pmatrix}$$

Energy inner products

$$i := 1..3$$
$$j := 1..3$$

$$Q_{i,j} := k \cdot u(L)_i \cdot u(L)_j + \int_0^L E \cdot I(x) \cdot \frac{d^2}{dx^2} u(x)_i \cdot \frac{d^2}{dx^2} u(x)_j \, dx$$

$$Q = \begin{pmatrix} 1.765 \times 10^8 & 3.704 \times 10^8 & 7.757 \times 10^8 \\ 3.704 \times 10^8 & 8.435 \times 10^8 & 1.894 \times 10^9 \\ 7.757 \times 10^8 & 1.894 \times 10^9 & 4.535 \times 10^9 \end{pmatrix}$$

$$f_i := \int_0^L f(x) u(x)_i dx$$

$$f = \begin{pmatrix} 40 \\ 53.333 \\ 76.19 \end{pmatrix}$$

Solution

$$\alpha := Q^{-1} \cdot f$$

$$\alpha = \begin{pmatrix} 2.983 \times 10^{-6} \\ -2.24 \times 10^{-6} \\ 4.422 \times 10^{-7} \end{pmatrix}$$

Plot of solution

$$w(x) := \sum_{i=1}^{3} (\alpha_i \cdot u(x)_i)$$

$$x := 0, .01 .. 1$$

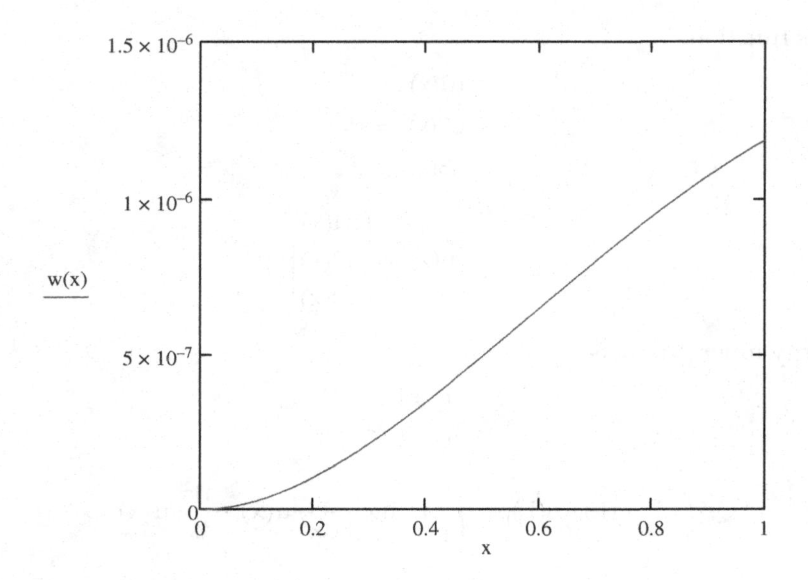

4.4 Finite-element method

The possibility of using admissible functions as basis functions for the application of the Rayleigh-Ritz method is suggested by Equation 4.29, which results from application of integration by parts to the energy inner product, $(f, g)_L$. This formulation illustrates that $(w, w)_L$ is proportional to the total potential energy if w is the displacement function, and that use of the energy inner product in this form requires a lower order of differentiability than the differential operator. Thus, admissible basis functions are required to satisfy only geometric boundary conditions and have a level of differentiability specified by the energy inner product when written in the form of Equation 4.29.

The relaxation of the requirement that the basis functions satisfy the level of differentiability specified by the differential operator enables piecewise-defined basis functions with a lower level of differentiability to be chosen. Such basis functions are called interpolating splines. A Rayleigh-Ritz method using splines chosen such that all geometric boundary conditions are satisfied and an appropriate level of differentiability is satisfied is called a finite-element method.

A set of piecewise linear splines which can be used as basis functions for the Rayleigh-Ritz method applied to second-order differential operators is illustrated in Figure 4.4. The interval $0 \leq x \leq 1$ is divided into n subintervals (a finite number), each of length ℓ. Thus, $n = 1/\ell$. The mathematical forms of the $n+1$ basis functions are

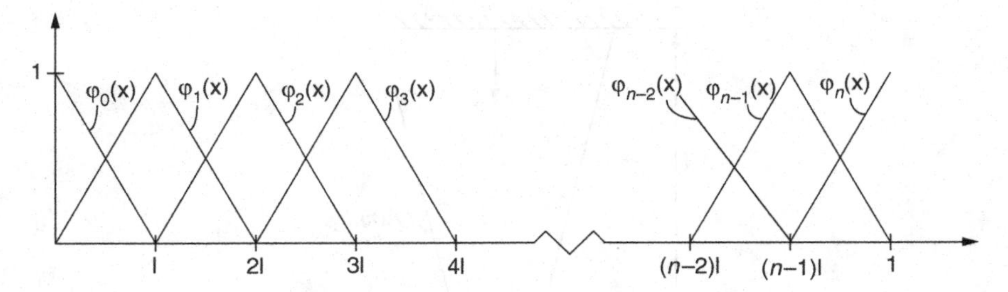

Figure 4.4 Piecewise-defined basis elements for use in the finite-element method to approximate the solution of a second-order differential equation.

$$
\begin{aligned}
\phi_0(x) &= \left(1-\frac{x}{\ell}\right)[1-u(x-\ell)] \\
\phi_1(x) &= \left(\frac{x}{\ell}\right)[u(x)-u(x-\ell)]+\left(2-\frac{x}{\ell}\right)[u(x-\ell)-u(x-2\ell)] \\
\phi_2(x) &= \phi_1(x-\ell) \\
&\vdots \\
\phi_k &= \phi_1(x-k\ell) \\
&\vdots \\
\phi_n(x) &= \left[\frac{x}{\ell}+(1-n)\right]u[x-(n-1)\ell]
\end{aligned}
\tag{4.30}
$$

where $u(z)$ is the unit step function defined such that

$$
u(z)=\begin{cases} 0 & z<0 \\ 1 & z>0 \end{cases} \tag{4.31}
$$

The Rayleigh-Ritz formulation is

$$
u=\sum_{i=0}^{n} c_i\phi_i(x) \tag{4.32}
$$

The geometric boundary condition $u(0)=u_0$ is applied by choosing $C_0=u_0$, and the boundary condition $u(1)=u_1$ is applied by requiring $C_n=u_1$.

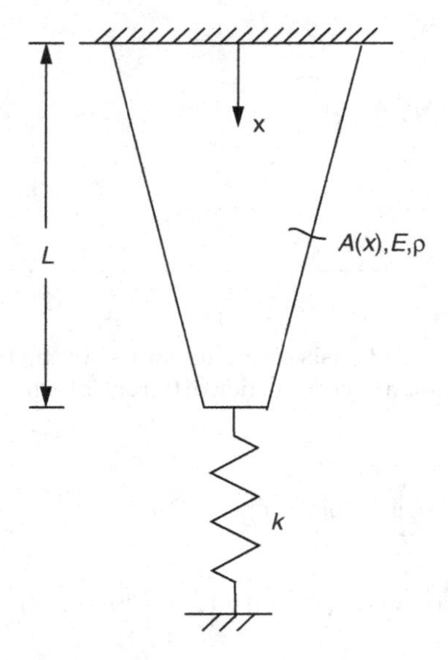

Figure 4.5 System of Example 4.5.

Example 4.5 The differential equation governing the vertical displacement of the non-uniform bar of Figure 4.5 under its own weight is

$$\frac{d}{dx}\left(EA\frac{du}{dx}\right)=\int_{0}^{x}\rho g A(\xi)d\xi \tag{a}$$

The bar is fixed at $x=0$ and attached to a linear spring of stiffness k at $x=L$. The resulting boundary conditions are

$$u(0)=0 \tag{b}$$

$$EA(L)\frac{du}{dx}(L)+ku(L)=0 \tag{c}$$

Nondimensional variables are introduced by

$$x^{*}=\frac{x}{L} \tag{d}$$

$$w^{*}=\frac{w}{L} \tag{e}$$

$$\alpha(x^{*})=\frac{A}{A_0} \tag{f}$$

Substitution of Equation d, Equation e, and Equation f into Equation a, assuming that E is constant, rearranging, and dropping the*'s leads to

$$\frac{d}{dx}\left(\alpha(x)\frac{du}{dx}\right)=\mu\int_0^x \alpha(\xi)d\xi \tag{g}$$

where

$$\mu=\frac{\rho g L^2}{E} \tag{h}$$

The nondimensional boundary conditions are of the form

$$u(0)=0 \tag{i}$$

$$\frac{du}{dx}(1)+\nu u(1)=0 \tag{j}$$

where

$$\nu=\frac{kL}{EA(L)} \tag{k}$$

Develop a finite-element approximation for the vertical displacement of a bar of length 30.5 m, made from a material of elastic modulus 200×10^9 N/m^2 and mass density 7000 kg/m^3, if (a) the bar is uniform with area 2×10^{-4} m^2, and (b) the area varies linearly according to $A(x)=2 \times 10^{-4}\ (1+.02x)$. Divide the interval from 0 to 1 into five segments of equal length.

Solution (a) If A is constant, then $\alpha=1$. For the numerical values given, the nondimensional parameters are $\mu=3.194 \times 10^{-4}$ and $\mathrm{v}=0.153$. The right-hand side of Equation g is simply μx. The piecewise-defined basis functions are illustrated in the MATHCAD file which follows the Rayleigh-Ritz approximation is

$$u(x)=\sum_{i=0}^{5} a_i\phi_i(x) \tag{l}$$

The geometric boundary condition at $x=0$ is imposed by setting $a_0=0$. The natural boundary condition at $x=1$ is not satisfied by the basis functions. Instead, the potential energy of the spring is included in the formulation of the energy inner product. The piecewise-defined set of functions can be used as a basis when the energy inner product is written in the form

$$(f,g)=\int_0^1 \alpha(x)\frac{df}{dx}\frac{dg}{dx}dx+\nu g(1)g(1) \tag{m}$$

A MATHCAD file is developed to obtain the solution. The file is written in general so that only the description of the variation of the area with x needs to be changed. The MATHCAD file set up for part (b) is illustrated below.

Example 4.5
Finite element approximation for displacement of hanging bar with attached spring. The interval from x = 0 to x = 1 is divided into five elements of equal length leading to definition of the basis functions as

$$\phi0(x) := (1 - 5 \cdot x)(\Phi(x) - \Phi(x - 0.2))$$

$$\phi1(x) := 5 \cdot x \cdot (\Phi(x) - \Phi(x - 0.2)) + (2 - 5 \cdot x) \cdot (\Phi(x - 0.4))$$

$$\phi2(x) := \phi1(x - 0.2)$$

$$\phi3(x) := \phi1(x - 0.4)$$

$$\phi4(x) := \phi1(x - 0.6)$$

$$\Phi5(x) := (5 \cdot x - 4) \cdot \Phi(x - 0.8)$$

where $\Phi(x)$ is the Heaviside function defined as 0 for $x < 0$ and 1 for $x > 0$. it is also called the unit step function $u(x)$

$x := 0, .01 .. 1$

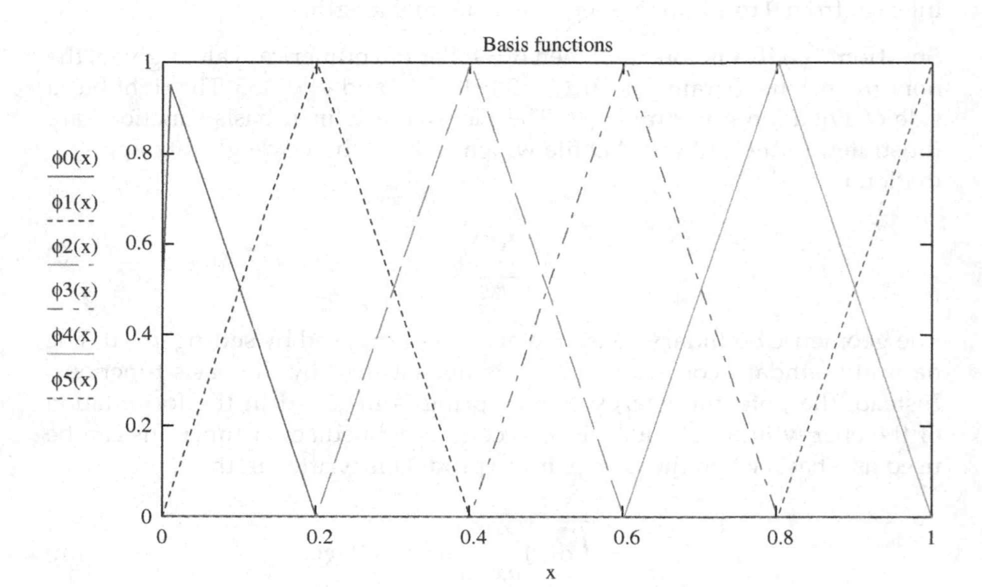

Dimensional Parameters

$\rho := 7000$ Mass density in kg/m^3

$g := 9.81$ Acceleration due to gravity in m/s^2

$L := 30.5$ Length of bar in m

$k := 200000$ Stiffness of spring in N/m

$E := 200 \cdot 10^9$ Elastic modulus in N/m^2

$A0 := 2 \cdot 10^{-4}$ Area at x = 0 in m^2

Variation of area

$\text{Area}(xd) := A0 \cdot (1 + 0.00xd)$ Variation of area over length of bar, xd is the dimensional coordinate along the length of the bar

$AL := \text{Area}(L)$ Calculated area at x = L in m^2

$AL = 2 \times 10^{-4}$

Nondimensional Parameters

$$\mu := \frac{\rho \cdot g \cdot L^2}{E} \qquad \mu = 3.194 \times 10^{-4}$$

$$\nu := \frac{k \cdot L}{E \cdot AL} \qquad \nu = 0.153$$

Nondimensional functions

$$\alpha(x) := \frac{\text{Area}(x \cdot L)}{A0}$$ Nondimensional area in terms of nondimensional variable x

$$f(x) := \mu \cdot \int_0^x \alpha(y)dy$$ Nondimensional loading due to gravity

Inner product evaluations

$$A_{1,1} := \int_0^1 \left(\frac{d}{dx}\phi 1(x)\right) \cdot \left(\frac{d}{dx}\phi 1(x)\right) \cdot \alpha(x)dx + \nu \cdot \phi 1(1) \cdot \phi 1(1) \quad A_{1,1} = 10$$

$$A_{1,2} := \int_0^1 \left(\frac{d}{dx}\phi 1(x)\right) \cdot \left(\frac{d}{dx}\phi 2(x)\right) \cdot \alpha(x)dx + \nu \cdot \phi 1(1) \cdot \phi 2(1)$$

$$A_{1,3} := \int_0^1 \left(\frac{d}{dx}\phi 1(x)\right) \cdot \left(\frac{d}{dx}\phi 3(x)\right) \cdot \alpha(x)dx + \nu \cdot \phi 1(1) \cdot \phi 3(1)$$

$$A_{1,4} := \int_0^1 \left(\frac{d}{dx}\phi 1(x)\right) \cdot \left(\frac{d}{dx}\phi 4(x)\right) \cdot \alpha(x)dx + \nu \cdot \phi 1(1) \cdot \phi 4(1)$$

$$A_{1,5} := \int_0^1 \left(\frac{d}{dx}\phi 1(x)\right) \cdot \left(\frac{d}{dx}\phi 5(x)\right) \cdot \alpha(x)dx + \nu \cdot \phi 1(1) \cdot \phi 5(1)$$

$$A_{2,2} := \int_0^1 \left(\frac{d}{dx}\phi2(x)\right)\cdot\left(\frac{d}{dx}\phi2(x)\right)\cdot\alpha(x)dx + \nu\cdot\phi2(1)\cdot\phi2(1)$$

$$A_{2,3} := \int_0^1 \left(\frac{d}{dx}\phi2(x)\right)\cdot\left(\frac{d}{dx}\phi3(x)\right)\cdot\alpha(x)dx + \nu\cdot\phi3(1)\cdot\phi2(1)$$

$$A_{2,4} := \int_0^1 \left(\frac{d}{dx}\phi2(x)\right)\cdot\left(\frac{d}{dx}\phi4(x)\right)\cdot\alpha(x)dx + \nu\cdot\phi4(1)\cdot\phi2(1)$$

$$A_{2,5} := \int_0^1 \left(\frac{d}{dx}\phi2(x)\right)\cdot\left(\frac{d}{dx}\phi5(x)\right)\cdot\alpha(x)dx + \nu\cdot\phi5(1)\cdot\phi2(1)$$

$$A_{3,3} := \int_0^1 \left(\frac{d}{dx}\phi3(x)\right)\cdot\left(\frac{d}{dx}\phi3(x)\right)\cdot\alpha(x)dx + \nu\cdot\phi3(1)\cdot\phi3(1)$$

$$A_{3,4} := \int_0^1 \left(\frac{d}{dx}\phi3(x)\right)\cdot\left(\frac{d}{dx}\phi4(x)\right)\cdot\alpha(x)dx + \nu\cdot\phi3(1)\cdot\phi4(1)$$

$$A_{3,5} := \int_0^1 \left(\frac{d}{dx}\phi3(x)\right)\cdot\left(\frac{d}{dx}\phi5(x)\right)\cdot\alpha(x)dx + \nu\cdot\phi3(1)\cdot\phi5(1)$$

$$A_{4,4} := \int_0^1 \left(\frac{d}{dx}\phi4(x)\right)\cdot\left(\frac{d}{dx}\phi4(x)\right)\cdot\alpha(x)dx + \nu\cdot\phi4(1)\cdot\phi4(1)$$

$$A_{4,5} := \int_0^1 \left(\frac{d}{dx}\phi4(x)\right)\cdot\left(\frac{d}{dx}\phi5(x)\right)\cdot\alpha(x)dx + \nu\cdot\phi4(1)\cdot\phi5(1)$$

$$A_{5,5} := \int_0^1 \left(\frac{d}{dx}\phi5(x)\right)\cdot\left(\frac{d}{dx}\phi5(x)\right)\cdot\alpha(x)dx + \nu\cdot\phi5(1)\cdot\phi5(1)$$

$$A_{2,1} := A_{1,2} \qquad A_{3,1} := A_{1,3}$$

$$A_{4,1} := A_{1,4} \qquad A_{5,1} := A_{1,5}$$

$$A_{3,2} := A_{2,3} \qquad A_{4,2} := A_{2,4}$$

$$A_{5,2} := A_{2,5} \qquad A_{4,3} := A_{3,4}$$

$$A_{5,3} := A_{3,5} \qquad A_{5,4} := A_{4,5}$$

$$A = \begin{pmatrix} 10 & -5 & 0 & 0 & 0 \\ -5 & 10 & -5 & 0 & 0 \\ 0 & -5 & 10 & -5 & 0 \\ 0 & 0 & -5 & 10 & -4.987 \\ 0 & 0 & 0 & -4.987 & 5.159 \end{pmatrix}$$

$$b_1 := \int_0^1 f(x)\cdot\phi 1(x)dx \qquad b_2 := \int_0^1 f(x)\cdot\phi 2(x)dx$$

$$b_3 := \int_0^1 f(x)\cdot\phi 3(x)dx \qquad b_4 := \int_0^1 f(x)\cdot\phi 4(x)dx$$

$$b_5 := \int_0^1 f(x)\cdot\phi 5(x)dx$$

$$b = \begin{pmatrix} 1.276\times10^{-5} \\ 2.556\times10^{-5} \\ 3.836\times10^{-5} \\ 5.102\times10^{-5} \\ 2.986\times10^{-5} \end{pmatrix}$$

$$C := A^{-1}\cdot b$$

$$C = \begin{pmatrix} 2.818\times10^{-5} \\ 5.381\times10^{-5} \\ 7.433\times10^{-5} \\ 8.717\times10^{-5} \\ 9.005\times10^{-5} \end{pmatrix}$$

$$w(x) := C_1\cdot\phi 1(x) + C_2\cdot\phi 2(x) + C_3\cdot\phi 3(x) + (C_4\cdot\phi 4(x)) + C_5\cdot\phi 5(x)$$

$z := 0,.01..1$

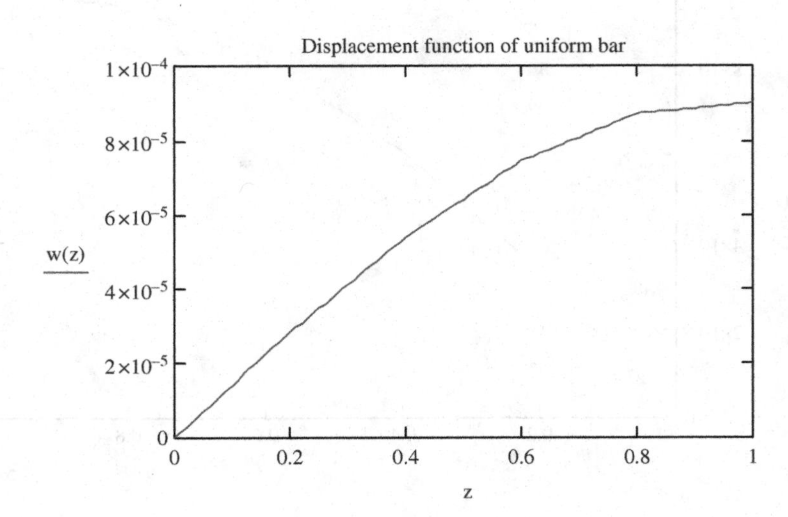

(b) For the nonuniform bar described,

$$\alpha(x) = \frac{2\times10^{-4}\left[1+(0.02)(30.5)x\right]}{2\times10^{-4}}$$

$$= 1+0.61x \tag{n}$$

The right-hand side of Equation g is evaluated as

$$\mu\int_0^x \alpha(\xi)d\xi = \mu\int_0^x (1+0.61\xi)d\xi$$

$$= \mu\left(x+0.305x^2\right) \tag{o}$$

The values of the nondimensional parameters are $\mu = 3.194 \times 10^{-4}$ and $v = 0.095$. The solution is obtained by modifying the MATHCAD file such that xd = 0.02. The resulting finite-element solution is illustrated.

A set of basis functions for a Rayleigh-Ritz approximation for the solution of a boundary-value problem involving a fourth-order operator must be twice differentiable, implying that the first derivative of the interpolating splines must be continuous.

Consider an interval $0 \le \xi \le 4\ell$. An interpolating spline $\psi(\xi)$ is determined such that $\psi(0) = 0$, $\psi'(0) = 0$, $\psi(4\ell) = 0$, $\psi'(4\ell) = 0$ and $\psi(2\ell) = 1$. Assuming that $\psi(\xi) = C_0 + C_1\xi + C_2\xi^2 + C_3\xi^3 + C_4\xi^4$, an appropriate spline can be determined as

$$\psi(\xi) = \frac{\xi^2}{\ell^2} - \frac{\xi^3}{2\ell^3} + \frac{\xi^4}{16\ell^4} \tag{4.33}$$

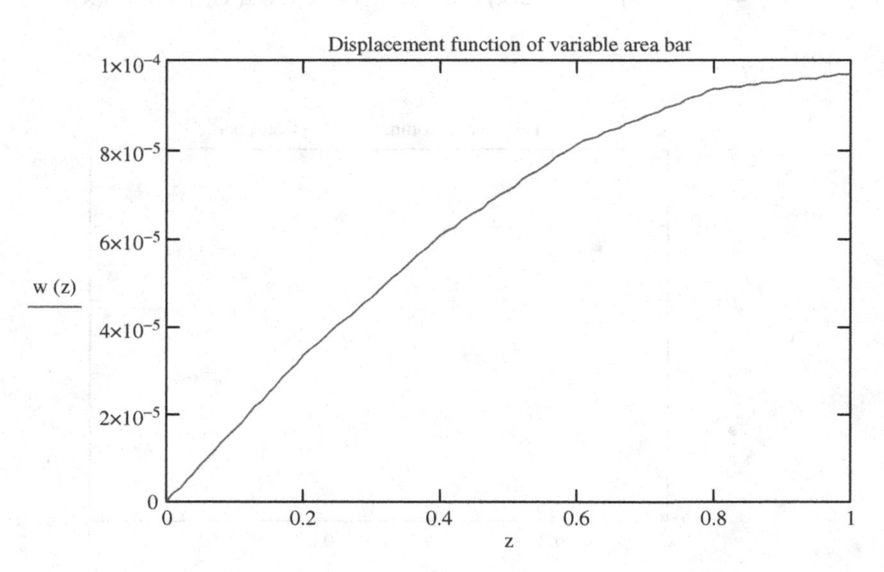

The function given in Equation 4.33 is called the generating spline because all basis functions can be defined from it.

The nondimensional interval $0 \leq x \leq 1$ is divided into n intervals, each of length ℓ. A set of basis functions is defined by

$$\phi_i(x) = \psi(x-(i-3)\ell)[u(x-(i-3)\ell) - u(x-(i+1)\ell] \quad i = 0,1,\ldots n+2$$

Example 4.6 Reconsider Example 4.3 in which a uniform Euler-Bernoulli beam rests on an elastic foundation and is subject to a transverse load. The nondimensional formulation of the governing differential equation is

$$\frac{d^4 w}{dx^4} + \phi w = \lambda f(x) \tag{a}$$

The nondimensional parameters are defined as

$$\phi = \frac{kL^3}{EI} \tag{b}$$

$$\lambda = \frac{F_{\max} L^3}{EI} \tag{c}$$

The beam is loaded so that

$$f(x) = x - x^2 \tag{d}$$

Dividing the interval from 0 to 1 into six segments of equal length, use the finite-element method to approximate the transverse deflection of a beam with (a) both ends free, (b) both ends fixed, and (c) the end at $x=0$ fixed and the end at $x=1$ pinned. In each case, use $\phi=1$ and $\lambda=1$.

Solution The basis elements are those shown in the MATHCAD file used to solve the problem. The appropriate energy-inner-product formulation is

$$(f,g)_L = \int_0^1 \frac{d^2 f}{dx^2}\frac{d^2 g}{dx^2}dx + \phi\int_0^1 f(x)g(x)dx \tag{e}$$

The general form of the Rayleigh-Ritz approximation is

$$w(x) = \sum_{i=0}^{8} a_i \phi_i(x) \tag{f}$$

(a) All boundary conditions for a beam which is free at both ends are natural (bending moment and shear force are both zero). Thus, the set of eight

interpolating splines serves as basis functions. The results are shown in the attached MATHCAD file.
(b) Since the beam is fixed at both ends, all boundary conditions are geometric and must be satisfied. This could be accomplished at $x=0$ by setting $a_0=0$, $a_1=0$, and $a_2=0$. Satisfaction of geometric boundary conditions at $x=1$ could be accomplished by setting $a_6=0$, $a_7=0$, and $a_8=0$. Thus, the appropriate Rayleigh-Ritz approximation is

$$w(x)=a_3\phi_3(x)+a_4\phi_4(x)+a_5\phi_5(x) \tag{g}$$

However, when only three basis functions are used, the geometric quantities of deflection and slope are not independent over the length of the beam. Upon inspection, it can be noted that $w'(2q)=-w'(4q)$ independently of $f(x)$. This is an artificial condition enforced by the choice of basis functions. An expanded set of basis functions can be chosen from the interpolating splines by requiring that

$$\begin{aligned} w(0)=0&=\sum_{i=0}^{8} a_i\phi_i(0) \\ &=a_0\phi_0(0)+a_1\phi_1(0)+a_2\phi_2(0) \\ &=a_0\psi(3\ell)+a_1\psi(2\ell)+a_2\psi(\ell) \\ &=\frac{9}{16}a_0+a_1+\frac{9}{16}a_2 \end{aligned} \tag{h}$$

$$\begin{aligned} w'(0)=0&=\sum_{i=0}^{8} a_i\phi_i'(0) \\ &=a_0\psi'(3\ell)+a_2\psi'(\ell) \\ &=\frac{3}{4}a_0-\frac{3}{4}a_2 \end{aligned} \tag{i}$$

Equation h and Equation i are used to determine $a_0=a_2=-8/9a_1$. Thus, a basis function satisfying the geometric boundary conditions at $x=0$ is

$$u(x)=-\frac{8}{9}\phi_0(x)+\phi_1(x)-\frac{8}{9}\phi_2(x) \tag{j}$$

A similar application of the boundary conditions at $x=1$ leads to a basis function of

$$v(x)=-\frac{8}{9}\phi_6(x)+\phi_7(x)-\frac{8}{9}\phi_8(x) \tag{k}$$

An appropriate set of basis functions for a Rayleigh-Ritz approximation is $\{u(x), \phi_3(x), \phi_4(x), \phi_5(x), v(x)\}$. MATHCAD results can be seen as below.
(c) The boundary conditions at $x=0$ for a fixed-pinned beam are the same as those in part (b), leading to the choice of $u(x)$ in Equation j as a basis function. The geometric boundary condition at $x=1$ is satisfied by requiring that

$$w(1) = 0 = a_6\psi(3\ell) + a_7\psi(2\ell) + a_8\psi(\ell)$$
$$= \frac{9}{16}a_6 + a_7 + \frac{9}{16}a_8 \tag{l}$$

Equation l is the only condition which must be satisfied by the interpolating splines at $x=1$. This leads to $a_7 = -9/16\ (a_6 + a_8)$ and a choice of a set of six basis functions, namely $\{u(x), \phi_3(x), \phi_4(x), \phi_5(x), \phi_6(x) - (9/16)\phi_7(x), \phi_8 - (9/16)\phi_7(x)\}$. The MATHCAD file for the finite-element approximation of the displacement of a fixed-pinned beam on an elastic foundation is shown in Figure 4.11.

Example 4.6 Finite element modeling of beam on an elastic foundation

$\mathrm{n} := 6$ Number of intervals

$\mathrm{q} := \frac{1}{\mathrm{n}}$ Length of an interval

$\psi(\xi) := \frac{\xi^2}{q^2} - \frac{\xi^3}{2 \cdot q^3} + \frac{\xi^4}{16 \cdot q^4}$ Generating spline

Interpolating splines

$$\phi0(x) := \psi(x + 3 \cdot q) \cdot (\Phi(x) - \Phi(x - q))$$

$$\phi1(x) := \psi(x + 2 \cdot q) \cdot (\Phi(x) - \Phi(x - 2q))$$

$$\phi2(x) := \psi(x + q) \cdot (\Phi(x) - \Phi(x - 3 \cdot q))$$

$$\phi3(x) := \psi(x) \cdot (\Phi(x) - \Phi(x - 4 \cdot q))$$

$$\phi4(x) := \psi(x - q) \cdot (\Phi(x - q) - \Phi(x - 5 \cdot q))$$

$$\phi5(x) := \psi(x - 2 \cdot q) \cdot (\Phi(x - 2 \cdot q) - \Phi(x - 6 \cdot q))$$

$$\phi6(x) := \psi(x - 3 \cdot q) \cdot (\Phi(x - 3 \cdot q) - \Phi(x - 7 \cdot q))$$

$$\phi7(x) := \psi(x - 4 \cdot q) \cdot (\Phi(x - 4 \cdot q) - \Phi(x - 8 \cdot q))$$

$$\phi8(x) := \psi(x - 5 \cdot q) \cdot (\Phi(x - 5 \cdot q) - \Phi(x - 9 \cdot q))$$

Plot interpolating splines

$x := 0, .01 .. 1$

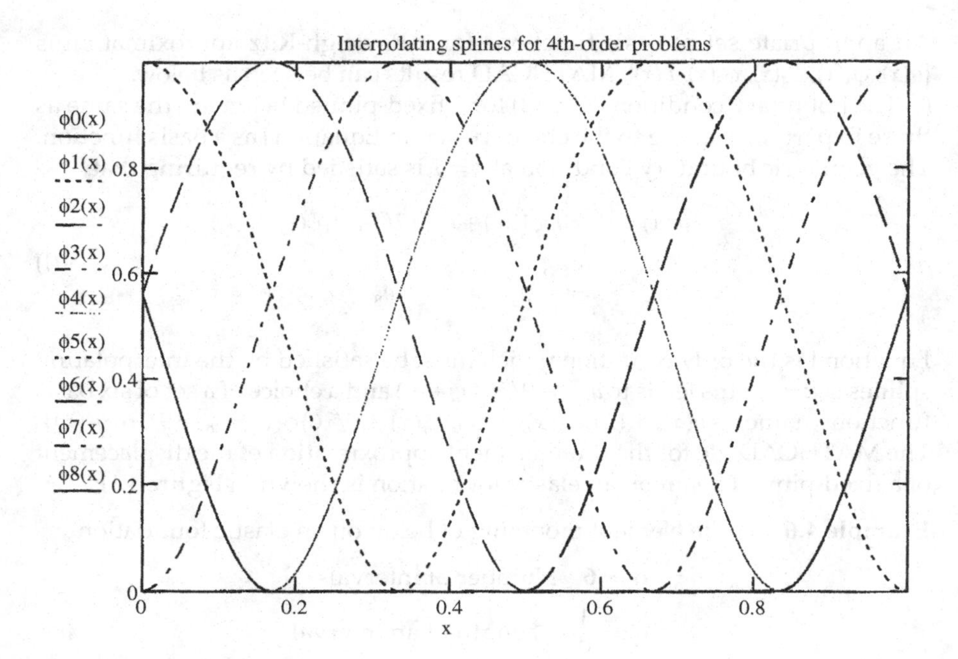

Nonhomogeneous term

$$f(x) := x - x^2$$

$$\phi := 1$$

a. Free free beam

Basis functions

$$A(x) := \begin{pmatrix} \phi0(x) \\ \phi1(x) \\ \phi2(x) \\ \phi3(x) \\ \phi4(x) \\ \phi5(x) \\ \phi6(x) \\ \phi7(x) \\ \phi8(x) \end{pmatrix}$$

$i := 1..9$

$j := 1..9$

Matrix of energy inner products

$$Q_{i,j} := \left(\int_0^1 \frac{d^2}{dx^2}A(x)_i \cdot \frac{d^2}{dx^2}A(x)_j dx\right) + \phi \cdot \int_0^1 A(x)_i \cdot A(x)_j dx$$

Vector of inner products

$$b_i := \int_0^1 f(x) \cdot A(x)_i \, dx$$

$$Q = \begin{pmatrix} 213.313 & -90.416 & -151.173 & 31.053 & 0 & 0 & 0 & 0 & 0 \\ -90.416 & 345.735 & 21.751 & -302.346 & 31.053 & 0 & 0 & 0 & 0 \\ -151.173 & 21.751 & 478.158 & -68.664 & -302.345 & 31.053 & 0 & 0 & 0 \\ 31.053 & -302.346 & -68.664 & 691.471 & -68.664 & -302.345 & 31.053 & 0 & 0 \\ 0 & 31.053 & -302.345 & -68.664 & 691.471 & -68.664 & -302.345 & 31.053 & 0 \\ 0 & 0 & 31.053 & -302.345 & -68.664 & 691.471 & -68.664 & -302.346 & 31.053 \\ 0 & 0 & 0 & 31.053 & -302.345 & -68.664 & 478.158 & 21.751 & -151.173 \\ 0 & 0 & 0 & 0 & 31.053 & -302.346 & 21.751 & 345.735 & -90.416 \\ 0 & 0 & 0 & 0 & 0 & 31.053 & -151.173 & -90.416 & 213.313 \end{pmatrix}$$

$$b = \begin{pmatrix} 1.559 \times 10^{-3} \\ 0.016 \\ 0.046 \\ 0.073 \\ 0.083 \\ 0.073 \\ 0.046 \\ 0.016 \\ 1.559 \times 10^{-3} \end{pmatrix}$$

Solution for coefficients

$W := Q^{-1} \cdot b$

$$W = \begin{pmatrix} 5.053 \times 10^{-3} \\ 5.068 \times 10^{-3} \\ 5.143 \times 10^{-3} \\ 5.135 \times 10^{-3} \\ 5.183 \times 10^{-3} \\ 5.135 \times 10^{-3} \\ 5.143 \times 10^{-3} \\ 5.068 \times 10^{-3} \\ 5.053 \times 10^{-3} \end{pmatrix}$$

Rayleigh Ritz Approximation

$$\text{wfree}(x) := \sum_{i=1}^{9} (W_1 \cdot A(x)_i)$$

$x := 0, .01 .. 1$

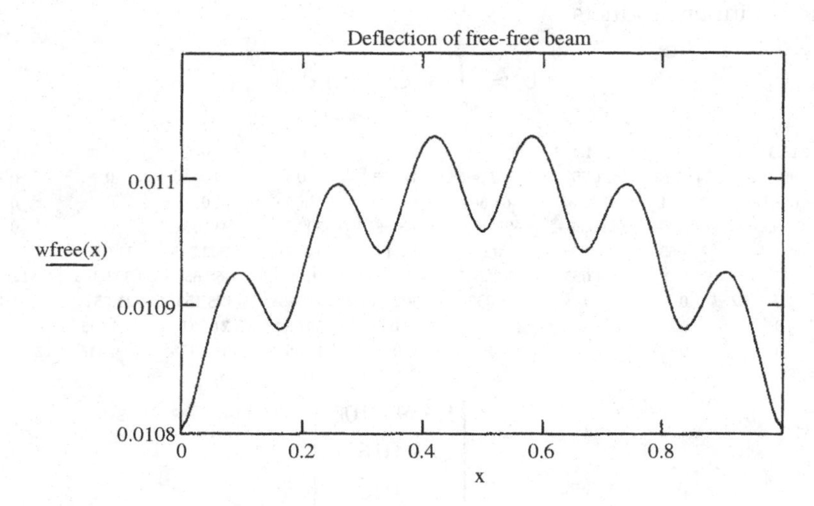

Fixed fixed beam

Case 1 Use of three basis functions

$$B(x) := \begin{pmatrix} \phi 3(x) \\ \phi 4(x) \\ \phi 5(x) \end{pmatrix}$$

$$i := 1..3$$
$$j := 1..3$$

$$R_{i,j} := \left(\int_0^1 \frac{d^2}{dx^2} B(x)_i \cdot \frac{d^2}{dx^2} B(x)_j dx \right) + \phi \cdot \int_0^1 B(x)_i \cdot B(x)_j dx$$

$$r_i := \int_0^1 f(x) \cdot B(x)_i dx$$

$$R = \begin{pmatrix} 691.471 & -68.664 & -302.345 \\ -68.664 & 691.471 & -68.664 \\ -302.345 & -68.664 & 691.471 \end{pmatrix}$$

$$r = \begin{pmatrix} 0.073 \\ 0.083 \\ 0.073 \end{pmatrix}$$

$\text{Wfixed} := R^{-1} \cdot r$

$$\text{Wfixed} = \begin{pmatrix} 2.174 \times 10^{-4} \\ 1.636 \times 10^{-4} \\ 2.174 \times 10^{-4} \end{pmatrix}$$

$$\text{wfixed}(x) := \sum_{i=1}^{3} (\text{Wfixed}_i \cdot B(x)_i)$$

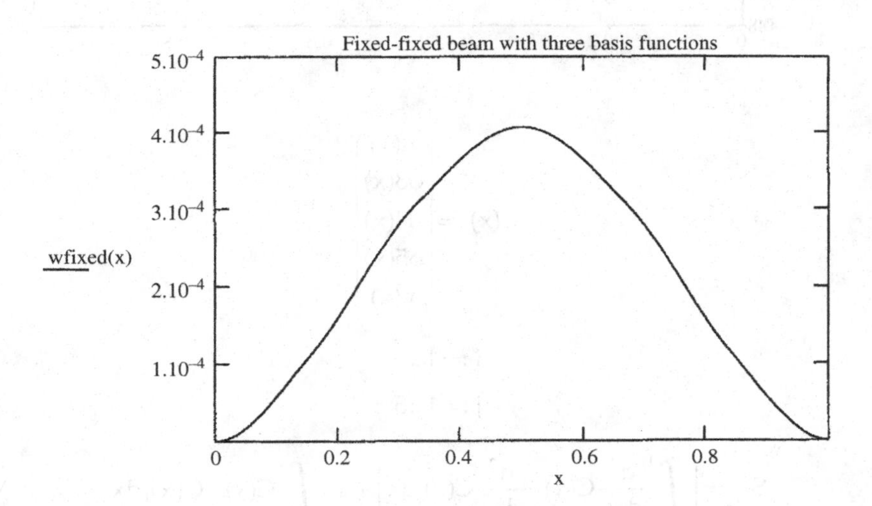

Fixed fixed beam with five basis functions

$$u(x) := \left(\frac{-8}{9} \cdot \psi(x + 3 \cdot q) + \psi(x + 2 \cdot q) - \frac{8}{9} \cdot \psi(x + q)\right) \cdot (\Phi(x) - \Phi(x - q))$$

$$u(x) := u(x) + \left(\psi(x + 2 \cdot q) - \frac{8}{9} \cdot \psi(x + q)\right) \cdot (\Phi(x - q) - \Phi(x - 2 \cdot q))$$

$$u(x) := u(x) - \frac{8}{9} \cdot \psi(x + q) \cdot (\Phi(x - 2 \cdot q) - \Phi(x - 3 \cdot q))$$

$$v(x) := \left(\frac{-8}{9} \cdot \psi(x - 3 \cdot q) + \psi(x - 4 \cdot q) - \frac{8}{9} \cdot \psi(x - 5 \cdot q)\right) \cdot (\Phi(x - 5 \cdot q))$$

$$v(x) := v(x) + \left(\psi(x - 4 \cdot q) - \frac{8}{9} \cdot \psi(x - 3 \cdot q)\right) \cdot (\Phi(x - 4q) - \Phi(x - 5 \cdot q))$$

$$v(x) := v(x) - \frac{8}{9} \cdot \psi(x - 3q) \cdot (\Phi(x - 3 \cdot q) - \Phi(x - 4 \cdot q))$$

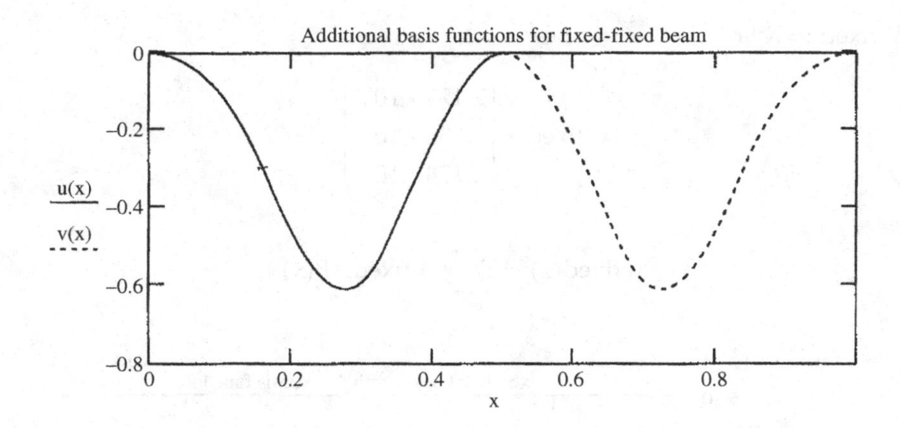

$$C(x) := \begin{pmatrix} u(x) \\ \phi 3(x) \\ \phi 4(x) \\ \phi 5(x) \\ v(x) \end{pmatrix}$$

$$i := 1..5$$

$$j := 1..5$$

$$S_{i,j} := \left(\int_0^1 \frac{d^2}{dx^2} C(x)_i \cdot \frac{d^2}{dx^2} C(x)_j dx \right) + \phi \cdot \int_0^1 C(x)_i \cdot C(x)_j \, dx$$

$$s_i := \int_0^1 f(x) \cdot C(x)_i dx$$

$$S = \begin{pmatrix} 775.284 & -268.931 & 299.756 & -27.603 & -73.91 \\ -268.931 & 691.471 & -68.664 & -302.345 & -38.772 \\ 299.756 & -68.664 & 691.471 & -68.664 & 255.623 \\ -27.603 & -302.345 & -68.664 & 691.471 & -277.889 \\ -73.91 & -38.772 & 255.623 & -277.889 & 988.055 \end{pmatrix}$$

$$s = \begin{pmatrix} -0.026 \\ 0.073 \\ 0.083 \\ 0.073 \\ -0.026 \end{pmatrix}$$

$\text{Wfixeda} := S^{-1} \cdot s$

$$\text{Wfixeda} = \begin{pmatrix} -2.24\times10^{-5} \\ 2.063\times10^{-4} \\ 1.736\times10^{-4} \\ 2.103\times10^{-4} \\ -5.839\times10^{-6} \end{pmatrix}$$

$$\text{wfixeda}(x) := \sum_{i=1}^{5} (\text{Wfixeda}_i \cdot C(x)_i)$$

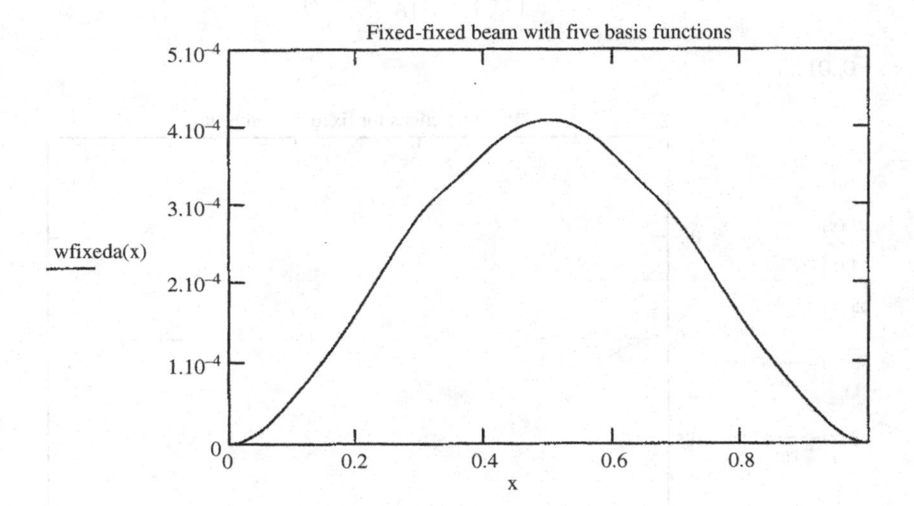

$$q1(x) := \frac{d}{dx}\text{wfixeda}(x)$$

$q1(2\cdot q) = 6.917\times10^{-4}$ $\quad$ $q1(4\cdot q) = -7.58\times10^{-4}$

Fixed Pinned beam

$$u(x) := \left(\frac{-8}{9}\cdot\Psi(x+3\cdot q) + \Psi(x+2\cdot q) - \frac{8}{9}\cdot\Psi(x+q)\right)\cdot(\Phi(x) - \Phi(x-q))$$

$$u(x) := u(x) + \left(\Psi(x+2\cdot q) - \frac{8}{9}\cdot\Psi(x+q)\right)\cdot(\Phi(x-q) - \Phi(x-2\cdot q))$$

$$u(x) := u(x) - \frac{8}{9}\cdot\Psi(x+q)\cdot(\Phi(x-2\cdot q) - \Phi(x-3\cdot q))$$

Basis functions

$$T(x) := \begin{pmatrix} u(x) \\ \phi 3(x) \\ \phi 4(x) \\ \phi 5(x) \\ \phi 6(x) - \frac{9}{16} \cdot \phi 7(x) \\ \phi 8(x) - \frac{9}{16} \cdot \phi 7(x) \end{pmatrix}$$

$x := 0, .01 .. 1$

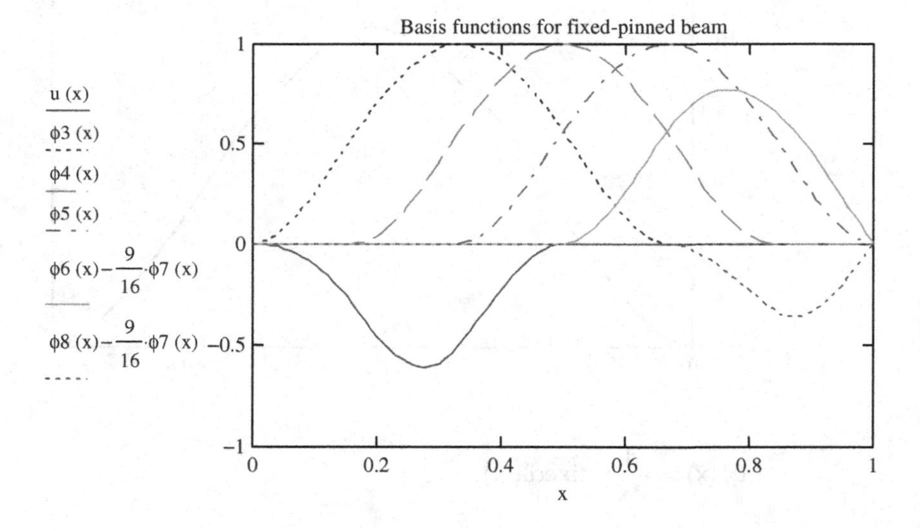

$$i := 1..6$$

$$h_i := \int_0^1 f(x) \cdot T(x)_i \, dx$$

$$j := 1..6$$

$$H_{i,j} := \sum_{k=1}^{6} \int_{(k-.99)\cdot q}^{(k-.01)\cdot q} \frac{d^2}{dx^2} T(x)_i \frac{d^2}{dx^2} T(x)_j \, dx$$

$$H_{i,j} := H_{i,j} + \phi \cdot \int_0^1 T(x)_i \cdot T(x)_j \, dx$$

$$H=\begin{pmatrix} 753.208 & -262.878 & 294.095 & -29.258 & 0 & 0 \\ -262.878 & 669.588 & -68.616 & -294.04 & 33.133 & 0 \\ 294.095 & -68.616 & 669.588 & -68.616 & -312.677 & -18.637 \\ -29.258 & -294.04 & -68.616 & 669.588 & 96.782 & 198.53 \\ 0 & 33.133 & -312.677 & 96.782 & 546.972 & -2.493 \\ 0 & 0 & -18.637 & 198.53 & -2.493 & 411.67 \end{pmatrix}$$

$$h=\begin{pmatrix} -0.026 \\ 0.073 \\ 0.083 \\ 0.073 \\ 0.037 \\ -7.308\times10^{-3} \end{pmatrix}$$

$$\text{Wpinned} := H^{-1}\cdot h$$

$$\text{Wpinned}=\begin{pmatrix} -5.663\times10^{-5} \\ 2.119\times10^{-4} \\ 2.693\times10^{-4} \\ 2.394\times10^{-4} \\ 1.653\times10^{-4} \\ -1.2\times10^{-4} \end{pmatrix}$$

$$\text{wpinned}(x) := \sum_{i=1}^{6}\left(\text{Wpinned}_i \cdot T(x)_i\right)$$

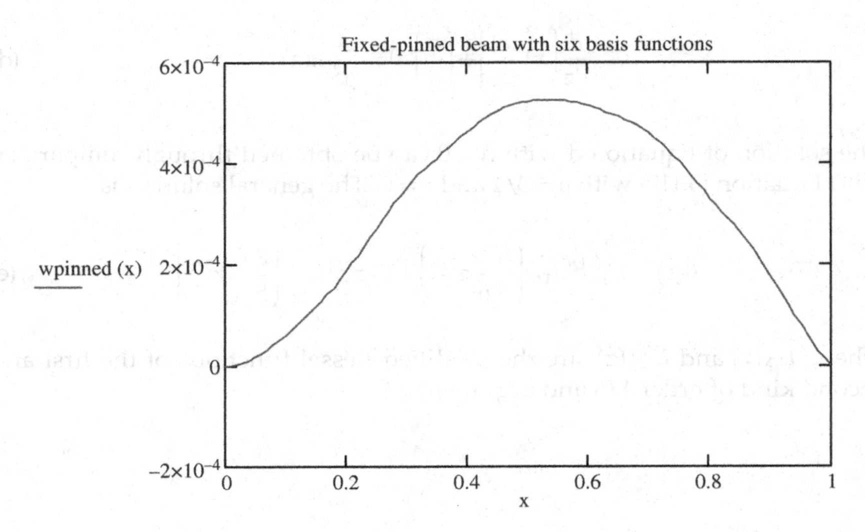

Example 4.7 The differential equation governing the nondimensional temperature in a variable-area fin is

$$\frac{d}{dx}\left[(1-\mu x)^{\frac{3}{4}}\frac{d\theta}{dx}\right]-\phi^2\theta=\Lambda u(x) \tag{a}$$

where $\phi^2 = hL^2/kb$ and $\Lambda = u_{\max}L^2/2kb\ell(T_0-T_\infty)$. The surface has a fixed temperature at $x=0$ and is insulated at $x=1$, leading to the boundary conditions

$$\theta(0)=1 \tag{b}$$

$$\frac{d\theta}{dx}(1)=0 \tag{c}$$

a. Obtain the exact solution for Equation a, Equation b, and Equation c with $\Lambda=0$.
b. Use the method of variation of parameters to determine an exact solution of Equation a, Equation b, and Equation c with $u(x)=1-x$.
c. Use the Rayleigh-Ritz method using fourth-order polynomials which satisfy only the geometric boundary conditions as basis functions to approximate the temperature distribution in the surface for (i) $\Lambda=0$ and (ii) $u(x)=1-x$. Use $\phi=1.5$ and $\mu=0.5$.
d. Use a finite-element method with five elements to approximate the temperature distribution over the surface for (i) $\Lambda=0$ and (ii) $u(x)=1-x$. Use $\phi=1.5$ and $\mu=0.5$.

Solution (a) Let $z=1-\mu x$. Changing the independent variable in Equation a from x to z leads to

$$\frac{d}{dz}\left(z^{\frac{3}{4}}\frac{d\theta}{dz}\right)-\left(\frac{\phi}{\mu}\right)^2\theta=\frac{\Lambda}{\mu^2}u(z) \tag{d}$$

The solution of Equation d with $\Lambda=0$ can be obtained through comparison with Equation (3.110) with $\mu=3/4$ and $v=0$. The general solution is

$$\theta(z)=C_1z^{1/8}I_{1/5}\left(\frac{8}{5}\frac{\phi}{\mu}z^{5/8}\right)+C_2z^{1/8}K_{1/5}\left(\frac{8}{5}\frac{\phi}{\mu}z^{5/8}\right) \tag{e}$$

where $I_{1/5}(z)$ and $K_{1/5}(z)$ are the modified Bessel functions of the first and second kind of order 1/5 and argument z.

The boundary condition, Equation, c, can be written in terms of z as $d\theta/dz\,(1-\mu)=0$, which when applied to Equation e using Equation f, leads to $C_2 = -\beta C_1$, where

$$\beta = \frac{I_{1/5}\left(\frac{8\phi}{5\mu}(1-\mu)^{5/8}\right) + \frac{8\phi}{\mu} I'_{1/5}\left(\frac{8\phi}{5\mu}(1-\mu)^{5/8}\right)}{K_{1/5}\left(\frac{8\phi}{5\mu}(1-\mu)^{5/8}\right) + \frac{8\phi}{\mu} K'_{1/5}\left(\frac{8\phi}{5\mu}(1-\mu)^{5/8}\right)} \tag{f}$$

The boundary condition, Equation b, can be written in terms of z as $\theta(z=1)=1$, which when applied to Equation e and using Equation f, leads to

$$\theta(z) = \frac{z^{1/8} I_{1/5}\left(\frac{8\phi}{5\mu} z^{5/8}\right) - \beta z^{1/8} K_{1/5}\left(\frac{8\phi}{5\mu} z^{5/8}\right)}{I_{1/5}\left(\frac{8\phi}{5\mu}\right) - \beta K_{1/5}\left(\frac{8\phi}{5\mu}\right)} \tag{g}$$

Returning to the original independent variable,

$$\theta(x) = \frac{(1-\mu x)^{1/8} I_{1/5}\left(\frac{8\phi}{5\mu}(1-\mu x)^{5/8}\right) - \beta(1-\mu x)^{1/8} K_{1/5}\left(\frac{8\phi}{5\mu}(1-\mu x)^{5/8}\right)}{I_{1/5}\left(\frac{8\phi}{5\mu}\right) - \beta K_{1/5}\left(\frac{8\phi}{5\mu}\right)} \tag{h}$$

(b) The particular solution of Equation a is assumed to be of the form

$$\theta(x) = A(x)(1-\mu x)^{1/8} I_{1/5}\left(\frac{8\phi}{5\mu}(1-\mu x)^{5/8}\right) + B(x)(1-\mu x)^{1/8} K_{1/5}\left(\frac{8\phi}{5\mu}(1-\mu x)^{5/8}\right) \tag{i}$$

Application of the method of variation of parameters leads to

$$\frac{dA}{dx} = \frac{u(x)}{W(x)}(1-\mu x)^{1/8} K_{1/5}\left(\frac{8\phi}{5\mu}(1-\mu x)^{5/8}\right) \tag{j}$$

$$\frac{dB}{dx} = \frac{u(x)}{W(x)}(1-\mu x)^{1/8} I_{1/5}\left(\frac{8\phi}{5\mu}(1-\mu x)^{5/8}\right) \tag{k}$$

where $W(x)$ is the Wronskian of the homogeneous solutions. Application of Abel's formula leads to $W(x) = e^{-(1-\mu x)^{3/4}}$. Solving Equation i and Equation j for $A(x)$ and $B(x)$ and using the solutions in Equation i leads to the general solution of Equation a as

$$\theta(x) = C_1(1-\mu x)^{1/8} I_{1/5}\left(\frac{8\phi}{5\mu}(1-\mu x)^{5/8}\right) + C_2(1-\mu x)^{1/8} K_{1/5}\left(\frac{8\phi}{5\mu}(1-\mu x)^{5/8}\right)$$

$$+ (1-\mu x)^{1/8} I_{1/5}\left(\frac{8\phi}{5\mu}(1-\mu x)^{5/8}\right)\int_0^x u(\xi)e^{(1-\mu\xi)^{3/4}}(1-\mu\xi)^{1/8} K_{1/5}\left(\frac{8\phi}{5\mu}(1-\mu\xi)^{5/8}\right)d\xi$$

$$+ (1-\mu x)^{1/8} K_{1/5}\left(\frac{8\phi}{5\mu}(1-\mu x)^{5/8}\right)\int_0^x u(\xi)e^{(1-\mu\xi)^{3/4}}(1-\mu\xi)^{1/8} I_{1/5}\left(\frac{8\phi}{5\mu}(1-\mu\xi)^{5/8}\right)d\xi \tag{l}$$

Application of Equation b to Equation l leads to

$$I_{1/5}\left(\frac{8\phi}{5\mu}\right)C_1 + K_{1/5}\left(\frac{8\phi}{5\mu}\right)C_2 = 1 \tag{m}$$

Application of Equation c to Equation l leads to

$$\left[\frac{1}{8}(1-\mu)^{-7/8} I_{1/5}\left(\frac{8\phi}{5\mu}(1-\mu)^{5/8}\right) + \frac{\phi}{\mu}(1-\mu)^{-1/4} I'_{1/5}\left(\frac{8\phi}{5\mu}(1-\mu)^{5/8}\right)\right]C_1$$

$$+ \left[\frac{1}{8}(1-\mu)^{-7/8} K_{1/5}\left(\frac{8\phi}{5\mu}(1-\mu)^{5/8}\right) + \frac{\phi}{\mu}(1-\mu)^{-1/4} K'_{1/5}\left(\frac{8\phi}{5\mu}(1-\mu)^{5/8}\right)\right]C_2$$

$$= -\left[\frac{1}{8}(1-\mu)^{-7/8} I_{1/5}\left(\frac{8\phi}{5\mu}(1-\mu)^{5/8}\right) + \frac{\phi}{\mu}(1-\mu)^{-1/4} I'_{1/5}\left(\frac{8\phi}{5\mu}(1-\mu)^{5/8}\right)\right]$$

$$\int_0^1 u(\xi)e^{(1-\mu\xi)^{3/4}}(1-\mu\xi)^{1/8} K_{1/5}\left(\frac{8\phi}{5\mu}(1-\mu\xi)^{5/8}\right)d\xi$$

$$-\left[\frac{1}{8}(1-\mu)^{-7/8} K_{1/5}\left(\frac{8\phi}{5\mu}(1-\mu)^{5/8}\right) + \frac{\phi}{\mu}(1-\mu)^{-1/4} K'_{1/5}\left(\frac{8\phi}{5\mu}(1-\mu)^{5/8}\right)\right]$$

$$\int_0^1 u(\xi)e^{(1-\mu\xi)^{3/4}}(1-\mu\xi)^{1/8} I_{1/5}\left(\frac{8\phi}{5\mu}(1-\mu\xi)^{5/8}\right)d\xi$$

$$- 2u(1)e^{(1-\mu)^{3/4}}(1-\mu)^{1/4} K_{1/5}\left(\frac{8\phi}{5\mu}(1-\mu)^{5/8}\right) I_{1/5}\left(\frac{8\phi}{5\mu}(1-\mu)^{5/8}\right) \tag{n}$$

Simultaneous solution of Equation m and Equation n leads to C_1 and C_2.

(c) There are two methods to develop a Rayleigh-Ritz approximation to a homogeneous differential equation with a nonhomogeneous boundary condition. The nonhomogeneity can be transferred to the differential equation through a change of dependent variable. Defining $\Psi(x) = \theta(x) - 1$, Equation a, Equation b, and Equation c can be rewritten as

$$\frac{d}{dx}\left[(1-\mu x)^{3/4}\frac{d\Psi}{dx}\right] - \phi^2\Psi = \Lambda u(x) - \phi^2 \tag{o}$$

$$\Psi(0) = 0 \tag{p}$$

$$\frac{d\Psi}{dx}(1) = 0 \tag{q}$$

A set of basis functions in $P^4[0,1]$ which all satisfy the geometric boundary condition of Equation q is $\phi_1(x) = x, \phi_2(x) = x^2, \phi_3(x) = x^3$ and $\phi_4(x) = x^4$. The Rayleigh-Ritz solution using these basis functions for $\Lambda = 0$ and $\Lambda = 1$ with $u(x) = 1 - x$ is given in the MATHCAD output as shown below.

Rayleigh Ritz approximations for Example 4.7

Basis functions

$$\varphi 1(x) := x$$

$$\varphi 2(x) := x^2$$

$$\varphi 3(x) := x^3$$

$$\varphi 4(x) := x^4$$

$$\varphi(x) := \begin{pmatrix} \varphi 1(x) \\ \varphi 2(x) \\ \varphi 3(x) \\ \varphi 4(x) \end{pmatrix}$$

Parameters

$$\Lambda := 0$$

$$\eta := 1.5$$

$$\mu := 0.5$$

$\alpha(x) := (1 - \mu \cdot x)^{\frac{3}{4}}$ Nondimensional cross sectional area

$u(x) := 1 - x$

$f(x) := \Lambda \cdot u(x) - \eta^2$

Matrix and vector calculations

$$i := 1, 2 .. 4$$

$$g := \int_0^1 f(x) \cdot \varphi(x)_i \, dx$$

$$j := 1, 2 .. 4$$

$$K_{i,j} := \left[\int_0^1 \alpha(x) \cdot \left(\frac{d}{dx} \varphi(x)_i \right) \cdot \left(\frac{d}{dx} \varphi(x)_j \right) dx \right]$$

$$K = \begin{pmatrix} 0.803 & 0.736 & 0.701 & 0.681 \\ 0.736 & 0.935 & 1.021 & 1.067 \\ 0.701 & 0.021 & 1.2 & 1.313 \\ 0.681 & 1.067 & 1.313 & 1.483 \end{pmatrix}$$

$$g = \begin{pmatrix} -1.125 \\ -0.75 \\ -0.563 \\ -0.45 \end{pmatrix}$$

Solution of equations

$$W := K^{-1} \cdot g$$

$$W = \begin{pmatrix} -2.254 \\ 0.744 \\ 0.047 \\ 0.154 \end{pmatrix}$$

Rayleigh Ritz approximation

$$\Psi(x) := \sum_{i=1}^{4} (W_i \cdot \varphi(x)_i)$$

$$x := 0, .01 .. 1$$

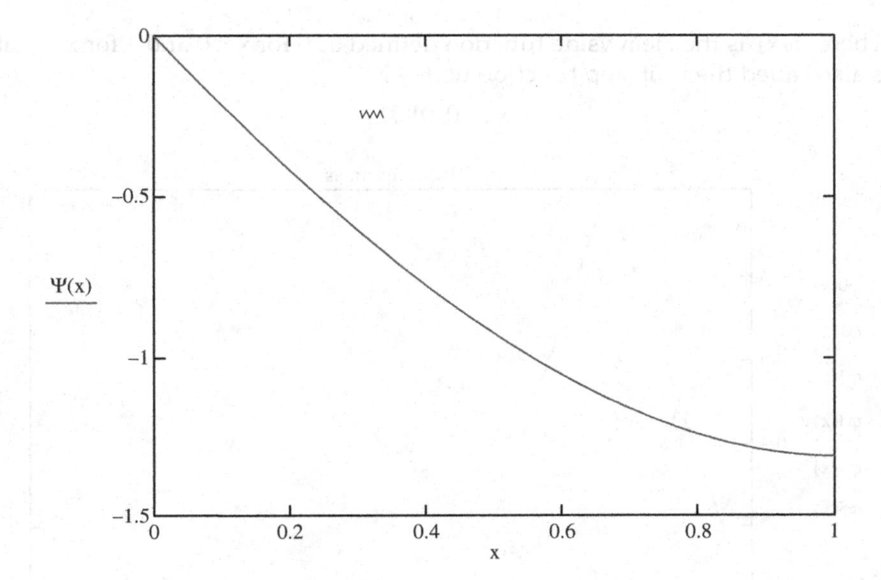

An alternative to a change of dependent variable is to assume a Rayleigh-Ritz approximation of

$$\theta(x) = 1 + \alpha_1 x + \alpha_2 x^2 + \alpha_3 x^3 + \alpha_4 x^4 \tag{r}$$

Use of Equation r leads to the same result, but requires a reformulation of the sets of algebraic equations.

(d) The finite-element response is obtained using the MATHCAD file shown in below.

Finite-element approximations for Example 4.7

Basis functions
The interval from x=0 to x=1 is divided into five elements of equal length leading to definition of the basis functions as

$$\varphi 0(x) := (1 - 5\cdot x)(\Phi(x) - \Phi(x - 0.2))$$

$$\varphi 1(x) := 5\cdot x\cdot(\Phi(x) - \Phi(x - 0.2)) + (2 - 5\cdot x)\cdot(\Phi(x - 0.2) - \Phi(x - 0.4))$$

$$\varphi 2(x) := \varphi 1(x - 0.2)$$

$$\varphi 3(x) := \varphi 1(x - 0.4)$$

$$\varphi 4(x) := \varphi 1(x - 0.6)$$

$$\varphi 5(x) := (5\cdot x - 4)\cdot \Phi(x - 0.8)$$

where $\Phi(x)$ is the Heavyside function defined as 0 for $x < 0$ and 1 for $x > 0$. it is also called the unit step function u(x)

$$x := 0, .01 .. 1$$

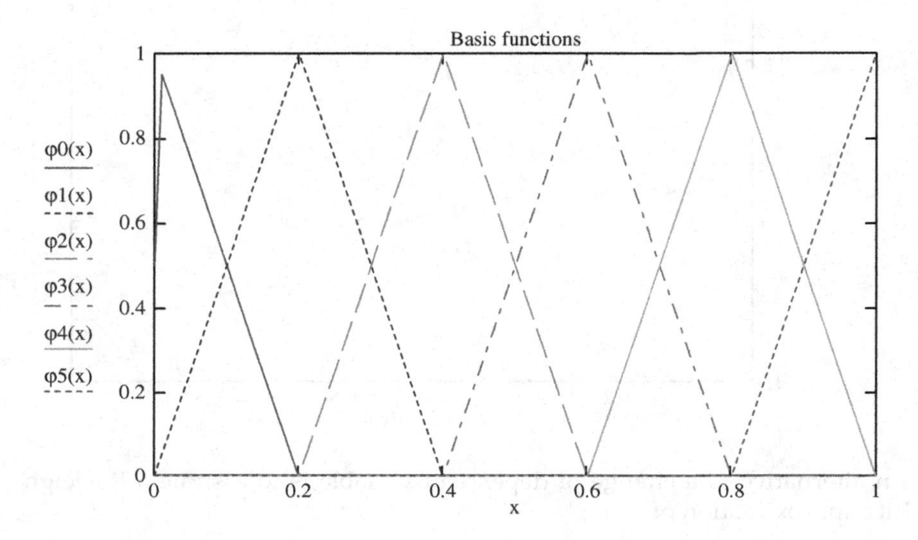

$$\varphi(x) := \begin{pmatrix} \varphi1(x) \\ \varphi2(x) \\ \varphi3(x) \\ \varphi4(x) \\ \varphi5(x) \end{pmatrix}$$

Parameters

$$\Lambda := 0$$

$$\eta := 1.5$$

$$\mu := 1.5$$

$\alpha(x) := (1 - \mu x)^{\frac{3}{4}}$ Nondimensional cross sectional area

$$u(x) := 1 - x$$

$$f(x) := \Lambda \cdot u(x) - \eta^2$$

Matrix and vector calculations

$$i := 1, 2 .. 5$$

$$g := \int_0^1 f(x) \cdot \varphi(x)_i \cdot dx$$

$$j := 1,2..5$$

$$K_{i,j} := \left[\int_0^1 \alpha(x) \cdot \left(\frac{d}{dx}\varphi(x)_j\right) dx\right]$$

$$K = \begin{pmatrix} 9.237 & -4.426 & 0 & 0 & 0 \\ -4.426 & 8.455 & -4.029 & 0 & 0 \\ 0 & -4.029 & 7.648 & -3.619 & 0 \\ 0 & 0 & -3.619 & 6.811 & -3.184 \\ 0 & 0 & 0 & -3.184 & 3.197 \end{pmatrix}$$

$$g = \begin{pmatrix} -0.45 \\ -0.45 \\ -0.45 \\ -0.45 \\ -0.225 \end{pmatrix}$$

Solution of equations

$$W := K^{-1} \cdot g$$

$$W = \begin{pmatrix} -0.415 \\ -0.765 \\ -1.037 \\ -1.216 \\ -1.282 \end{pmatrix}$$

Finite element approximation

$$\Psi(x) := \sum_{i=1}^{5} (W_i \cdot \varphi(x)_i)$$

$$x := 0, .01 .. 1$$

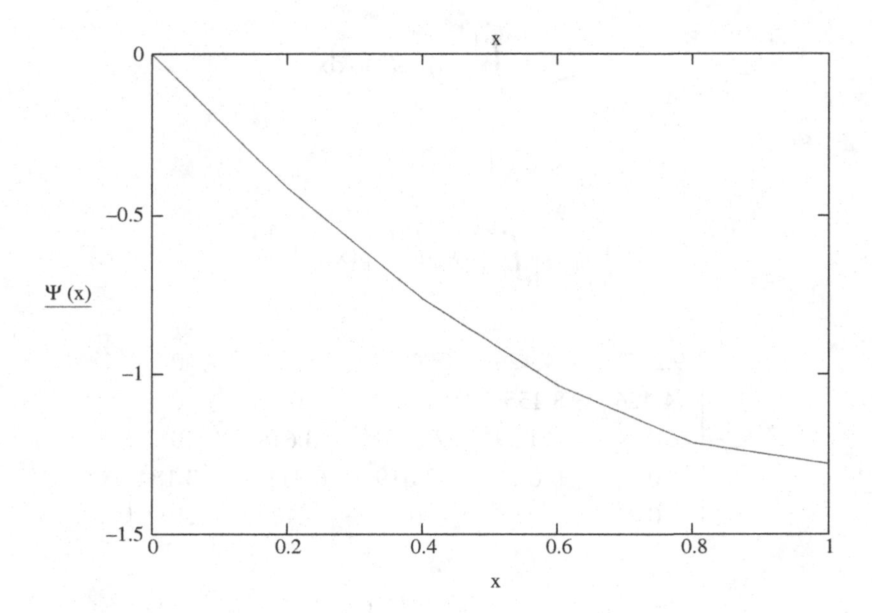

4.5 Galerkin's method

The least-squares method provides a best approximation in the variational sense for the solution of $\mathbf{Lu}=\mathbf{f}$ when $\mathbf{L}$ is a linear operator by minimizing the distance between $\mathbf{Lq}$ and $\mathbf{f}$, where $\mathbf{q}$ is a vector from the span of the basis functions. The Rayleigh-Ritz method provides a best approximation in the variational sense for the solution when $\mathbf{L}$ is positive definite and self-adjoint with respect to a defined inner product. The Rayleigh-Ritz approximation minimizes the distance between $\mathbf{u}$ and $\mathbf{q}$ with respect to the energy norm.

Galerkin's method is an alternate method for formulating an approximation to the solution of $\mathbf{Lu}=\mathbf{f}$, which may be used when $\mathbf{L}$ does not satisfy the constraints of the Rayleigh-Ritz method or perhaps the least-squares method.

Let $S_D=\{\phi_1,\phi_2,\ldots,\phi_n\}$ be a set of linearly independent vectors in the domain of $\mathbf{L}$, and let $S_R=\{\psi_1,\psi_2,\ldots,\psi_n\}$ be a set of linearly independent vectors in the range of $\mathbf{L}$. Let (f,g) represent a valid inner product defined in R. The Galerkin approximation is of the form

$$\hat{\mathbf{u}}=\sum_{i=1}^{n}\alpha_i\phi_i \tag{4.36}$$

The error in the approximation is

$$\begin{aligned}\mathbf{e}&=\mathbf{L}\hat{\mathbf{u}}-\mathbf{f}\\&=\mathbf{L}\left(\sum_{i=1}^{n}\alpha_i\phi_i\right)-\mathbf{f}\end{aligned} \tag{4.37}$$

The Galerkin method requires that e be orthogonal to every member of the set S_R. That is,

$$(\mathbf{e},\psi_k)=0 \quad \mathrm{k}=1,2,\ldots,n \tag{4.38}$$

Equation 4.38 represents a set of simultaneous algebraic equations to solve for the coefficients in the Galerkin expansion, Equation 4.38. Enforcement of Equation 4.38 implies that the error vector has no projection onto any of the vectors in the set S_R.

If **L** is a linear operator, then Equation 4.38 can be rearranged using the following steps allowed by the linearity of **L** and properties of inner products:

$$0=\left(\sum_{i=1}^{n}\alpha_i\mathbf{L}\phi_i-\mathbf{f},\psi_k\right)$$

$$=\sum_{i=1}^{n}\alpha_i(\mathbf{L}\phi_i,\psi_k)-(\mathbf{f},\psi_k) \tag{4.39}$$

Equation 4.39 can be rearranged into the following set of simultaneous linear algebraic equations:

$$\sum_{i=1}^{n}(\mathbf{L}\phi_i,\psi_k)\alpha_i=(\mathbf{f},\psi_\mathbf{k}) \quad k=1,2,\ldots,n \tag{4.40}$$

If the sets of vectors S_R and S_D are identical, then Equation 4.40 becomes

$$\sum_{i=1}^{n}(\mathbf{L}\phi_i,\phi_k)\alpha_i=(\mathbf{f},\phi_k) \quad k=1,2,\ldots,n \tag{4.41}$$

Equation 4.41 is the same set of equations as that developed from the Rayleigh-Ritz formulation when **L** is a positive definite and self-adjoint operator.

Example 4.9 A first-order model of the nondimensional deflection of a microscale cantilever beam due to electrostatic actuation is

$$\frac{d^4u}{dx^4}=-\frac{\phi}{(1+u)^2} \tag{a}$$

The cantilever beam is subject to the boundary conditions $u(0)=1, du/dx(0)=0, d^2u/dx^2(1)=0$ and $d^3u/dx^3(1)=0$. Two linearly independent functions which satisfy these conditions are

$$v_1(x)=x^2-\frac{3}{4}x^4+\frac{1}{5}x^5 \tag{b}$$

$$v_2(x) = x^3 - \frac{3}{2}x^4 + \frac{3}{10}x^5 \tag{c}$$

Determine a Galerkin approximation for the solution of Equation a with $\phi = 1$, using Equation b and Equation c as basis functions for both S_D and S_R.

Solution A Galerkin approximation is assumed in the form of

$$\hat{u}(x) = \alpha_1 v_1(x) + \alpha_2 v_2(x) \tag{d}$$

Equation a can be rewritten as

$$(1+u)^2 \frac{d^4u}{dx^4} = -1 \tag{e}$$

Equation e is nonlinear, and therefore the formulation of Equation 4.41 cannot be used. The error in using Equation d as an approximation is

$$\begin{aligned} e(x) &= (1+\hat{u}^2)\frac{d^4\hat{u}}{dx^4} + 1 \\ &= [1 + \alpha_1 v_1 + \alpha_2 v_2]^2[(-12 + 24x)\alpha_1 + (-36 + 36x)\alpha_2] + 1 \end{aligned} \tag{f}$$

Requiring $e(x)$ to be orthogonal to $v_1(x)$ with respect to the standard inner product on $C^4[0,1]$ leads to

$$\int_0^1 e(x) v_1(x) dx = 0$$

$$\int_0^1 \{[1 + \alpha_1 v_1 + \alpha_2 v_2]^2[(-12 + 24x)\alpha_1 + (-36 + 36x)\alpha_2] + 1\} v_1 dx = 0 \tag{g}$$

Integration and simplification of Equation g lead to

$$\begin{aligned} &1.23\alpha_1 - 2.275\alpha_2 + 0.987\alpha_1^2 - 1.292\alpha_1\alpha_2 - 0.087\alpha_2^2 \\ &\quad + 0.202\alpha_1^3 - 0.177\alpha_1^2\alpha_2 + 0.011\alpha_1\alpha_2^2 - 3.606\times 10^{-3}\alpha_2^3 = -0.217 \end{aligned} \tag{h}$$

Requiring $e(x)$ to be orthogonal to $v_2(x)$ with respect to the standard inner product on $C^4[0,1]$ leads to

$$\int_0^1 e(x)v_2(x)dx = 0$$

$$\int_0^1 \{[1+\alpha_1 v_1+\alpha_2 v_2]^2[(-12+24x)\alpha_1+(-36+36x)\alpha_2]+1\}v_2 dx = 0 \quad \text{(i)}$$

Integration and simplification of Equation i lead to

$$\begin{aligned}&-0.685\alpha_1-0.254\alpha_2-0.036\alpha_1\alpha_2-0.148\alpha_1^2-0.026\alpha_2^2\\&+0.029\alpha_1^2\alpha_2-1.11x10^{-3}\alpha_1\alpha_2^3-0.037\alpha_2^3=0\end{aligned} \quad \text{(j)}$$

The solutions of the nonlinear algebraic equations Equation h and Equation j are not unique. One solution is $\alpha_1 = 0.032$ and $\alpha_2 = 0.080$.

Problems

4.1. a. Find the least squares approximation to $f(x) = xe^{-2x}$ over the interval $-1 \le x \le 1$ from the span of $1, x, x^2$ using the standard inner product on C[–1,1].
b. Calculate the error in the approximation using the inner product generated norm.
c. Determine the Taylor series expansion for f(x) through the quadratic term. Calculate the error in this approximation and compare to the error in the least squares approximation.

4.2. Repeat Problem 4.1 using the Legendre polynomials $P_0(x)$, $P_1(x)$, $P_2(x)$ as basis functions.

4.3. Find the least squares approximation for $f(x) = \cos^2(2x)$ over the interval $0 \le x \le 2\pi$ using $1, x^2, x^4$ as basis functions and using the standard inner product on C[0,2π].

4.4. Repeat Problem 4.3 using a set of orthonormal functions which have the same span as $1, x^2, x^4$ as basis functions.

4.5. Find the least squares approximation to $J_0(x)$ using $1, x^2, x^4$ as basis functions and using the inner product defined by $(f,g) = \int_0^1 f(x)g(x)xdx$.

4.6. Find the least squares approximation to $J_1(x)$ using x, x^3, x^5 as basis functions and using the inner product defined by $(f,g) = \int_0^1 f(x)g(x)xdx$.

4.6. Determine the least squares quadratic fit for y as a function of x for the data

X	0	0.25	0.5	0.75	1.0	1.25	1.5	1.75	2.0
Y	3.0	2.2	2.0	1.9	2.5	3.4	5.1	7.5	10.0

4.7. The following table presents the data for lattice sums of an atom above a carbon

X (nm)	1.0	1.5	2.0	2.5	3.0	3.5	4.0
S	2.40	0.50	0.17	0.076	0.040	0.025	0.015

By observation it is determined that the data may fir a curve of the form $S=ax^b$. Taking the natural logarithm of the proposed regression equation leads to $\ln(s)=\ln(a)+b\ \ln(x)$. Determine the linear least squares approximation for ln(s) as a function of ln(x). Use the regression to approximate S for x = 2.8 nm and extrapolate S for x = 5.0 nm.

4.8. The first ten zeroes of the Bessel function of the first kind of order zero $J_0\ (x)$ are 2.405, 5.530, 8.654, 11.792, 14.931, 18.071, 21.212, 24.352, 27.493 and 30.653. Use a quadratic least squares regression to predict the zeroes as a function of the number of the zero.

Problems 4.9–4.17 refer to the system of Figure P4.9. The differential equation for the non-dimensional steady-state temperature in the bar which is subject to an external heat generation over its length is

$$\frac{d}{dx}\left[\alpha(x)\frac{d\theta}{dx}\right]-\beta(x)\theta=q(x) \tag{a}$$

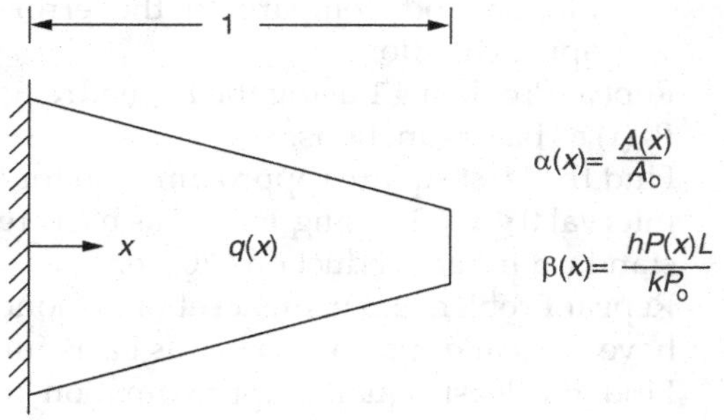

Figure P4.9

4.9. Consider a uniform bar such that $\alpha(x)=1$, $\beta(x)=0.5$ with $q(x)=x(1-x)$. The bar is subject to a constant temperature at x = 0 and is insulated at x = 1 such that $\theta(0)=0$, $d\theta/dx\ (1)=0$. Let S be the vector space which is the intersection of $P^3[0,1]$ with the subspace of C[0,1] of all functions satisfying the homogeneous boundary conditions. Find the best least squares approximation for the temperature distribution in the bar from S.

4.10. Consider a uniform bar such that $\alpha(x)=1, \beta(x)=0.5$ with $q(x)=x(1-x)$. The bar is subject to a constant temperature at $x=0$ and is insulated at $x=1$ such that $\theta(0)=0$, $d\theta/dx\ (1)=0$. Let S be the vector space which is the intersection of $P^3\ [0,1]$ with the subspace of C[0,1] of all functions satisfying the homogeneous boundary conditions. Fin the best Rayleigh-Ritz approximation for the temperature distribution from S.

4.11. Repeat Problem 4.10 but using an orthonormal basis for S with respect to the energy inner product.

4.12. Consider a uniform bar such that $\alpha(x)=1, \beta(x)=0.5$ with $q(x)=x(1-x)$. The bar is subject to a constant temperature at $x=0$ and is open for heat transfer at $x=1$ such that $\theta(0)=0$, $d\theta/dx\ (1)+0.5\theta(1)=0$. Let S be the vector space which is the intersection of $P^3\ [0,1]$ with the subspace of C[0,1] of all functions satisfying the homogeneous boundary conditions. Find the best Rayleigh-Ritz approximation for the temperature distribution in the bar from S.

4.13. Use the Rayleigh-Ritz method to approximate the temperature distribution in a surface with $\alpha(x)=(1-0.1x)^{3/2}$, and $q(x)=x(1-x)$. The end $x=0$ is maintained at a constant temperature, $\theta(0)=0$ while its other end is insulated $d\theta/dx\ (1)=0$. Use a basis for the intersection of $P^4\ [0,1]$ with the subspace of C[0,1] of all functions which satisfy the boundary conditions. Compare with the exact solution.

4.14. Use the Rayleigh-Ritz method to approximate the temperature distribution in a surface with $\alpha(x)=(1-0.1x)^3$, $\beta(x)=(1-0.1x)^{3/2}$ and $q(x)=x(1-x)$. The end $x=0$ is maintained at a constant temperature, $\theta(0)=0$ while its other end is insulated $d\theta/dx\ (1)=0$. Use a basis for the intersection of $P^4\ [0,1]$ with the subspace of C[0,1] of all functions which satisfy the boundary conditions. Compare with the exact solution.

4.15. Use the Rayleigh-Ritz method to approximate the temperature distribution in a surface with $\alpha(x)=(1-0.1x)$, $\beta(x)=1$ and $q(x)=x(1-x)$. The end $x=0$ is maintained at a constant temperature, $\theta(0)=0$ while its other end is insulated $d\theta/dx\ (1)=0$. Use a basis for the intersection of $P^4\ [0,1]$ with the subspace of C[0,1] of all functions which satisfy the geometric boundary condition only. Use the energy formulation of the inner products. Compare with the exact solution.

4.16. Use the finite-element method to solve Problem 4.10. Use five equally spaced elements between $x=0$ and $x=1$.

4.17. Use the finite-element method to solve Problem 4.15. Use five equally spaced elements between $x=0$ and $x=1$.

Problems 4.18–4.19 refer to the approximate solution of the nondimensional differential equation for the static deflection of a stretched beam on an elastic foundation and subject to transverse loading, as illustrated in Figure P4.18. In all problems the subspace S refers to the intersection of P^6 [0,1] with a subspace of C^4 [0,1] which consists of all functions which satisfy the homogeneous boundary conditions,

$$\frac{d^2}{dx^2}\left[\alpha(x)\frac{d^2w}{dx^2}\right]-\varepsilon\frac{d^2w}{dx^2}+\eta w=f(x) \tag{a}$$

Use $f(x)=2x^2(1-x)^2$ for all problems.

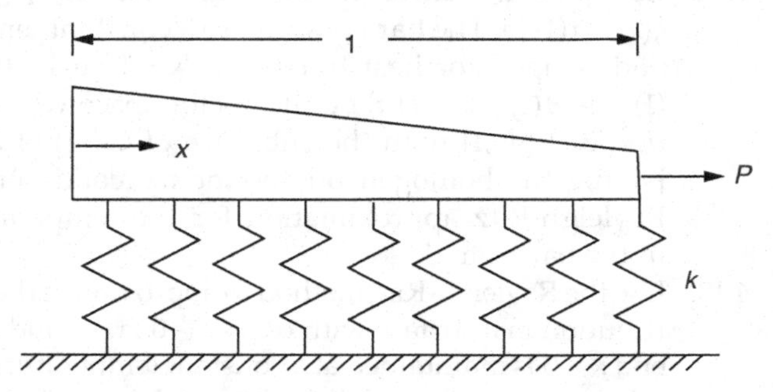

Figure P4.18

4.18. Use the least squares method to approximate the deflection of a uniform, $\alpha(x)=1$ fixed-pinned beam $w(0)=0$, $dw/dx\ (0)=0$, $w(1)=0$, $d^2w/dx^2\ (1)=0$ with $\varepsilon=0$ and $\eta=2$. Use basis functions for S as trial functions.

4.19. Use the Rayleigh-Ritz method to approximate the deflection of a uniform, $\alpha(x)=1$ fixed-pinned beam $w(0)=0$, $dw/dx\ (0)=0$, $w(1)=0, d^2w/dx^2\ (1)=0$ with $\varepsilon=0$ and $\eta=2$. Use basis functions for S as trial functions.

4.20. Use the Rayleigh-Ritz method to approximate the deflection of a uniform, $\alpha(x)=1$ fixed-free beam $w(0)=0$, $dw/dx\ (0)=0$, d^2w/dx^2 $(1)=0$, $d^3w/dx^3(1)=0$ with $\varepsilon=1$ and $\eta=2$. Use basis functions for S as trial functions.

4.21. Use the Rayleigh-Ritz method to approximate the deflection of a uniform, $\alpha(x)=1$ fixed-free beam $w(0)=0$, $dw/dx\ (0)=0$, d^2w/dx^2 $(1)=0$, $d^3w/dx^3\ (1)=0$ with $\varepsilon=1$ and $\eta=2$. Use basis functions from P^5 [0,1] which satisfy only the geometric boundary conditions as basis functions.

4.22. Repeat Problem 4.20, but with $\alpha(x)=1+0.2x$.

4.23. Solve Problem 4.19 using the finite element method. Use five equally spaced elements.

4.24. Use the finite-element method to solve Problem 4.20, but with $\alpha(x)=(1+0.25x)^2$, Use five equally spaced elements.

Chapter 5

Eigenvalue problems

5.1 Eigenvalue and eigenvector problems

An eigenvalue of an operator $\mathbf{L}$ is a scalar λ for which a nontrivial solution ($\mathbf{u} \neq \mathbf{0}$) of

$$\mathbf{L}\mathbf{u} = \lambda\mathbf{u} \tag{5.1}$$

exists. The resulting nontrivial solution is called an eigenvector of $\mathbf{L}$ corresponding to the eigenvalue λ.

An eigenvector corresponding to an eigenvalue of a linear operator is not unique. Let λ be an eigenvalue of a linear operator $\mathbf{L}$, and let $\mathbf{u} \neq \mathbf{0}$ be a corresponding eigenvector satisfying Equation 5.1. Define $\mathbf{v} = c\mathbf{u}$ for an arbitrary $c \neq 0$. Consider

$$\begin{aligned} \mathbf{L}\mathbf{v} &= \mathbf{L}(c\mathbf{u}) \\ &= c\mathbf{L}\mathbf{u} \\ &= c\lambda\mathbf{u} \\ &= \lambda\mathbf{v} \end{aligned} \tag{5.2}$$

Equation 5.2 illustrates that $\mathbf{v}$ is also an eigenvector of $\mathbf{L}$ corresponding to the eigenvalue λ.

Thus, an eigenvector of a linear operator is unique to at best a multiplicative constant.

If all the eigenvectors of an operator corresponding to an eigenvalue are multiples of one another, the eigenvalue is nondegenerate. If two or more linearly independent vectors exist for an eigenvalue, that eigenvalue is said to be degenerate.

Let λ_1 and λ_2 be distinct eigenvalues of a linear operator $\mathbf{L}$ with corresponding eigenvectors $\mathbf{u}_1$ and $\mathbf{u}_2$, respectively. If the eigenvectors are linearly dependent, then there is some $c \neq 0$ such that $\mathbf{u}_2 = c\mathbf{u}_1$. Then

$$\mathbf{L}\mathbf{u}_2 = \mathbf{L}(c\mathbf{u}_1) \tag{5.3}$$

Using the linearity of $\mathbf{L}$ and the definition of the eigenvalue-eigenvector problem, Equation 5.3 becomes

$$\begin{aligned}\lambda_2\mathbf{u}_2 &= c\lambda_1\mathbf{u}_1\\ &= \lambda_1\mathbf{u}_2\end{aligned} \tag{5.4}$$

Equation 5.5 implies that $\lambda_1 = \lambda_2$, which contradicts the assumption that the eigenvalues are distinct. Thus, the assumption that the vectors are linearly dependent is false. Since the pair of eigenvectors chosen is arbitrary, it is easily shown that if all pairs of vectors in a set are linearly independent, then all vectors in the set are linearly independent. Since the pair of eigenvectors chosen is arbitrary, all pairs of eigenvectors corresponding to distinct eigenvalues are linearly independent. Thus the set of eigenvectors corresponding to distinct eigenvalues of a linear operator are linearly independent.

5.2 Eigenvalues of adjoint operators

Let **L** be a linear operator whose domain is the vector space S_D. Let (u, v) constitute a valid inner product on S_D. Let $\mathbf{L}^*$ be the adjoint of **L**, whose domain is S_R, the range of **L**, with respect to the inner product (u, v). Thus

$$(\mathbf{Lu},\mathbf{v}) = (\mathbf{u},\mathbf{L}^*\mathbf{v}) \tag{5.5}$$

for all **u** in S_D and all **v** in S_R. The eigenvalue-eigenvector problem for **L** is of the form of Equation 5.1, and the eigenvalue-eigenvector problem for $\mathbf{L}^*$ is

$$\mathbf{L}^*\mathbf{v} = \mu\mathbf{v} \tag{5.6}$$

where μ is an eigenvalue of $\mathbf{L}^*$ and **v** is an eigenvector of $\mathbf{L}^*$ corresponding to μ.

Taking the inner product of Equation 5.1 with **v** and using properties of the inner product leads to

$$\begin{aligned}(\mathbf{Lu},\mathbf{v}) &= (\lambda\mathbf{u},\mathbf{v})\\ &= \lambda(\mathbf{u},\mathbf{v})\end{aligned} \tag{5.7}$$

Using the definition of the adjoint on the left-hand side of Equation 5.7 gives

$$\left(\mathbf{u},\mathbf{L}^*\mathbf{v}\right) = \lambda(\mathbf{u},\mathbf{v}) \tag{5.8}$$

However, since **v** is the eigenvector of $\mathbf{L}^*$ corresponding to the eigenvalue μ, Equation 5.8 can be rewritten as

$$(\mathbf{u},\mu\mathbf{v}) = \lambda(\mathbf{u},\mathbf{v}) \tag{5.9}$$

Noting that the properties of a valid inner product imply that $(\mathbf{u},\mu\mathbf{v})=\bar{\mu}(\mathbf{u},\mathbf{v})$, Equation 5.9 can be rearranged to yield

$$(\bar{\mu}-\lambda)(\mathbf{u},\mathbf{v})=0 \tag{5.10}$$

Equation 5.10 implies that either $(\mathbf{u}, \mathbf{v})=0$ or $\lambda=\bar{\mu}$. This conclusion leads to the following:

Theorem 5.1 (Biorthogonality) If λ is an eigenvalue of a linear operator **L** with a corresponding eigenvector **u**, and if $\mu \neq \bar{\lambda}$ is an eigenvalue of $\mathbf{L}^*$ with corresponding eigenvector **v**, then $(\mathbf{u}, \mathbf{v})=0$.

Proof The theorem follows directly form Equation 5.11.

Theorem 5.2 If $\mathbf{L}^*$ is the adjoint of a linear operator **L** with respect to a valid inner product, then the eigenvalues of $\mathbf{L}^*$ are the complex conjugates of the eigenvalues of **L**.

Proof Equation 5.11 implies that unless $\mu=\bar{\lambda}$, the eigenvectors of **L** and $\mathbf{L}^*$ are mutually orthogonal. Suppose that λ is an eigenvalue of **L**, but that $\bar{\lambda}$ is not an eigenvalue of $\mathbf{L}^*$. Then the eigenvector **u** is orthogonal to all eigenvectors of $\mathbf{L}^*$. It is mentioned in Section 5.6 that the set of eigenvectors of $\mathbf{L}^*$ is complete on S_R. The statement that **u** is orthogonal to all eigenvectors of $\mathbf{L}^*$ is thus contradictory, and the theorem is proved by contradiction.

Theorem 5.3 Eigenvalues of a self-adjoint operator are real.

Proof From Theorem 5.2, every eigenvalue of an adjoint operator $\mathbf{L}^*$ is the complex conjugate of an eigenvalue of the operator **L**. If λ is an eigenvalue of a self-adjoint operator **L**, then $\lambda=\bar{\lambda}$.

Theorem 5.4 Eigenvectors corresponding to distinct eigenvalues of a self-adjoint operator are mutually orthogonal with respect to the inner product for which the operator is self-adjoint.

Proof If the operator is self-adjoint, then **u** and **v** are eigenvectors corresponding to eigenvalues of **L**. The theorem follows directly from Equation 5.10.

Theorem 5.3 implies that if **L** is self-adjoint with respect to any inner product, then its eigenvalues are real. Theorem 5.4 then implies that if **L** is self-adjoint with respect to an inner product, then the eigenvectors are mutually orthogonal with respect to that inner product.

5.3 *Eigenvalues of positive definite operators*

Suppose that **L** is positive definite with respect to a defined inner product $(\mathbf{u}, \mathbf{v})$. That is,

$$(\mathbf{L}\mathbf{u},\mathbf{u})\geq 0 \tag{5.11}$$

for all **u** in S_D, and (**Lu**, **u**) = 0 if and only if **u** = **0**. Let λ be an eigenvalue of a positive definite operator with an eigenvector **u**, **Lu** = λ**u**. Taking the inner product of this equation with **u** leads to (**Lu**, **u**) = λ(**u**, **u**) or

$$\lambda = \frac{(\mathbf{Lu},\mathbf{u})}{\|\mathbf{u}\|^2} \tag{5.12}$$

where the norm is the inner-product-generated norm. Since **L** is positive definite, the right-hand side of Equation 5.12 is that of two positive quantities. This proves

Theorem 5.5 All eigenvalues of a positive definite operator are positive.

Consider an operator **L** which has an eigenvalue $\lambda = 0$. Then there exists $\mathbf{u} \neq \mathbf{0}$ such that **Lu** = **0.** For this **u**, (**Lu**,**u**) = (**0**,**u**) = 0. Hence the operator is, at best, non-negative definite. If an operator is non-negative definite, then Equation 5.13 shows that all its eigenvalues are non-negative. To show that $\lambda = 0$ must be an eigenvalue of a non-negative definite operator, it is necessary to show that (**Lu**, **u**) = 0 implies that **u** = **0** or **Lu** = **0**. It is shown in Section 5.6 that this is true for self-adjoint operators.

Theorem 5.6 If **L** is self-adjoint and non-negative definite with respect to an inner product (**u**, **v**), then $\lambda = 0$ is an eigenvalue of **L**.

5.4 *Eigenvalue problems for operators in finite-dimensional vector spaces*

Let **A** be a real $n \times n$ matrix. The eigenvalue-eigenvector problem for **A** is of the form

$$\mathbf{Au} = \lambda\mathbf{u} \tag{5.13}$$

where **u** is a vector in R^n. Let **I** be the $n \times n$ identity operator, **Iu** = **u**. Equation 5.13 can be rewritten as

$$(\mathbf{A} - \lambda\mathbf{I})\mathbf{u} = \mathbf{0} \tag{5.14}$$

Equation 5.14 can be expanded into

$$\begin{bmatrix} a_{1,1}-\lambda & a_{1,2} & a_{1,3} & \cdots & a_{1,n} \\ a_{2,1} & a_{2,2}-\lambda & a_{2,3} & \cdots & a_{2,n} \\ a_{3,1} & a_{3,2} & a_{3,3}-\lambda & \cdots & a_{3,n} \\ \vdots & \vdots & \vdots & \ddots & \vdots \\ a_{n,1} & a_{n,2} & a_{n,3} & \cdots & a_{n.n} \end{bmatrix} \begin{bmatrix} u_1 \\ u_2 \\ u_3 \\ \vdots \\ u_n \end{bmatrix} = \begin{bmatrix} 0 \\ 0 \\ 0 \\ \vdots \\ 0 \end{bmatrix} \tag{5.15}$$

Equation 5.15 represents a system of n simultaneous linear equations to solve for the components of the eigenvector. A nontrivial solution exists if and only if the determinant of the coefficient matrix is zero, that is,

$$\begin{vmatrix} a_{1,1}-\lambda & a_{1,2} & a_{1,3} & \cdots & a_{1,n} \\ a_{2,1} & a_{2,2}-\lambda & a_{2,3} & \cdots & a_{2,n} \\ a_{3,1} & a_{3,2} & a_{3,3}-\lambda & \cdots & a_{3,n} \\ \vdots & \vdots & \vdots & \ddots & \vdots \\ a_{n,1} & a_{n,2} & a_{n,3} & \cdots & a_{n.n} \end{vmatrix} = 0 \tag{5.16}$$

The determinant of a square matrix is a combination of permutations of elements of the matrix. Each permutation is a product of one element from each column. One permutation is the product of the diagonal elements. Thus it is clear that evaluation of the right-hand side of Equation 5.16 yields a polynomial in powers of λ with highest power n. The eigenvalues are the roots of this polynomial, called the characteristic polynomial.

Since the eigenvalues are the roots of a polynomial of order n, an $n \times n$ matrix has n eigenvalues. All eigenvalues may not be distinct; some may be repeated. If **A** is a real matrix, then all coefficients of its characteristic polynomial are real. This implies that if an eigenvalue is complex, that it has a nonzero imaginary part, then its complex conjugate is also an eigenvalue. The eigenvectors are the nontrivial solutions of Equation 5.15 corresponding to each eigenvalue.

It has been shown in chapter 2 that the adjoint of a real $n \times n$ matrix **A** with respect to the standard inner product on R^n is the transpose of the matrix, $\mathbf{A}^* = \mathbf{A}^T$. A matrix is self-adjoint with respect to the standard inner product if it is symmetric. Thus, from Theorems 5.3 and 5.4, all eigenvalues of a symmetric matrix are real, and the corresponding eigenvectors are mutually orthogonal with respect to the standard inner product on R^n.

Example 5.1 The stress tensor $\underset{\approx}{\sigma}$ defines the state of stress at a point in a continuum:

$$\underset{\approx}{\sigma} = \begin{pmatrix} \sigma_{11} & \sigma_{12} & \sigma_{13} \\ \sigma_{21} & \sigma_{22} & \sigma_{23} \\ \sigma_{31} & \sigma_{32} & \sigma_{33} \end{pmatrix} \tag{a}$$

The stress tensor represents the stress vector acting on three mutually perpendicular planes passing through the point. The stress vector acting on a plane can be resolved into a component normal to the plane (the normal stress) and a component that lies in the plane (the shearing stress), as illustrated in Figure 5.1. The diagonal elements are the normal stresses, the component of stress normal to the planes whose normals are in the x-, y-, and z-directions. The off-diagonal elements are the shear stresses; σ_{ij} is the stress acting in

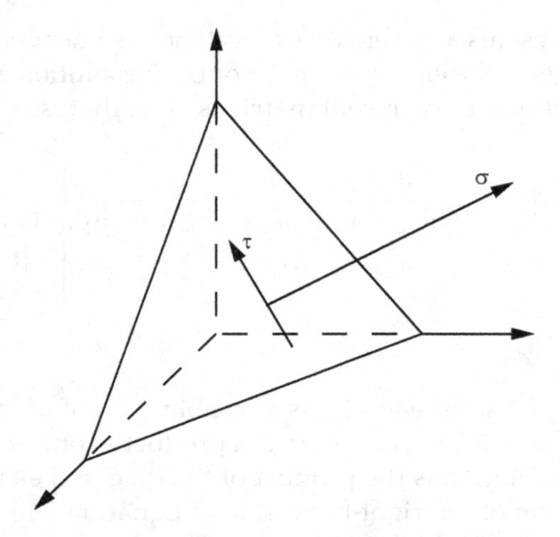

Figure 5.1 The stress vector acting on a plane is resolved into a component normal to the plane (the normal stress) and a component in the plane (the shear stress).

the *j*-direction on the plane whose normal is in the *i*-direction. The angular momentum equation can be used to show that the stress tensor is symmetric ($\sigma_{ij} = \sigma_{ji}$) in the absence of body coupling or intrinsic angular momentum.

The stress vector acting on a plane can be written in the form

$$\boldsymbol{\sigma} = \underset{\approx}{\sigma}\mathbf{n} \tag{b}$$

where **n** is a unit vector normal to the plane. The stress vector can be resolved into two components: a normal stress perpendicular to the plane and a shear stress in the plane:

$$\boldsymbol{\sigma} = \sigma\mathbf{n} + \tau\mathbf{m} \tag{c}$$

where σ is the normal stress, τ is the shear stress, and **m** is a unit vector in the plane.

A principal stress is defined as the normal stress on a plane on which the shear stress is zero:

$$\boldsymbol{\sigma} = \sigma\mathbf{n} \tag{d}$$

Equating Equation b and Equation d shows that the stress vector acting on a plane of principal stress is

$$\underset{\approx}{\sigma}\mathbf{n} = \sigma\mathbf{n} \tag{e}$$

Equation e can be written in matrix form as

$$\begin{pmatrix} \sigma_{11} & \sigma_{12} & \sigma_{13} \\ \sigma_{21} & \sigma_{22} & \sigma_{23} \\ \sigma_{31} & \sigma_{32} & \sigma_{33} \end{pmatrix} \begin{pmatrix} n_1 \\ n_2 \\ n_3 \end{pmatrix} = \sigma \begin{pmatrix} n_1 \\ n_2 \\ n_3 \end{pmatrix} \tag{f}$$

Thus, the principal stresses are the eigenvalues of the stress tensor, and their corresponding eigenvectors are vectors normal to the principal planes.

The principal stresses are eigenvalues of a 3×3 symmetric matrix. Thus there are three principal stresses. Since the stress tensor is symmetric, Theorems 5.3 and 5.4 imply that all principal stresses are real and that the planes on which the principal stresses act are mutually orthogonal. Note that the stress tensor is not necessarily positive definite, and that hence the principal stresses can be of either sign.

The eigenvectors are unique only to a multiplicative constant. Unit vectors normal to the principal planes are obtained by normalizing the eigenvectors, requiring that their standard norm be one:

$$\sqrt{n_1^2 + n_2^2 + n_3^2} = 1 \tag{f}$$

Example 5.2 Natural frequencies of discrete systems: The differential equations governing the free vibration of an undamped n-degree-of-freedom mechanical linear system such as that shown in Figure 5.2 are of the form

$$\mathbf{M}\ddot{\mathbf{x}} + \mathbf{K}\mathbf{x} = \mathbf{0} \tag{a}$$

where $\mathbf{x}$ is the n-dimensional displacement vector, the vector of displacements of designated particles, $\mathbf{M}$ is the $n \times n$ mass matrix and $\mathbf{K}$ is the $n \times n$ stiffness matrix. As an example, the differential equations of the system shown in Figure 5.2 are

$$\begin{bmatrix} m & 0 & 0 \\ 0 & m & 0 \\ 0 & 0 & 2m \end{bmatrix} \begin{bmatrix} \ddot{x}_1 \\ \ddot{x}_2 \\ \ddot{x}_3 \end{bmatrix} + \begin{bmatrix} 3k & -2k & 0 \\ -2k & 4k & -2k \\ 0 & -2k & 2k \end{bmatrix} \begin{bmatrix} x_1 \\ x_2 \\ x_3 \end{bmatrix} = \begin{bmatrix} 0 \\ 0 \\ 0 \end{bmatrix} \tag{b}$$

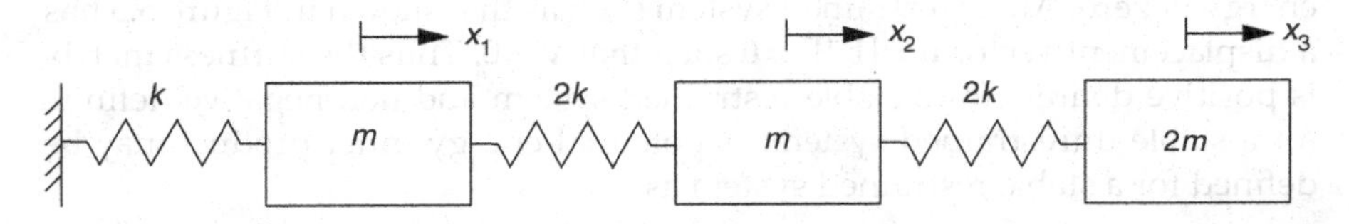

Figure 5.2 The natural frequencies of a multi-degree-of-freedom mechanical system are the square roots of the inverse of the system's mass matrix multiplied by its stiffness matrix.

When an energy formulation is used to derive the differential equations, both the mass and stiffness matrices are symmetric.

A normal-mode solution of Equation a can be assumed as

$$\mathbf{x} = \mathbf{X}e^{i\omega t} \tag{c}$$

where ω is a natural frequency of the system and $\mathbf{X}$ is its mode shape vector. Substitution of Equation c into Equation a leads to

$$\mathbf{KX} = \omega^2\mathbf{MX} \tag{d}$$

The mass matrix cannot be singular, hence its inverse $\mathbf{M}^{-1}$ exists, and one can premultiply Equation d by $\mathbf{M}^{-1}$ to obtain

$$\mathbf{M}^{-1}\mathbf{KX} = \omega^2\mathbf{X} \tag{e}$$

It can be concluded from Equation e that the natural frequencies of an n-degree-of-freedom system are the square roots of the eigenvalues of $\mathbf{M}^{-1}\mathbf{K}$ and that the mode shapes are their corresponding eigenvectors.

While it is guaranteed that $\mathbf{M}$ and $\mathbf{K}$ are symmetric matrices, $\mathbf{M}^{-1}\mathbf{K}$ is not necessarily symmetric. Thus $\mathbf{M}^{-1}\mathbf{K}$ is not necessarily self-adjoint with respect to the standard inner product on R^n. If $\mathbf{v}$ is the velocity vector for an n-degree-of-freedom system at any instant, then it can be shown that $(\mathbf{Mv},\mathbf{v}) = 2T$, where T is the kinetic energy of the system at that instant. However, the system's kinetic energy is always positive unless its velocity vector is zero. Thus, since $\mathbf{v}$ is defined arbitrarily, the mass matrix is positive definite with respect to the standard inner product, since it is also symmetric a kinetic-energy inner product is thus defined as

$$(\mathbf{u},\mathbf{v})_M = (\mathbf{Mu},\mathbf{v}) \tag{f}$$

If $\mathbf{u}$ is the displacement vector for an n-degree-of-freedom linear system at any instant, it can be shown that $(\mathbf{Ku},\mathbf{u}) = 2V$, where V is the potential energy of the system at that instant. The potential energy function for a stable system is always positive. Furthermore, if the system is restrained from rigid-body motion, then only a state of zero displacement leads to a potential energy of zero. An unrestrained system such as that shown in Figure 5.3 has a displacement vector $\mathbf{u} = [1\ 1]^T \neq \mathbf{0}$ such that $V = 0$. Thus the stiffness matrix is positive definite for a stable restrained system and non-negative definite for a stable unrestrained system. A potential-energy inner product may be defined for a stable restrained system as

$$(\mathbf{u},\mathbf{v})_K = (\mathbf{Ku},\mathbf{v}) \tag{g}$$

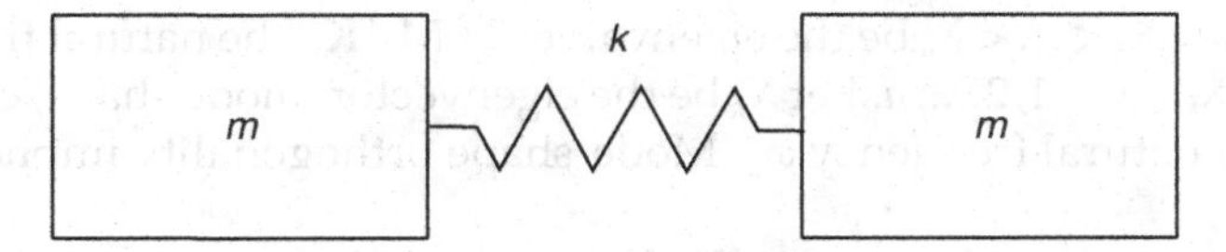

Figure 5.3 The stiffness matrix of an unrestrained system is singular. The lowest natural frequency of the system is zero, which corresponds to a rigid-body mode of vibration.

Consider the following derivation in which the definitions of energy inner products, the self-adjointness of **K** and **M** with respect to the standard inner product, and the properties of inner products are used:

$$\begin{aligned}(\mathbf{M}^{-1}\mathbf{K}\mathbf{u},\mathbf{v})_M &= (\mathbf{M}\mathbf{M}^{-1}\mathbf{K}\mathbf{u},\mathbf{v})\\ &= (\mathbf{K}\mathbf{u},\mathbf{v})\\ &= (\mathbf{u},\mathbf{K}\mathbf{v})\\ &= (\mathbf{u},\mathbf{M}\mathbf{M}^{-1}\mathbf{K}\mathbf{v})\\ &= (\mathbf{M}\mathbf{u},\mathbf{M}^{-1}\mathbf{K}\mathbf{v})\\ &= (\mathbf{u},\mathbf{M}^{-1}\mathbf{K}\mathbf{v})_M \end{aligned} \tag{h}$$

It can be concluded from Equation h that $\mathbf{M}^{-1}\mathbf{K}$ is self-adjoint with respect to the kinetic-energy inner product. A similar procedure is used to show that $\mathbf{M}^{-1}\mathbf{K}$ is self-adjoint with respect to the potential-energy inner product. The positive-definiteness of $\mathbf{M}^{-1}\mathbf{K}$ with respect to the kinetic-energy inner product is shown by

$$\begin{aligned}(\mathbf{M}^{-1}\mathbf{K}\mathbf{u},\mathbf{u})_M &= (\mathbf{M}\mathbf{M}^{-1}\mathbf{K}\mathbf{u},\mathbf{u})\\ &= (\mathbf{K}\mathbf{u},\mathbf{u})\end{aligned} \tag{i}$$

Equation i implies that if **K** is positive definite with respect to the standard inner product, then $\mathbf{M}^{-1}\mathbf{K}$ is positive definite with respect to the kinetic-energy inner product.

Since $\mathbf{M}^{-1}\mathbf{K}$ is self-adjoint with respect to the kinetic-energy inner product, Theorem 5.3 guarantees that all its eigenvalues are real. Theorem 5.5 and the positive-definiteness of $\mathbf{M}^{-1}\mathbf{K}$ guarantee that all its eigenvalues are positive. Theorem 5.4 and the self-adjointness of $\mathbf{M}^{-1}\mathbf{K}$ lead to the conclusion that eigenvectors corresponding to distinct eigenvalues of $\mathbf{M}^{-1}\mathbf{K}$ are mutually orthogonal with respect to the kinetic-energy inner product.

Let $0<\lambda<\lambda_2<\ldots<\lambda_n$ be the eigenvalues of $\mathbf{M}^{-1}\mathbf{K}$. The natural frequencies are $\omega_k=\sqrt{\lambda_k}\quad k=1,2,\ldots,n$. Let $\mathbf{X}_k$ be the eigenvector (mode shape) corresponding to the natural frequency ω_k. Mode-shape orthogonality implies that

$$0=(\mathbf{X}_i,\mathbf{X}_j)_{\mathrm{M}}\quad i\neq j \tag{j}$$

The mode shapes are normalized by requiring that

$$(\mathbf{X}_i,\mathbf{X}_i)_{\mathrm{M}}=1 \tag{k}$$

If the mode shapes are normalized according to Equation k, then

$$\begin{aligned}(\mathbf{X}_i,\mathbf{X}_i)_{\mathrm{K}}&=(\mathbf{K}\mathbf{X}_i,\mathbf{X}_i)\\&=\left(\omega_i^2\mathbf{M}\mathbf{X}_i,\mathbf{X}_i\right)\\&=\omega_i^2(\mathbf{X}_i,\mathbf{X}_i)_{\mathrm{M}}\\&=\omega_i^2\end{aligned} \tag{l}$$

Example 5.3 Determine the natural frequencies and normalized mode-shape vectors for the system shown in Figure 5.2 and demonstrate orthogonality of the mode shapes.

Solution Using the results of Example 5.2, the natural frequencies are the square roots of the eigenvalues of $\mathbf{M}^{-1}\mathbf{K}$. Using the mass and stiffness matrices from Equation b of Example 5.2,

$$\mathbf{M}^{-1}\mathbf{K}=\phi\begin{bmatrix}3&-2&0\\-2&4&-2\\0&-1&1\end{bmatrix} \tag{a}$$

where $\phi=k/m$. The eigenvalues of $\mathbf{M}^{-1}\mathbf{K}$ are determined from

$$\begin{aligned}0&=\left|\mathbf{M}^{-1}\mathbf{K}-\lambda\mathbf{I}\right|\\&=\begin{vmatrix}3\phi-\lambda&-2\phi&0\\-2\phi&4\phi-\lambda&-2\phi\\0&-\phi&\phi-\lambda\end{vmatrix}\end{aligned} \tag{b}$$

The characteristic equation obtained by expanding the determinant of Equation b is

$$-\lambda^3+8\phi\lambda^2-13\phi^2\lambda+2\phi^3=0 \tag{c}$$

The roots of Equation c are 0.1716 ϕ, 2 ϕ, and 5.8284 ϕ. The natural frequencies, the square roots of the eigenvalues, are

$$\omega_1 = 0.4142\sqrt{\frac{k}{m}} \qquad \omega_2 = 1.4142\sqrt{\frac{k}{m}} \qquad \omega_3 = 2.4142\sqrt{\frac{k}{m}} \tag{d}$$

The eigenvector corresponding to the middle eigenvalue is obtained from

$$\begin{bmatrix} 3\phi-\lambda_2 & -2\phi & 0 \\ -2\phi & 4\phi-\lambda_2 & -2\phi \\ 0 & -\phi & \phi-\lambda_2 \end{bmatrix}\begin{bmatrix} X_1 \\ X_2 \\ X_3 \end{bmatrix} = \begin{bmatrix} 0 \\ 0 \\ 0 \end{bmatrix}$$
$$\begin{bmatrix} \phi & -2\phi & 0 \\ -2\phi & 2\phi & -2\phi \\ 0 & -\phi & -\phi \end{bmatrix}\begin{bmatrix} X_1 \\ X_2 \\ X_3 \end{bmatrix} = \begin{bmatrix} 0 \\ 0 \\ 0 \end{bmatrix} \tag{e}$$

The first line of the matrix system of Equation e leads to $X_1 = 2X_2$, while the third line leads to $X_3 = -X_2$. Letting X_2 be an arbitrary constant C, an eigenvector corresponding to λ_2 is $\mathbf{X}_2 = C[2 \quad 1 \quad -1]^T$. The eigenvector is normalized by requiring that $(\mathbf{X}_2, \mathbf{X}_2)_M = 1$. To this end,

$$C^2 m[2 \quad 1 \quad -1]\begin{bmatrix} 1 & 0 & 0 \\ 0 & 1 & 0 \\ 0 & 0 & 2 \end{bmatrix}\begin{bmatrix} 2 \\ 1 \\ -1 \end{bmatrix} = 1 \tag{f}$$

Evaluation of the inner product in Equation f and solving for C gives $C = 0.3780/\sqrt{m}$, leading to a normalized eigenvector, $\mathbf{X}_2 = (1/\sqrt{m})[0.7559 \quad 0.3980 \quad -0.3980]$. The normalized eigenvectors corresponding to natural frequencies of ω_1 and ω_3 are obtained using similar procedures. The results are $\mathbf{X}_1 = 1/\sqrt{m}[0.3366 \quad 0.4760 \quad 0.5745]^T$ are $\mathbf{X}_3 = (1/\sqrt{m})[0.5615 \quad -0.7941 \quad 0.1645]^T$.

5.5 Second-order differential operators

The general form of a linear second-order differential operator is

$$Ly = a_2(x)\frac{d^2y}{dx^2} + a_1(x)\frac{dy}{dx} + a_0(x)y \tag{5.17}$$

The right-hand side of Equation 5.17 can be manipulated to yield an alternate representation,

$$Ly = \frac{1}{r(x)}\left[\frac{d}{dx}\left(p(x)\frac{dy}{dx}\right) + q(x)y\right] \tag{5.18}$$

The functions in Equation 5.18 are defined such that $r(x)>0$. Application of the product rule for differentiation to the right-hand side of Equation 5.18 leads to

$$Ly=\frac{1}{r(x)}\left[p(x)\frac{d^2y}{dx^2}+\frac{dp}{dx}\frac{dy}{dx}+q(x)y\right] \tag{5.19}$$

The operator in Equation 5.19 is equivalent to the operator in Equation 5.17 if

$$a_2(x)=\frac{p(x)}{r(x)} \tag{5.20}$$

$$a_1(x)=\frac{1}{r(x)}\frac{dp}{dx} \tag{5.21}$$

$$a_0=\frac{q(x)}{r(x)} \tag{5.22}$$

Dividing Equation 5.21 by Equation 5.20 leads to

$$\frac{a_1(x)}{a_2(x)}=\frac{1}{p(x)}\frac{dp}{dx} \tag{5.23}$$

which when integrated gives

$$\ln p=\int\frac{a_1}{a_2}dx+\hat{C} \tag{5.24}$$

or

$$p(x)=Ce^{\int\frac{a_1}{a_2}dx} \tag{5.25}$$

Without loss of generality, $C=1$.

Equation 5.20 implies that

$$r(x)=\frac{p(x)}{a_2(x)} \tag{5.26}$$

and from Equation 5.22

$$q(x)=a_0(x)r(x) \tag{5.27}$$

If Equation 5.26 leads to $r(x)<0$, the negative sign is shifted to $p(x)$ to ensure that $r(x)$ is always positive.

The operator L is defined for all y(x) in S_D, a subspace of $C^2[a, b]$. Define an inner product on $C^2[a, b]$ as

$$(f,g)_r = \int_a^b f(x)g(x)r(x)dx \tag{5.28}$$

Consider

$$(Lf,g)_r = \int_a^b \frac{1}{r(x)}\left[\frac{d}{dx}\left(p(x)\frac{dt}{dx}\right) + q(x)f(x)\right]g(x)r(x)dx$$

$$= \int_a^b \frac{d}{dx}\left[p(x)\frac{dt}{dx}\right]g(x)dx + \int_a^b q(x)f(x)g(x)dx \tag{5.29}$$

Using integration by parts on the first integral of Equation 5.29 and choosing $u = g(x)$ and $dv = (d/dx)[p(x)(df/dx)]dx$ such that $du = (dg/dx)dx$ and $v = p(x)(df/dx)$ leads to

$$(Lf,g)_r = p(x)g(x)\frac{df}{dx}\bigg|_a^b - \int_a^b p(x)\frac{df}{dx}\frac{dg}{dx}dx + \int_a^b q(x)f(x)g(x)dx \tag{5.30}$$

Using integration by parts on the first integral of Equation 5.30 and choosing $u = p(x)(dy/dx)$ and $dv = (df/dx)dx$ such that $du = d/dx[p(x)(dg/dx)]\,dx$ and $v = f(x)$ leads to

$$(Lf,g)_r = p(x)g(x)\frac{df}{dx}\bigg|_a^b - p(x)\frac{dg}{dx}f(x)\bigg|_a^b + \int_a^b \left[\frac{d}{dx}\left[p(x)\frac{dg}{dx}\right]f(x) + q(x)f(x)g(x)\right]dx$$

$$= p(x)g(x)\frac{df}{dx}\bigg|_a^b - p(x)\frac{dg}{dx}f(x)\bigg|_a^b + \int_a^b \frac{1}{r(x)}\left[\left(\frac{d}{dx}p(x)\frac{dg}{dx}\right) + q(x)g(x)\right]f(x)r(x)dx \tag{5.31}$$

$$= p(x)g(x)\frac{df}{dx}\bigg|_a^b - p(x)\frac{dg}{dx}f(x)\bigg|_a^b + (f,Lg)_r$$

It is clear from Equation 5.31 that **L** is self-adjoint with respect to the inner product defined in Equation 5.28 if

$$p(x)g(x)\frac{df}{dx}\bigg|_a^b - p(x)\frac{dg}{dx}f(x)\bigg|_a^b = 0 \tag{5.32}$$

for all $f(x)$, $g(x)$ in S_D. Note that

$$\begin{aligned} &p(x)g(x)\frac{df}{dx}\bigg|_a^b - p(x)\frac{dg}{dx}f(x)\bigg|_a^b \\ &= p(b)g(b)\frac{df}{dx}(b) - p(a)g(a)\frac{df}{dx}(a) - p(b)\frac{dg}{dx}(b)f(b) + p(a)\frac{dg}{dx}(a)f(a) \\ &= p(b)\left[g(b)\frac{df}{dx}(b) - \frac{dg}{dx}(b)f(b)\right] - p(a)\left[g(a)\frac{df}{dx}(a) - \frac{dg}{dx}(a)f(a)\right] \end{aligned} \tag{5.33}$$

Clearly the right-hand side of Equation 5.33 is zero if $p(b) = p(a) = 0$.
The general form of the boundary conditions at $x = a$ and $x = b$ is

$$y(a) + \alpha y'(a) = 0 \tag{5.34}$$

$$y(b) + \beta y'(b) = 0 \tag{5.35}$$

Since $f(x)$ and $g(x)$ satisfy these conditions, substitution of Equation 5.34 and Equation 5.35 into Equation 5.33 leads to

$$\begin{aligned} &p(b)\left[g(b)\frac{df}{dx}(b) - \frac{dg}{dx}(b)f(b)\right] - p(a)\left[g(a)\frac{df}{dx}(a) - \frac{dg}{dx}(a)f(a)\right] \\ &\quad = p(b)[-\beta g'(b)f'(b) - g'(b)(-\beta f'(b))] \\ &\quad - p(a)[-\alpha g'(a)f'(a) - g'(a)(-\alpha f'(a))] = 0 \end{aligned} \tag{5.36}$$

If $\alpha = 0$ in Equation 5.33, then the terms in Equation 5.33 evaluated at $x = a$ are identically zero. If $\alpha = \infty$, such that Equation 5.34 is $y'(a) = 0$ then the terms in Equation 5.33 evaluated at $x = a$ are identically zero.

Consider next the case where $p(a) = p(b)$ and where S_D is defined to be composed of all functions $y(x)$ such that $y(a) = y(b)$ $y'(a) = y'(b)$ In this case, the right-hand side of Equation 5.32 becomes

$$\begin{aligned} &p(b)[g(b)f'(b) - g'(b)f(b)] - p(a)[g(a)f'(a) - g'(a)f(a)] \\ &\quad - p(b)[g(b)f'(b) - g'(b)f(b)] + p(a)[g(a)f'(a) - g'(a)f(a)] = 0 \end{aligned} \tag{5.37}$$

The following theorem has been proved:

Theorem 5.7 Any linear second-order differential operator can be written as $Ly = 1/r(x)[d/dx(p(x)(dy/dx)) + q(x)y]$, which is self-adjoint with respect to the inner product $(f,g)_r = \int_a^b f(x)g(x)r(x)dx$, provided that

(i) either $p(a) = 0$ or S_D is defined such that $y(a) + \alpha y'(a) = 0$ for all y in S_D and either $p(b) = 0$ or S_D is defined such that $y(b) + \beta y'(b) = 0$ for all y in S_D
or
(ii) $p(a) = p(b)$ and S_D is defined by

$$y(a) = y(b)$$

and

$$y'(a) = y'(b)$$

Now consider the positive definiteness of **L**. Using Equation 5.30, after one application of integration by parts,

$$(Lf, f)_r = p(x)f(x)\frac{df}{dx}\bigg|_a^b - \int_a^b p(x)\frac{df}{dx}\frac{df}{dx}dx + \int_a^b q(x)f(x)f(x)dx \tag{5.38}$$

If $p(x) \leq 0$, then $-\int_a^b p(x)[df/dx]^2\,dx \geq 0$. Now if $p(a) = p(b) = 0$, then the boundary terms are zero. Also if $p(a) = p(b)$ and if $f(a) = f(b)$, and $f'(a) = f'(b)$, then the boundary terms are zero.

Now consider boundary conditions of the form of Equation 5.34 and Equation 5.35. Then from Equation 5.37, the boundary terms simplify to

$$\begin{aligned} &p(b)y'(b)(-\beta y'(b)) - p(a)y'(a)(-\alpha y'(a)) \\ &= -\beta p(b)y'(b)^2 + \alpha p(a)y'(a)^2 \end{aligned} \tag{5.39}$$

If $p(a) < 0$ and $p(b) < 0$, then the boundary terms are non-negative if $\beta \geq 0$ and $\alpha \leq 0$. If a boundary condition is simply $y(a) = 0$ or $y(b) = 0$, then the corresponding boundary term is zero.

From the above and from Equation 5.39, it is clear that if $p(x) \leq 0$ and $q(x) \geq 0$ for all x, $a \leq x \leq b$, and if boundary conditions are of the form of Equation 5.35 and Equation 5.36 with $\beta \geq 0$ and $\alpha \leq 0$, then **L** is at least non-negative definite. Furthermore, Equation 5.39 shows that if $q(x) > 0$ for some x, then **L** is positive definite. Thus it only remains to determine the conditions under which, if $q(x) = 0$, there exists an $f(x) \neq 0$ such that $(\mathbf{L}f,f)_r = 0$. From Equation 5.39 and Equation 5.40, this is possible only if $df/dx = 0$ for all x, $a \leq x \leq b$, which requires that $f(x) = c_0 + c_1x$. If the boundary conditions are of the form of Equation 5.35 and Equation 5.36, then application of Equation 5.35 leads to $c_1(1 + \alpha a) + \alpha c_0 = 0$ and $c_1(1 + \beta b) + \beta c_0 = 0$. The only solution of these equations is $c_1 = c_0 = 0$.

However, if the boundary conditions are such $y'(a)=0$ and $y'(b)=0$, then $f(x)=c_0$ satisfies both conditions and is in the domain of **L**. In this latter case, **L** is non-negative definite.

If the domain of **L** is specified by boundary conditions of the form $y(a)=y(b)$ and $y'(a)=y'(b)$, then $f(x)=c_0$ is in the domain of **L**, and **L** is non-negative definite.

The above results can be summarized in the following theorem:

Theorem 5.8 A second-order differential operator defined by $Ly=1/r(x)\, d[p(x)dy/dx]/dx+[q(x)y]$ is positive definite with respect to the inner product $(f,y)=\int_a^b f(x)g(x)r(x)dx$ if $p(x)\leq 0$ and $q(x)>0$ for all x, $a\leq x\leq b$ and S_D is defined by the conditions of Theorem 5.7 with $\alpha\leq 0$ and $\beta\geq 0$. If $q(x)=0$ and S_D is defined by the boundary conditions of Equation 5.35, then **L** is positive definite only if either α or β has a finite value. Also, if $q(x)=0$, $p(a)=p(b)=0$, and if S_D is defined such that $y(a)=y(b)$ and $y'(a)=y'(b)$. then **L** is non-negative definite.

A second-order differential operator with boundary conditions such that it is self-adjoint and non-negative definite with respect to the inner product $(f,g)_r=\int_a^b f(x)g(x)r(x)dx$ constitutes a Sturm-Liouville system. The eigenvalue problem for a Sturm-Liouville system, called a Sturm-Liouville problem, is of the form

$$\frac{1}{r(x)}\left[\frac{d}{dx}\left(p(x)\frac{dy}{dx}\right)+q(x)y\right]=\lambda y(x) \tag{5.40}$$

with boundary conditions as specified in Theorems 5.7 and 5.8.

The domain of a Sturm-Liouville operator **L** is specified by boundary conditions derived from the geometry and physics of a physical problem. Note that if $p(a)=0$, then the only condition required at $x=a$ is that $y(a)$ and $y'(a)$ remain finite. Often Sturm-Liouville operators obtained from modeling physical systems using cylindrical coordinates have $p(r)=r$, hence $p(0)=0$. There is no boundary in a circular cylinder at $r=0$; however, it is expected that the dependent variable remains finite at the center of the cylinder.

The conditions $y(a)=y(b)$ and $y'(a)=y'(b)$ are often called periodicity conditions in that they require the dependent variable and its spatial derivative to have the same values at both ends of the interval. These conditions are often enforced in the circumferential direction in cylindrical coordinates using the angular coordinate θ, $0\leq\theta\leq 2\pi$. For the solution to be single-valued at $x=0$, it is required $y(0)=y(2\pi)$ and $y'(0)=y'(2\pi)$.

The following theorem is a summary of theorems 5.3, 5.4, 5.7, and 5.8:

Theorem 5.9 The eigenvalues of a Strum-Liouville problem are real and non-negative, $0\leq\lambda_1\leq\lambda_2\leq\lambda_3\leq\cdots\leq\lambda_{k-1}\leq\lambda_k\leq\lambda_{k+1}\leq\cdots$. If conditions, as described in Theorem 5.8, are such that **L** is positive definite on its domain, then the smallest eigenvalue is positive. Furthermore, eigenvectors, $w_i(x)$ and $w_j(x)$

corresponding to distinct eigenvalues λ_i and λ_j of a Sturm-Liouville problem are orthogonal such that $(w_i, w_j)_r = 0$.

Example 5.4 A MEMS actuator is modeled as a simply supported uniform column of length L, pinned at both ends, made from a material of elastic modulus E, and having a uniform cross-section of area A and area moment of inertia I. During operation, the column is subject to resistive heating which changes the temperature of the column from its installed temperature T_0. The increase in temperature induces thermal stresses in the column leading to a uniform compressive axial load,

$$P = \alpha AE(T - T_0) \tag{a}$$

where α is the coefficient of thermal expansion for the column. Let $w(x)$ represent the transverse deflection of the column. A free-body diagram of a segment of the column of length x is illustrated in Figure 5.4. The internal bending moment is related to the deflection by

$$M = -EI\frac{d^2w}{dx^2} \tag{b}$$

Summation of moments about the left support leads to

$$M + Pw = 0 \tag{c}$$

which, when substituted into Equation b, leads to

$$-EI\frac{d^2w}{dx^2} = Pw \tag{d}$$

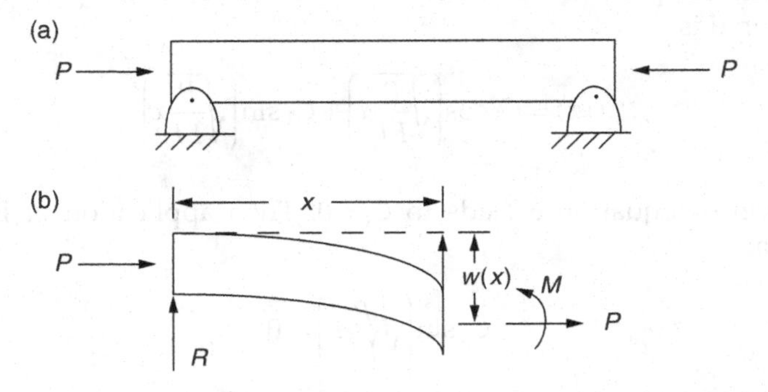

Figure 5.4 (a) The column is subject to a uniform axial load which develops due to change in temperature from an initial temperature. (b) If the column has a transverse deflection, a free-body diagram of a segment of the column reveals an internal bending moment and axial load.

Since the column is restrained from displacement at each end,

$$w(0)=0 \tag{e}$$

$$w(L)=0 \tag{f}$$

An obvious solution of Equation d, Equation e, and Equation f is $w(x)=0$. The critical buckling load is the smallest value of P for which the transverse displacement of the column is nonzero. Determine the critical buckling temperature, the smallest temperature at which the column buckles.

Solution It is clear from Equation d that the critical buckling load is the smallest eigenvalue of the operator

$$Lw=-EI\frac{d^2w}{dx^2} \tag{g}$$

with a domain defined by the boundary conditions of Equation e and Equation f. Comparison of the form of **L** defined in Equation g with the self-adjoint form of Equation 5.18 reveals that $p(x)=-EI$, $r(x)=1$ and $q(x)=0$. The operator of Equation d is in the form of a Sturm-Liouville operator; thus, from Theorem 5.7, the operator is self-adjoint with respect to the standard inner product on $C^2[0, L]$. The form of the operator and the boundary conditions are such that the conditions of Theorem 5.8 are satisfied and the operator is positive definite.

Since the operator is self-adjoint, all eigenvalues must be real, which implies that a critical buckling load exists. The positive definiteness of the operator confirms the obvious, that the critical buckling load is positive.

Assuming a solution of the form $w(x)=e^{\beta x}$ and substituting into the Sturm-Liouville problems leads to $\beta=\pm i\sqrt{P/EI}$. Thus the general solution of Equation d is

$$w(x)=C_1\cos\left(\sqrt{\frac{P}{EI}}x\right)+C_2\sin\left(\sqrt{\frac{P}{EI}}x\right) \tag{h}$$

Application of Equation e leads to $C_1=0$. Then application of Equation f shows that

$$C_2\sin\left(\sqrt{\frac{P}{EI}}L\right)=0 \tag{i}$$

A nontrivial solution exists only if

$$\sin\left(\sqrt{\frac{P}{EI}}L\right)=0 \tag{j}$$

The critical buckling load, the smallest value of P which satisfies Equation j, is

$$P_{cr} = \frac{\pi^2 EI}{L^2} \tag{k}$$

The critical temperature, the temperature at which buckling occurs, can be obtained from Equation a and Equation k as

$$\begin{aligned} T_{cr} &= T_0 + \frac{P_{cr}}{\alpha AE} \\ &= T_0 + \frac{\pi^2 I}{\alpha AL^2} \end{aligned} \tag{l}$$

The resulting buckled shape of the column is

$$w(x) = C\sin\left(\frac{\pi x}{L}\right) \tag{m}$$

Example 5.5 Consider the boundary-value problem

$$\frac{d^2y}{dx^2} + 3\frac{dy}{dx} + \lambda y = 0 \tag{a}$$

$$y(0) = 0 \tag{b}$$

$$y'(1) + 3y(1) = 0 \tag{c}$$

a. Determine the values of λ for which a nontrivial solution of Equation a, Equation b, and Equation c exists.
b. For each eigenvalue, determine its normalized mode shape, and
c. Demonstrate orthogonality of the eigenvectors corresponding to distinct eigenvalues.

Solution The eigenvalue problem of Equation a, Equation b, and Equation c is in the form of a Sturm-Liouville problem with

$$Ly = -\frac{d^2y}{dx^2} - 3\frac{dy}{dx} \tag{d}$$

The operator of Equation e is of the form of Equation 5.18 with $a_2(x) = -1$, $a_1(x) = -3$, and $a_0(x) = 0$. The operator can be rewritten in the form of Equation 5.18 using Equation 5.25, Equation 5.26, and Equation 5.27:

$$p(x) = e^{\int \frac{a_1(x)}{a_2(x)}dx}$$

$$= e^{\int \frac{(-3)}{(-1)}dx} \tag{e}$$

$$= e^{3x}$$

$$r(x) = \frac{p(x)}{a_2(x)} \tag{f}$$

$$= -e^{3x}$$

The function $r(x)$ as determined in Equation f is negative; therefore $p(x)$ and $r(x)$ can be redefined as

$$p(x) = -e^{3x} \tag{g}$$

$$r(x) = e^{3x} \tag{h}$$

Since $a_0(x) = 0$, Equation 5.28 leads to $q(x) = 0$. The operator can be written in the form of Equation 5.19 as

$$L = \frac{1}{e^{3x}}\frac{d}{dx}\left(-e^{3x}\frac{dy}{dx}\right) \tag{i}$$

The boundary conditions are of the form of Equation 5.34 and Equation 5.35 with $\alpha = 0$ and $\beta = 3$. Thus the conditions of Theorem 5.8 are satisfied, and **L** is a positive definite operator.

A solution of Equation a is assumed to be of the form

$$y(x) = e^{\alpha x} \tag{j}$$

Substitution of Equation j into Equation a leads to

$$\alpha^2 + 3\alpha + \lambda = 0 \tag{k}$$

Application of the quadratic formula to Equation k gives

$$\alpha = \frac{1}{2}(-3 \pm \sqrt{9 - 4\lambda}) \tag{l}$$

Since **L** is self-adjoint and positive definite with respect to an inner product, the values of λ are all real and positive. The mathematical form of the solution depends on the values of α; if the values of α are real and distinct, $(\lambda < 9/4)$

exponential solutions are obtained, whereas if the values of α are complex conjugates, $(\lambda > 9/4)$ trigonometric solutions are obtained. The case when the values α are real but not distinct, $(\lambda = 9/4)$, is a special case.

First, suppose that $(\lambda < 9/4)$, thus defining $\alpha_1 = 1/2\left(-3-\sqrt{9-4\lambda}\right)$ and $\alpha_2 = 1/2\left(-3+\sqrt{9-4\lambda}\right)$; the general solution of Equation a is then

$$y(x) = C_1 e^{\alpha_1 x} + C_2 e^{\alpha_2 x} \tag{m}$$

where C_1 and C_2 are constants of integration. Requiring the solution to satisfy the boundary conditions, Equation b and Equation c, leads to

$$C_1 + C_2 = 0 \tag{n}$$

$$(\alpha_1 + 3)C_1 + (\alpha_2 + 3)C_2 = 0 \tag{o}$$

Equation n and Equation o have only the trivial solution unless the determinant of the coefficient matrix obtained when formulating the equations in a matrix form is singular. This would require $\alpha_1 - \alpha_2 = 0$, which is contrary to the hypothesis of $\lambda < 9/4$. Thus there are no eigenvalues such that $\lambda < 9/4$.

Assume that $\lambda > 9/4$. The general solution of Equation a is

$$y(x) = C_1 e^{-\frac{3}{2}x} \cos\left(\frac{1}{2}\sqrt{4\lambda - 9}x\right) + C_2 e^{-\frac{3}{2}x} \sin\left(\frac{1}{2}\sqrt{4\lambda - 9}x\right) \tag{p}$$

Application of Equation b to Equation p gives $C_1 = 0$. Application of Equation c to the resulting form of Equation p leads to

$$C_2\left[\frac{3}{2}\sin\left(\sqrt{4\lambda-9}\right) + \sqrt{4\lambda-9}\cos\left(\sqrt{4\lambda-9}\right)\right] = 0 \tag{q}$$

A nontrivial solution exists only if $C_2 = 0$. Thus, Equation q implies that the eigenvalues are solutions of

$$-\frac{2}{3}\sqrt{4\lambda-9} = \tan\left(\sqrt{4\lambda-9}\right) \tag{r}$$

Defining $q = \sqrt{4\lambda - 9}$, Equation r is rewritten as

$$-\frac{2}{3}q = \tan q \tag{s}$$

Equation s is a transcendental equation whose solutions are obtained numerically. The solutions correspond to the points of intersection of the graphs of $-(2/3)q$ and $\tan(q)$ as illustrated in Figure 5.5. The five smallest values of q which satisfy Equation s and their corresponding values of λ are listed in Table 5.1.

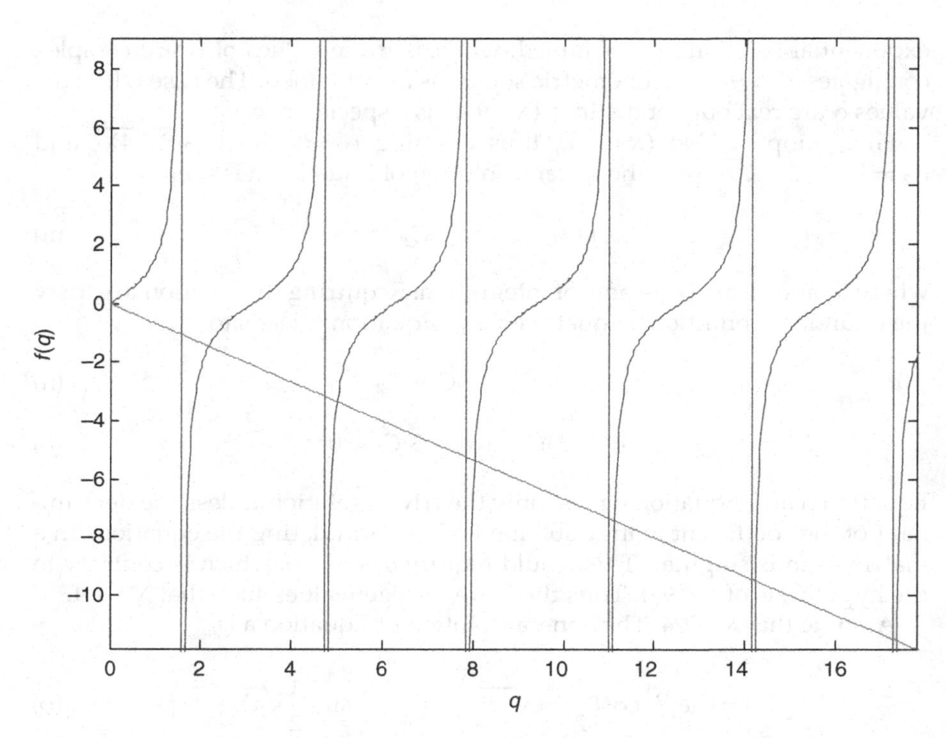

Figure 5.5 The transcendental equation tan $q = -(2/3)q$ has an infinite, but countable, number of positive solutions. As q gets large, the solutions approach odd multiples of $\pi/2$.

Table 5.1 Eigenvalues for Example 5.5

i	1	2	3	4	5
q	2.175	5.004	8.038	11.130	14.242
λ	3.433	8.510	18.403	33.219	52.959

There are an infinite, but countable, number of eigenvalues which can be indexed as

$$0 < \lambda_1 < \lambda_2 < \cdots < \lambda_{k-1} < \lambda_k < \lambda_{k+1} < \cdots \tag{t}$$

The corresponding eigenvectors are

$$y_k(x) = C_k e^{-\frac{3}{2}x} \sin\left(\sqrt{4\lambda_k - 9}\,x\right) \tag{u}$$

Eigenvectors corresponding to distinct eigenvalues satisfy the orthogonality conditions $(y_i, y_j)_r = 0$, which leads to

$$
\begin{aligned}
0 &= \int_0^1 \left[C_i e^{-\frac{3}{2}x} \sin\left(\sqrt{4\lambda_i - 9}x\right)\right]\left[C_j e^{-\frac{3}{2}x} \sin\left(\sqrt{4\lambda_j - 9}x\right)\right] e^{3x} dx \\
&= \int_0^1 C_i C_j \sin\left(\sqrt{4\lambda_i - 9}x\right)\sin\left(\sqrt{4\lambda_j - 9}x\right) dx
\end{aligned} \tag{v}
$$

The eigenvectors are normalized by requiring

$$
\begin{aligned}
1 &= (y_k, y_k)_r \\
&= \int_0^1 \left[C_k e^{-\frac{3}{2}x} \sin\left(\sqrt{4\lambda_k - 9}x\right)\right]^2 e^{3x} dx \\
&= C_k^2 \int_0^1 \left[\sin\left(\sqrt{4\lambda_k - 9}x\right)\right]^2 dx \\
&= C_k^2 \int_0^1 \frac{1}{2}\left[1 - \cos\left(2\sqrt{4\lambda_k - 9}x\right)\right] dx \\
&= \frac{C_k^2}{2}\left[x + \frac{1}{2\sqrt{4\lambda_k - 9}} \sin\left(2\sqrt{4\lambda_k - 1}x\right)\right]_{x=0}^{x=1} \\
&= \frac{C_k^2}{2}\left[1 + \frac{1}{2\sqrt{4\lambda_k - 9}} \sin\left(2\sqrt{4\lambda_k - 1}\right)\right]
\end{aligned} \tag{w}
$$

Substitution of the double-angle formula for sines and Equation r into Equation w leads to

$$
\begin{aligned}
1 &= \frac{C_k^2}{2}\left[1 + \frac{2\sin\left(\sqrt{4\lambda_k - 9}\right)\cos\left(\sqrt{4\lambda_k - 9}\right)}{2\sqrt{4\lambda_k - 9}}\right] \\
&= \frac{C_k^2}{2}\left[1 - \frac{2}{3}\cos^2\left(\sqrt{4\lambda_k - 9}\right)\right]
\end{aligned} \tag{x}
$$

Equation x can be rearranged to yield

$$
C_k = \left[\frac{2}{1 - \frac{2}{3}\cos^2\left(\sqrt{4\lambda}\right)}\right]^{\frac{1}{2}} \tag{y}
$$

The Sturm-Liouville problem of Example 5.5 has an infinite, but countable, number of eigenvalues and eigenvectors. It can be shown, but the proof is lengthy and noninstructive, that these observations hold true for all Sturm-Liouville problems.

Theorem 5.10 There are an infinite, but countable, number of eigenvalues of a Sturm-Liouville problem $0,<\lambda_1<\lambda_2<\cdots<\lambda_{k-1}<\lambda_k<\lambda_{k+1}<\cdots$.

An analysis of the results of example 5.5 illustrates that the conditions of Theorem 5.8 are sufficient, but not necessary, conditions for positive definiteness. Consider the substitution, $\mu=4\lambda-9$. The results of Example 5.5 show that all values of μ are positive. Using this substitution in the eigenvalue problem $\mathbf{L}y=\lambda y$ leads to $4\mathbf{L}y-9y=\mu y$. Thus the values of μ are the eigenvalues of the operator $4\mathbf{L}-9$. When put into the form of Equation 5.19, $q(x)=-9e^{-3/2x}$. Thus, $q(x)<0$ and the conditions of Theorem 5.8 are not met, yet all of the eigenvalues of the operator are positive.

Example 5.6 An eigenvalue problem often encountered during the solution of problems using a polar, cylindrical, or spherical coordinate system is

$$\frac{d^2w}{d\theta^2}+\lambda w=0 \tag{a}$$

$$w(0)=w(2\pi) \tag{b}$$

$$\frac{dw}{dx}(0)=\frac{dw}{dx}(2\pi) \tag{c}$$

The independent variable θ is a circumferential coordinate whose range is $0\leq\theta\leq 2\pi$. Equation b and Equation c enforce the conditions that the dependent variable w is single-valued at $\theta=0$ and $\theta=2\pi$. These are called periodicity conditions because their application leads to a solution that is periodic in θ and of period 2π. Determine all eigenvalues and their corresponding normalized eigenvectors.

Solution Equation a can be rewritten in the form of Equation 5.18 with

$$\mathbf{L}w=-\frac{d^2w}{dx^2} \tag{d}$$

Equation d is in the form of a Sturm-Liouville operator with $p(x)=-1$, $r(x)=1$, and $q(x)=0$. The boundary conditions are the periodicity conditions of case (iii) of Theorem 5.7. Thus all eigenvalues are real, and eigenvectors corresponding to distinct eigenvalues are self-adjoint with respect to the standard inner product for $C^2[0,2\pi]$.

Application of Theorem 5.8 shows that the operator is non-negative definite on the subspace of $C^2[0,2\pi]$ defined by Equation b and Equation c. Thus, all eigenvalues are non-negative, but the system may have an eigenvalue of zero.

Assuming a solution to Equation a of the form $y(x)=e^{\beta x}$ leads to $\beta=\pm\sqrt{-\lambda}$. Since the operator is non-negative definite, only non-negative values of λ need to be considered. First consider $\lambda=0$, for which the general solution of Equation a is

$$y(x)=C_1+C_2x \tag{e}$$

Application of Equation b gives $C_1=C_1+2\pi C_2$, which implies that $C_2=0$. Then using $y(x)=C_1$, Equation c is satisfied for all values of C_1. Thus $\lambda=0$ is an eigenvalue of **L** with a corresponding eigenvector of $y_0(x)=C_1$, a constant. Normalization of the eigenvector leads to

$$\begin{aligned} 1&=\int_0^{2\pi}[y_0(x)]^2\,dx \\ &=\int_0^{2\pi}C_1^2dx \\ &=2\pi C_1^2 \end{aligned} \tag{f}$$

Equation f can be solved to give $C_1=1/\sqrt{2\pi}$, leading to $y_0(x)=1/\sqrt{2\pi}$.

If $\lambda>0$, the general solution of Equation a is

$$y(x)=C_1\cos\left(\sqrt{\lambda}x\right)+C_2\sin\left(\sqrt{\lambda}x\right) \tag{g}$$

Application of Equation b to Equation g leads to

$$C_1=C_1\cos\left(2\pi\sqrt{\lambda}\right)+C_2\sin\left(2\pi\sqrt{\lambda}\right) \tag{h}$$

Application of Equation c to Equation g gives

$$\sqrt{\lambda}C_2=-C_1\sqrt{\lambda}\sin\left(2\pi\sqrt{\lambda}\right)+C_2\sqrt{\lambda}\cos\left(2\pi\sqrt{\lambda}\right) \tag{i}$$

Equation h and Equation i can be summarized in matrix form as

$$\begin{bmatrix}\cos\left(2\pi\sqrt{\lambda}\right)-1 & \sin\left(2\pi\sqrt{\lambda}\right) \\ -\sin\left(2\pi\sqrt{\lambda}\right) & \cos\left(2\pi\sqrt{\lambda}\right)-1\end{bmatrix}\begin{bmatrix}C_1\\C_2\end{bmatrix}=\begin{bmatrix}0\\0\end{bmatrix} \tag{j}$$

A nontrivial solution of Equation j exists if and only if the determinant of the coefficient matrix is zero. To this end,

$$\begin{aligned}
&\left[\cos\left(2\pi\sqrt{\lambda}\right)-1\right]^2+\left[\sin\left(2\pi\sqrt{\lambda}\right)\right]^2=0\\
&\left[\cos\left(2\pi\sqrt{\lambda}\right)\right]^2-2\cos\left(2\pi\sqrt{\lambda}\right)+1+\left[\sin\left(2\pi\sqrt{\lambda}\right)\right]^2=0\\
&2\left[1-\cos\left(2\pi\sqrt{\lambda}\right)\right]=0
\end{aligned} \tag{k}$$

The values of λ which solve Equation k are

$$\lambda=n^2 \quad n=1,2,... \tag{l}$$

Substitution of Equation l into Equation j gives

$$\begin{bmatrix} 0 & 0 \\ 0 & 0 \end{bmatrix}\begin{bmatrix} C_1 \\ C_2 \end{bmatrix}=\begin{bmatrix} 0 \\ 0 \end{bmatrix} \tag{m}$$

Since Equation m is identically satisfied for all values of C_1 and C_2, the constants are independent, and the general form of the eigenvector corresponding to $\lambda=n^2$ is

$$y_n(x)=C_{1,n}\cos(nx)+C_{2,n}\sin(nx) \tag{n}$$

Algebraic calculation reveals (cos(nx), sin(nx)$=0$). Also

$$\begin{aligned}
\int_0^{2\pi}[\cos(nx)]^2\,dx &= \int_0^{2\pi}\frac{1}{2}[1+\cos(2nx)]dx\\
&=\frac{1}{2}\left[x+\frac{1}{2n}\sin(2nx)\right]_{x=0}^{x=2\pi}\\
&=\pi
\end{aligned} \tag{o}$$

Equation o leads to a normalization constant of $C_{1,n}=1/\sqrt{\pi}$. A similar calculation leads to $C_{2,n}=1/\sqrt{\pi}$.

The eigenvalues and normalized eigenvectors for the Sturm-Liouville problem of Equation a, Equation b, and Equation c are

$$\lambda_n=n^2 \quad n=0,1,2,... \tag{p}$$

$$y_0(x)=\frac{1}{\sqrt{2\pi}} \tag{q}$$

$$\left.\begin{aligned} y_{n,1} &= \frac{1}{\sqrt{\pi}}\cos(nx) \\ y_{n,2} &= \frac{1}{\sqrt{\pi}}\sin(nx) \end{aligned}\right\} \quad n = 1,2,... \tag{r}$$

The eigenvalue problem described by Equation a, Equation b and Equation c is a positive semi-definite Sturm-Liouville problem. The smallest eigenvalue for the system is zero. The eigenvector corresponding to the zero eigenvalue is given by Equation q. As with all Sturm-Liouville problems, the system has an infinite, but countable, number of eigenvalues. However, all non-zero eigenvalues are degenerate. That is, there is more than one linearly independent eigenvector corresponding to each eigenvalue. The two normalized linearly independent eigenvectors corresponding to an eigenvalue of $\lambda = n^2$, for $n = 1,2,...$ are given in Equation r. For each integer $n > 0$, the most general solution to Equation a subject to Equation b and Equation c is

$$y_n(x) = A_n \cos(nx)/\sqrt{\pi} + B_n \sin(nx)/\sqrt{\pi}.$$

5.6 *Eigenvector expansions*

5.6.1 *Completeness theorem*

It is shown in Section 2.9 that if $\phi_1, \phi_2, \ldots, \phi_{k-1}, \phi_k, \phi_{k+1}, \ldots$ is a complete orthonormal set with respect to a valid inner product defined on a vector space V, then *f*, an arbitrary element of V, has a representation of the form

$$f = \sum_i (f, \phi_i)\phi_i \tag{5.41}$$

The eigenvectors of a self-adjoint operator defined on a vector space of finite dimension *n* are orthogonal with respect to a defined inner product and span R^n. The eigenvectors form a basis for R^n and are complete in R^n. Thus any vector in R^n can be represented by an eigenvector expansion of the form of Equation 5.41.

Completeness of a set of vectors in an infinite-dimensional vector space is more difficult to prove. A comprehensive general theory of the subject is outside the objectives of this book. However, some basic results are listed and proved, followed by a general discussion which will guide the development and use of eigenvector expansions.

Theorem 5.10 Let $\mathbf{u}_1, \mathbf{u}_2, \ldots$ be a set of orthonormal vectors with respect to an inner product $(\mathbf{u}, \mathbf{v})$ defined on an infinite-dimensional vector space V. Then for some $\mathbf{u}$ in V, if $\mathbf{u} = \sum_{i=1}^{\infty} \alpha_i \mathbf{u}_i$ is a convergent series, then $\alpha_i = (\mathbf{u}, \mathbf{u}_i)$.

Proof Assume that $\mathbf{u} = \sum_{i=1}^{\infty} \alpha_i \mathbf{u}_i$ is a convergent series. Taking the inner product of both sides of this equation with $\mathbf{u}_k$ for an arbitrary *k* leads to$(\mathbf{u},\mathbf{u}_k)=(\sum_{i=1}^{\infty}\alpha_i\mathbf{u}_i,\mathbf{u}_k)$. Since the series is convergent, the order of summation and of taking the inner product may be interchanged, leading to $(\mathbf{u},\mathbf{u}_k)=\sum_{i=1}^{\infty}\alpha_i(\mathbf{u}_i,\mathbf{u}_k)$. Using the orthonormality of the set of vectors leads to the desired result.

Theorem 5.11 **(Bessel's inequality)** Let $\mathbf{u}_1,\mathbf{u}_2,\ldots$ be a set of orthonormal vectors with respect to an inner product **(u,v)** defined on an infinite-dimensional vector space V. Then for any **u** in V,

$$\sum_{i=1}^{\infty}(\mathbf{u},\mathbf{u}_i)\leq\|\mathbf{u}\|^2 \tag{5.42}$$

where the norm is the inner-product-generated norm.

Proof The square of any real number is non-negative, thus

$$\left\|\mathbf{u}-\sum_{i=1}^{n}(\mathbf{u},\mathbf{u}_i)\mathbf{u}_i\right\|^2\geq 0 \tag{a}$$

The definition of the inner-product-generated norm and the properties of inner products are used on Equation a, leading to

$$\begin{aligned}
0&\leq\left(\mathbf{u}-\sum_{i=1}^{n}(\mathbf{u},\mathbf{u}_i)\mathbf{u}_i,\mathbf{u}-\sum_{j=1}^{n}(\mathbf{u},\mathbf{u}_j)\mathbf{u}_j\right)\\
&=(\mathbf{u},\mathbf{u})-2\left(\mathbf{u},\sum_{i=1}^{n}(\mathbf{u},\mathbf{u}_i)\mathbf{u}_i\right)+\left(\sum_{i=1}^{n}(\mathbf{u},\mathbf{u}_i)\mathbf{u}_i,\sum_{j=1}^{n}(\mathbf{u},\mathbf{u}_j)\mathbf{u}_j\right)\\
&=\|\mathbf{u}\|^2-2\sum_{i=1}^{n}(\mathbf{u},\mathbf{u}_i)(\mathbf{u},\mathbf{u}_i)+\sum_{i=1}^{n}\sum_{j=1}^{n}(\mathbf{u},\mathbf{u}_i)(\mathbf{u},\mathbf{u}_j)(\mathbf{u}_i,\mathbf{u}_j)
\end{aligned} \tag{b}$$

The set $\mathbf{u}_1,\mathbf{u}_2,\ldots$ is assumed to be orthonormal; thus, $(\mathbf{u}_i,\mathbf{u}_j)=\delta_{i,j}$, which when used in Equation b, leads to

$$0\leq\|\mathbf{u}\|^2-2\sum_{i=1}^{n}(\mathbf{u},\mathbf{u}_i)^2+\sum_{i=1}^{n}(\mathbf{u},\mathbf{u}_i)^2 \tag{c}$$

Rearrangement of Equation c leads to

$$\sum_{i=1}^{n}(\mathbf{u},\mathbf{u}_i)^2\leq\|\mathbf{u}^2\| \tag{d}$$

The sequence of partial sums $s_n = \sum_{i=1}^{n}(\mathbf{u},\mathbf{u}_i)^2$ is a bounded sequence of positive numbers which is not decreasing. Thus, $\lim_{n\to\infty} s_n \le \|\mathbf{u}\|^2$, and the theorem is proved.

Theorem 5.12 (Parseval's Identity) Let $\mathbf{u}_1,\mathbf{u}_2,\ldots$ be a set of orthonormal vectors with respect to an inner product **(u,v)** defined on an infinite-dimensional vector space V. Then $\mathbf{u}_1,\mathbf{u}_2,\ldots$ is complete on V if and only if

$$\|\mathbf{u}\|^2 = \sum_{i=1}^{\infty}(\mathbf{u},\mathbf{u}_i)^2 \tag{5.43}$$

for all **u** in V.

Proof If $\mathbf{u}_1,\mathbf{u}_2,\ldots$ is complete in V, then every **u** in V can be represented by $\mathbf{u} = \sum_{i=1}^{\infty}\alpha_i\mathbf{u}_i$, where from Theorem 5.10, $\alpha_i = (\mathbf{u},\mathbf{u}_i)$. Thus

$$\begin{aligned}\|\mathbf{u}\|^2 &= \left(\sum_{i=1}^{\infty}(\mathbf{u},\mathbf{u}_i)\mathbf{u}_i, \sum_{j=1}^{\infty}(\mathbf{u},\mathbf{u}_j)\mathbf{u}_j\right) \\ &= \sum_{i=1}^{\infty}\sum_{j=1}^{\infty}(\mathbf{u},\mathbf{u}_i)(\mathbf{u},\mathbf{u}_j)(\mathbf{u}_i,\mathbf{u}_j)\end{aligned} \tag{a}$$

Since $\mathbf{u}_1,\mathbf{u}_2,\ldots$ are an orthonormal set, Equation a reduces to

$$\|\mathbf{u}\|^2 = \sum_{i=1}^{\infty}(\mathbf{u},\mathbf{u}_i)^2 \tag{b}$$

The series on the right-hand side of Equation b is bounded by Bessel's inequality. Thus Parseval's identity is established.

If Parseval's identity is satisfied for all **u** in V, then **u** has an expansion as $\mathbf{u} = \sum_{i=1}^{\infty}\alpha_i\mathbf{u}_i$, and thus the set $\mathbf{u}_1,\mathbf{u}_2,\ldots$ is complete in V.

Theorem 5.13 Let $\mathbf{u}_1,\mathbf{u}_2,\ldots$ be a set of orthonormal vectors with respect to an inner product **(u,v)** defined on an infinite-dimensional vector space V. The set is complete in V if and only if $\mathbf{v} = \mathbf{0}$ is the only vector orthogonal to every vector in the set.

Proof First assume the set is complete in V, and let $\mathbf{w} \ne 0$ be a vector in V orthogonal to all vectors in the set. Then from Parseval's identity,

$$\|\mathbf{w}\|^2 = \sum_{i=1}^{\infty}(\mathbf{w},\mathbf{u}_i) \tag{a}$$

However, since **w** is orthogonal to all vectors in the set, $(\mathbf{w}, \mathbf{u}_i) = 0$, and Equation a leads to $\|\mathbf{w}\| = 0$. However, the only vector which has a norm of zero is the zero vector. The assumption that $\mathbf{w} \ne 0$ is thus contradicted.

The converse can be proved by contradiction. Assume that $\mathbf{v}=0$ is orthogonal to all members of the set, but that the set is not complete in V. Then there exists a vector $\mathbf{y}$ in V such that Parseval's identity is not satisfied, and Bessel's inequality is actually an inequality. The completion is left as an exercise.

Parseval's identity provides a criterion to determine whether a set of orthonormal vectors is complete in a vector space. It, along with Theorem 5.13, can be used to prove the following:

Theorem 5.14 Let $\mathbf{L}$ be a proper Sturm-Liouville operator with eigenvalues $0 \leq \lambda_1 \leq \lambda_2 \leq \cdots \leq \lambda_{k-1} \leq \lambda_k \leq \lambda_{k+1} \leq \cdots$ and corresponding normalized eigenvector $\phi_1, \phi_2, \ldots, \phi_{k-1}, \phi_k, \phi_{k+1}, \ldots$. The set of eigenvectors is complete in S_D, the domain of $\mathbf{L}$.

Proof The full proof of this theorem is beyond the objective of this study. Using Theorem 5.13, it suffices to show that $\mathbf{v}=0$ is the only vector orthogonal to all elements in the set $\phi_1, \phi_2, \ldots, \phi_{k-1}, \phi_k, \phi_{k+1}, \ldots$. Assume that $\mathbf{w} \neq 0$ is orthogonal to each vector in the set. Then for each k, $(\mathbf{w}, \phi_k) \neq 0$. Noting that $\mathbf{L}\phi_k = \lambda_k \phi_k$, then for $\lambda_k \neq 0$ $(\mathbf{w}, \phi_k) = 1/\lambda_k(\mathbf{w}, \mathbf{L}\phi_k)$. The self-adjointness of $\mathbf{L}$ implies that $(\mathbf{w}, \mathbf{L}\phi_k) = (\mathbf{Lw}, \phi_k)$. Thus $(\mathbf{Lw} - \lambda_k \mathbf{w}, \phi_k) = 0$. It then remains to show that this condition is satisfied for all k only if $\mathbf{w}=0$.

Theorem 5.14 implies that any vector f in S_D has an eigenvector expansion of the form of Equation 5.41. An eigenvector expansion for an infinite-dimensional vector space is called a Fourier series. Note that the convergence of the series is convergence with respect to the definition of the inner-product-generated norm. For a Sturm-Liouville problem, convergence is with respect to the norm $\|f-g\| = \left[\int_a^b [f(x)-g(x)]^2 r(x)dx\right]^{1/2}$. This is a mean convergence. Pointwise convergence can be considered for each specific case.

Example 5.7 (a) Expand $f(x) = x\left(-(1/4)e^{-3} + e^{-3x}\right)$ into a series of eigenvectors as in Example 5.5. Note that $f(x)$ is in S_D. (b) Apply Parseval's identity to the expansion.

Solution The eigenvectors of Example 5.5 are

$$y_k(x) = C_k e^{-\frac{3}{2}x} \sin(q_k x) \tag{a}$$

where q_k is the kth positive solution of

$$-\frac{2}{3} q_k = \tan(q_k) \tag{b}$$

The eigenvectors are normalized by requiring that

$$(y_{k=}, y_k)_r = 1$$

$$\int_0^1 \left[C_k e^{-\frac{3}{2}x} \sin(q_k x)\right]^2 e^{3x} dx = 1$$

$$C_k^2 \int_0^1 [\sin(q_k x)]^2 dx = 1$$

$$C_k^2 \int_0^1 \frac{1}{2}(1-\cos(2q_k x))dx = 1$$

$$\frac{C_k^2}{2}\left[x - \frac{1}{2q_k}\sin(2q_k x)\right]_{x=0}^{x=1} = 1$$

$$\frac{C_k^2}{2}\left[1 - \frac{1}{2q_k}\sin(2q_k)\right] = 1$$

$$\frac{C_k^2}{2}\left[1 - \frac{1}{q_k}\sin(q_k)\cos(q_k)\right] = 1 \qquad \text{(c)}$$

Equation b is used to simplify Equation c, leading to

$$C_k = \sqrt{\frac{2}{1+\frac{2}{3}[\cos(q_k)]^2}} \qquad \text{(d)}$$

The eigenvector expansion for $f(x)$ is

$$f(x) = \sum_{k=1}^{\infty}\left(f, C_k e^{-\frac{3}{2}x}\sin(q_k x)\right)_r C_k e^{-\frac{3}{2}x}\sin(q_k x) \qquad \text{(e)}$$

The inner products are evaluated as

$$\begin{aligned}\left(f, C_k e^{-\frac{3}{2}x}\sin(q_k x)\right)_r &= \int_0^1 x\left(-\frac{1}{4}e^{-3} + e^{-3x}\right)\left[C_k e^{-\frac{3}{2}x}\sin(q_k x)\right]e^{3x}dx \\ &= C_k \int_0^1 x e^{\frac{3}{2}x}\left(-\frac{1}{4}e^{-3} + e^{-3x}\right)\sin(q_k x)dx \\ &= C_k \frac{e^{-\frac{3}{2}}\left[\left(\frac{171}{8} - q_k^2\right)\sin(q_k) - q_k\left(\frac{71}{4} - 3q_k^2\right)\cos(q_k)\right] - 9q_k}{4\left(\frac{9}{4} + q_k^2\right)^2}\end{aligned} \qquad \text{(f)}$$

(b) Application of Parseval's identity leads to

$$\|f\|^2 = \sum_{k=1}^{\infty} \alpha_k^2$$

$$\int_0^1 x^2\left(-\frac{1}{4}e^{-3} + e^{-3x}\right)^2 e^{\frac{3}{2}x}\,dx = \sum_{k=1}^{\infty}\left\{C_k \frac{e^{-\frac{3}{2}}\left[\left(\frac{171}{8} - q_k^2\right)\sin(q_k) - q_k\left(\frac{71}{4} - 3q_k^2\right)\cos(q_k)\right] - 9q_k}{4\left(\frac{9}{4} + q_k^2\right)^2}\right\}^2 \quad \text{(g)}$$

$$0.709 = \sum_{k=1}^{\infty}\left\{C_k \frac{e^{-\frac{3}{2}}\left[\left(\frac{171}{8} - q_k^2\right)\sin(q_k) - q_k\left(\frac{71}{4} - 3q_k^2\right)\cos(q_k)\right] - 9q_k}{4\left(\frac{9}{4} + q_k^2\right)^2}\right\}^2$$

Example 5.8 Expand the function $f(x) = 1 + (4/3\pi^2)x^2 - (1/3\pi^4)x^4$ in terms of the eigenvectors of Example 5.4.

Solution The eigenvalues and normalized eigenvectors from Example 5.4 are

$$\lambda_n = n^2 \quad n = 0,1,2,\ldots \quad \text{(a)}$$

$$y_0(x) = \frac{1}{\sqrt{2\pi}} \quad \text{(b)}$$

$$\left.\begin{aligned} y_{n,1} &= \frac{1}{\sqrt{\pi}}\cos(nx) \\ y_{n,2} &= \frac{1}{\sqrt{\pi}}\sin(nx) \end{aligned}\right\} \quad n = 1,2,\ldots \quad \text{(c)}$$

The eigenvector expansion is of the form

$$f(x) = (f, y_0)y_0 + \sum_{n=1}^{\infty}[(f, y_{n,1})y_{n,1} + (f, y_{n,2})y_{n,2}] \quad \text{(d)}$$

where

$$\begin{aligned} (f, y_0) &= \int_0^{2\pi}\left(1 + \frac{4}{3\pi^2}x^2 - \frac{1}{3\pi^4}x^4\right)\frac{1}{\sqrt{2\pi}}\,dx \\ &= \frac{128\pi + 15}{15\sqrt{2\pi}} \end{aligned} \quad \text{(e)}$$

$$(f,y_{n,1}) = \int_0^{2\pi}\left(1+\frac{4}{3\pi^2}x^2 - \frac{1}{3\pi^4}x^4\right)\frac{1}{\sqrt{\pi}}\cos(nx)dx$$
$$= \frac{6-16n^2\pi^3}{3\sqrt{\pi}n^5\pi^4} \tag{f}$$

and

$$(f,y_{n,2}) = \int_0^{2\pi}\left(1+\frac{4}{3\pi^2}x^2 - \frac{1}{3\pi^4}x^4\right)\frac{1}{\sqrt{\pi}}\sin(nx)dx$$
$$= -\frac{56}{3\sqrt{\pi}\pi^2n^3} \tag{g}$$

Substitution of Equation e, Equation f, and Equation g into Equation d leads to

$$f(x) = \frac{128\pi+15}{30\pi} + \sum_{n=1}^{\infty}\left[\frac{6-16n^2\pi^3}{3\pi^5n^5}\cos(\text{nx}) - \frac{56}{3\pi^3n^3}\sin(nx)\right] \tag{h}$$

5.6.2 Trigonometric fourier series

The eigenvectors of the Sturm-Liouville problem introduced in Example 5.5 are the basis functions used in a trigonometric Fourier series expansion. If $f(x)$ is a function of period 2π, its Fourier series expansion generated from the eigenvectors is

$$f(x) = (f,y_0)y_0 + \sum_{n=1}^{\infty}[(f,y_{n,1})y_{n,1} + (f,y_{n,2})y_{n,2}] \tag{5.44}$$

where

$$(f,y_0) = \frac{1}{\sqrt{2\pi}}\int_0^{2\pi} f(x)dx \tag{5.45}$$

$$(f,y_{n,1}) = \frac{1}{\sqrt{\pi}}\int_0^{2\pi} f(x)\cos(nx)dx \tag{5.46}$$

$$(f,y_{n,2}) = \frac{1}{\sqrt{\pi}}\int_0^{2\pi} f(x)\sin(nx)dx \tag{5.47}$$

Substitution of Equation 5.45, Equation 5.46, and Equation 5.47 into Equation 5.44 leads to

$$f(x)=\frac{a_0}{2}+\sum_{n=1}^{\infty}[a_n\cos(nx)+b_n\sin(nx)] \tag{5.48}$$

where the Fourier coefficients are defined by

$$a_n=\frac{1}{\pi}\int_0^{2\pi} f(x)\cos(nx)dx \quad n=0,1,2,... \tag{5.49}$$

$$b_n=\frac{1}{\pi}\int_0^{2\pi} f(x)\sin(nx)dx \quad n=1,2,... \tag{5.50}$$

Theorems 5.10–5.14 show that the trigonometric Fourier series for $f(x)$ converges in the mean to $f(x)$. A general theorem regarding pointwise convergence of the Fourier series can be proved.

Theorem 5.15 The trigonometric Fourier series representation of a piecewise continuous function $f(x)$ defined for $0\leq x\leq 2\pi$ is that of Equation 5.48, where the Fourier coefficients are as calculated in Equation 5.49 and Equation 5.50. The trigonometric Fourier series representation converges to a function which is periodic with period 2π and converges pointwise to $f(x)$ at every x where $f(x)$ is continuous. If $f(x)$ has a jump discontinuity at x_0, the Fourier series converges to $1/2\left[f(x_0^-)+f(x_0^+)\right]$ at $x=x_0$.

Theorem 5.15 implies that the trigonometric Fourier series representation for $f(x)$ converges to the periodic extension for $f(x)$ outside the interval $0\leq x\leq 2\pi$. The periodic extension $f_p(x)$ of $f(x)$ is defined such that $f_p(x\pm 2n\pi)$ $=f(x)$ for any integer $n=0,1,2,...$ and for all x, $0\leq x\leq 2\pi$.

The Fourier series reduces to simpler forms if $f(x)$ is an even function or an odd function.

Theorem 5.16 If $f(x)$ is an even function such that $f_p(-x)=f_p(x)$ for all x, $0\leq x\leq 2\pi$, then its trigonometric Fourier series representation is given by Equation 5.48 with $b_n=0$, $n=1,2,...$

Theorem 5.17 If $f(x)$is an odd function such that $f_p(-x)=-f_p(x)$ for all x, $0\leq x\leq 2\pi$, then its trigonometric Fourier series representation is given by Equation 5.48 with $a_n=0$, $n=0,1,2,...$

Example 5.9 (a) Develop the trigonometric Fourier series representation for the function shown in Figure 5.6. (b) Apply Parseval's identity to the result. (c) Draw the function to which the Fourier series representation converges for $-6\pi\leq x\leq 6\pi$.

Solution (a) The function shown in Figure 5.6 is an odd function. Therefore, according to Theorem 5.17, $a_n=0$, $n=0,1,2,\ldots$ The coefficients of the sine terms can be calculated as

$$b_n = \int_0^{\pi} (1)\sin(nx)dx + \int_{\pi}^{2\pi} (-1)\sin(nx)dx \tag{a}$$

$$= \frac{2}{n}[1-(-1)^n]$$

The trigonometric Fourier series representation of the function shown in Figure 5.6 is

$$f(t) = 2\sum_{n=1}^{\infty} \frac{1-(-1)^n}{n}\sin(nx)$$

$$= 4\sum_{n=1,3,5}^{\infty} \frac{1}{n}\sin(nx) \tag{b}$$

(b) Application of Parseval's identity leads to

$$\int_0^{2\pi} [f(x)]^2\,dx = \sum_{n=1,3,5}^{\infty}\left(\frac{4}{n}\right)^2$$

$$2\pi = \sum_{n=1,3,5}^{\infty}\left(\frac{4}{n}\right)^2$$

$$\frac{\pi}{8} = \sum_{n=1,3,5}^{\infty}\frac{1}{n^2} \tag{c}$$

(c) The trigonometric Fourier series for $f(x)$ converges to the function illustrated in Figure 5.7. The Fourier series converges to the periodic extension of $f(x)$, $f_p(x)$, and converges pointwise to $f_p(x)$ at all x except $x=\pm n\pi$, at which it converges to zero.

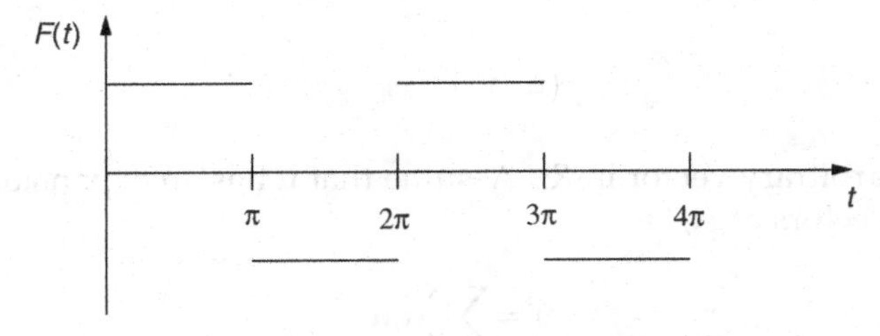

Figure 5.6 The periodic function of the system of Example 5.9 is illustrated over two periods.

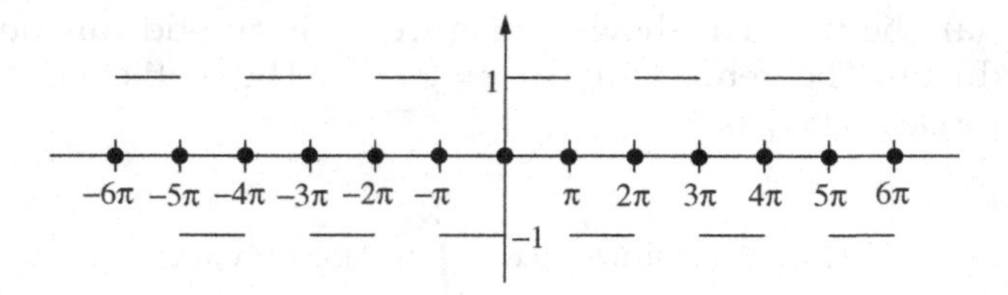

Figure 5.7 The Fourier series for the function of Example 5.9 converges to a periodic function of period 2π, which converges pointwise to the periodic extension of $F(t)$ at all t where it is continuous, and to the average value at values of t where it has a jump discontinuity.

5.6.3 Completeness of eigenvectors for self-adjoint and non-self-adjoint operators

Let $\mathbf{L}$ be an $n \times n$ matrix operator. If $\mathbf{L}$ is symmetric, it is self-adjoint with respect to the standard inner product for R^n, and all its eigenvalues and eigenvectors are real. If $\mathbf{L}$ is not symmetric, it may still have real eigenvalues and eigenvectors. For example, $\mathbf{L}$ could be self-adjoint with respect to an inner product other than the standard inner product. However, not all matrices have real eigenvalues. Such matrices are not self-adjoint with respect to any valid inner product. The associated scalar field for a matrix with complex eigenvalues is the set of complex numbers.

The eigenvectors of a non-self-adjoint matrix are linearly independent and therefore form a basis for C^n, the space of n-dimensional vectors whose elements are complex numbers. The space R^n is a subspace of C^n. Any vector in R^n can be written as a linear combination of eigenvectors of the matrix. However, the coefficients in the expansion can be complex.

Recall that the eigenvalues of the adjoint are complex conjugates of the eigenvalues of the matrix and that the eigenvectors satisfy biorthogonality. If $\mathbf{u}$ is an eigenvector of a matrix $\mathbf{A}$ corresponding to an eigenvector λ and $\mathbf{v}$ is an eigenvector of $\mathbf{A}^*$ corresponding to an eigenvalue $\mu \neq \bar{\lambda}$, then $(\mathbf{u},\mathbf{v}) = 0$. Let $\lambda_1, \lambda_2, \ldots, \lambda_n$ be the eigenvalues of a matrix $\mathbf{A}$ with corresponding eigenvectors $\mathbf{u}_1, \mathbf{u}_2, \ldots, \mathbf{u}_n$. The eigenvalues of $\mathbf{A}^*$ are $\bar{\lambda}_1, \bar{\lambda}_2, \ldots, \bar{\lambda}_n$ which have corresponding eigenvectors $\mathbf{v}_1, \mathbf{v}_2, \ldots, \mathbf{v}_n$. The eigenvectors satisfy the biorthonormality condition,

$$(\mathbf{u}_i, \mathbf{v}_j) = \delta_{i,j} \tag{5.51}$$

Let $\mathbf{u}$ be an arbitrary vector in R^n. Assume that $\mathbf{u}$ has an expansion in terms of the eigenvectors of $\mathbf{A}$,

$$\mathbf{u} = \sum_{i=1}^{n} \alpha_i \mathbf{u}_i \tag{5.52}$$

Taking the inner product of both sides of Equation 5.52 with $\mathbf{v}_j$ leads to

$$(\mathbf{u},\mathbf{v}_j)=\sum_{i=1}^{n}\alpha_i(\mathbf{u}_i,\mathbf{v}_j) \tag{5.53}$$

Due to the biorthogonality condition, Equation 5.51, the only nonzero term in the summation on the right-hand side of Equation 5.53 corresponds to $i=j$, leading to

$$\alpha_j=(\mathbf{u},\mathbf{v}_j) \tag{5.54}$$

Thus, if the operator is not self-adjoint, the coefficients in the eigenvector expansion are the inner products with the vector and the eigenvectors of the adjoint operator.

Theorems have been presented showing that the eigenvectors of a Sturm-Liouville operator are complete in the vector space defined as the domain of the operator. Because of this completeness, any vector in the domain of the operator has a convergent eigenvector expansion. The convergence is with respect to the inner-product-generated norm. Without further proof, it is assumed that eigenvectors of all self-adjoint operators whose eigenvalues are countable are complete in the domain of the operator and that a convergent eigenvector expansion exists for any vector in the domain of the operator. This is true for fourth-order differential operators, integral operators, partial differential operators, operators defined for a set of coupled differential equations, and operators defined for a differential equation coupled with an algebraic equation.

There are self-adjoint operators whose eigenvalues are not countable. In such cases, the eigenvectors corresponding to discrete eigenvalues are not complete. An example is an operator of the form $Ly=-d/dx(p(x)dy/dx)+q(x)y$ with boundary conditions at $x=a$ and $x=b$, but $p(x)=0$ for some x, $a\leq x\leq b$. Such a problem is called a turning-point problem and may not have countable eigenvalues.

The proof of Theorem 5.6 from Section 5.3 can now be completed. It is left to show that if $(\mathbf{Lu},\mathbf{u})=0$ for some $\mathbf{u}\neq 0$, then $\lambda=0$ is an eigenvalue of $\mathbf{L}$ if $\mathbf{L}$ is self-adjoint and all eigenvalues are non-negative.

Since $\mathbf{L}$ is self-adjoint, any vector in the domain of $\mathbf{L}$ has an eigenvector expansion $\mathbf{u}=\sum_i\alpha_i\mathbf{u}_i$. Assume that for some $\mathbf{u}$, $(\mathbf{Lu},\mathbf{u})=0$; then

$$\begin{aligned}0&=(\mathbf{Lu},\mathbf{u})\\&=\left(\mathbf{L}\left(\sum_i\alpha_i\mathbf{u}_i\right),\sum_j\alpha_j\mathbf{u}_j\right)\end{aligned} \tag{5.55}$$

Using the linearity of **L**, the properties of inner products, and the orthonormality of the eigenvectors, Equation 5.55 reduces to

$$\sum_i \alpha_i^2 \lambda_i = 0 \tag{5.56}$$

Since the operator is non-negative definite, all nonzero eigenvalues must be positive. Thus, Equation 5.56 is true only if one of the eigenvalues is zero and all the coefficients in the eigenvector expansion are zero except for the one multiplying the eigenvector corresponding to $\lambda = 0$.

5.6.4 *Solution of nonhomogeneous equations using eigenvector expansions*

Let **L** be a linear operator defined on a domain S_D. Assume **L** is self-adjoint and positive definite with respect to a valid inner product $(f, g)_r$ on S_D. Let $0 < \lambda_1 \le \lambda_2 \le \cdots \le \lambda_{k-1} \le \lambda_k \le \lambda_{k+1} \le \cdots$ be the eigenvalues of **L** with corresponding normalized eigenvectors $\phi_1, \phi_2, \ldots, \phi_{k-1}, \phi_k, \phi_{k+1}, \ldots$. The eigenvectors are complete in S_D.

Consider the nonhomogeneous problem

$$\mathbf{Lu} = \mathbf{f} \tag{5.57}$$

where the domain of **L** is S_D and **f** is an element of the range of **L**. Since the eigenvectors of **L** are complete in S_D and **u** must be in S_D, it has an eigenvector expansion of the form

$$\mathbf{u} = \sum_i \alpha_i \phi_i \tag{5.58}$$

Substitution of Equation 5.58 into Equation 5.57 leads to

$$\mathbf{L}\left(\sum_i \alpha_i \phi_i\right) = \mathbf{f} \tag{5.59}$$

Since **L** is a linear operator and by definition $\mathbf{L}\phi_i = \lambda_i \phi_i$, Equation 5.59 can be rewritten as

$$\sum_i \alpha_i \lambda_i \phi_i = \mathbf{f} \tag{5.60}$$

Taking the inner product of Equation 5.60 with ϕ_j for an arbitrary j leads to

$$\begin{aligned} \left(\sum_i \alpha_i \lambda_i \phi_i, \phi_j\right)_r &= (\mathbf{f}, \phi_j)_r \\ \sum_i \alpha_i \lambda_i (\phi_i, \phi_j)_r &= (\mathbf{f}, \phi_j)_r \end{aligned} \tag{5.61}$$

The orthonormality condition for a set of normalized eigenvectors of a self-adjoint operator is $(\phi_i, \phi_j)_r = \delta_{i,j}$. Its application in Equation 5.61 leads to

$$\alpha_j = \frac{1}{\lambda_j}(\mathbf{f}, \phi_j)_r \tag{5.62}$$

The solution of Equation 5.57 can be written as

$$\mathbf{u} = \sum_i \frac{1}{\lambda_i}(\mathbf{f}, \phi_i)_r \phi_i \tag{5.63}$$

Equation 5.63 provides the description of the inverse operator of a self-adjoint operator,

$$\mathbf{L}^{-1}\mathbf{f} = \sum_i \frac{1}{\lambda_i}(\mathbf{f}, \phi_i)_r \phi_i \tag{5.64}$$

Note that the inverse operator does not exist if $\lambda = 0$ is an eigenvalue of **L**.

Example 5.10 The flexibility matrix **A** for a discrete mechanical system is the inverse of the stiffness matrix. The stiffness matrix of the system is

$$\mathbf{K} = \begin{bmatrix} 4 & -2 & 0 \\ -2 & 3 & -1 \\ 0 & -1 & 2 \end{bmatrix} \tag{a}$$

Use Equation 5.63 to determine **A**.

Solution The stiffness matrix is symmetric, and therefore it is self-adjoint with respect to the standard inner product on R^3. The eigenvalues and normalized eigenvectors of **K** can be determined as

$$\lambda_1 = 0.8549 \quad \mathbf{u}_1 = \begin{bmatrix} 0.4320 \\ 0.6793 \\ 0.5932 \end{bmatrix} \tag{b}$$

$$\lambda_2 = 2.4760 \quad \mathbf{u}_2 = \begin{bmatrix} 0.4913 \\ 0.3744 \\ -0.7864 \end{bmatrix} \tag{c}$$

$$\lambda_3 = 5.6691 \quad \mathbf{u}_3 = \begin{bmatrix} 0.7563 \\ -0.6312 \\ 0.1720 \end{bmatrix} \tag{d}$$

Defining $\mathbf{f} = [f_1 f_2 f_3]^T$, application of Equation 5.64 leads to

$$\mathbf{Af} = \frac{1}{0.8549}\left\{[0.4320\ \ 0.6973\ \ 0.5932]\begin{bmatrix} f_1 \\ f_2 \\ f_3 \end{bmatrix}\right\}\begin{bmatrix} 0.4320 \\ 0.6793 \\ 0.5932 \end{bmatrix}$$

$$+\frac{1}{2.4670}\left\{[0.4913\ \ 0.3744\ \ -0.7864]\begin{bmatrix} f_1 \\ f_2 \\ f_3 \end{bmatrix}\right\}\begin{bmatrix} 0.4913 \\ 0.3744 \\ -0.7864 \end{bmatrix}$$

$$+\frac{1}{5.6691}\left\{[0.7563\ \ -0.6312\ \ 0.1720]\begin{bmatrix} f_1 \\ f_2 \\ f_3 \end{bmatrix}\right\}\begin{bmatrix} 0.7563 \\ -0.6312 \\ 0.1720 \end{bmatrix}$$

$$= \begin{bmatrix} 0.4167 & 0.3333 & 0.1667 \\ 0.3333 & 0.6667 & 0.3333 \\ 0.1667 & 0.3333 & 0.6667 \end{bmatrix}\begin{bmatrix} f_1 \\ f_2 \\ f_3 \end{bmatrix} \tag{e}$$

Example 5.11 The differential equations governing the forced response of a multi-degree-of-freedom discrete system are of the form

$$\mathbf{M}\ddot{\mathbf{x}} + \mathbf{Kx} = \mathbf{F} \tag{a}$$

where **M** is the mass matrix, **K** is the stiffness matrix, **x** is the displacement vector, and **F** is the force vector. Recall from Example 5.2 that the natural frequencies are the square roots of the eigenvalues of $\mathbf{M}^{-1}\mathbf{K}$ and the mode-shape vectors are the corresponding eigenvectors. Also recall that $\mathbf{M}^{-1}\mathbf{K}$ is self-adjoint with respect to the kinetic- and potential-energy inner products and that the eigenvectors are normalized with respect to the kinetic-energy inner product. Let $\omega_1 \leq \omega_2 \leq \cdots \leq \omega_n$ be the natural frequencies of an n-degree-of-freedom system, and let $\mathbf{X}_1, \mathbf{X}_2, \ldots, \mathbf{X}_n$ be the corresponding normalized mode-shape vectors. The forced response, at any time, is in R^n. Therefore, it has an eigenvector expansion of the form

$$\mathbf{x} = \sum_{i=1}^{n} c_i(t)\mathbf{X}_i \tag{b}$$

Show that substitution of Equation b into Equation a leads to a decoupling of the differential equations and a process called modal analysis.

Solution Substitution of Equation b into Equation a and the linearity of the mass and stiffness matrices lead to

$$\sum_{i=1}^{n} \ddot{c}_i \mathbf{M}\mathbf{X}_\mathrm{i} + \sum_{i=1}^{n} c_i \mathbf{K}\mathbf{X}_\mathrm{i} = \mathbf{F} \tag{c}$$

Taking the standard inner product of both sides of Equation a with $\mathbf{X}_j$ for arbitrary j and using the definitions of energy inner products and properties of inner products

$$\sum_{i=1}^{n} \ddot{c}_i (\mathbf{M}\mathbf{X}_\mathrm{i}, \mathbf{X}_\mathrm{j}) + \sum_{i=1}^{n} c_i (\mathbf{K}\mathbf{X}_\mathrm{i}, \mathbf{X}_\mathrm{j}) = (\mathbf{F}, \mathbf{X}_\mathrm{j})$$

$$\sum_{i=1}^{n} \ddot{c}_i (\mathbf{X}_\mathrm{i}, \mathbf{X}_\mathrm{j})_M + \sum_{i=1}^{n} c_i (\mathbf{X}_\mathrm{i}, \mathbf{X}_\mathrm{j})_K = (\mathbf{F}, \mathbf{X}_\mathrm{j}) \tag{d}$$

Orthonormality of the mode shapes implies $(\mathbf{X}_\mathrm{i}, \mathbf{X}_\mathrm{j})_M = \delta_{i,j}$ and $(\mathbf{X}_\mathrm{i}, \mathbf{X}_\mathrm{j})_K = \delta_{i,j}\omega_i^2$. Using these in Equation d results in

$$\ddot{c}_j + \omega_j^2 = (\mathbf{F}, \mathbf{X}_\mathrm{j}) \tag{e}$$

Equation e represents a set of uncoupled differential equations to solve for the values of each c_i.

The procedure in which the forced response of a dynamic system is determined as an eigenvector expansion is called modal analysis. The eigenvalues are the squares of the natural frequencies, and the eigenvectors are their corresponding mode shapes. The forced response is written as an eigenvector expansion with the coefficients as functions of time. Orthogonality properties of the mode shapes are used to derive a set of uncoupled differential equations for the time-dependent coefficients. Each differential equation may be solved independently and the forced response obtained from the eigenvector expansion.

The time-dependent coordinates are called principal coordinates, a set of coordinates defined so that, when they are used as dependent variables in the differential equations, the equations are uncoupled. They are a set of coordinates because they can be obtained by a linear transformation from the original set of coordinates (called generalized coordinates) represented by the vector $\mathbf{x}$. Define $\mathbf{P}$, the $n \times n$ modal matrix as the matrix whose columns are the normalized mode-shape vectors. Then

$$\mathbf{x} = \mathbf{P}\mathbf{c} \tag{5.65}$$

where $\mathbf{c}$ is the vector of principal coordinates. Since its columns are the normalized mode-shape vectors, which must be linearly independent, $\mathbf{P}$ is invertible, and therefore the transformation expressed in Equation 5.65 is one-to-one.

5.7 Fourth-order differential operators

Consider the following form of a fourth-order differential operator:

$$\mathbf{L}y = \frac{1}{r(x)}\left\{\frac{d^2}{dx^2}\left[s(x)\frac{d^2y}{dx^2}\right] + \frac{d}{dx}\left[p(x)\frac{dy}{dx}\right] + q(x)y\right\} \tag{5.66}$$

Let $f(x)$ and $g(x)$ be arbitrary elements of S_D, which is defined as all functions in $C^4[a,b]$ which satisfy all boundary conditions. Consider the inner product

$$\begin{aligned}(\mathbf{L}f, g)_r &= \int_a^b f(x)g(x)r(x)dx \\ &= \int_a^b \frac{d^2}{dx^2}\left[s(x)\frac{d^2f}{dx^2}\right]g(x)dx + \int_a^b \frac{d}{dx}\left[p(x)\frac{df}{dx}\right] + g(x)dx \\ &\quad + \int_a^b f(x)g(x)dx\end{aligned} \tag{5.67}$$

Applying integration by parts twice to the first integral in Equation 5.67 and once to the second integral in the equation leads to

$$\begin{aligned}(\mathbf{L}f, g)_r &= \left[g(x)\frac{d}{dx}\left(s(x)\frac{d^2f}{dx^2}\right)\right]_a^b - \left[s(x)\frac{dg}{dx}\frac{d^2f}{dx^2}\right]_a^b + \left[p(x)g(x)\frac{df}{dx}\right]_a^b \\ &\quad + \int_a^b \left[s(x)\frac{d^2f}{dx^2}\frac{d^2g}{dx^2} - p(x)\frac{df}{dx}\frac{dg}{dx} + q(x)f(x)g(x)\right]dx\end{aligned} \tag{5.68}$$

It can be similarly shown that

$$\begin{aligned}(f, \mathbf{L}g)_r &= \left[f(x)\frac{d}{dx}\left(s(x)\frac{d^2g}{dx^2}\right)\right]_a^b - \left[s(x)\frac{df}{dx}\frac{d^2g}{dx^2}\right]_a^b + \left[p(x)f(x)\frac{dg}{dx}\right]_a^b \\ &\quad + \int_a^b \left[s(x)\frac{d^2f}{dx^2}\frac{d^2g}{dx^2} - p(x)\frac{df}{dx}\frac{dg}{dx} + q(x)f(x)g(x)\right]dx\end{aligned} \tag{5.69}$$

The operator $\mathbf{L}$ is self-adjoint if the right-hand side of Equation 5.68 is equal to the right-hand side of Equation 5.69. The requirement for self-adjointness becomes

$$\left[g(x)\frac{d}{dx}\left(s(x)\frac{d^2f}{dx^2}\right)\right]_a^b - \left[s(x)\frac{dg}{dx}\frac{d^2f}{dx^2}\right]_a^b + \left[p(x)g(x)\frac{df}{dx}\right]_a^b$$

$$= \left[f(x)\frac{d}{dx}\left(s(x)\frac{d^2g}{dx^2}\right)\right]_a^b - \left[s(x)\frac{df}{dx}\frac{d^2g}{dx^2}\right]_a^b + \left[p(x)f(x)\frac{dg}{dx}\right]_a^b \qquad (5.70)$$

If the boundary conditions are such that Equation 5.70 is satisfied for all f and g in S_D, then $\mathbf{L}$ is self-adjoint with respect to the inner product of Equation 5.67.

Boundary conditions such that the operator is self-adjoint with respect to the inner product of Equation 5.67 are of the form

$$a_1y(a) - b_1\frac{d}{dx}\left(s\frac{d^2y}{dx^2}\right)\bigg|_{x=a} = 0 \quad a_2y(b) - b_2\frac{d}{dx}\left(s\frac{d^2y}{dx^2}\right)\bigg|_{x=b} = 0$$

$$\alpha_1\frac{dy}{dx}(a) - \beta_1 s(a)\frac{d^2y}{dx^2}(a) + \gamma_1 p(a)y(a) = 0$$

$$\alpha_2\frac{dy}{dx}(b) - \beta_2 s(b)\frac{d^2y}{dx^2}(b) + \gamma_2 p(b)y(b) = 0 \qquad (5.71)$$

The positive-definiteness of $\mathbf{L}$ can be examined by substituting $g = f$ in Equation 5.68, leading to

$$(\mathbf{L}f, f)_r = \left[f(x)\frac{d}{dx}\left(s(x)\frac{d^2f}{dx^2}\right)\right]_a^b - \left[s(x)\frac{df}{dx}\frac{d^2f}{dx^2}\right]_a^b + \left[p(x)f(x)\frac{df}{dx}\right]_a^b$$

$$+ \int_a^b \left[s(x)\frac{d^2f}{dx^2}\frac{d^2f}{dx^2} - p(x)\frac{df}{dx}\frac{df}{dx} + q(x)f(x)g(x)\right]dx \qquad (5.72)$$

Equation 5.72 shows that if $s(x) \geq 0, p(x) \leq 0$, and $q(x) \geq 0$, then if all boundary terms are zero or positive, the operator is non-negative definite. $\mathbf{L}$ is positive definite if $q(x) > 0$ as long as the boundary terms are zero or positive, or if $q(x) = 0$ and the boundary terms are positive. If $q(x) = 0$ and the boundary terms are zero, then the operator is non-negative definite if the boundary conditions allow $f(x) = c_1 + c_2x$ to be an element of S_D.

The general form of a fourth-order differential operator is

$$\mathbf{L}y = a_4(x)\frac{d^4y}{dx^4} + a_3(x)\frac{d^2y}{dx^2} + a_2(x)\frac{d^2y}{dx^2} + a_1(x)\frac{dy}{dx} + a_0(x)y \qquad (5.73)$$

Expansion of derivatives in Equation 5.73 leads to

$$\mathbf{L}y = \frac{1}{r(x)}\left[s(x)\frac{d^4y}{dx^4} + 2\frac{ds}{dx}\frac{d^3y}{dx^3} + \left(\frac{d^2s}{dx^2} + p(x)\right)\frac{d^2y}{dx^2} + \frac{dp}{dx}\frac{dy}{dx} + q(x)y\right] \qquad (5.74)$$

Comparison of Equation 5.74 with Equation 5.73 shows that the fourth-order differential operator may be written in a self-adjoint form if functions $s(x)$, $p(x)$, and $q(x)$ can be found such that

$$a_4(x) = \frac{s(x)}{r(x)} \tag{5.75}$$

$$a_3(x) = 2\frac{s'(x)}{r'(x)} \tag{5.76}$$

$$a_2(x) = \frac{s''(x) + p(x)}{r(x)} \tag{5.77}$$

$$a_1(x) = \frac{p'(x)}{r(x)} \tag{5.78}$$

$$a_0(x) = \frac{q(x)}{r(x)} \tag{5.79}$$

Since there are five coefficients in the differential equation of Equation 5.73 and only three functions to be determined for the self-adjoint form of Equation 5.74, not all fourth-order differential operators can be rewritten in the self-adjoint form. However, this form is only a sufficient, not a necessary, condition for self-adjointness. It may be possible to determine an inner product for which the operator is self-adjoint without being able to rewrite it in the form of Equation 5.66.

As discussed subsequently, the inner product of Equation 5.67 is an energy inner product. Since **L** is self-adjoint with respect to this inner product, when it is also positive definite, an energy inner product exists of the form

$$(f,g)_L = \int_0^1 (\mathbf{L}f)g(x)r(x)dx \tag{5.80}$$

Theorem 5.18 A fourth-order differential operator of the form of Equation 5.66 is self-adjoint with respect to the inner product defined in Equation 5.67 if the domain of the operator S is defined such that all $f(x)$ in S satisfy conditions of the form of Equation 5.72. Furthermore, the operator is positive definite for appropriate choices of the constants $a_1, b_1, a_2, b_2, \alpha_1, \beta_1, \gamma_1, \alpha_2, \beta_2$ and γ_2 (see Problem 5.48).

When the operator is self-adjoint, Theorems 5.2–5.6 can be applied, leading to Theorem 5.19.

Theorem 5.19 If the domain of **L** is defined such that the operator of Equation 5.66 is self-adjoint with respect to the inner product defined in Equation 5.67, then all eigenvalues of **L** are real, and eigenvectors corresponding to distinct eigenvectors are orthogonal with respect to the inner product of Equation 5.67. The eigenvalues are infinitely countable and are distinct. Furthermore, if the domain is defined such that **L** is positive definite, then all eigenvalues of **L** are positive.

Theorems 5.13 (Bessel's inequality) and 5.14 (Parseval's identity) apply when the operator is self-adjoint. The self-adjointness is then used to show completeness of the eigenvectors and to develop an expansion theorem.

Theorem 5.20 Suppose that S, the domain of **L**, is defined such that the operator of Equation 5.66 is self-adjoint with respect to the inner product defined in Equation 5.67. Let $\lambda_1 < \lambda_2 < \cdots < \lambda_{k-1} < \lambda_k < \lambda_{k+1} < \cdots$ be the eigenvalues of **L**, with $\phi_k(x)$, the normalized eigenvector (normalized with respect to the inner product of Equation 5.67), corresponding to λ_k. The set of eigenvectors is complete on S, and if $f(x)$ is in S, then the expansion $\sum_{i=1}^{\infty}(f,\phi_i)_r\phi_i$ converges to $f(x)$.

The nondimensional partial differential equation governing the free vibrations of motion of a nonuniform stretched beam on an elastic foundation is

$$\frac{\partial^2}{\partial x^2}\left(\alpha(x)\frac{\partial^2 w}{\partial x^2}\right)-\frac{\partial}{\partial x}\left(\varepsilon(x)\frac{\partial w}{\partial x}\right)+\eta(x)w+\beta(x)\frac{\partial^2 w}{\partial t^2}=0 \tag{5.81}$$

where $\alpha(x)$ is a nondimensional function of the nonuniform bending stiffness, $\varepsilon(x)$ is a nondimensional axial load allowing for the load to vary over the span of the beam, $\eta(x)$ is the nondimensional stiffness per length of the elastic foundation, allowing it to vary over the span of the beam, and $\beta(x)$ is the nondimensional inertia of the beam. A normal-mode solution for Equation 5.81 can be assumed to be

$$w(x,t)=W(x)e^{i\omega t} \tag{5.82}$$

where ω is a natural frequency and $W(x)$ is its corresponding mode shape. Substitution of Equation 5.82 into Equation 5.81 leads to

$$\frac{d^2}{dx^2}\left(\alpha(x)\frac{d^2W}{dx^2}\right)-\varepsilon(x)\frac{d^2W}{dx^2}+\eta(x)W=\omega^2\beta(x)W \tag{5.83}$$

Dividing Equation 5.83 by $\beta(x)$ leads to an eigenvalue problem of the form of Equation 5.1 with ω^2 as the eigenvalue, $W(x)$ as the eigenvector, and

$$\mathbf{L}W=\frac{1}{\beta(x)}\left[\frac{d^2}{dx^2}\left(\alpha(x)\frac{d^2W}{dx^2}\right)-\frac{d}{dx}\left(\varepsilon(x)\frac{dW}{dx}\right)+\eta(x)W\right] \tag{5.84}$$

Equation 5.84 is in the self-adjoint form of Equation 5.66. The problem is self-adjoint if the boundary conditions are defined such that Equation 5.70

is satisfied by $s(x)=\alpha(x), p(x)=\varepsilon(x), q(x)=\eta(x)$ and $r(x)=\beta(x)$. Assuming that the problem is nondimensionalized, $a=0$ and $b=1$. Consider, for convenience, a uniform beam in which all properties are constant. Using these simplifications, the boundary terms at $x=1$ in Equation 5.70 become

$$\alpha g(1)\frac{d^3 f}{dx^3}(1)-\alpha\frac{dg}{dx}(1)\frac{d^2 f}{dx^2}(1)+\varepsilon g(1)\frac{df}{dx}(1)=\alpha f(1)\frac{d^3 g}{dx^3}(1)$$
$$-\alpha\frac{df}{dx}(1)\frac{d^3 g}{dx^3}(1)+\varepsilon f(1)\frac{dg}{dx}(1) \qquad (5.85)$$

If the end at $x=1$ is fixed, then if f and g are in S_D, then $f(1)=g(1)=0$ and $df/dx(1)=dg/dx(1)=0$. In this case, both sides of Equation 5.85 are identically zero, and the equation is identically satisfied. Such is also the case if the end is pinned. If the end at $x=0$ is free, the end moment is zero, and thus $d^2f/dx^2=d^2g/dx^2=0$. The applied load is assumed to remain horizontal. The end of the beam has a slope equal to $dw/dx(1)$. The component of the axial force in the shear direction, the direction perpendicular to the neutral axis, is (nondimensionally) $\varepsilon(dw/dx(1))$. Thus the appropriate boundary condition for a free end at $x=1$ is $\alpha w(1)+\varepsilon(d^3w/dx^3)=0$. It is then clear that Equation 5.85 is identically satisfied when the end is free.

Thus, the fourth-order differential operator is self-adjoint, for a stretched-beam operator is self-adjoint on S_D with respect to the inner product

$$(\mathbf{L}f,g)_\beta=\int_0^1 f(x)g(x)\beta(x)dx \qquad (5.86)$$

Noting that $\beta(x)$ is the nondimensional inertia of the beam, the inner product of Equation 5.86 is a kinetic-energy inner product in that $(\mathbf{L}(\partial w/\partial t), \partial w/\partial t)_\beta=2T$.

Example 5.12 Determine the natural frequencies and normalized mode shapes of a uniform pinned-fixed beam subject to a constant axial load, $\varepsilon=2$, and on a uniform elastic foundation with $\eta=1$. Demonstrate mode-shape orthogonality.

Solution The eigenvalue problem is

$$\frac{d^4W}{dx^4}-2\frac{d^2W}{dx^2}+W=\lambda W \qquad (a)$$

$$W(0)=0 \qquad (b)$$

$$\frac{d^2W}{dx^2}(0)=0 \qquad (c)$$

$$W(1) = 0 \tag{d}$$

$$\frac{dW}{dx}(1) = 0 \tag{e}$$

A solution of Equation a can be assumed to be the form $W(x) = e^{\gamma x}$, which when substituted into Equation a, leads to

$$\gamma^4 - 2\gamma^2 + 1 - \lambda = 0 \tag{f}$$

The solution of Equation f is

$$\gamma = \pm\left[1 \pm \sqrt{\lambda}\right]^{\frac{1}{2}} \tag{g}$$

If $\lambda < 1$, the values of λ obtained from Equation g are all real, and a nontrivial solution cannot be obtained. The same is true for $\lambda = 0$. For $\lambda > 1$, Equation g has two real roots, $\pm u = \pm\sqrt{1+\sqrt{\lambda}}$, and two complex roots, $\pm iv = \pm i\sqrt{\sqrt{\lambda} - 1}$. The general solution of Equation a can be written as

$$W(x) = C_1 \cosh(ux) + C_2 \sinh(ux) + C_3 \cos(vx) + C_4 \sin(vx) \tag{h}$$

Application of boundary conditions Equation b, Equation c, Equation d, and Equation e gives

$$W(0) = 0 = C_1 + C_3 \tag{i}$$

$$\frac{d^2W}{dx^2}(0) = 0 = u^2C_1 - v^2C_3 \tag{j}$$

$$W(1) = 0 = C_1 \cosh(u) + C_2 \sinh(u) + C_3 \cos(v) + C_4 \sin(v) \tag{k}$$

$$\frac{dW}{dx}(1) = 0 = uC_1 \sinh(u) + uC_2 \cosh(u) - vC_3 \sin(v) + vC_4 \cos(v) \tag{l}$$

Equation i and Equation j are satisfied only by $C_1 = C_3 = 0$. Equation k and Equation l can then be written in matrix form as

$$\begin{bmatrix} \sinh(u) & \sin(v) \\ u\cosh(u) & v\cos(v) \end{bmatrix} \begin{bmatrix} C_2 \\ C_4 \end{bmatrix} = \begin{bmatrix} 0 \\ 0 \end{bmatrix} \tag{m}$$

A nontrivial solution of Equation m exists only if

$$v\cos(v)\sinh(u) - u\cosh(u)\sin(v) = 0 \tag{n}$$

Equation n can be rearranged to yield

$$v \tanh(u) = u \tan(v) \tag{o}$$

Equation o is a transcendental equation with an infinite, but countable, number of solutions: $0 < \lambda_1 < \lambda_2 < \cdots < \lambda_{k-1} < \lambda_k < \lambda_{k+1} < \cdots$. As k gets large, u_k/v_k approaches 1 and $\tanh(u_k)$ approaches 1. Thus the larger eigenvalues are such that v_k approaches $(4k-3)\pi/4$ with $\lambda_k = \left(v_k^2 - 1\right)^2$. The five lowest eigenvalues and nondimensional natural frequencies are given in Table 5.2.

The eigenvectors can be determined from Equation h and Equation m as

$$W_k(x) = C_k \left[\sinh(u_k x) - \frac{\sinh(u_k)}{\sin(v_k)} \sin(v_k x)\right] \tag{p}$$

The mode shapes are normalized by requiring

$$(W_k, W_k) = 1$$

$$\int_0^1 C_k^2 \left[\sinh(u_k x) - \frac{u_k \cosh(u_k)}{v_k \cos(v_k)} \sin(v_k x)\right]^2 dx = 1$$

$$C_k = 2\left\{2\left[\left(\frac{u_k \cosh u_k}{v_k \cos v_k}\right)^2 - \frac{\sin(2v_k)}{v_k}\right] - \frac{8u_k \cosh u_k}{(u_k^2 + v_k^2) v_k \cos v_k} (\cosh u_k \sin v_k - \cos v_k \sinh u_k) - 2 + \frac{\sinh(2u_k)}{u_k}\right\}^{-\frac{1}{2}}$$

Mode-shape orthogonality is demonstrated by

Table 5.2 Eigenvalues and Natural Frequencies for Example 5.12

k	1	2	3	4	5
λ_k	2.62×10^2	2.583×10^3	1.106×10^4	3.211×10^4	7.451×10^4
u_k	4.15×10^0	7.20×10^0	1.03×10^1	1.34×10^1	1.66×10^1
v_k	3.90×10^0	7.06×10^0	1.02×10^1	1.32×10^1	1.65×10^1
ω_k	1.62×10^1	5.08×10^1	1.05×10^2	1.79×10^2	2.73×10^2
C_k	3.29×10^{-2}	1.54×10^{-3}	6.85×10^{-5}	3.09×10^{-6}	1.31×10^{-7}

$$(W_k, W_m) = \int_0^1 C_k \left[\sinh(u_k x) - \frac{\sinh(u_k)}{\sin(v_k)} \sin(v_k x)\right] C_m$$

$$\left[\sinh(u_m x) - \frac{\sinh(u_m)}{\sin(v_m)} \sin(v_m x)\right] dx$$

$$= C_k C_m \int_0^1 \left[\sinh(u_k x)\sinh(u_m x) - \frac{\sinh(u_k)}{\sin(v_k)} \sin(v_k x)\sinh(u_m x)\right.$$

$$- \frac{\sinh(u_m)}{\sin(v_m)} \sinh(u_k x)\sin(v_m x)$$

$$\left. + \frac{\sinh(u_k)}{\sin(v_k)} \frac{\sinh(u_m)}{\sin(v_m)} \sin(v_k x)\sin(v_m x)\right] dx$$

The normalized mode shapes are illustrated in Figure 5.8.

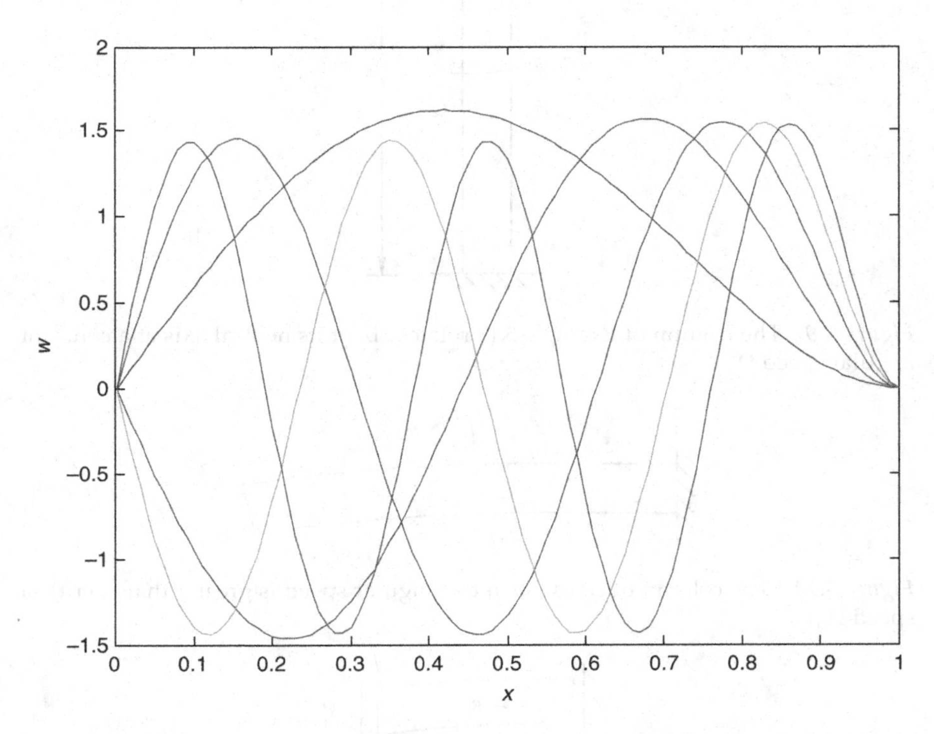

Figure 5.8 Mode shapes for Example 5.12 are orthonormal with respect to the standard inner product for $C^4[0,1]$.

Example 5.13 A uniform column free at $x=0$, but fixed at $x=L$, rotates at a constant angular speed Ω about its central axis, as illustrated in Figure 5.9. If the column is stable, it remains vertical, but if it is unstable, it will deform as shown in Figure 5.10.

Let s be a coordinate along the axis of the column, measured from its free end. If the column is stable, then $s=x$. Define θ as the slope of the elastic curve; then $\cos\theta=dx/ds$ and $\sin\theta=dr/ds$, where $r(x)$ is the distance from the vertical position to the center of the column. Figure 5.11 shows a sectional cut of the column, illustrating the internal shear force which is $V=EI(d^2\theta/ds^2)$. This section of the column also has a centripetal force due to its rotation, $F=-\int_s^0 r(s)\Omega^2\rho gA ds$. Application of Newton's second law to this section of the column leads to

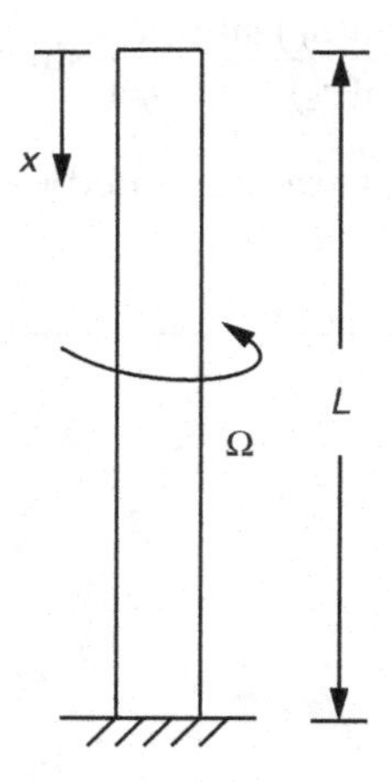

Figure 5.9 The column of Example 5.13 rotates about its neutral axis at a constant angular speed Ω.

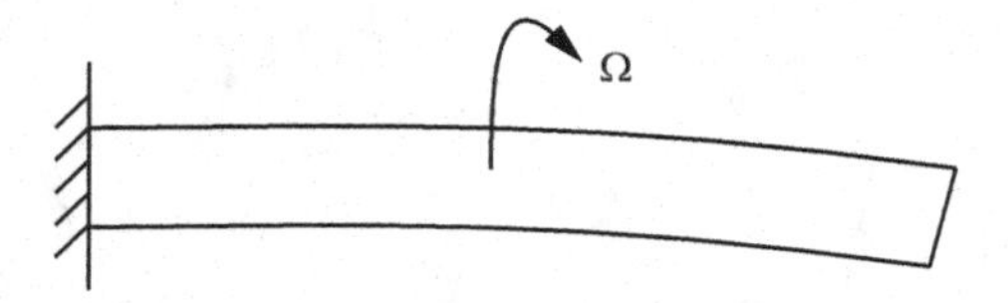

Figure 5.10 The column buckles when the angular speed is greater than a critical speed Ω_{cr}.

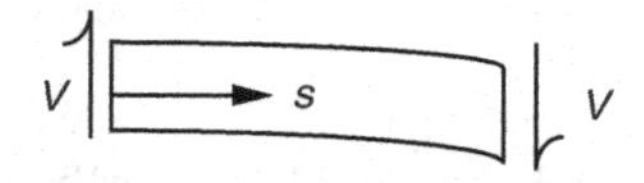

Figure 5.11 The free-body diagram when the column reaches a critical speed shows only an internal shear force. A centripetal inertial force leads to buckling.

$$EI\frac{d^2\theta}{ds^2} = -\rho g A\Omega^2 \int_s^0 r(s)ds \tag{a}$$

Define $u(s) = \int_0^s r(s)ds$. Then $du/ds = r(s)$. Since the goal is to obtain the speed at which the column buckles, it can be reasonably be assumed that at the onset of buckling, θ is small, and therefore $\sin\theta \approx \theta$. Combining the above leads to $\theta = d^2u/ds^2$. Substitution into Equation a gives

$$EI\frac{d^4u}{ds^4} = \rho g A\Omega^2 u \tag{b}$$

The appropriate boundary conditions specified for a column free at $x=0$ and fixed at $x=\mathrm{L}$ are

$$u(0) = 0 \tag{c}$$

$$\frac{d^3u}{ds^3}(0) = 0 \tag{d}$$

$$\frac{du}{ds}(L) = 0 \tag{e}$$

$$\frac{d^2u}{ds^2}(L) = 0 \tag{f}$$

The critical speed is the speed at which instability occurs. It is calculated from

$$\Omega_c = \left(\frac{\lambda EI}{\rho g A}\right)^{\frac{1}{2}} \tag{g}$$

where λ is the smallest eigenvalue of $Lu = d^4u/ds^4$ defined on S_D whose members are elements of $C^4[0,L]$ which satisfy the boundary conditions of Equation c, Equation d, Equation e, and Equation f. (a) Determine whether or not L is self-adjoint. If not, what is the adjoint problem? (b) Determine the critical buckling speed.

Solution (a) The operator is self-adjoint on S_D if Equation 5.69 is true with $p(x)=0$ and $s(x)=1$. Requiring $f(x)$ and $g(x)$ each to satisfy Equation c, Equation d, Equation e, and Equation f leads to

$$g(L)\frac{d^3f}{dx^3}(L) - \frac{dg}{dx}(0)\frac{d^2f}{dx^2}(0) = f(L)\frac{d^3g}{dx^3}(L) - \frac{df}{dx}(0)\frac{d^2g}{dx^2}(0) \tag{h}$$

However, Equation h is not satisfied for all $f(x)$ and $g(x)$ in S_D, and therefore L is not self-adjoint.

Let $f(x)$ be an element of S_D, and let $g(x)$ be an element of S_D^*, the domain of L^*, the adjoint of L. Since only the boundary terms did not vanish when testing for self-adjointness, it is suspected that L^* has the same differential form as L, but has a different domain. The appropriate boundary conditions for L^* are chosen so that Equation h is satisfied for all $f(x)$ in S_D and all $g(x)$ in S_D^*. This is true only if

$$\frac{dg}{dx}(0)=0 \tag{i}$$

$$\frac{d^2g}{dx^2}(0)=0 \tag{j}$$

$$g(L)=0 \tag{k}$$

$$\frac{d^3g}{dx^3}(L)=0 \tag{l}$$

(b) Defining $\lambda=q^4$, the general solution of $d^4u/ds^4=q^4u$ is

$$u(s)=C_1\cos(qs)+C_2\sin(qs)+C_3\cosh(qs)+C_4\sinh(qs) \tag{m}$$

Application of Equation c, Equation d, Equation e, and Equation f to Equation m leads to

$$C_1+C_3=0 \tag{n}$$

$$-C_2+C_4=0 \tag{o}$$

$$-\sin(qL)C_1+\cos(qL)C_2+\sinh(qL)C_3+\cosh(qL)C_4=0 \tag{p}$$

$$-\cos(qL)C_1-\sin(qL)C_2+\cosh(qL)C_3+\sinh(qL)C_4=0 \tag{q}$$

Equation n, Equation o, Equation p, and Equation q have a nontrivial solution if and only if

$$\cos(qL)\cosh(qL)=-1 \tag{r}$$

The smallest value of q which satisfies Equation r is 1.875/L. Thus $\lambda=q^4=12.360L^4$. The critical speed is calculated from Equation g as

$$\Omega_c=3.516\left(\frac{EI}{\rho gAL^4}\right)^{\frac{1}{2}} \tag{s}$$

5.8 Differential operators with eigenvalues in boundary conditions

A general formulation of an eigenvalue problem for a second-order differential operator is

$$Ly = \lambda y \tag{5.87}$$

$$L_a y(a) = \lambda c_a y(a) \tag{5.88}$$

$$L_b y(b) = \lambda c_b y(b) \tag{5.89}$$

where **L** is in the self-adjoint form of Equation 5.18,

$$Ly = 1/r(x)[d(p(x)dy/dx)/dx] + [q(x)y]$$

The boundary operators are of the form

$$L_a y(a) = \alpha_a y'(a) + \beta_a y(a) \tag{5.90}$$

$$L_b y(b) = \alpha_b y'(b) + \beta_b y(b) \tag{5.91}$$

and c_a and c_b are non-negative constants. Consider the inner product defined by

$$(f, g)_L = \int_a^b Lf(x)g(x)r(s)dx + [L_a f(a)]g(a) + [L_b f(b)]g(b) \tag{5.92}$$

It is possible to demonstrate that under the conditions required in Section 5.5 for L to be self-adjoint and positive definite with respect to the inner product of Equation 5.92, and with $\alpha_a\beta_a \leq 0$ and $\alpha_b\beta_b \geq 0$, Equation 5.92 represents a valid inner product. Let λ_i and λ_j be distinct eigenvalues of the system defined by Equation 5.87, Equation 5.88, and Equation 5.89, with corresponding eigenvectors y_i and y_j. Thus

$$Ly_i = \lambda_i y_i \tag{5.93}$$

$$L_a y_i(a) = \lambda_i c_a y_i(a) \tag{5.94}$$

$$L_b y_i(b) = \lambda_i c_b y_i(b) \tag{5.95}$$

Using the inner-product definition of Equation 5.92 and Equation 5.93, Equation 5.94, and Equation 5.95.

$$(y_i, y_j)_L = \int_a^b (Ly_i)y_j(x)r(x)dx + [L_a y_i(a)]y_j(a) + [L_b y_i(b)]y_j(b)$$

$$= \int_a^b \lambda_i y_i(x)y_j(x)r(x)dx + \lambda_i c_a y_i(a)y_j(a) + \lambda_i c_b y_i(b)y_j(b) \tag{5.96}$$

In a similar fashion, it can be determined that

$$(y_j, y_i)_L = \int_a^b \lambda_j y_i(x) y_j(x) r(x) dx + \lambda_j c_a y_i(a) y_j(a) + \lambda_j c_b y_i(b) y_j(b) \quad (5.97)$$

Subtracting Equation 5.97 from Equation 5.96 leads to

$$(y_i y_j)_L - (y_j, y_i)_L$$
$$= (\lambda_i - \lambda_j) \left[\int_a^b y_i(x) y_j(x) r(x) dx + c_a y_i(a) y_j(a) + c_b y_i(b) y_j(b) \right] \quad (5.98)$$

Since the inner product is commutative, Equation 5.98 implies that

$$(\lambda_i - \lambda_j) \left[\int_a^b y_i(x) y_j(x) r(x) dx + c_a y_i(a) y_j(a) + c_b y_i(b) y_j(b) \right] = 0 \quad (5.99)$$

Since λ_i and λ_j are distinct,

$$\int_a^b y_i(x) y_j(x) r(x) dx + c_a y_i(a) y_j(a) + c_b y_i(b) y_j(b) = 0 \quad (5.100)$$

Equation 5.100 shows that the eigenvectors corresponding to distinct eigenvalues of an eigenvalue problem of the form of Equation 5.87, Equation 5.88, and Equation 5.89 are orthogonal with respect to the inner product of Equation 5.92.

The development of the orthogonality condition leads to the definition of the kinetic-energy inner product defined by

$$(f, g)_M = \int_a^b f(x) g(x) r(x) dx + c_a f(a) g(a) + c_b f(b) g(b) \quad (5.101)$$

Eigenvectors corresponding to distinct eigenvalues are also orthogonal with respect to the inner product of Equation 5.101.

If the inner product of Equation 5.101 is used to normalize eigenvectors such that

$$1=(y_i,y_i)_M$$
$$=\int_a^b [y_i(x)]^2 r(x)dx + c_a[f(a)]^2 + c_b[f(b)]^2 \tag{5.102}$$

then

$$(y_i,y_i)_L = \lambda_i \tag{5.103}$$

Example 5.14 The differential equation governing the torsional oscillations of the nonuniform circular shaft shown in Figure 5.12 is

$$\frac{\partial}{\partial x}\left(JG\frac{\partial\theta}{\partial x}\right)=\rho J\frac{\partial^2\theta}{\partial t^2} \tag{a}$$

where G is the shaft's shear modulus, ρ is its mass density, and J is the cross-sectional area polar moment of inertia. Since the shaft is fixed at $x=0$,

$$\theta(0,t)=0 \tag{b}$$

Application of the moment equation to the thin rigid disk at $x=\mathrm{L}$ leads to

$$-J(L)G\frac{\partial\theta}{\partial x}(L,t)=I\frac{\partial^2\theta}{\partial t^2}(L,t) \tag{c}$$

where I is the mass moment of inertia of the disk about its centroidal axis.

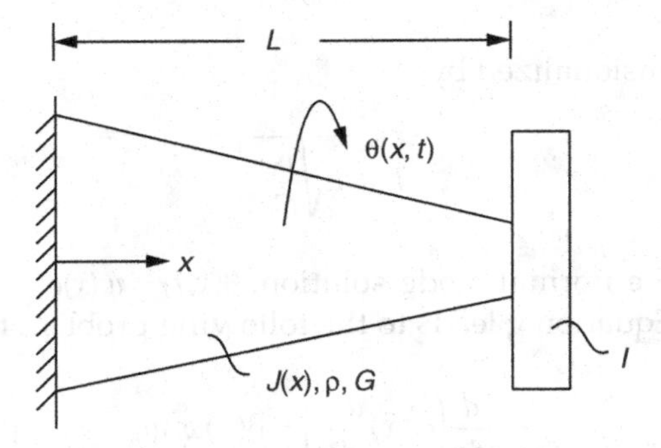

Figure 5.12 The shaft of Example 5.14 has a thin disk attached at its end. Upon application of the normal-mode solution, an eigenvalue problem is obtained in which the eigenvalue appears in the boundary condition as well as the differential equation.

Nondimensional variables are introduced according to

$$x^* = \frac{x}{L}$$

$$t^* = \frac{t}{T}$$

$$\alpha(x^*) = \frac{JG}{J_0G_0} \tag{d}$$

$$\beta(x) = \frac{\rho J}{\rho_0 J_0}$$

where the subscript "0" refers to a reference value of the quantity. Substitution of Equation d into Equation a, Equation b, and Equation c leads to the nondimensional problem formulation,

$$\frac{\partial}{\partial x}\left(\alpha\frac{\partial\theta}{\partial x}\right) = \beta\frac{\partial^2\theta}{\partial t^2} \tag{e}$$

$$\theta(0,t) = 0 \tag{f}$$

$$\alpha(1)\frac{\partial\theta}{\partial x}(1,t) = -\mu\frac{\partial^2\theta}{\partial t^2}(1,t) \tag{g}$$

where

$$\mu = \frac{I}{\rho_0 J_0 L} \tag{h}$$

and t is nondimensionalized by

$$T = L\sqrt{\frac{\rho_0}{G_0}} \tag{i}$$

Substitution of the normal-mode solution, $\theta(x,t) = w(x)e^{i\omega t}$, into Equation e, Equation f, and Equation g leads to the following problem for $w(x)$:

$$-\frac{d}{dx}\left(\alpha(x)\frac{dw}{dx}\right) = \beta(x)\omega^2 w \tag{j}$$

$$w(0) = 0 \tag{k}$$

$$\alpha(1)\frac{dw}{dx}(1) = \mu\omega^2 w(1) \tag{l}$$

The eigenvalue problem of Equation j, Equation k, and Equation l is of the form of Equation 5.87, Equation 5.88, and Equation 5.89 with

$$Lw = -\frac{1}{\beta(x)}\frac{d}{dx}\left(\alpha(x)\frac{dw}{dx}\right) \quad \lambda = \omega^2 \tag{m}$$

$$L_a w(a) = w(a) \quad c_a = 0 \tag{n}$$

$$L_b w(b) = \alpha(1)\frac{dw}{dx}(L) \quad c_b = \mu \tag{o}$$

The inner product of Equation 5.101 becomes

$$(f,g)_M = \int_0^L f(x)g(x)\beta(x)dx + \mu f(a)g(a) \tag{p}$$

The total kinetic energy of the system is the kinetic energy of the shaft plus the kinetic energy of the disk,

$$T = \frac{1}{2}\int_0^L \left(\frac{\partial\theta}{\partial t}\right)^2 \rho J dx + \frac{1}{2} I\left(\frac{\partial\theta}{\partial t}(L,t)\right)^2 \tag{q}$$

Assuming that $\theta(x, t) = w(x)\,\mathrm{Re}[e^{i\omega t}] = w(x)\cos(\omega t)$ and transforming to nondimensional variables using Equation d, Equation h, and Equation i leads to

$$\begin{aligned} T &= \frac{1}{2}\omega^2\cos^2(\omega t)\left(\frac{G_0 J_0}{L}\right)\left[\int_0^1 [w(x)]^2\beta(x)dx + \mu[w(1)]^2\right] \\ &= \frac{1}{2}\omega^2\cos^2(\omega t)\left(\frac{G_0 J_0}{L}\right)(w,w)_M \end{aligned} \tag{r}$$

Equation r shows that the kinetic energy is proportional to the energy inner product of the system response with itself. Determine the nondimensional natural frequencies and mode shapes of a uniform shaft with a thin disk at the end such that $\mu = 0.75$.

Solution The eigenvalue problem for a uniform shaft is

$$\frac{d^2w}{dx^2} + \lambda w = 0 \tag{s}$$

$$w(0) = 0 \tag{t}$$

$$\frac{dw}{dx}(1) = \mu\lambda w(1) \tag{u}$$

The general solution of Equation s is

$$w(x) = C_1\cos(\sqrt{\lambda}x) + C_2\sin(\sqrt{\lambda}x) \tag{v}$$

Application of Equation t to Equation v leads to $C_1 = 0$. Application of Equation u to the resulting form of Equation v leads to

$$\begin{aligned} \sqrt{\lambda}C_2\cos(\sqrt{\lambda}) &= \mu\lambda C_2\sin(\sqrt{\lambda}) \\ \frac{1}{\mu\sqrt{\lambda}} &= \tan(\sqrt{\lambda}) \end{aligned} \tag{w}$$

Equation w is a transcendental equation whose solutions for λ are the eigenvalues of the system. The solutions correspond to the values of the points of intersection of the curves shown in Figure 5.13. As expected, there are an infinite, but countable, number of eigenvalues, $0 < \lambda_1 < \lambda_2 < \cdots < \lambda_{k-1} < \lambda_k < \lambda_{k+1} < \cdots$.

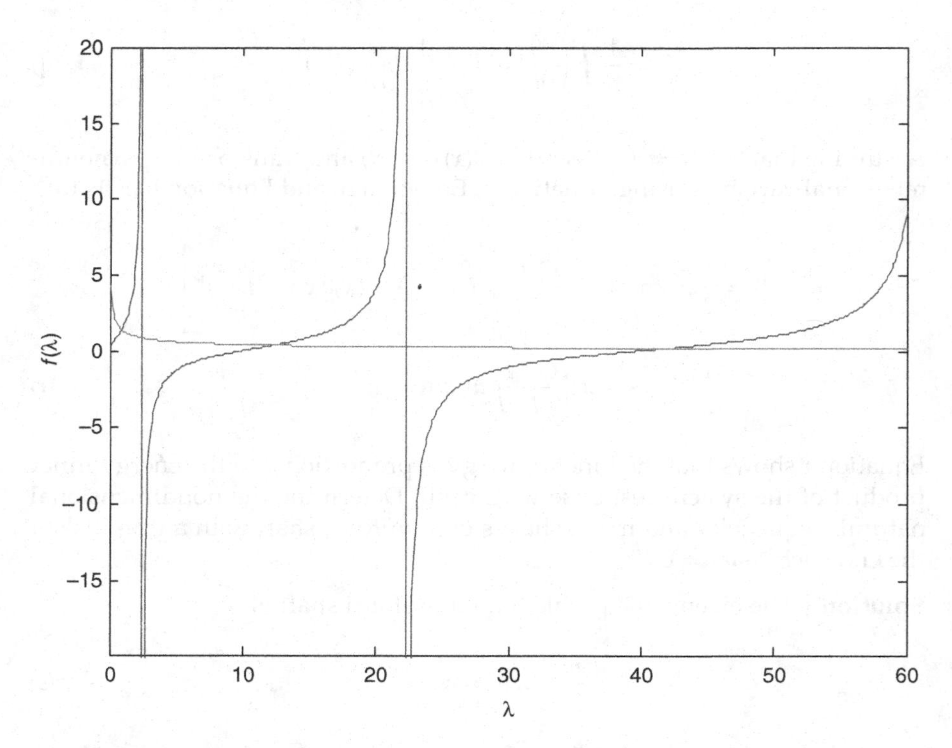

Figure 5.13 The points of intersection are the eigenvalues of Example 5.14.

As the eigenvalues grow large, the points of intersection of the curves approach integer multiples of π such that $\lim_{k\to\infty}\lambda_k = [(k-1)\pi]^2$. The five lowest values of the eigenvalues and the nondimensional natural frequencies for several values of μ are given in Table 5.3.

The mode shape corresponding to the eigenvalue λ_k is

$$w_k(x) = C_k \sin(\sqrt{\lambda_k}\,x) \tag{x}$$

Mode-shape orthogonality is demonstrated by

$$\begin{aligned}
(w_i, w_j)_M &= \int_0^1 w_i(x) w_j(x)\,dx + \mu w_i(1) w_j(1) \\
&= C_i C_j \left[\int_0^1 \sin(\sqrt{\lambda_i}\,x)\sin(\sqrt{\lambda_j}\,x)\,dx + \mu \sin(\sqrt{\lambda_i})\sin(\sqrt{\lambda_j}) \right] \\
&= C_i C_j \left\{ \frac{\sin[(\sqrt{\lambda_i}-\sqrt{\lambda_j})x]}{2(\sqrt{\lambda_i}-\sqrt{\lambda_j})} - \frac{\sin[(\sqrt{\lambda_i}+\sqrt{\lambda_j})x]}{2(\sqrt{\lambda_i}+\sqrt{\lambda_j})} \right\} \Bigg|_{x=0}^{x=1} \\
&\quad + C_i C_j \mu \sin(\sqrt{\lambda_i})\sin(\sqrt{\lambda_j}) \\
&= C_i C_j \left\{ \frac{\sin(\sqrt{\lambda_i})\cos(\sqrt{\lambda_j}) - \cos(\sqrt{\lambda_i})\sin(\sqrt{\lambda_j})}{2(\sqrt{\lambda_i}-\sqrt{\lambda_j})} \right. \\
&\quad - \frac{\sin(\sqrt{\lambda_i})\cos(\sqrt{\lambda_j}) + \cos(\sqrt{\lambda_i})\sin(\sqrt{\lambda_j})}{2(\sqrt{\lambda_i}+\sqrt{\lambda_j})} \\
&\quad \left. + \mu \sin(\sqrt{\lambda_i})\sin(\sqrt{\lambda_j}) \right\}
\end{aligned} \tag{y}$$

Table 5.3 Natural Frequencies of Example 5.14 for Various Values of μ

μ	ω_1	ω_2	ω_3	ω_4	ω_5
0.1	1.429	4.306	7.182	10.200	13.241
0.25	1.265	3.935	6.814	9.812	12.869
0.5	1.077	3.649	6.578	9.630	12.721
0.75	0.951	3.505	6.486	9.563	12.671
1.0	0.860	3.426	6.437	9.529	12.640
2.0	0.653	3.292	6.362	9.477	12.606
10.0	0.311	3.173	6.292	9.435	12.574

Obtaining a common denominator of $\sqrt{\lambda_i}-\sqrt{\lambda_j}$ and using Equation w to substitute for $\sin(\sqrt{\lambda_i})$ and $\sin(\sqrt{\lambda_j})$ in the last term of Equation y leads to $(w_i, w_j)_M = 0$.

The eigenvectors are normalized by requiring

$$\begin{aligned}
1 &= (w_i, w_i)_M \\
&= C_i^2\left\{\int_0^1 [\sin(\sqrt{\lambda_i}x)]^2\,dx + \mu[\sin(\sqrt{\lambda_i})]^2\right\} \\
&= C_i^2\left\{\frac{1}{2}[x-] + \mu[\sin(\sqrt{\lambda_i})]^2\right\}_{x=0}^{x=1} \\
&= C_i^2\left\{\frac{1}{2}\left[1-\frac{1}{2\sqrt{\lambda_i}}\sin(2\sqrt{\lambda_i})\right] + \mu[\sin(\sqrt{\lambda_i})]^2\right\} \\
&= C_i^2\left\{\frac{1}{2}\left[1-\frac{1}{\sqrt{\lambda_i}}\sin(\sqrt{\lambda_i})\cos(\sqrt{\lambda_i})\right] + \mu[\sin(\sqrt{\lambda_i})]^2\right\} \qquad \text{(z)}
\end{aligned}$$

Noting from Equation x that $\cos(\sqrt{\lambda_i}) = \mu\sqrt{\lambda_i}\sin(\sqrt{\lambda_i})$, Equation z can be used to determine

$$C_i = \sqrt{\frac{2}{1+\mu[\sin\sqrt{\lambda_i}]^2}} \qquad \text{(aa)}$$

The normalized mode shapes corresponding to the five lowest natural frequencies for several values of μ are given in Figure 5.14.

5.9 *Eigenvalue problems involving Bessel functions*

Many applications lead to eigenvalue problems involving Bessel functions. Consider the eigenvalue problem $\mathbf{L}\mathbf{y} = \lambda\mathbf{y}$, where $\mathbf{L}$ is Bessel's operator of order n,

$$\mathbf{L}y = -\frac{1}{x}\frac{d}{dx}\left(x\frac{dy}{dx}\right) + n^2\frac{y}{x^2} \qquad (5.104)$$

Equation 5.104 is of the form of the general Sturm-Liouville operator, Equation 5.18, with $p(x) = -x$, $q(x) = n^2/x$ and $r(x) = x$. If the eigenvalue problem is to be solved over a domain $a \le x \le b$, then $p(x) < 0$ and $q(x) > 0$ are such that the operator is positive definite and self-adjoint with respect to the inner product,

$$(f, g) = \int_a^b f(x)g(x)x\,dx \qquad (5.105)$$

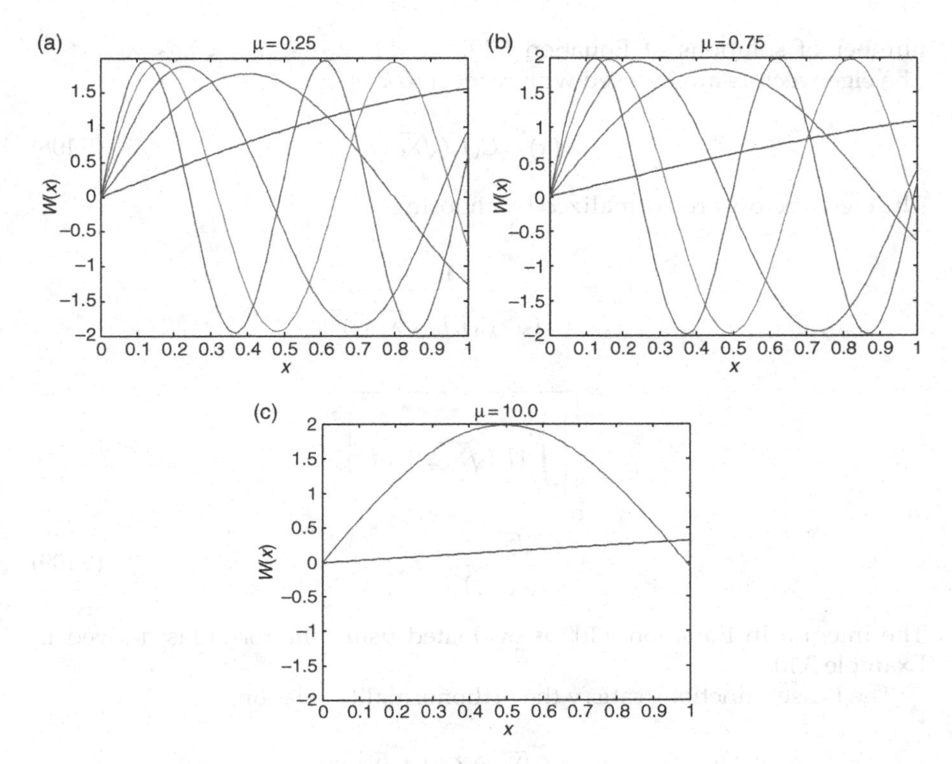

Figure 5.14 Five lowest-mode shapes for the system of Example 5.14 with (a) $\mu=0.25$, (b) $\mu=0.75$, and (c) two lowest-mode shapes for $\mu=10.0$.

If $a=0$, then since $p(0)=0$, it is required only at $x=0$ that the solution be finite.

Consider the eigenvalue problem defined by the operator of Equation 5.104 defined for $0\leq x\leq 1$ with the condition $y(1)=0$. The general solution of the eigenvalue problem is

$$y(x)=C_1J_n(\sqrt{\lambda}x)+C_2Y_n(\sqrt{\lambda}x) \tag{5.106}$$

Recalling that $J_n(0)$ is finite, but $Y_n(0)$ does not exist, it is required that $C_2=0$ for the solution to remain finite at $x=0$. Application of the boundary condition at $x=1$ leads to the equation defining the eigenvalues:

$$J_n(\sqrt{\lambda})=0 \tag{5.107}$$

Since the operator is self-adjoint with respect to the inner product of Equation 5.105, it can be concluded that there are an infinite, but countable, number of solutions of Equation 5.107, $\lambda_1 < \lambda_2 < \cdots < \lambda_{k-1} < \lambda_k < \lambda_{k+1} < \cdots$. The eigenvectors are indexed with respect to k as

$$y_k(x) = C_k J_n(\sqrt{\lambda_k}\, x) \tag{5.108}$$

The eigenvectors are normalized by choosing

$$\begin{aligned} C_k &= \frac{1}{(J_n(\sqrt{\lambda_k}\, x), J_n(\sqrt{\lambda_k}\, x))^{\frac{1}{2}}} \\ &= \frac{1}{\left\{\int_0^1 [J_n(\sqrt{\lambda_k}\, x)]^2 x dx\right\}^{\frac{1}{2}}} \\ &= \frac{\sqrt{2}}{J_{n+1}(\sqrt{\lambda_k})} \end{aligned} \tag{5.109}$$

The integral in Equation 5.109 is evaluated using the formulas derived in Example 3.10.

The Bessel functions satisfy the orthonormality relation,

$$\begin{aligned} \delta_{k,m} &= (C_k J_n(\sqrt{\lambda_k}\, x), C_m J_n(\sqrt{\lambda_m}\, x)) \\ &= C_k C_m \int_0^1 J_n(\sqrt{\lambda_k}\, x) J_n(\sqrt{\lambda_m}\, x) dx \end{aligned} \tag{5.110}$$

Noting that the eigenvalues are the square roots of the zeroes of the Bessel functions, and defining $\alpha_{k,n}$ as the kth positive zero of $J_n(x)$, the normalized eigenvectors can be written as

$$y_{k,n} = \sqrt{2}\,\frac{J_n(\alpha_{k,n} x)}{J_{n+1}(\alpha_{k,n})} \tag{5.111}$$

The first ten zeroes for the first five Bessel functions are given in Table 5.4.

The Bessel functions $J_n(\alpha_{k,n} x)$ for $k = 1, 2, \ldots$ are complete in the subspace V of $C^2[0,1]$, defined such that if $f(x)$ is in V, then $f(0) = 1$ and $f(x)$ is finite at $x = 0$. Thus any function in V has an expansion of the form

$$f(x) = \sum_{k=0}^{\infty} \sqrt{2} A_k \frac{J_n(\alpha_{k,n} x)}{J_{n+1}(\alpha_{k,n})} \tag{5.112}$$

Table 5.4 Roots of Bessel Functions of Integer Order. $\alpha_{n,m}$ is the mth Positive Zero of $J_n(x)$

n	0	1	2	3	4
$\alpha_{n,1}$	2.405	3.832	5.131	6.380	7.558
$\alpha_{n,2}$	5.520	7.016	8.417	9.761	11.065
$\alpha_{n,3}$	8.6754	10.173	11.620	13.015	14.373
$\alpha_{n,4}$	11.792	13.324	14.796	16.223	17.617
$\alpha_{n,5}$	14.931	16.471	17.940	19.409	20.827
$\alpha_{n,6}$	18.071	19.616	21.117	22.583	24.019
$\alpha_{n,7}$	21.212	22.760	24.270	25.748	27.199
$\alpha_{n,8}$	24.352	25.904	27.421	28.908	30.371
$\alpha_{n,9}$	27.493	29.047	30.569	32.065	33.537
$\alpha_{n,10}$	30.653	32.190	33.217	35.219	36.699

where

$$A_k = \frac{\sqrt{2}}{J_{n+1}(\alpha_{k,n})} \int_0^1 J_n(\alpha_{k,n}x) f(x) x dx \tag{5.113}$$

A series of the form of Equation 5.112 is called a Fourier-Bessel expansion. As with the trigonometric Fourier series, it converges pointwise to $f(x)$ for all x at which $f(x)$ is continuous.

Example 5.15 Expand $f(x) = 1 - x$ in a Fourier-Bessel series. Use $n = 0, 1, 2, 3, 4, 5$ and compare results.

Solution The coefficients in the Fourier-Bessel series for $f(x)$ are

$$\begin{aligned} A_k &= \left(f(x), \sqrt{2}\frac{J_n(\alpha_{k,n}x)}{J_{n+1}(\alpha_{k,n})} \right)_x \\ &= \frac{\sqrt{2}}{J_{n+1}(\alpha_{k,n})} \int_0^1 J_n(\alpha_{k,n}x)(1-x) x dx \end{aligned} \tag{a}$$

The first six Fourier coefficients for $n = 0, 1, 2, 3, 4$ are listed in Table 5.5. For example, the Fourier-Bessel expansion for $n = 2$ is

$$\begin{aligned} f(x) = \sqrt{2}\Bigg[& 0.2302\frac{J_2(5.131x)}{J_3(5.131)} - 0.1888\frac{J_2(8.417x)}{J_3(8.417)} + 0.0793\frac{J_2(11.62x)}{J_3(11.62)} \\ & - 0.0558\frac{J_2(14.796x)}{J_3(14.796)} + 0.0431\frac{J_2(17.96x)}{J_3(17.96)} - 0.0339\frac{J_2(21.117x)}{J_3(21.117)} \Bigg] + \cdots \end{aligned} \tag{b}$$

Table 5.5 Coefficients in Fourier-Bessel Series for $f(x)=1-x$. A_k^n the kth Coefficient in the Expansion Using $J_n(\alpha_{k,n}x)$ as Basis Functions

n	0	1	2	3	4
A_1^n	0.288	0.261	0.230	0.206	0.181
A_2^n	−0.0165	−0.0918	−0.119	−0.129	−0.131
A_3^n	0.0102	0.0560	0.0793	0.0934	0.0985
A_4^n	−0.00285	−0.00359	−0.0558	−0.0683	−0.0762
A_5^n	0.00248	0.0269	0.0431	0.0541	0.0619
A_6^n	−0.00138	−0.0202	−0.0339	−0.0438	−0.0511

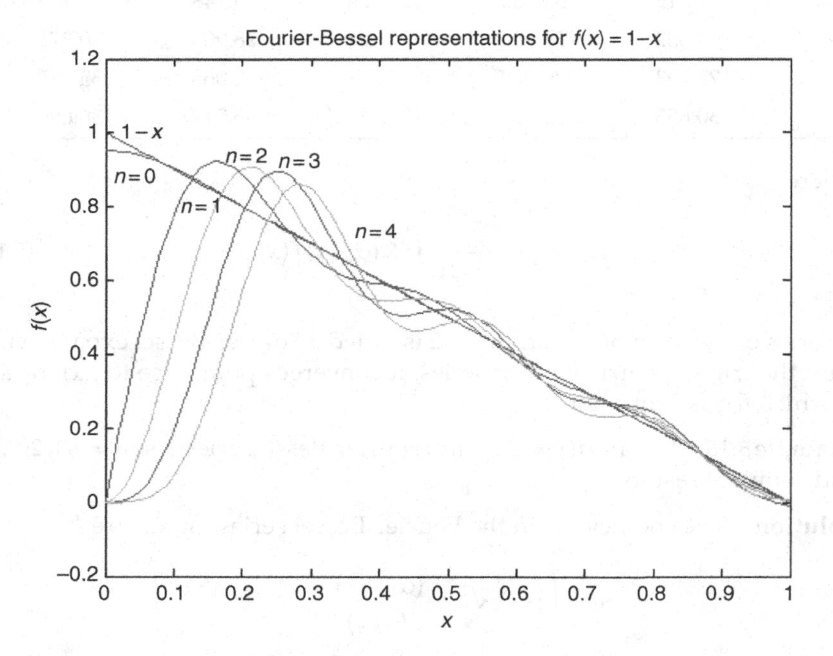

Figure 5.15 The Fourier-Bessel series representation for $f(x)=1-x$ is better for smaller n.

The Fourier-Bessel approximations for $f(x)$ are illustrated in Figure 5.15. The approximation for $n=0$ is best near $x=0$ because $f(0)=1$, whereas $J_n(0)=0$ for $n>1$. The Fourier series converges pointwise to $f(x)$ for all x except at $x=0$ for $n>0$.

There are many applications which lead to eigenvalue problems involving Bessel's equation or equations whose solution involves Bessel functions. The examples below include a buckling problem and a problem to determine natural frequencies and mode shapes for a nonuniform distributed-parameter system.

Example 5.16 The differential equation for the slope of the elastic curve of a vertical column under its own weight

$$\frac{d}{dx}\left(EI\frac{d\theta}{dx}\right)+f(x)\theta=0 \tag{a}$$

where

$$f(x)=\rho g\int_0^x A(\xi)d\xi \tag{b}$$

Introducing nondimensional variables,

$$x^*=\frac{x}{L}$$

$$w^*=\frac{w}{L}$$

$$\alpha(x)=\frac{EI}{E_0I_0}$$

$$\beta(x)=\frac{\rho g\int_0^L A(\xi)d\xi}{\rho_0 g A_0 L} \tag{c}$$

into Equation b leads to

$$\frac{d}{dx}\left(\alpha(x)\frac{d\theta}{dx}\right)+\lambda f(x)\theta=0 \tag{d}$$

where

$$\lambda=\frac{\rho_0 g A_0 L^3}{E_0 I_0} \tag{e}$$

The critical buckling length, L_{cr}, is the length at which the column buckles under its own weight. Determine the critical buckling length of a fixed-free column when the column is uniform.

Solution The appropriate boundary conditions for a fixed-free column are

$$\frac{d\theta}{dx}(0)=0 \tag{f}$$

$$\theta(1)=0 \tag{g}$$

For a uniform column, $\alpha(x)=1$, and therefore Equation d simplifies to

$$\frac{d^2\theta}{dx^2}+\lambda x\theta=0 \tag{h}$$

Equation h is of the form of Equation 3.110 with $r=0$ and $s=1$. The solution of Equation h is of the form of Equation 3.111. The resulting general solution of Equation h is

$$\theta(x)=C_1x^{1/2}J_{1/3}\left(\frac{2}{3}\sqrt{\lambda}x^{3/2}\right)+C_2x^{1/2}Y_{1/3}\left(\frac{2}{3}\sqrt{\lambda}x^{3/2}\right) \tag{i}$$

Since the order of the Bessel functions is not an integer or a half-integer an alternate representation of Equation i is

$$\theta(x)=C_1x^{1/2}J_{1/3}\left(\frac{2}{3}\sqrt{\lambda}x^{3/2}\right)+C_2x^{1/2}J_{-1/3}\left(\frac{2}{3}\sqrt{\lambda}x^{3/2}\right) \tag{j}$$

The attempted application of Equation g meets with difficulty. Whereas $J_{-1/3}(0)$ is undefined, it is not clear what the value of $d/dx(x^{1/2}J_{-1/3}(x))$ is at $x=0$. It is convenient to use the series expansions of the Bessel functions to apply the boundary condition at $x=0$. To this end,

$$x^{1/2}J_{1/3}\left(\frac{2}{3}\sqrt{\lambda}x^{3/2}\right)=x^{1/2}\sum_{k=0}^{\infty}\frac{(-1)^k}{\Gamma(k+1)\Gamma\left(k+\frac{4}{3}\right)}\left(\frac{1}{3}\sqrt{\lambda}x^{3/2}\right)^{2k+1/3}$$

$$=\sum_{k=0}^{\infty}\frac{(-1)^k\left(\frac{1}{3}\sqrt{\lambda}\right)^{2k+1/3}}{\Gamma(k+1)\Gamma\left(k+\frac{4}{3}\right)}x^{3k+1} \tag{k}$$

Then

$$\frac{d}{dx}\left[x^{1/2}J_{1/3}\left(\frac{2}{3}\sqrt{\lambda}x^{3/2}\right)\right]=\sum_{k=0}^{\infty}\frac{(-1)^k(3k+1)\left(\frac{1}{3}\sqrt{\lambda}\right)^{2k+1/3}}{\Gamma(k+1)\Gamma\left(k+\frac{4}{3}\right)}x^{3k} \tag{l}$$

Equation l is used to determine

$$\frac{d}{dx}\left[x^{1/2}J_{1/3}\left(\frac{2}{3}\sqrt{\lambda}x^{3/2}\right)\right]_{x=0}=\frac{\left(\frac{1}{3}\sqrt{\lambda}\right)^{1/3}}{\Gamma\left(\frac{4}{3}\right)} \tag{m}$$

Similarly,

$$x^{1/2}J_{-1/3}\left(\frac{2}{3}\sqrt{\lambda}x^{3/2}\right)=x^{1/2}\sum_{k=0}^{\infty}\frac{(-1)^k}{\Gamma(k+1)\Gamma\left(k+\frac{2}{3}\right)}\left(\frac{1}{3}\sqrt{\lambda}x^{3/2}\right)^{2k-1/3}$$
$$=\sum_{k=0}^{\infty}\frac{(-1)^k\left(\frac{1}{3}\sqrt{\lambda}\right)^{2k-1/3}}{\Gamma(k+1)\Gamma\left(k+\frac{2}{3}\right)}x^{3k} \tag{n}$$

$$\frac{d}{dx}\left[x^{1/2}J_{-1/3}\left(\frac{2}{3}\sqrt{\lambda}x^{3/2}\right)\right]=\sum_{k=0}^{\infty}\frac{(-1)^k(3k)\left(\frac{1}{3}\sqrt{\lambda}\right)^{2k-1/3}}{\Gamma(k+1)\Gamma\left(k+\frac{2}{3}\right)}x^{3k-1} \tag{o}$$

$$\frac{d}{dx}\left[x^{1/2}J_{-1/3}\left(\frac{2}{3}\sqrt{\lambda}x^{3/2}\right)\right]_{x=0}=0 \tag{p}$$

Thus, in view of Equation k and Equation p, application of Equation g to Equation j leads to $C_1=0$ and

$$\theta(x)=C_2x^{1/2}J_{-1/3}\left(\frac{2}{3}\sqrt{\lambda}x^{3/2}\right) \tag{q}$$

Application of Equation h to Equation q leads to the transcendental equation

$$J_{-1/3}\left(\frac{2}{3}\sqrt{\lambda}\right)=0 \tag{r}$$

The smallest value of λ which satisfies Equation r is $\lambda=7.833$, and therefore, from Equation e, the critical buckling length of a uniform fixed-free column is

$$L_{cr}=1.986\left(\frac{EI}{\rho GA_0}\right)^{1/3} \tag{s}$$

Example 5.17 Consider the nonuniform shaft with a thin disk and torsional spring attached at one end ($x=1$) and fixed at its other end ($x=0$). The nondimensional problem governing the natural frequencies and mode shapes of the shaft is

$$\frac{d}{dx}\left(\alpha(x)\frac{d\Theta}{dx}\right)+\omega^2\alpha(x)\Theta=0 \tag{a}$$

$$\Theta(0) = 0 \tag{b}$$

$$\frac{d\Theta}{dx}(1) + \kappa\Theta(1) = \mu\omega^2\Theta(1) \tag{c}$$

Determine the first five natural frequencies and mode shapes when $\kappa = 1.5$ and $\mu = 1$ for $\alpha(x) = (1 - 0.2x)^2$.

Solution Substitution of Equation d into Equation a leads to

$$\frac{d}{dx}\left[(1 - 0.2x)^2 \frac{d\Theta}{dx}\right] + \omega^2(1 - 0.2x)^2\Theta = 0 \tag{d}$$

Define

$$z = 1 - 0.2x \tag{e}$$

Equation d is rewritten using z as the independent variable as

$$\frac{d}{dz}\left(z^2 \frac{d\Theta}{dz}\right) + \lambda z^2\Theta = 0 \tag{f}$$

where

$$\lambda = 25\omega^2 \tag{g}$$

Using z as the independent variable, the boundary conditions, Equation b and Equation c, become

$$\Theta(z = 1) = 0 \tag{h}$$

$$\frac{d\Theta}{dz}(0.8) + 5\kappa\Theta(0.8) = \frac{\mu}{5}\lambda\Theta(0.8) \tag{i}$$

Equation f is of the form of Equation 3.108 with $r = 2$ and $s = 2$. The solution to Equation f is obtained by comparison with Equation 3.108 as

$$\Theta(z) = C_1 z^{-1/2} J_{-1/2}(\sqrt{\lambda}z) + C_2 z^{-1/2} Y_{-1/2}(\sqrt{\lambda}z) \tag{j}$$

Application of Equation h to Equation j gives

$$C_2 = -\frac{J_{-1/2}(\sqrt{\lambda})}{Y_{-1/2}(\sqrt{\lambda})} C_1 \tag{k}$$

which when used in Equation j leads to

$$\Theta(z) = z^{1/2} C_1 \left[J_{-1/2}(\sqrt{\lambda}z) - \frac{J_{-1/2}(\sqrt{\lambda})}{Y_{-1/2}(\sqrt{\lambda})} Y_{-1/2}(\sqrt{\lambda}z)\right] \tag{l}$$

Application of Equation i to Equation l results in

$$(0.2\lambda-0.6940)\left[J_{-1/2}(0.8\sqrt{\lambda})-\frac{J_{-1/2}(\sqrt{\lambda})}{Y_{-1/2}(\sqrt{\lambda})}Y_{-1/2}(0.8\sqrt{\lambda})\right]$$

$$=0.8922\sqrt{\lambda}\left[J'_{-1/2}(0.8\sqrt{\lambda})-\frac{J_{-1/2}(\sqrt{\lambda})}{Y_{-1/2}(\sqrt{\lambda})}Y'_{-1/2}(0.8\sqrt{\lambda})\right] \quad \text{(m)}$$

The kinetic-energy inner product for this system is

$$(f,g)_K=\int_0^1 f(x)g(x)(1-0.2x)^2\,dx+f(1)g(1) \quad \text{(n)}$$

The mode shapes satisfy the orthogonality relation, $(\Theta_i(x),\Theta_j(x))_M=0$ for $i\neq j$. The mode-shape vectors are normalized by requiring that $(\Theta_i(x),\Theta_i(x))_M=1$.

An alternate representation of the solution of Equation f is in terms of spherical Bessel functions. Equation f is in the form of Equation 3.15 with $n=0$, whose solution in terms of spherical Bessel functions is given in terms of Equation 3.116. Thus, the solution of Equation f can be expressed as

$$\Theta(z)=C_1 j_0\left(\sqrt{\lambda z}\right)+C_2 y_0\left(\sqrt{\lambda z}\right) \quad \text{(o)}$$

Application of boundary conditions, Equation b and Equation c, to Equation o leads to the transcendental equation for the natural frequencies as

$$-\omega j_1(4\omega)y_0(5\omega)+\omega j_0(5\omega)y_1(5\omega)+(1.5-\omega^2)[j_0(4\omega)y_0(5\omega)-j_0(5\omega)y_0(4\omega)]=0 \quad \text{(p)}$$

The differentiation relations $j_0'(x)=j_1(x)$ and $yj_0(x)=y_1(x)$ are used in the development of Equation p. The mode shapes are of the form

$$\Theta_k(x)=C_k\left[j_0(5\omega_k(1-0.2x))-\frac{j_0(5\omega_k)}{y_0(5\omega_k)}y_0(5\omega_k(1-0.2x))\right] \quad \text{(q)}$$

The mode shapes are normalized by choosing

$$C_k=\left\{\int_0^1\left[j_0(5\omega_k(1-0.2x))-\frac{j_0(5\omega_k)}{y_0(5\omega_k)}y_0(5\omega_k(1-0.2x))\right]^2(1-0.2x)^2\,dx\right.$$

$$\left.+\left[j_0(4\omega_k)-\frac{j_0(5\omega_k)}{y_0(5\omega_k)}y_0(4\omega_k))\right]^2\right\}^{-\frac{1}{2}} \quad \text{(r)}$$

The following MATHCAD file is used to determine the five lowest natural frequencies for the system using Equation p and to plot the corresponding mode shapes.

$f1(\omega) := js(0,5\cdot\omega)$ The function js(n,x) is the spherical Bessel function of the first kind of order n and argument x

$f2(\omega) := js(0,4\cdot\omega)$

$f3(\omega) := js(1,4\cdot\omega)$

$g1(\omega) := ys(0,5\cdot\omega)$ The function ys(n,x) is the spherical Bessel function of the second kind of order n and argument x

$g2(\omega) := ys(0,4\cdot\omega)$

$g3(\omega) := ys(1,4\cdot\omega)$

Transcendental function used to determine natural frequencies

$$h(\omega) := -\omega\cdot f3(\omega)\cdot g1(\omega) + f1(\omega)\cdot\omega\cdot g3(\omega) + (1.5-\omega^2)\cdot(f2(\omega)\cdot g1(\omega) - f1(\omega)\cdot g2(\omega))$$

$x := 0.4, 0.41..\ 1C$

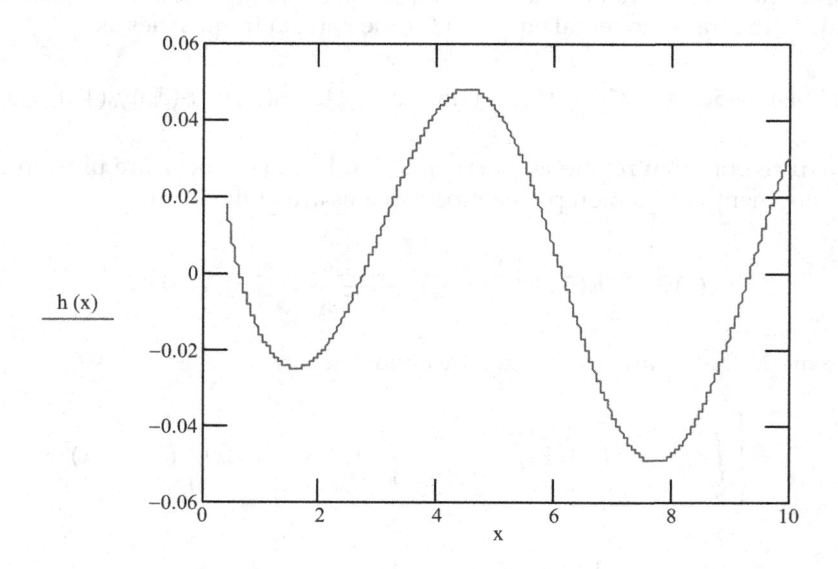

A graph of h over a small region helps to locate range of roots and to make good initial guesses for iterations

$\omega := 2$ Guess for natural frequency obtained from graph

$a := \text{root}(h(\omega),\omega)$ Internal root finding program given function and initial guess

$a := 2.726$ Value of natural frequency

Function defining mode shape

$$f(x) := js[0,5\cdot a\cdot(1-0.2\cdot x)] - \frac{js(0,5\cdot a)}{ys(0,5\cdot a)}\cdot ys[0,5\cdot a\cdot(1-0.2\cdot x)]$$

Plotting mode shape

$x := 0,.01..1$

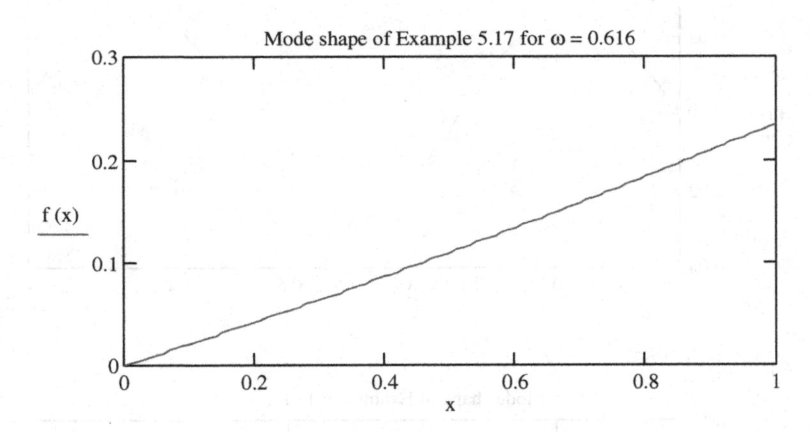

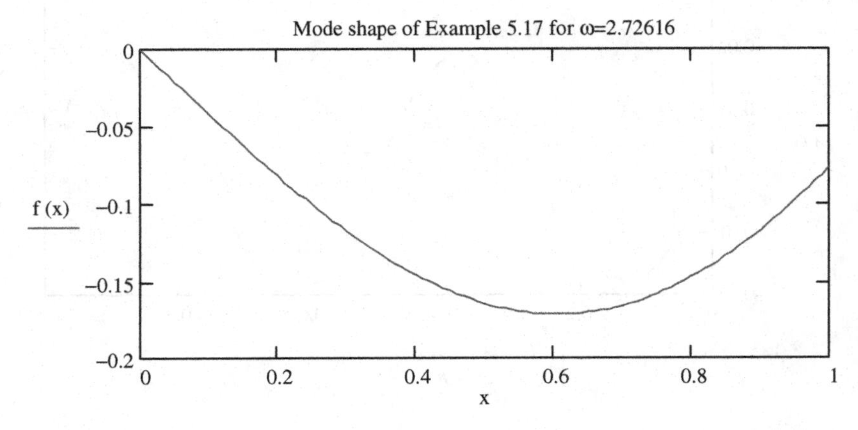

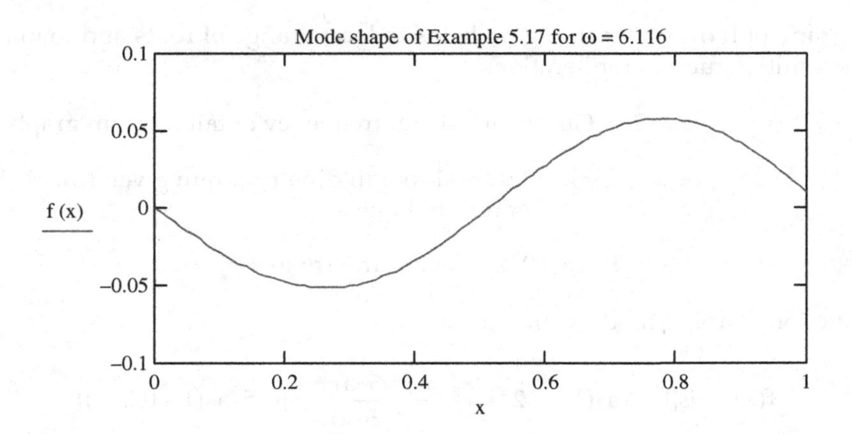
Mode shape of Example 5.17 for ω = 6.116
0.1
0.05
0
−0.05
−0.1
f (x)
0
0.2
0.4
0.6
0.8
1
x

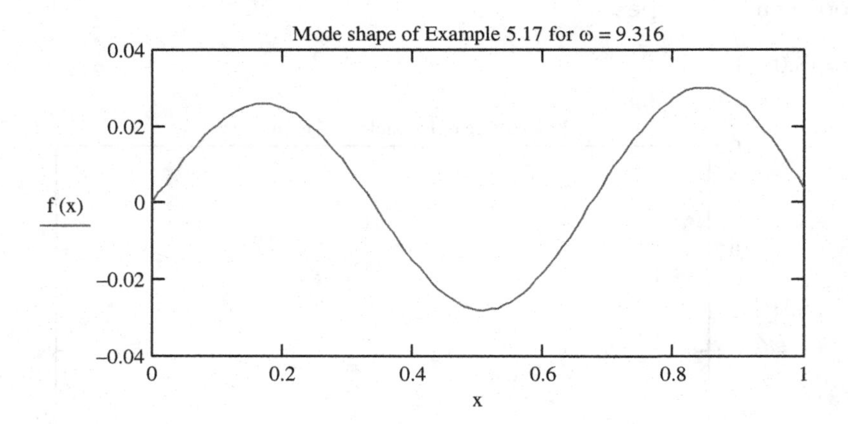
Mode shape of Example 5.17 for ω = 9.316
0.04
0.02
0
−0.02
−0.04
f (x)
0
0.2
0.4
0.6
0.8
1
x

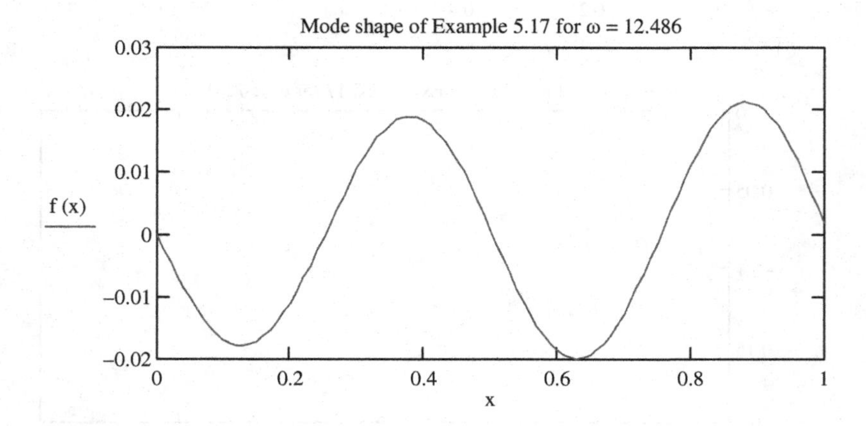
Mode shape of Example 5.17 for ω = 12.486
0.03
0.02
0.01
0
−0.01
−0.02
f (x)
0
0.2
0.4
0.6
0.8
1
x

5.10 Eigenvalue problems in other infinite-dimensional vector spaces

A model for vibrations of double-walled nanotubes is presented in Example 1.4. The model is that of concentric beams connected by an elastic layer. A generalization of the model is that of a series of beams, not necessarily concentric, connected by elastic layers. The outermost beams are connected to rigid foundations through elastic layers. The nondimensional differential equations for such a system can be summarized as

$$\mu_i \frac{\partial^4 w_i}{\partial x^4} - \varepsilon \frac{\partial^2 w_i}{\partial x_i^2} + \eta_{i-1}(w_i - w_{i-1}) + \eta_i(w_i - w_{i+1}) + \beta_i \frac{\partial^2 w_i}{\partial t^2} \quad i = 1,2,\ldots n \tag{5.114}$$

The first and nth equations of the form of Equation 5.114 should be modified to take into account that w_0 and w_{n+1} are zero. A normal-mode solution is assumed to be

$$w_i(x,t) = u_i(x)e^{i\omega t} \tag{5.115}$$

Use of Equation 5.115 in the equations summarized by Equation 5.114 leads to

$$\mu_i \frac{d^4 u_i}{dx^4} - \varepsilon \frac{d^2 u_i}{dx^2} - \eta_{i-1}u_{i-1} + (\eta_{i-1} + \eta_i)u_i - \eta_i u_{i+1} = \beta_i \omega^2 u_i \quad i = 1,2,\ldots,n \tag{5.116}$$

The differential equations of Equation 5.116 can be summarized in matrix form as

$$\mathbf{K} + \mathbf{K}_c\mathbf{u} = \omega^2 \mathbf{M}\mathbf{u} \tag{5.117}$$

where $\mathbf{K}$ is the structural-stiffness operator matrix which can be written as

$$\mathbf{K} = \mathbf{K}_b + \mathbf{K}_a \tag{5.118}$$

in which $\mathbf{K}_b$ is a diagonal operator matrix representing the bending stiffness with

$$(k_b)_{i,i} = \mu_i \frac{\partial^4}{\partial x^4} \tag{5.119}$$

$\mathbf{K}_a$ is a diagonal operator matrix representing the axial stiffness as

$$(k_a)_{i,i} = -\varepsilon \frac{\partial^2}{\partial x^2} \tag{5.120}$$

$\mathbf{K}_c$ is a tridiagonal matrix of coupling stiffnesses due to the elastic layers with

$$(k_c)_{i,i+1} = -\eta_i \quad i = 2,3,\ldots,n \tag{5.121a}$$

$$(k_c)_i = \eta_{i-1} + \eta_i \quad i = 1,2,\ldots,n \tag{5.121b}$$

$$(k_c)_{i-1,i} = -\eta_{i-1} \quad i = 1,2,\ldots,n-1 \tag{5.121c}$$

and $\mathbf{M}$ is a diagonal mass matrix with

$$(m)_{i,i} = \beta_i \quad i = 1,2,\ldots,n \tag{5.122}$$

Multiplying by $\mathbf{M}^{-1}$, Equation 5.117 becomes

$$\mathbf{M}^{-1}(\mathbf{K} + \mathbf{K}_c)\mathbf{u} = \omega^2\mathbf{u} \tag{5.123}$$

It is clear from Equation 5.123 that the natural frequencies are the square roots of the eigenvalues of the linear operator defined by $\mathbf{L} = \mathbf{M}^{-1}(\mathbf{K} + \mathbf{K}_c)$. The domain of $\mathbf{L}$ is Q, the space of vectors of the form $\begin{bmatrix} f_1(x) & f_2(x) & \ldots & f_n(x) \end{bmatrix}^T$, where each $f_i(x)$ is an element of S, the subspace of $C^4[0,1]$ of functions that satisfy all boundary conditions for the individual beams. The standard inner product on Q is defined by

$$(\mathbf{u},\mathbf{v}) = \int_0^1 \mathbf{v}^T\mathbf{u}dx \tag{5.124}$$

The kinetic-energy inner product on Q is defined as

$$(\mathbf{u},\mathbf{v})_M = \int_0^1 \mathbf{v}^T\mathbf{M}\mathbf{u}dx \tag{5.125}$$

A potential-energy inner product is defined by

$$(\mathbf{u},\mathbf{v})_L = \int_0^1 \mathbf{v}^T(\mathbf{L}\mathbf{u})dx \tag{5.126}$$

It is easy to show that since $\mathbf{K}_c$ and $\mathbf{M}$ are symmetric matrices and the stretched-beam operator defined by $\mathbf{L}_b u = d^4u/dx^4 - \varepsilon(d^2u/dx^2)$ is self-adjoint with respect to the standard inner product on S, then $\mathbf{L}$ is self-adjoint with respect to the kinetic-energy inner product of Equation 5.125 and the potential-energy inner product of Equation 5.126. Furthermore, if either $\mathbf{K}_c$ or $\mathbf{L}_b$ is positive definite, then $\mathbf{L}$ is positive definite. In this case, Theorems 5.18 and 5.19 imply that all eigenvalues of $\mathbf{L}$ are real and positive and that eigenvectors corresponding to distinct eigenvalues are orthogonal with respect to the kinetic-energy inner product.

It is difficult to obtain numerical values for the natural frequencies for the general stretched problem. A Rayleigh-Ritz method is presented in Section 5.15. However, it is easy to obtain the natural frequencies and mode shapes for unstretched beams. Thus, for the remainder of this section, set $\varepsilon=0$, in which case $\mathbf{L}_b=\mathbf{M}^{-1}(\mathbf{K}_b+\mathbf{K}_c)$.

Let δ_k, for $k=1,2,\ldots$ be the eigenvalues of the problem $d^4\phi/dx^4=\delta^4\phi$ subject to the boundary conditions satisfied by each of the beams. Let $\phi_k(x)$ be the normalized eigenvectors corresponding to δ_k. A solution of Equation 5.117 is assumed to be in the form of

$$\mathbf{u}_k=\mathbf{a}_k\phi_k(x) \tag{5.127}$$

where $\mathbf{a}_k$ is a vector of constants. Substitution of Equation 5.127 into Equation 5.117 leads to

$$\mathbf{M}^{-1}(\hat{\mathbf{K}}_k+\mathbf{K}_c)\mathbf{a}_k=\omega^2\mathbf{a}_k \tag{5.128}$$

where $\hat{\mathbf{K}}_k$ is a diagonal matrix with

$$\left(\hat{k}\right)_{i,i}=\delta_k^4\mu_i \tag{5.129}$$

Equation 5.129 is in the form of an eigenvalue problem for an $n\times n$ matrix. Since $\mathbf{M}$ and $\hat{\mathbf{K}}_k+\mathbf{K}_c$ are symmetric, the eigenvalues are all real, and if $\hat{\mathbf{K}}_k+\mathbf{K}_c$ is positive definite, all eigenvalues are positive.

For each k, there are n natural frequencies obtained from the solution of Equation 5.129. Thus the natural frequencies are represented by $\omega_{k,j}$ for $k=1,2,\ldots$ and $j=1,2,\ldots,n$. Each natural frequency has a corresponding mode shape of $\mathbf{u}_{k,j}=\mathbf{a}_{k,j}\phi_k(x)$, where $\mathbf{a}_{k,j}$ is the eigenvector of Equation 5.128 corresponding to the natural frequency $\omega_{k,j}$.

Example 5.18 Consider three unstretched, elastically connected beams with the following mass, bending-stiffness operator, and stiffness-coupling matrices:

$$\mathbf{M}=\begin{bmatrix}1&0&0\\0&1.5&0\\0&0&2\end{bmatrix}\qquad \mathbf{K}_b=\begin{bmatrix}\dfrac{\partial^4}{\partial x^4}&0&0\\0&2\dfrac{\partial^4}{\partial x^4}&0\\0&0&3\dfrac{\partial^4}{\partial x^4}\end{bmatrix}$$

$$\mathbf{K}_c=\begin{bmatrix}10000&-10000&0\\-10000&30000&-20000\\0&-20000&30000\end{bmatrix} \tag{a}$$

The normalized mode shape corresponding to a natural frequency ω_k of a uniform fixed-fixed Euler-Bernoulli beam is

$$\phi_k(x) = C_k\{\cosh(\sqrt{\omega_k}\,x) - \cos(\sqrt{\omega_k}\,x) - \alpha_k[\sinh(\sqrt{\omega_k}\,x) - \sin(\sqrt{\omega_k}\,x)]\} \quad \text{(b)}$$

where

$$\alpha_k = \frac{\cosh(\sqrt{\omega_k}) - \cos(\sqrt{\omega_k})}{\sinh(\sqrt{\omega_k}) - \sin(\sqrt{\omega_k})} \quad \text{(c)}$$

and

$$C_k = \left[\int_0^1 \{\cosh(\sqrt{\omega_k}\,x) - \cos(\sqrt{\omega_k}\,x) - \alpha_k[\sinh(\sqrt{\omega_k}\,x) - \sin(\sqrt{\omega_k}\,x)]\}^2\,dx\right]^{-1/2} \quad \text{(d)}$$

The lowest natural frequencies and values of δ for a fixed-fixed beam are

$$\begin{aligned}
\omega_1 &= 22.37 & \delta_1 &= 4.724\\
\omega_2 &= 61.66 & \delta_2 &= 7.852\\
\omega_3 &= 120.9 & \delta_3 &= 10.99\\
\omega_4 &= 199.9 & \delta_4 &= 14.14\\
\omega_5 &= 298.9 & \delta_5 &= 17.29
\end{aligned} \quad \text{(e)}$$

The matrices $\hat{\mathbf{K}}_k$ for $k = 1, 2, \ldots, 5$ are calculated as

$$\hat{\mathbf{K}}_1 = \begin{bmatrix} 500.41 & 0 & 0 \\ 0 & 1.00\times 10^3 & 0 \\ 0 & 0 & 1.50\times 10^3 \end{bmatrix}$$

$$\hat{\mathbf{K}}_2 = \begin{bmatrix} 3.80\times 10^3 & 0 & 0 \\ 0 & 7.60\times 10^3 & 0 \\ 0 & 0 & 1.14\times 10^4 \end{bmatrix}$$

$$\hat{\mathbf{K}}_3 = \begin{bmatrix} 1.46\times 10^4 & 0 & 0 \\ 0 & 2.92\times 10^4 & 0 \\ 0 & 0 & 4.38\times 10^4 \end{bmatrix} \quad \text{(f)}$$

$$\hat{\mathbf{K}}_4 = \begin{bmatrix} 3.99\times 10^4 & 0 & 0 \\ 0 & 7.99\times 10^4 & 0 \\ 0 & 0 & 1.20\times 10^5 \end{bmatrix}$$

$$\hat{\mathbf{K}}_3 = \begin{bmatrix} 8.92\times 10^4 & 0 & 0 \\ 0 & 1.78\times 10^5 & 0 \\ 0 & 0 & 2.67\times 10^5 \end{bmatrix}$$

The natural frequencies of the first set of modes for the elastically connected beams are the square roots of the eigenvalues of $\mathbf{M}^{-1}(\hat{\mathbf{K}}_1 + \mathbf{K}_c)$. The mode shapes are of the form $\mathbf{a}_{1,j}\,\phi_1(x)$ for $j = 1, 2, 3$, where $\mathbf{a}_{1,j}$ are the eigenvectors of $\mathbf{M}^{-1}\left(\hat{\mathbf{K}}_1 + \mathbf{K}_c\right)$ normalized with respect to the kinetic-energy inner product $(\mathbf{u}, \mathbf{v})_M = (\mathbf{Mu}, \mathbf{v})$. The first set of natural frequencies and normalized mode shapes is

$$\omega_{1,1} = 49.17 \qquad \mathbf{w}_{1,1} = \begin{bmatrix} 0.6069 \\ 0.4905 \\ 0.3479 \end{bmatrix} \phi_1(x)$$

$$\omega_{1,2} = 111.8 \qquad \mathbf{w}_{1,2} = \begin{bmatrix} 0.7377 \\ -0.1487 \\ -0.4597 \end{bmatrix} \phi_1(x) \tag{g}$$

$$\omega_{1,3} = 178.8 \qquad \mathbf{w}_{1,3} = \begin{bmatrix} 0.2958 \\ -0.6356 \\ 0.3915 \end{bmatrix} \phi_1(x)$$

The set $\omega_{1,1}, \omega_{1,2}$, and $\omega_{1,3}$ is a set of intramodal frequencies. The first five sets of intramodal frequencies are listed in Table 5.6, while the corresponding normalized mode shapes are given in Table 5.7.

Table 5.6 Sets of Intramodal Frequencies for Example 5.18

	$k=1$	$k=2$	$k=3$	$k=4$	$k=5$
$\omega_{k,1}$	49.17	80.73	139.01	216.52	312.16
$\omega_{k,2}$	111.88	128.79	174.44	254.42	364.15
$\omega_{k,3}$	178.84	190.97	226.45	294.22	395.89

Table 5.7 Normalized Mode-Shape Vectors $\mathbf{a}_{k,j}$, for Example 5.18

k/j	1	2	3	4	5
1	$\begin{bmatrix} 0.6069 \\ 0.4905 \\ 0.3679 \end{bmatrix}$	$\begin{bmatrix} 0.6550 \\ 0.4779 \\ 0.3369 \end{bmatrix}$	$\begin{bmatrix} 0.7902 \\ 0.4184 \\ 0.2377 \end{bmatrix}$	$\begin{bmatrix} 0.9259 \\ 0.2852 \\ 0.1016 \end{bmatrix}$	$\begin{bmatrix} 0.9776 \\ 0.1678 \\ 0.0804 \end{bmatrix}$
2	$\begin{bmatrix} 0.7377 \\ -0.1487 \\ 0.4597 \end{bmatrix}$	$\begin{bmatrix} 0.7012 \\ -0.1954 \\ -0.4749 \end{bmatrix}$	$\begin{bmatrix} 0.5572 \\ -0.3297 \\ -0.5076 \end{bmatrix}$	$\begin{bmatrix} 0.3438 \\ -0.5078 \\ -0.4975 \end{bmatrix}$	$\begin{bmatrix} 0.1947 \\ -0.6512 \\ -0.4037 \end{bmatrix}$
3	$\begin{bmatrix} 0.2958 \\ -0.6356 \\ 0.3915 \end{bmatrix}$	$\begin{bmatrix} 0.2791 \\ -0.6325 \\ 0.4012 \end{bmatrix}$	$\begin{bmatrix} 0.2321 \\ -0.6188 \\ 0.4311 \end{bmatrix}$	$\begin{bmatrix} 0.1563 \\ -0.5722 \\ 0.4921 \end{bmatrix}$	$\begin{bmatrix} 0.0804 \\ -0.4630 \\ 0.5796 \end{bmatrix}$

5.11 Solvability conditions

Let **L** be a linear operator defined on S_D such that $\lambda = 0$ is an eigenvalue of **L**. Then $\mathbf{Lu} = 0$ has a nontrivial (and nonunique) solution. Now consider

$$\mathbf{Lu} = \mathbf{f} \tag{5.130}$$

for a given **f**. Let $\mathbf{u}_1$ be a nontrivial solution to $\mathbf{Lu} = 0$. Then if **u** is a solution of Equation 5.130,

$$\begin{aligned} \mathbf{L}(\mathbf{u} + \mathbf{c}\mathbf{u}_1) &= \mathbf{Lu} + \mathbf{L}(\mathbf{c}\mathbf{u}_1) \\ &= \mathbf{Lu} + \mathbf{c}\mathbf{L}\mathbf{u}_1 \\ &= \mathbf{f} + 0 \\ &= \mathbf{f} \end{aligned} \tag{5.131}$$

and thus, $\mathbf{u} + \mathbf{c}\mathbf{u}_1$ is also a solution of Equation 5.130 for any scalar c.

The above shows that if a solution to Equation 5.130 exists, then it is not unique. However, it is necessary to consider under what circumstances a solution of Equation 5.130 exists. Equation (5.64) presents a representation of the inverse operator for a linear self-adjoint operator constructed from its eigenvalues and eigenvectors. Equation (5.64) shows that if the operator has a zero eigenvalue, then its inverse operator does not exist. In such a case, there is not a one-to-one correspondence between the domain of **L** and the range of **L**. Equation 5.131 shows that more than one element of the domain is mapped into an element of the range. Conversely, not every element of the range has a corresponding element in the domain. That is, a solution of Equation 5.130 does not exist for all **f** seemingly in the range of **L**.

Consider the differential equation

$$\frac{d^2u}{dx^2} + n^2\pi^2 u = 2x \tag{5.132}$$

where n is an integer, and subject to the boundary conditions

$$\frac{du}{dx}(0) = 0 \tag{5.133}$$

$$\frac{du}{dx}(1) = 0 \tag{5.134}$$

The homogeneous solution of Equation 5.132 is

$$u_h(x) = C_1 \cos(n\pi x) + C_2 \sin(n\pi x) \tag{5.135}$$

Note that if the boundary conditions of Equation 5.133 and Equation 5.134 are applied to Equation 5.135, $C_2=0$, but C_1 remains arbitrary, resulting in a nontrivial homogeneous solution of $C_1\cos(n\pi x)$. Thus $\lambda=0$ is an eigenvalue of the operator $d^2u/dx^2+n^2\pi^2u=\lambda u$ subject to $du/dx(0)=0$ and $du/dx(1)=0$.

The particular solution of Equation 5.132 is $u_p(x)=(2/n^2\pi^2)x$, resulting in a general solution of

$$u(x)=C_1\cos(n\pi x)+C_2\sin(n\pi x)+\frac{2}{n^2\pi^2}x \tag{5.136}$$

Application of Equation 5.133 to Equation 5.136 leads to $C_2=-(2/n^3\pi^3)$. Subsequent application of Equation 5.134 to Equation 5.136 then leads to $C_2=(-1)^{n+1}(2/n^3\pi^3)$. The results for C_2 are consistent for even n, but are contradictory for odd n. This implies that a nonunique solution of

$$u(x)=C_1\cos(n\pi x)-\frac{2}{n^3\pi^3}\sin(n\pi x)+\frac{2}{n^2\pi^2}x \tag{5.137}$$

exists for even values of n, but that no solution exists for odd values of n.

The above example illustrates that if $\lambda=0$ is an eigenvalue of an operator **L**, then a solution of the nonhomogenous equation of the form of Equation 5.130 exists under certain conditions.

Consider the case when **L** is self-adjoint with respect to an inner product, (**u**, **v**). Assume that $\mathbf{u}_1$ is a nontrivial solution of $\mathbf{Lu}=\mathbf{0}$. Taking the inner product of both sides of Equation 5.130 with $\mathbf{u}_1$ gives

$$(\mathbf{Lu},\mathbf{u}_1)=(\mathbf{f},\mathbf{u}_1) \tag{5.138}$$

However, since **L** is self-adjoint and $\mathbf{Lu}_1=0$, $(\mathbf{Lu},\mathbf{u}_1)=(\mathbf{u},\mathbf{Lu}_1)=(\mathbf{u},0)=0$. Thus

$$(\mathbf{f},\mathbf{u}_1)=0 \tag{5.139}$$

If for a specific **f**, Equation 5.139 is not true, then a contradiction has been obtained, and the assumption that $\mathbf{u}_1$ exists is false. Hence, if **L** is self-adjoint with respect to an inner product, the existence of a solution of $\mathbf{Lu}=\mathbf{f}$ requires that **f** be orthogonal to all nontrivial solutions of $\mathbf{Lu}=0$. This is a solvability condition, a condition which must be satisfied for a solution of Equation 5.130 to exist.

Suppose L is not self-adjoint, let $\mathbf{L}^*$ be the adjoint of **L**, and let **v** be a nontrivial solution of

$$\mathbf{L}^*\mathbf{v}=0 \tag{5.140}$$

Taking the inner product of both sides of Equation 5.130 with **v**, using the definition of the adjoint operator, and then using Equation 5.140 results in

$$(\mathbf{Lu},\mathbf{v}) = (\mathbf{f},\mathbf{v})$$

$$(\mathbf{u},\mathbf{Lv}^*) = (\mathbf{f},\mathbf{v})$$

$$(\mathbf{u},0) = (\mathbf{f},\mathbf{v})$$

$$0 = (\mathbf{f},\mathbf{v}) \tag{5.141}$$

Equation 5.141 is the appropriate solvability condition when the operator is not self-adjoint. The nonhomogeneous term must be orthogonal to all nontrivial homogeneous solutions of the adjoint equation.

Theorem 5.20 (Fredholm Alternative) If $\mathbf{Lu}=\mathbf{0}$ has a nontrivial solution, call it $\mathbf{u}_1$ ($\lambda=0$ is an eigenvalue of **L**), then $\mathbf{Lu}=\mathbf{f}$ has a solution if and only if $(\mathbf{f},\mathbf{v})=0$, where **v** is the nontrivial solution of $\mathbf{L}^*\mathbf{v}=0$. When a solution of $\mathbf{Lu}=\mathbf{f}$ exists, it is not unique, because $\mathbf{u}+c\mathbf{u}_1$ is also a solution for any scalar c.

Let **A** be an $n\times n$ matrix. The system of equations $\mathbf{Au}=\mathbf{0}$ has a nontrivial solution if and only if **A** is singular $|\mathbf{A}|=0$. Thus the solution of $\mathbf{Au}=\mathbf{f}$ exists and is unique if **A** is nonsingular. However, if **A** is singular, a solution of $\mathbf{Au}=\mathbf{f}$ exists if and only if **f** is orthogonal to the nontrivial solutions of $\mathbf{A}^*\mathbf{v}=0$ or, since the adjoint of a real matrix with respect to the standard inner product is its transpose, **f** must be orthogonal to nontrivial solutions of $\mathbf{A}^T\mathbf{u}=0$.

Example 5.19 Consider the nonhomogeneous system of algebraic equations defined by

$$\begin{bmatrix}1 & 2\\3 & 6\end{bmatrix}\begin{bmatrix}u_1\\u_2\end{bmatrix}=\begin{bmatrix}f_1\\f_2\end{bmatrix} \tag{a}$$

The coefficient matrix in this system is singular, and therefore one of the eigenvalues of matrix is zero. The adjoint of a real $n\times n$ matrix with respect to the standard inner product on R^n is its transpose. Thus, the homogeneous adjoint problem is

$$\begin{bmatrix}1 & 3\\2 & 6\end{bmatrix}\begin{bmatrix}v_1\\v_2\end{bmatrix}=\begin{bmatrix}0\\0\end{bmatrix} \tag{b}$$

which has a nontrivial solution, $v=\begin{bmatrix}3\\-1\end{bmatrix}$. The solvability condition that must be satisfied for a solution of $\mathbf{Au}=\mathbf{f}$ to exist is $\left([3 \quad -1]^T, [f_1 \quad f_2]^T\right)=0$, or $3f_1-f_2=0$.

A vector for which the solvability condition is satisfied is $f=\begin{bmatrix}1\\3\end{bmatrix}$, for which Equation a leads to

$$u_1+2u_2=1 \tag{c}$$

$$3u_1 + 6u_2 = 3 \tag{d}$$

Equation c can be rearranged to give $u_1 = 1 - 2u_2$. Equation d is simply three times Equation c. Thus u_2 is arbitrary, call it c, and $u_1 = 1 - 2c$. The solution can be summarized in vector form as

$$\begin{bmatrix} u_1 \\ u_2 \end{bmatrix} = \begin{bmatrix} 1 \\ 0 \end{bmatrix} + c \begin{bmatrix} -2 \\ 1 \end{bmatrix} \tag{e}$$

Example 5.20 Example 2.21 illustrates the use of a system of equations to determine the point of intersection of three planes. Consider three planes whose equations are

$$x + y - z = d_1 \tag{a}$$

$$2x - y + 2z = d_2 \tag{b}$$

$$-3x + 3y - 5z = d_3 \tag{c}$$

The matrix formulation of Equation a, Equation b, and Equation c is

$$\begin{bmatrix} 1 & 1 & -1 \\ 2 & -1 & 2 \\ -3 & 3 & -5 \end{bmatrix} \begin{bmatrix} x \\ y \\ z \end{bmatrix} = \begin{bmatrix} d_1 \\ d_2 \\ d_3 \end{bmatrix} \tag{d}$$

The matrix in Equation d is singular. Determine the conditions under which the three planes intersect.

Solution Since the matrix is singular, it has an eigenvalue of zero. The adjoint of the matrix is its transpose. The nontrivial solution of the homogeneous adjoint problem can be obtained from

$$\begin{bmatrix} 1 & 2 & -3 \\ 1 & -1 & 3 \\ -1 & 2 & -5 \end{bmatrix} \begin{bmatrix} x \\ y \\ z \end{bmatrix} = \begin{bmatrix} 0 \\ 0 \\ 0 \end{bmatrix} \tag{e}$$

A solution of Equation e is

$$\mathbf{v} = \begin{bmatrix} 1 \\ -2 \\ -1 \end{bmatrix} \tag{f}$$

The solvability condition is

$$(\mathbf{d},\mathbf{v})=0$$

$$\begin{bmatrix}1 & -2 & -1\end{bmatrix}\begin{bmatrix}d_1\\ d_2\\ d_3\end{bmatrix}=0$$

$$d_1-2d_2-d_3=0 \tag{g}$$

The general solution of Equation e for this choice of d_1, d_2 and $d_3$is

$$\begin{bmatrix}x\\ y\\ z\end{bmatrix}=\begin{bmatrix}0\\ 5\\ 3\end{bmatrix}+c\begin{bmatrix}1\\ -4\\ -3\end{bmatrix} \tag{h}$$

for any value of c. Equation h is the vector representation of a line in three-dimensional space represented by the parametric equations

$$y=5-4x$$

$$z=3-3x \tag{i}$$

For example, the planes have a line of intersection when $d_1=2$, $d_2=1$ and $d_3=0$.

5.12 *Asymptotic approximations to solutions of eigenvalue problems*

Consider a problem in which it is desired to determine the eigenvalues of a linear operator of the form $\hat{\mathbf{L}}=\mathbf{L}+\varepsilon\mathbf{L}_1$, where ε is a small nondimensional parameter. Let λ be an eigenvalue of $\mathbf{L}$ with corresponding eigenvector $\mathbf{y}$, $\mathbf{Ly}=\lambda\mathbf{y}$. The eigenvalue problem for the perturbed operator is

$$\hat{\mathbf{L}}\hat{\mathbf{y}}=\hat{\lambda}\hat{\mathbf{y}} \tag{5.142}$$

Approximate solutions to equations containing small parameters are often attempted using asymptotic methods. The solution to an equation containing a small dimensionless parameter is attempted by expanding the unknown in an expansion in terms of functions, often powers, of the small parameter. The expansion is not assumed to be convergent, but asymptotic in which each term is smaller in magnitude than its preceding term. Only a few of the terms in the expansion are determined and the truncation error noted as a function of the dimensionless parameter. The resulting expansion must be uniform for all values of the independent variables and all parameters. Only problems that lead to uniform expansions are considered in this study.

Asymptotic expansions for the perturbed eigenvalue and eigenvector are assumed to be

$$\hat{\lambda} = \lambda + \varepsilon\lambda_1 + \varepsilon^2\lambda_2 + \cdots \tag{5.143}$$

and

$$\hat{y} = y + \varepsilon y_1 + \varepsilon^2 y_2 + \cdots \tag{5.144}$$

Substituting Equation 5.144 and Equation 5.143 into Equation 5.142 leads to

$$(\mathbf{L} + \varepsilon\mathbf{L}_1)(\mathbf{y} + \varepsilon\mathbf{y}_1 + \cdots) + \cdots = (\lambda + \varepsilon\lambda_1 + \cdots)(\mathbf{y} + \varepsilon\mathbf{y}_1 + \cdots) + \cdots \tag{5.145}$$

The operators are linear, thus

$$\mathbf{Ly} + \varepsilon\mathbf{Ly}_1 + \varepsilon\mathbf{L}_1\mathbf{y} + O(\varepsilon^2) = \lambda\mathbf{y} + \varepsilon\lambda\mathbf{y}_1 + \varepsilon\lambda_1\mathbf{y} + O(\varepsilon^2) \tag{5.146}$$

where $O(\varepsilon^2)$ represents all terms of order ε^2 and higher. Collecting coefficients of like powers of ε gives

$$\mathbf{Ly} - \lambda\mathbf{y} + \varepsilon(\mathbf{Ly}_1 + \mathbf{L}_1\mathbf{y} - \lambda\mathbf{y}_1 - \lambda_1\mathbf{y}) + O(\varepsilon^2) = 0 \tag{5.147}$$

Since ε is a small independent dimensionless parameter, powers of ε are linearly independent, and therefore coefficients of powers of ε must vanish independently. This leads to a set of hierarchal equations:

$$O(1) \quad \mathbf{Ly} = \lambda\mathbf{y} \tag{5.148}$$

$$O(\varepsilon) \quad \mathbf{Ly}_1 = \lambda\mathbf{y}_1 - \mathbf{L}_1\mathbf{y} + \lambda_1\mathbf{y} \tag{5.149}$$

Let $\lambda_{k,0}$ and $\mathbf{y}_{k,0}$ constitute an eigenvalue-eigenvector pair which satisfies Equation 5.148. Define $\lambda_{k,1}$ as the perturbation of the eigenvalue and $\mathbf{y}_{1,k}$ the perturbation in the corresponding eigenvector. The eigenvalue perturbation is the value of $\lambda_{k,1}$ such that a solution, $\mathbf{y}_{1,k}$, exists of

$$\mathbf{Ly}_{k,1} = \lambda_{k,0}\mathbf{y}_{k,1} - \mathbf{L}_1\mathbf{y}_{k,0} + \lambda_{k,1}\mathbf{y}_{k,0} \tag{5.150}$$

Let $\mathbf{L}^*$ be the adjoint of $\mathbf{L}$ with respect to an inner product $(\mathbf{u},\mathbf{v})$. Let $\mathbf{W}_{k,0}$ be the eigenvector of $\mathbf{L}^*$ corresponding to its eigenvalue $\overline{\lambda}_{k,0}$. Application of the Fredholm alternative to Equation 5.150 leads to

$$(-\mathbf{L}_1\mathbf{y}_{k,0} + \lambda_{k,1}\mathbf{y}_{k,0}, \mathbf{w}_{k,0}) = 0 \tag{5.151}$$

which in turn leads to

$$\lambda_{k,1} = \frac{(\mathbf{L_1 y_{k,0}}, \mathbf{w_{k,0}})}{(\mathbf{y_{k,0}}, \mathbf{w_{k,0}})} \tag{5.152}$$

If the eigenvectors are normalized, then Equation 5.152 reduces to $\lambda_{k,1} = (\mathbf{L_1 y_{k,0}}, \mathbf{w_{k,0}})$.

The perturbation in the eigenvector is obtained using the expansion theorem. If the boundary conditions satisfied by $\mathbf{y_{k,1}}$ are the same as those satisfied by $\mathbf{y_{k,0}}$ then

$$\mathbf{y_{k,1}} = \sum_{\substack{j=1 \\ j\neq k}} \alpha_{k,j} \mathbf{y_{j,0}} \tag{5.153}$$

Substitution of Equation 5.153 into Equation 5.150 leads to

$$\mathbf{L}\left(\sum_{\substack{j=1 \\ j\neq k}} \alpha_{k,j} \mathbf{y}_{j,0}\right) - \lambda_{k,0} \sum_{\substack{j=1 \\ j\neq k}} \alpha_{k,j} \mathbf{y}_{j,0} = \lambda_{k,1} \mathbf{y_{k,0}} - \mathbf{L_1 y_{k,0}} \tag{5.154}$$

Taking the inner product of both sides of Equation 5.154 with $\mathbf{W}_{\ell,0}$, with $\ell \neq j$ leads to

$$\begin{gathered}\sum_{\substack{j=1 \\ j\neq k}} \alpha_{k,j} (\mathbf{L y_{j,0}}, \mathbf{w}_{\ell,0}) - \lambda_{k,0} \sum_{\substack{j=1 \\ j\neq k}} \alpha_{k,j} (\mathbf{y_{j,0}}, \mathbf{w}_{\ell,0}) = (\lambda_{k,1} \mathbf{y_{k,0}} - \mathbf{L_1 y_{k,0}}, \mathbf{w}_{\ell,0}) \\ \sum_{\substack{j=1 \\ j\neq k}} (\bar{\lambda}_{\ell,0} - \lambda_{k,0}) \alpha_{k,j} (\mathbf{y_{j,0}}, \mathbf{w}_{\ell,0}) = (\lambda_{k,1} \mathbf{y_{k,0}} - \mathbf{L_1 y_{k,0}}, \mathbf{w}_{\ell,0})\end{gathered} \tag{5.155}$$

The principle of bi-orthonormality between the eigenvectors of **L** and the eigenvectors of **L*** requires $(\mathbf{y_{j,0}}, \mathbf{w}_{\ell,0}) = \delta_{j,\ell}$. Equation 5.155 simplifies to

$$\alpha_{k,\ell} = -\frac{(\mathbf{L_1 y_{k,0}}, \mathbf{w}_{\ell,0})}{\bar{\lambda}_{\ell,0} - \lambda_{k,0}} \tag{5.156}$$

Example 5.21 Consider the eigenvalue problem

$$-\frac{d^2 y}{dx^2} + \varepsilon\left(\frac{dy}{dx} + 2y\right) = \lambda y \tag{a}$$

$$y(0) = 0 \tag{b}$$

$$y(1) = 0 \tag{c}$$

Noting that the eigenvalues and normalized eigenvectors of the problem $-d^2y/dx^2 = \lambda y$ subject to Equation b and Equation c are

$$\lambda_{n,0} = (n\pi)^2 \tag{d}$$

$$y_{n,0} = \frac{1}{\sqrt{2}}\sin(n\pi x) \tag{e}$$

determine corrections to the eigenvalues and eigenvectors for small ε.

Solution Equation a can be written as $\mathbf{L}\mathbf{y} + \varepsilon\mathbf{L}_1\mathbf{y} = \lambda\mathbf{y}$, where $\mathbf{L}_1\mathbf{y} = dy/dx + 2y$. Asymptotic expansions for the eigenvalues and eigenvectors are assumed to be

$$\lambda_n = \lambda_{n,0} + \varepsilon\lambda_{n,1} + \cdots \tag{f}$$

$$y_n = y_{n,0} + \varepsilon\lambda_{n,1} + \cdots \tag{g}$$

The first-order correction for the eigenvalue is obtained using Equation 5.151. Since the $O(1)$ problem is self-adjoint, $\mathbf{y} = \mathbf{w} = y_{n,0}$, and since the eigenvectors have been normalized, $(y_{n,0}, y_{n,0}) = 1$. Thus

$$\begin{aligned}
\lambda_{n,1} &= (\mathbf{L}_1 y_{n,0}, y_{n,0}) \\
&= \int_0^1 \left(\frac{dy_{n,0}}{dx} + 2y_{n,0}\right) y_{n,0}\,dx \\
&= \frac{1}{2}\int_0^1 [n\pi\cos(n\pi x) + 2\sin(n\pi x)]\sin(n\pi x)dx \\
&= \frac{1}{2}
\end{aligned} \tag{h}$$

An approximation for the eigenvalues is

$$\lambda_n = (\eta\pi)^2 + \frac{\varepsilon}{2} + O(\varepsilon^2) \tag{i}$$

The perturbation in the eigenvectors is obtained using Equation 5.152:

$$y_{n,1} = \sum_{k \neq n} \alpha_{n,k} y_{k,0} \tag{j}$$

where from Equation 5.155,

$$\alpha_{n,k} = \frac{(\mathbf{L}_1 y_{n,0}, y_{k,0})}{(k\pi)^2 - (n\pi)^2}$$

$$= \frac{\frac{1}{2}\int_0^1 [n\pi\cos(n\pi x) + 2\sin(n\pi x)]\sin(k\pi x)dx}{(k^2 - n^2)\pi^2}$$

$$= \begin{cases} 0 & n \text{ even and } k \text{ even or } n \text{ odd and } k \text{ odd} \\ \dfrac{2nk}{(k^2 - n^2)\pi} & n \text{ even and } k \text{ odd or } n \text{ odd and } k \text{ even} \end{cases} \tag{k}$$

Thus, for even values of n,

$$y_{n,1} = \frac{\sqrt{2}n}{\pi} \sum_{k=1,3,5} \frac{k}{(k^2 - n^2)^2} \sin(k\pi x) \tag{l}$$

and for odd values of n,

$$y_{n,1} = \frac{\sqrt{2}n}{\pi} \sum_{k=2,4,6} \frac{k}{(k^2 - n^2)^2} \sin(k\pi x) \tag{m}$$

Occasionally the boundary conditions for the $O(\varepsilon)$ problem are different than those for the $O(1)$ problem. This difference is usually in the form of a nonhomogeneous boundary condition developed at $O(\varepsilon)$. A transformation of the dependent variable can be performed to remove the nonhomogeneity from the boundary condition to the differential equation. An alternative is to take the inner product of both sides of Equation 5.149 with the homogeneous solution of the adjoint problem, as illustrated above. However, the evaluation on the left-hand side leads to a boundary term that does not vanish, which becomes part of the solvability condition. This method is illustrated in the following example.

Example 5.22 The nondimensional natural frequencies of a longitudinal motion of the nonuniform bar shown in Figure 5.16 are obtained from solving the differential equation,

$$\frac{d}{dx}\left(\alpha(x)\frac{du}{dx}\right) + \alpha(x)\omega^2 u = 0 \tag{a}$$

subject to the boundary conditions

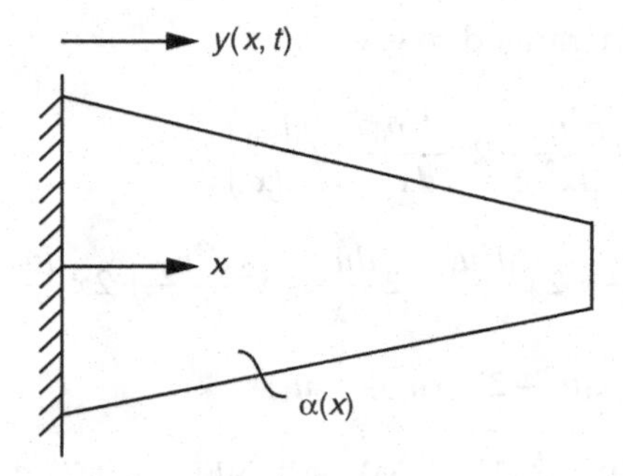

Figure 5.16 The natural frequencies of the bar of Example 5.22 are approximated using an asymptotic expansion in terms of the taper rate of the bar.

$$u(0) = 0 \tag{b}$$

$$\alpha(1)\frac{du}{dx}(1) + \eta u(1) = 0 \tag{c}$$

Consider a circular bar with a slight linear taper such that

$$\alpha(x) = (1 - \varepsilon x)^2 \tag{d}$$

where ε is a small dimensionless parameter.

The natural frequencies are the square roots of the eigenvalues of the operator $Lu = -1/\alpha(x)[d(\alpha(x)du/dx)/dx]$. Use an asymptotic method to determine a two-term expansion for the natural frequencies.

Solution Equation a can be written as

$$-\frac{d}{dx}\left[(1 - 2\varepsilon x + \varepsilon^2 x^2)\frac{du}{dx}\right] = \lambda(1 - 2\varepsilon x + \varepsilon^2 x^2)u \tag{e}$$

Asymptotic expansions for the eigenvalues are obtained by assuming

$$u = u_0 + \varepsilon u_1 + \varepsilon^2 u_2 + \cdots \tag{f}$$

$$\lambda = \lambda_0 + \varepsilon\lambda_1 + \varepsilon^2\lambda_2 + \cdots \tag{g}$$

Substitution of Equation f and Equation g into Equation e leads to

$$-\frac{d}{dx}\left[(1 - 2\varepsilon x + \varepsilon^2 x^2)\frac{d}{dx}(u_0 + \varepsilon u_1 + \varepsilon^2 u_2 + \cdots)\right]$$

$$= (\lambda_0 + \varepsilon\lambda_1 + \varepsilon^2\lambda_2 + \cdots)(1 - 2\varepsilon x + \varepsilon^2 x^2)(u_0 + \varepsilon u_1 + \varepsilon^2 u_2 + \cdots) \tag{h}$$

Equation h can be rearranged to give

$$-\left[\frac{d^2u_0}{dx^2}+\varepsilon\left(\frac{d^2u_1}{dx^2}-2x\frac{d^2u_0}{dx^2}-2\frac{du_0}{dx}\right)\right.$$

$$\left.+\varepsilon^2\left(\frac{d^2u_2}{dx^2}-2x\frac{d^2u_1}{dx^2}-2\frac{du_1}{dx}+x^2\frac{d^2u_0}{dx^2}+2x\frac{du_0}{dx}\right)\right]$$

$$=\lambda_0u_0+\varepsilon(\lambda_1u_0-2\lambda_1xu_0+\lambda_0u_1)$$

$$+\varepsilon^2(\lambda_0u_2-2\lambda_0xu_1+\lambda_0x^2u_0+\lambda_1u_1-2\lambda_1xu_0+\lambda_2u_0)+\cdots \tag{i}$$

Setting coefficients of like powers of ε to zero leads to the hierarchical equations

$$O(1)\quad -\frac{d^2u_0}{dx^2}=\lambda_0u_0 \tag{j}$$

$$O(\varepsilon)\quad \frac{d^2u_1}{dx^2}+\lambda_0u_1=2x\frac{d^2u_0}{dx^2}+2\frac{du_0}{dx}-\lambda_1u_0+2\lambda_0xu_0 \tag{k}$$

Equation f and Equation g are used in the boundary conditions, Equation b and Equation c, leading to

$$u_0(0)=0 \tag{l}$$

$$u_1(0)=0 \tag{m}$$

$$u_0'(1)+\eta u_0(1)=0 \tag{n}$$

$$u_1'(1)+\eta u_1(1)=2u_0'(1) \tag{o}$$

The solution for $u_0(x)$ is obtained as

$$u_{k,0}(x)=c_k\sin\sqrt{\lambda_k}\,x \tag{p}$$

where $\sqrt{\lambda_k}=\beta_k$ and β_k solves

$$\eta\tan\beta=-\beta \tag{q}$$

Equation p is normalized so that

$$c_k=\left[\int_0^1\sin^2(\sqrt{\lambda_k}\,x)dx\right]^{-\frac{1}{2}}$$

$$=\sqrt{\frac{2\lambda_k}{\lambda_k+\sin^2\sqrt{\lambda_k}}} \tag{r}$$

For a solution of the $O(\varepsilon)$ problem to exist, a solvability condition must be satisfied. Taking the inner product of both sides of Equation k with $u_0(x)$ leads to

$$\left(u_0, \frac{d^2u_1}{dx^2} + \lambda_0 u_1\right) = (u_0, f(x)) \tag{s}$$

where $f(x) = 2x(d^2u_0/dx^2) + 2(du_0/dx) - \lambda_1 u_0 + 2\lambda_0 x u_0$. Focusing on the left-hand side of Equation s leads to

$$\left(u_0, \frac{d^2u_1}{dx^2} + \lambda_0 u_1\right) = \int_0^1 u_0\left(\frac{d^2u_1}{dx^2} + \lambda_0 u_1\right)dx = \int_0^1 u_0 \frac{d^2u_1}{dx^2}dx + \lambda_0 \int_0^1 u_0 u_1 dx$$

$$= u_0 \frac{du_1}{dx}\bigg|_0^1 - u_1 \frac{du_0}{dx}\bigg|_0^1 + \int_0^1 u_1 \frac{d^2u_0}{dx^2}dx + \lambda_0 \int_0^1 u_0 u_1 dx$$

$$= u_0(1)\frac{du_1(1)}{dx} - u_0(0)u_1'(0) - u_1(1)u_0'(1) + u_1(0)u_0'(0) + \int_0^1 u_1\left(\frac{d^2u_0}{dx^2} + \lambda_0 u_0\right)dx \tag{t}$$

Using boundary conditions for u_0 and u_1 at $x = 0$ and $x = 1$ and noting that the integrand of the integral is identically zero allows Equation t to be simplified to

$$\left(u_0, \frac{d^2u_1}{dx^2} + \lambda_0 u_1\right) = u_0(1)\left(\frac{du_1(1)}{dx} + \eta u_1(1)\right)$$

$$= u_0(1)(2u_0'(1))$$

$$= -2\eta(u_0(1))^2$$

$$= -2\eta(c_k \sin\sqrt{\lambda_k}\,)^2 \tag{u}$$

The solvability condition is obtained from Equation s and Equation u as

$$-2\eta c_k^2 \sin^2\sqrt{\lambda_k} = \left(c_k \sin(\sqrt{\lambda_k}x), 2x\frac{d^2u_0}{dx^2} + 2\frac{du_0}{dx} - \lambda k_{11} u_0 + 2\lambda_k x u_0\right) \tag{v}$$

Equation v is solved for λ_1, leading to

$$\lambda_1 = 2\eta c_k^2 \sin^2\sqrt{\lambda_k} + c_k^2 \int_0^1 \sin(\sqrt{\lambda_k}x)[2\sqrt{\lambda_k}\cos(\sqrt{\lambda_k}x)]\,dx$$

$$= c_k^2(1+2\eta)\sin^2\sqrt{\lambda_k}$$

$$= \frac{2\lambda_k(1+2\eta)\sin^2\sqrt{\lambda_k}}{\lambda_k + \sin^2\sqrt{\lambda_k}} \tag{w}$$

The first four eigenvalues corresponding to $\eta = 1.5$ are calculated from Equation q and Equation w as

$$\lambda_1 = 2.175 + 4.74\varepsilon$$

$$\lambda_2 = 5.004 + 7.079\varepsilon$$

$$\lambda_3 = 8.038 + 7.618\varepsilon$$

$$\lambda_4 = 11.13 + 7.795\varepsilon \tag{x}$$

Example 5.23 The stability of an elastic column has been considered in Example 5.4, where the critical buckling length is determined for a column subject to a compressive axial load, in Example 5.13 which dealt with the critical speed of a rotating column, and in Example 5.16 which dealt with the critical buckling length of a column under its own weight. Consider a column rotating with an angular speed Ω, subject to an axial load P, and to gravity. The nondimensional differential equation for a uniform column with these combined effects is

$$EI\frac{d^4u}{dx^4} + (P + \rho gALx)L^2\frac{d^2u}{dx^2} = \rho A\Omega^2 u \tag{a}$$

The boundary conditions for the column are

$$u(0) = 0 \tag{b}$$

$$\frac{d^3u}{dx^3}(0) = 0 \tag{c}$$

$$\frac{du}{dx}(1) = 0 \tag{d}$$

$$\frac{d^2u}{dx^2}(1) = 0 \tag{e}$$

It is desired to construct a stability surface, a surface in the parameter space P, Ω, L, which is a boundary between stability and instability. As illustrated in Figure 5.17, the intersection of this surface with the P-axis is at the critical buckling load, its intersection along the L-axis gives the critical buckling length, and its intersection along the Ω-axis gives the critical speed. The surface intersects the P-L plane, defining a stability curve, if $\Omega = 0$.

The critical values of the parameters are

$$P_{cr} = \frac{\pi^2 EI}{4L^2} \tag{f}$$

$$L_{cr} = 1.9866\left(\frac{EI}{\rho gA}\right)^{\frac{1}{3}} \tag{g}$$

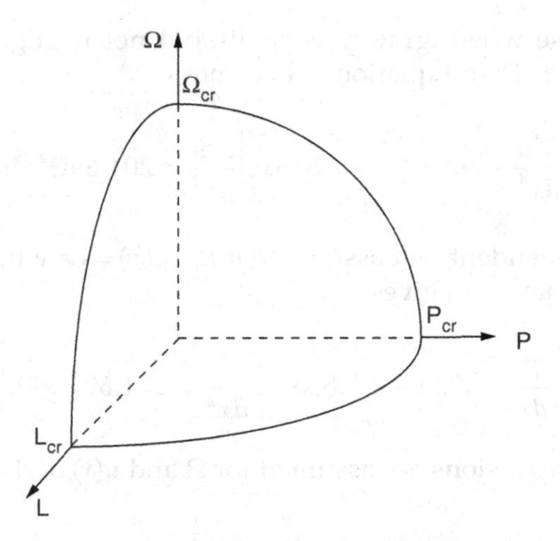

Figure 5.17 The buckling of a long rotating column subject to an axial load is a function of three parameters. The stability surface is the boundary between stable (inside the surface) and unstable (exterior to the surface) responses in the parameter space.

$$\Omega_{cr} = 3.516\left(\frac{EI}{\rho g A L^4}\right)^{\frac{1}{2}} \tag{h}$$

To define the surface better, the following nondimensional parameters are introduced:

$$L^* = \frac{L}{L_{cr}} \tag{i}$$

$$P^* = \frac{P}{P_{cr}\left(L = \frac{L_{cr}}{2}\right)} \tag{j}$$

$$\Omega^* = \frac{\Omega}{\Omega_{cr}\left(L = \frac{L_{cr}}{2}\right)} \tag{k}$$

Substitution of Equation i, Equation j, and Equation k into Equation a leads to

$$\frac{d^4u}{dx^4} + \left[9.86PL^2 + 7.833xL^3\right]\frac{d^2u}{dx^2} = 201.69\Omega^2L^4u \tag{l}$$

Consider the case where gravity is small, but not negligible. To this end, assume $L^3 = \varepsilon \ll 1$. Then Equation m becomes

$$\frac{d^4u}{dx^4} + \left[9.86P\varepsilon^{\frac{2}{3}} + 7.833x\varepsilon\right]\frac{d^2u}{dx^2} = 201.69\Omega^2\varepsilon^{\frac{4}{3}}u \tag{m}$$

P and L are independent, but assume that $P = O(\varepsilon) = \alpha\varepsilon$ where $\alpha = O(1)$. Substitution into Equation m gives

$$\frac{d^4u}{dx^4} + [9.86\alpha + 7.833x]\varepsilon\frac{d^2u}{dx^2} = 201.69\Omega^2\varepsilon^{\frac{4}{3}}u \tag{n}$$

Asymptotic expansions are assumed for Ω and $u(x)$ of the form

$$\Omega = \Omega_0\varepsilon^{-\frac{2}{3}} + \Omega_1\varepsilon^{\frac{1}{3}} + \cdots \tag{o}$$

$$u(x) = u_0(x) + \varepsilon u_1(x) + \cdots \tag{p}$$

Determine Ω_1.

Solution Substitution of Equation o and Equation p into Equation n, collecting coefficients of like powers of ε, and setting them to zero independently leads to

$$\frac{d^4u_0}{dx^4} = 201.69\Omega_0^2u_0 \tag{q}$$

$$\frac{d^4u_1}{dx^4} - 201.69\Omega_0^2u_1 = 403.38\Omega_0\Omega_1u_0 - (9.86\alpha + 7.833x)\frac{d^2u_0}{dx^2} \tag{r}$$

Both $u_0(x)$ and $u_1(x)$ satisfy boundary conditions of the form of Equation b, Equation c, Equation d, and Equation e.

Using the results of Example 5.13,

$$\Omega_0 = \frac{3.516}{(201.69)^{\frac{1}{2}}} = 0.2475 \tag{s}$$

and

$$\begin{aligned} u_0(x) = {} & \cos(3.768\sqrt{\Omega_0}x) + \frac{\sin(3.768\sqrt{\Omega_0}) + \sinh(3.768\sqrt{\Omega_0})}{\cos(3.768\sqrt{\Omega_0}) + \cosh(3.768\sqrt{\Omega_0})} \\ & \times \sin(3.768\sqrt{\Omega_0}x) - \cosh(3.768\sqrt{\Omega_0}x) \\ & + \frac{\sin(3.768\sqrt{\Omega_0}) + \sinh(3.768\sqrt{\Omega_0})}{\cos(3.768\sqrt{\Omega_0}) + \cosh(3.768\sqrt{\Omega_0})}\sinh(3.768\sqrt{\Omega_0}x) \end{aligned} \tag{t}$$

The system of Equation s subject to Equation b, Equation c, Equation d, and Equation e constitutes a nonhomogeneous problem for which a nontrivial homogeneous solution exists. Thus a solution exists if and only if a solvability condition is satisfied. It is shown in Example 5.16 that the system of Equation q subject to Equation b, Equation c, Equation d, and Equation e is not self-adjoint. Thus the appropriate solvability condition is that the nonhomogeneous terms of Equation s are orthogonal to the nontrivial solution of the adjoint problem.

The boundary conditions for the adjoint problem are $du/dx(0)=0, d^2u/dx^2(0)=0$, $u(L)=0$, and $d^3u/dx^3(L)=0$. The solution of Equation q subject to these boundary conditions is

$$\begin{aligned} u_0^*(x) &= \cos(3.768\sqrt{\Omega_0}x)+\frac{\sin(3.768\sqrt{\Omega_0})+\sinh(3.768\sqrt{\Omega_0})}{\cos(3.768\sqrt{\Omega_0})+\cosh(3.768\sqrt{\Omega_0})} \\ &\quad\times\sin(3.768\sqrt{\Omega_0}x)+\cosh(3.768\sqrt{\Omega_0}x) \\ &\quad-\frac{\sin(3.768\sqrt{\Omega_0})+\sinh(3.768\sqrt{\Omega_0})}{\cos(3.768\sqrt{\Omega_0})+\cosh(3.768\sqrt{\Omega_0})}\sinh(3.768\sqrt{\Omega_0}x) \end{aligned} \tag{u}$$

Requiring the nonhomogeneous term of Equation r to be orthogonal to $u_0^*(x)$ leads to

$$\left(403.38\Omega_0\Omega_1 u_0-(9.86\alpha+7.833x)\frac{d^2u_0}{dx^2},u_0^*\right)=0 \tag{v}$$

Equation w can be rearranged as

$$\begin{aligned} \Omega_1 &= \frac{\left((9.86\alpha+7.833x)\frac{d^2u_0}{dx^2},u_0^*\right)}{403.38\Omega_0(u_0,u_0^*)} \\ &= \frac{\int_0^1 (9.86\alpha+7.833x)\frac{d^2u_0}{dx^2}u_0^*dx}{403.38\Omega_0\int_0^1 u_0(x)u_0^*(x)dx} \\ &= -0.302\alpha-0.117 \end{aligned} \tag{w}$$

Thus,

$$\Omega=0.2475\varepsilon^{-\frac{2}{3}}-\varepsilon^{\frac{1}{3}}(0.302\alpha+0.117) \tag{x}$$

5.13 Rayleigh's quotient

A functional is a scalar function of a function of a vector. An energy inner product of a vector with itself is a functional. Suppose that **L** is a linear operator defined on a domain S_D and that **L** is a self-adjoint operator with respect to an inner product **(f,g)**. Let **y** be an arbitrary element of S_D. Rayleigh's quotient is the functional defined by

$$R(\mathbf{y}) = \frac{(\mathbf{Ly},\mathbf{y})}{(\mathbf{y},\mathbf{y})} \tag{5.156}$$

For each **y** in S_D, R(**y**) takes on a scalar value.

Let **u** be an eigenvector of **L** corresponding to an eigenvalue λ. Then

$$\begin{aligned} R(\mathbf{u}) &= \frac{(\mathbf{Lu},\mathbf{u})}{(\mathbf{u},\mathbf{u})} \\ &= \frac{(\lambda\mathbf{u},\mathbf{u})}{(\mathbf{u},\mathbf{u})} \\ &= \lambda \end{aligned} \tag{5.157}$$

Equation 5.157 shows that the value of Rayleigh's quotient for an operator **L** evaluated for an eigenvector of **L** is the eigenvalue corresponding to the eigenvector.

If **L** is self-adjoint and positive definite, then the expansion theorems imply that its eigenvectors $\mathbf{u}_1, \mathbf{u}_2, \ldots, \mathbf{u}_{k-1}, \mathbf{u}_k, \mathbf{u}_{k+1}, \ldots$ are complete in S_D. Thus any vector in S_D has an expansion of the form

$$\mathbf{y} = \sum_i \alpha_i \mathbf{u}_i \tag{5.158}$$

Substitution of Equation 5.158 into Equation 5.156 leads to

$$R(\mathbf{y}) = \frac{\left(\mathbf{L}\left(\sum_i \alpha_i \mathbf{u}_\mathrm{i}\right), \sum_j \alpha_j \mathbf{u}_\mathrm{j}\right)}{\left(\sum_i \alpha_i \mathbf{u}_\mathrm{j}, \sum_i \alpha_j \mathbf{u}_\mathrm{j}\right)} \tag{5.159}$$

Using the linearity of **L** and properties of inner products, Equation 5.159 becomes

$$R(\mathbf{y})=\frac{\sum_i\sum_j \alpha_i\alpha_j(\mathbf{Lu_i},\mathbf{u_j})}{\sum_i\sum_j \alpha_i\alpha_j(\mathbf{u_i},\mathbf{u_j})} \tag{5.160}$$

Noting that $\mathbf{u}_i$ is an eigenvector of $\mathbf{L}$ corresponding to the eigenvalue λ_i and using the orthonormality of the mode shapes, $(\mathbf{u}_i, \mathbf{u}_j)=\delta_{i,j}$, Equation 5.160 reduces to

$$R(\mathbf{y})=\frac{\sum_i \lambda_i\alpha_i^2}{\sum_i \alpha_i^2} \tag{5.161}$$

The vector $\mathbf{y}$ is dependent on the values of the coefficients in the eigenvector expansion thus the value of Rayleigh's quotient can be viewed as a function of the coefficients,

$$R(\mathbf{y})=R(\alpha_1,\alpha_2,\ldots,\alpha_{k-1},\alpha_k,\alpha_{k+1},\ldots) \tag{5.162}$$

Rayleigh's quotient is stationary when $dR=0$ or when

$$0=\frac{\partial R}{\partial\alpha_1}d\alpha_1+\frac{\partial R}{\partial\alpha_2}d\alpha_2+\cdots+\frac{\partial R}{\partial\alpha_k}d\alpha_k+\cdots \tag{5.163}$$

Since $\alpha_1, \alpha_2, \ldots$ are all independent parameters, Equation 5.163 shows that R is stationary if

$$0=\frac{\partial R}{\partial\alpha_1}=\frac{\partial R}{\partial\alpha_2}=\cdots=\frac{\partial R}{\partial\alpha_{k-1}}=\frac{\partial R}{\partial\alpha_k}=\frac{\partial R}{\partial\alpha_{k+1}}=\cdots \tag{5.164}$$

Setting $\partial R/\partial\alpha_k=0$ in Equation 5.161 gives

$$\frac{\left(\sum_i \lambda_i\frac{\partial\alpha_i^2}{\partial\alpha_k}\right)\left(\sum_j \alpha_j^2\right)-\left(\sum_i \lambda_i\alpha_i^2\right)\left(\sum_j \frac{\partial\alpha_j^2}{\partial\alpha_k}\right)}{\left(\sum_j \alpha_j^2\right)^2}=0 \tag{5.165}$$

The left-hand side of Equation 5.165 is zero if its numerator is zero. Simplifying leads to

$$\left(\sum_i 2\lambda_i\alpha_i\delta_{i,k}\right)\left(\sum_j \alpha_j^2\right)-\left(\sum_i \lambda_i\alpha_i^2\right)\left(\sum_j 2\alpha_j\delta_{j,k}\right)=0$$

$$\lambda_k\alpha_k\sum_j \alpha_j^2-\alpha_k\sum_j \lambda_j\alpha_j^2=0$$

$$\alpha_k\sum_j(\lambda_k-\lambda_j)\alpha_j^2=0 \tag{5.166}$$

An equation of the form of Equation 5.166 may be written for each k. There are as many independent solutions of the resulting set of equations as there are distinct eigenvalues. The kth solution is of the form $\alpha_j=\delta_{k,j}$, $j=1,2,\ldots$.

The uniqueness of these solutions is proven by assuming that the kth solution has more than just one α_k different from zero. Without loss of generality, assume α_i, α_ℓ and α_k are all nonzero. Then the ith, ℓth and kth equations lead to

$$(\lambda_i-\lambda_\ell)\alpha_\ell^2+(\lambda_i-\lambda_k)\alpha_k^2=0$$

$$(\lambda_\ell-\lambda_i)\alpha_i^2+(\lambda_\ell-\lambda_k)\alpha_k^2=0$$

$$(\lambda_k-\lambda_i)\alpha_i^2+(\lambda_k-\lambda_\ell)\alpha_\ell^2=0 \tag{5.167}$$

The determinant of the coefficient matrix for the system of equations in Equation 5.167 is $2(\lambda_i-\lambda_\ell)(\lambda_\ell-\lambda_k)(\lambda_k-\lambda_i)$. Thus, the only solution of Equation 5.167 is $\alpha_i=\alpha_\ell=\alpha_k=0$. This argument may be repeated with the assumption of any number of nonzero coefficients.

The above shows that the only vectors in S_D that render $R(\mathbf{y})$ stationary are eigenvectors of $\mathbf{L}$. Equation 5.157 shows that the value of Rayleigh's quotient of an eigenvector is the eigenvalue corresponding to the eigenvector. Thus the minimum value of Rayleigh's quotient for a self-adjoint operator is the operator's smallest eigenvalue.

Rayleigh's quotient can be used to determine an upper bound for the smallest eigenvalue of a self-adjoint operator. A set of trial vectors, each belonging to S_D, is developed. The smallest eigenvalue must be less than the smallest value of Rayleigh's quotient for all vectors in the set.

Now consider a subspace of S_D, call it S_{D1}, in which all vectors are orthogonal to $\mathbf{u}_1$, the eigenvector corresponding to the smallest eigenvalue of $\mathbf{L}$. All vectors in S_{D1} can be represented by Equation 5.158, but with $\alpha_1=0$. The process described in Equation 5.159, Equation 5.160, Equation 5.161, Equation 5.162, Equation 5.163, Equation 5.164, Equation 5.165, Equation 5.166, and Equation 5.167 are repeated for all vectors in S_{D1} with the result that the smallest value of R over all vectors in S_{D1} is λ_2. Similarly, it can be shown that the smallest value of R for all vectors in S_{D2}, the subspace of S_{D1} in which all elements are orthogonal to $\mathbf{u}_1$ and $\mathbf{u}_2$, is λ_3.

Example 5.24 The differential equation for the transverse deflection of a uniform column under an axial load has been shown to be

$$EI\frac{d^2y}{dx^2} + Py = 0 \tag{a}$$

Consider a long vertical column where the axial load is generated due to gravity. The axial load varies over the length of the column as

$$P = \rho g A x \tag{b}$$

where ρ is the mass density of the column and A is its cross-sectional area. Thus, Equation a becomes

$$EI\frac{d^2y}{dx^2} + \rho g A x y = 0 \tag{c}$$

The boundary conditions for a pinned-pinned column are

$$y(0) = 0 \tag{d}$$

$$y(L) = 0 \tag{e}$$

Nondimensional variables are introduced as

$$x^* = \frac{x}{L} \tag{f}$$

$$y^* = \frac{y}{L} \tag{g}$$

Substitution of Equation e and Equation f into Equation b, Equation c, and Equation d leads to

$$EI\frac{d^2y}{dx^2} + \lambda x y = 0 \tag{h}$$

$$\mathrm{y}(0) = 0 \tag{i}$$

$$y(1) = 0 \tag{j}$$

where

$$\lambda = \frac{\rho g A L^3}{EI} \tag{k}$$

The critical buckling length, L_{cr}, is the length at which the column will buckle under its own weight. Equation h and Equation i constitute a Sturm-Liouville problem with λ as the eigenvalue of the operator,

$$\mathbf{L}\mathbf{y} = -\frac{1}{x}\frac{d^2y}{dx^2} \tag{l}$$

The lowest eigenvalue of the operator defined by Equation l acting on the domain defined by Equation i and Equation j leads to the critical buckling length from Equation k.

Use Rayleigh's quotient to determine an upper bound on the critical buckling length. Use the trial functions $\sin\pi x$, $x(1-x)$ and $x(1-x^2)$.

Solution The inner product for which **L** is self-adjoint is

$$(f,g)_r = \int_0^1 f(x)g(x)x\,dx \tag{m}$$

Rayleigh's quotient is expressed as

$$\begin{aligned} R(\mathbf{y}) &= \frac{(\mathbf{L}y,y)_r}{(y,y)_r} \\ &= \frac{\int_0^1 \left(-\frac{1}{x}\frac{d^2y}{dx^2}\right) y(x)x\,dx}{\int_0^1 y(x)y(x)x\,dx} \\ &= \frac{-\int_0^1 y(x)\frac{d^2y}{dx^2}dx}{\int_0^1 [y(x)]^2 x\,dx} \end{aligned} \tag{n}$$

Application of Rayleigh's quotient to the suggested trial functions leads to

$$\begin{aligned} R(\sin(\pi x)) &= \frac{-\int_0^1 \sin(\pi x)[-\pi^2 \sin(\pi x)]dx}{\int_0^1 [\sin(\pi x)]^2 x\,dx} \\ &= 19.739 \end{aligned} \tag{o}$$

$$\begin{aligned} R(x(1-x)) &= \frac{-\int_0^1 x(1-x)(-2)dx}{\int_0^1 x^2(1-x)dx} \\ &= 4 \end{aligned} \tag{p}$$

$$R(x(1-x^2)) = \frac{-\int_0^1 x(1-x^2)(-6x)dx}{\int_0^1 x^2(1-x^2)dx}$$
$$= 6 \tag{q}$$

Thus, an upper bound on the critical buckling length can be determined such that

$$L_{cr} < \left(4\frac{EI}{\rho g A}\right)^{\frac{1}{3}} = 1.587\left(\frac{EI}{\rho g A}\right)^{\frac{1}{3}} \tag{r}$$

Example 5.25 The differential equations governing the motion of the five-degree-of-freedom system shown in Figure 5.18 are

$$\begin{bmatrix} 200 & 0 & 0 & 0 & 0 \\ 0 & 100 & 0 & 0 & 0 \\ 0 & 0 & 300 & 0 & 0 \\ 0 & 0 & 0 & 100 & 0 \\ 0 & 0 & 0 & 0 & 100 \end{bmatrix} \begin{bmatrix} \ddot{x}_1 \\ \ddot{x}_2 \\ \ddot{x}_3 \\ \ddot{x}_4 \\ \ddot{x}_5 \end{bmatrix} + \begin{bmatrix} 500 & -100 & 0 & 0 & 0 \\ -100 & 300 & -200 & 0 & 0 \\ 0 & -200 & 400 & -200 & 0 \\ 0 & 0 & -200 & 300 & -100 \\ 0 & 0 & 0 & -100 & 200 \end{bmatrix} \begin{bmatrix} x_1 \\ x_2 \\ x_3 \\ x_4 \\ x_5 \end{bmatrix} = \begin{bmatrix} 0 \\ 0 \\ 0 \\ 0 \\ 0 \end{bmatrix} \tag{a}$$

The mode shape corresponding to the lowest mode of such a discrete system has no nodes. That is, there is no point in the system except at a fixed support whose displacement is zero. This implies that all components of the mode-shape vector are of the same sign. Use Rayleigh's quotient with three trial vectors for the lowest mode shape to determine an upper bound on the lowest natural frequency.

400 N/m, x_1, 200 kg, 100 N/m, x_2, 100 kg, 200 N/m, x_3, 300 kg, 200 N/m, x_4, 100 kg, 100 N/m, x_5, 100 kg, 100 N/m

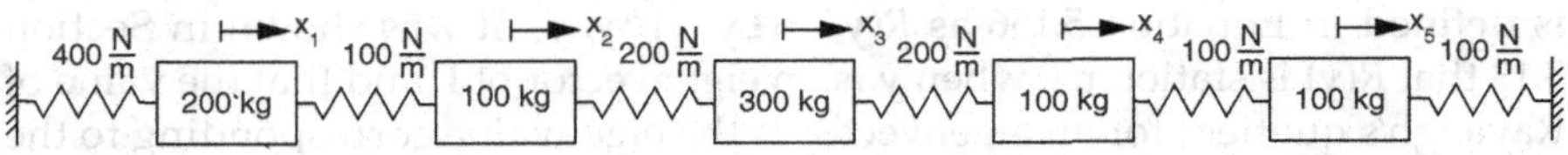

Figure 5.18 A Rayleigh-Ritz method is used to approximate the lowest natural frequency and mode shape for the five-degree-of-freedom system.

Solution It is shown in Section 5.4 that the natural frequencies are the square roots of the eigenvalues of $\mathbf{M}^{-1}\mathbf{K}$, which is self-adjoint with respect to the kinetic-energy inner product $(\mathbf{u}, \mathbf{v})_M = (\mathbf{Mu}, \mathbf{v})$. Thus the appropriate form of Rayleigh's quotient is

$$R(\mathbf{y}) = \frac{(\mathbf{M}^{-1}\mathbf{K}\mathbf{y}, \mathbf{y})_M}{(\mathbf{y}, \mathbf{y})_M} = \frac{(\mathbf{K}\mathbf{y}, \mathbf{y})}{(\mathbf{M}\mathbf{y}, \mathbf{y})} \tag{b}$$

Trial vectors with only positive components are chosen. Since the discrete system is constrained at both ends, the displacements of the masses closest to the support should be the smallest. The following trial vectors are used:

$$\mathbf{y}_1 = \begin{bmatrix} 1 \\ 2 \\ 3 \\ 2 \\ 1 \end{bmatrix} \quad \mathbf{y}_2 = \begin{bmatrix} 1 \\ 4 \\ 3 \\ 2 \\ 1 \end{bmatrix} \quad \mathbf{y}_3 = \begin{bmatrix} 2 \\ 3 \\ 4 \\ 2 \\ 1 \end{bmatrix} \tag{c}$$

Substitution into Equation b leads to $R(\mathbf{y}_1) = 0.2895$, $R(\mathbf{y}_2) = 0.3800$ and $R(\mathbf{y}_3) = 0.4143$. Thus an upper bound for the lowest natural frequency is

$$\omega_1 < \sqrt{0.2895} = 0.5380 \ \frac{\text{rad}}{\text{s}} \tag{d}$$

The exact value of the lowest natural frequency is 0.4633 rad/s.

5.14 Rayleigh-Ritz method

The Rayleigh-Ritz method used to approximate the solution of $\mathbf{Lu} = \mathbf{f}$ for a self-adjoint and positive definite operator $\mathbf{L}$ involves minimizing the energy norm of the difference between the exact solution and the Rayleigh-Ritz approximation. However, this approach does not work for eigenvalue problems because eigenvectors are unique to at most a multiplicative constant.

Let $\mathbf{L}$ be a linear operator, defined on a domain S_D, that is self-adjoint and positive definite with respect to an inner product $(\mathbf{u}, \mathbf{v})$. Rayleigh's quotient is defined in Equation 5.156 as $R(\mathbf{y}) = (\mathbf{Ly}, \mathbf{y})/(\mathbf{y}, \mathbf{y})$. It was shown in Section 5.13 that $R(\mathbf{y})$ is stationary when $\mathbf{y}$ is an eigenvector of $\mathbf{L}$ and that the value of Rayleigh's quotient for an eigenvector is the eigenvalue corresponding to the eigenvector. Let $\mathbf{u}_i$ $i = 1, 2, \ldots, n$ be a basis for a finite-dimensional subspace of S_D. A Rayleigh-Ritz approximation for an eigenvector is of the form

$$\mathbf{y} = \sum_{i=1}^{n} \alpha_i \mathbf{u}_i \tag{5.168}$$

The Rayleigh-Ritz approximation for the eigenvector is determined as the vector in S_D for which the difference between the exact eigenvalue and Rayleigh's quotient for the vector, $R(\mathbf{w}) - R(\mathbf{y})$, is minimized. This is equivalent to finding the vector in S_D for which Rayleigh's quotient is a minimum.

The equation defining Rayleigh's quotient can be rewritten as

$$(\mathbf{L}\mathbf{y}, \mathbf{y}) - R(\mathbf{y})(\mathbf{y}, \mathbf{y}) = 0 \tag{5.169}$$

Substitution of Equation 5.168 into Equation 5.169, using the linearity of $\mathbf{L}$ and the properties of inner products, leads to

$$\sum_{i=1}^{n}\sum_{j=1}^{n} \alpha_i\alpha_j(\mathbf{L}\mathbf{u}_i, \mathbf{u}_j) - R(\mathbf{y})\sum_{i=1}^{n}\sum_{j=1}^{n} \alpha_i\alpha_j(\mathbf{u}_i, \mathbf{u}_j) = 0 \tag{5.170}$$

Differentiating Equation 5.170 with respect to α_k for an arbitrary k, $k = 1, 2, \ldots, n$ leads to

$$\sum_{i=1}^{n}\sum_{j=1}^{n} \frac{\partial}{\partial \alpha_k}(\alpha_i\alpha_j)(\mathbf{L}\mathbf{u}_i, \mathbf{u}_j) - \frac{\partial R}{\partial \alpha_k}\sum_{i=1}^{n}\sum_{j=1}^{n} \alpha_i\alpha_j(\mathbf{u}_i \mathbf{u}_j)$$

$$- R(\mathbf{y})\sum_{i=1}^{n}\sum_{j=1}^{n} \frac{\partial}{\partial \alpha_k}(\alpha_i\alpha_j)(\mathbf{u}_i, \mathbf{u}_j) = 0 \tag{5.171}$$

Minimization of $R(\mathbf{y})$ requires $\partial R/\partial \alpha_k = 0$. Noting that $\partial/\partial\alpha_k(\alpha_i\alpha_j) = \alpha_i\delta_{j,k} + \alpha_j\delta_{i,k}$, using the self-adjointness of $\mathbf{L}$ and the properties of inner products, Equation 5.171 reduces to

$$\sum_{i=1}^{n} \alpha_i(\mathbf{L}\mathbf{u}_i, \mathbf{u}_k) = R(\mathbf{y})\sum_{i=1}^{n} \alpha_i(\mathbf{u}_i, \mathbf{u}_k) \quad k = 1, 2, \ldots, n \tag{5.172}$$

Equation 5.172 represents a system of n algebraic equations for the n coefficients in the Rayleigh-Ritz expansion and can be written in matrix form as

$$\mathbf{A}\alpha = R(\mathbf{y})\mathbf{B}\alpha \tag{5.173}$$

where α is the vector of coefficients,

$$(\mathbf{A})_{i,j} = (\mathbf{L}\mathbf{u}_i, \mathbf{u}_j) \tag{5.174}$$

and

$$(\mathbf{B})_{i,j} = (\mathbf{u}_i, \mathbf{u}_j) \tag{5.175}$$

Equation 5.173 is a homogeneous system of equations with $R(\mathbf{y})$ as a parameter. When this equation is rewritten as

$$\mathbf{B}^{-1}\mathbf{A}\alpha = R(\mathbf{y})\alpha \tag{5.176}$$

It is clear that the appropriate values of $R(\mathbf{y})$ are the eigenvalues of the matrix $\mathbf{B}^{-1}\mathbf{A}$ and the coefficients in the Rayleigh-Ritz expansion are the components of the corresponding eigenvectors.

Define $\mu_1, \mu_2, \ldots, \mu_n$ as the eigenvalues of $\mathbf{B}^{-1}\mathbf{A}$. An eigenvector of $\mathbf{B}^{-1}\mathbf{A}$ corresponding to μ_k is ϕ_k with components $\alpha_{i,k}$, $i=1,2,\ldots,n$. The matrix $\mathbf{B}$ is clearly symmetric due to the commutativity of the inner product. The matrix $\mathbf{A}$ is symmetric due to the self-adjointness of $\mathbf{L}$. Thus the eigenvalues are the eigenvalues of a matrix which is the product of two symmetric matrices. Whereas the product is not necessarily symmetric, it is self-adjoint with respect to an energy inner product defined as $(\mathbf{u}, \mathbf{v})_B = (\mathbf{Bu}, \mathbf{v})$ (Section 5.4). Then the eigenvectors are orthogonal with respect to this energy inner product, $(\phi_k, \phi_\ell)_B = 0 \quad k \neq \ell$. The eigenvectors can be normalized such that $(\phi_k, \phi_k)_B = 1$. The corresponding approximation to an eigenvector of $\mathbf{L}$ is

$$\mathbf{y}_\mathbf{k} = \sum_{i=1}^{n} \alpha_{i,k}\mathbf{u}_i \tag{5.177}$$

The Rayleigh-Ritz approximation for eigenvalues and eigenvectors is based on minimization of Rayleigh's quotient over all vectors in the span of the basis. Rayleigh's quotient is a minimum when the trial vector is the eigenvector corresponding to the lowest eigenvalue. Thus the approximate eigenvector of $\mathbf{L}$ obtained from the eigenvector corresponding to the smallest eigenvalue of $\mathbf{B}^{-1}\mathbf{A}$ approximates the eigenvector corresponding to the smallest eigenvector of $\mathbf{L}$. It is shown in Section 5.13 that if a trial vector is chosen orthogonal to the eigenvector corresponding to the lowest eigenvalue, then an upper bound is obtained for the next lowest eigenvalue. Thus it is reasonable to assume that if the approximate mode-shape vectors are mutually orthogonal, then they approximate eigenvectors for the larger eigenvalues, and the eigenvalues of $\mathbf{B}^{-1}\mathbf{A}$ are approximations of higher eigenvalues of $\mathbf{L}$. To this end, consider

$$(\mathbf{y}_\mathbf{k}, \mathbf{y}_\ell) = \left(\sum_{i=1}^{n} \alpha_{i,k}\mathbf{u}_i, \sum_{j=1}^{n} \alpha_{j,\ell}\, \mathbf{u}_\mathrm{j} \right)$$

$$= \sum_{i=1}^{n}\sum_{j=1}^{n} \alpha_{i,k}\alpha_{j,\ell}(\mathbf{u}_\mathrm{i}, \mathbf{u}_\mathrm{j}) \tag{5.178}$$

Recall that the elements of the matrix **B** are the inner products $(\mathbf{u}_i, \mathbf{u}_j)$. Then Equation 5.178 becomes

$$\begin{aligned}(\mathbf{y}_k, \mathbf{y}_\ell) &= (\phi_k, \phi_\ell)_B \\ &= \delta_{k,\ell}\end{aligned} \tag{5.179}$$

Equation 5.179 shows that the eigenvector approximations are orthogonal with respect to the inner product of which **L** is self-adjoint.

Theorem 5.21 Let **L** be a linear operator whose domain is D such that **L** is self-adjoint and positive definite with respect to an inner product (**u**,**v**). Let u_1, $u_2,\ldots,$ u_n be a basis for a finite-dimensional subspace of D. The Rayleigh-Ritz approximation for the lowest eigenvectors of **L**, calculated as $w_k = \sum_{i=1}^{n} \alpha_{i,k} u_i$ for k = 1,2,...,n, are mutually orthogonal such that $(w_i, w_j) = 0$ for $i, j = 1,2,\ldots,n$ and $i \neq j$.

Example 5.26 Consider again the five-degree-of-freedom system of Example 5.25. Use the Rayleigh-Ritz method to approximate the two lowest natural frequencies and mode-shape vectors from the subspace of R^5 spanned by $\mathbf{u}_1 = [1 \quad 2 \quad 3 \quad 2 \quad 1]^T$ and $\mathbf{u}_2 = [1 \quad 3 \quad 2 \quad -1 \quad -2]^T$.

Solution As explained in Example 5.25, $\mathbf{L} = \mathbf{M}^{-1}\mathbf{K}$ is self-adjoint with respect to the kinetic-energy inner product. Thus the inner products used in Rayleigh's quotient are $(\mathbf{u}, \mathbf{v}) = (\mathbf{Mu}, \mathbf{v})$ and $(\mathbf{Lu}, \mathbf{v}) = (\mathbf{Ku}, \mathbf{v})$. Then **A** is a 2×2 matrix whose elements are $a_{i,j} = (\mathbf{Ku}_i, \mathbf{u}_j)$, and **B** is a 2×2 matrix whose elements are $b_{i,j} = (\mathbf{Mu}_i, \mathbf{u}_j)$. Calculations lead to

$$\mathbf{A} = \begin{bmatrix} 1.1\times10^3 & 9\times10^2 \\ 9\times10^2 & 3.3\times10^3 \end{bmatrix} \tag{a}$$

$$\mathbf{B} = \begin{bmatrix} 3.8\times10^3 & 2.2\times10^3 \\ 2.2\times10^3 & 2.8\times10^3 \end{bmatrix} \tag{b}$$

$$\mathbf{B}^{-1}\mathbf{A} = \begin{bmatrix} 0.19 & -0.817 \\ 0.172 & 1.821 \end{bmatrix} \tag{c}$$

The eigenvalues and normalized eigenvectors of $\mathbf{B}^{-1}\mathbf{A}$ can be calculated as

$$\mu_1 = 0.281 \qquad \phi_1 = \begin{bmatrix} 1.7\times10^{-2} \\ -1.937\times10^{-3} \end{bmatrix} \tag{d}$$

$$\mu_2 = 1.729 \qquad \phi_2 = \begin{bmatrix} -1.4\times10^{-2} \\ 2.6\times10^{-2} \end{bmatrix} \tag{e}$$

The Rayleigh-Ritz approximations for the lowest natural frequencies are $\omega_1 = \sqrt{\mu_1} = 0.53$ rad/s and $\omega_2 = \sqrt{\mu_2} = 1.315$ rad/s. Approximations for the corresponding normalized mode-shape vectors are

$$\begin{aligned}\mathbf{y}_1 &= 1.7\times10^{-2}\mathbf{u}_1 - 1.937\times10^{-3}\mathbf{u}_2 \\ &= [0.015 \quad 0.029 \quad 0.048 \quad 0.037 \quad 0.021]^{\mathrm{T}}\end{aligned} \tag{f}$$

$$\begin{aligned}\mathbf{y}_2 &= -1.4\times10^{-2}\mathbf{u}_1 + 2.6\times10^{-2}\mathbf{u}_2 \\ &= [0.012 \quad 0.049 \quad 0.010 \quad -0.053 \quad -0.065]^{\mathrm{T}}\end{aligned} \tag{g}$$

The inner products developed for eigenvalue problems involving self-adjoint operators whose domains are $C^n[a,b]$ are energy inner products. For example, the Sturm-Liouville operator is self-adjoint with respect to an inner product $(f,g)_r = \int_a^b f(x)g(x)r(x)dx$, where $r(x)$ is determined from the operator. This is usually a kinetic-energy inner product. Since $\mathbf{L}$ is self-adjoint with respect to this inner product, when $\mathbf{L}$ is positive definite, the inner product $(f,g)_L = (\mathbf{L}f,g)_r$ is an energy product, usually a potential-energy inner product. Thus Rayleigh's quotient,

$$\begin{aligned}R(\mathbf{y}) &= \frac{(\mathbf{Ly},\mathbf{y})_r}{(\mathbf{y},\mathbf{y})_r} \\ &= \frac{(\mathbf{y},\mathbf{y})_K}{(\mathbf{y},\mathbf{y})_M}\end{aligned} \tag{5.180}$$

is the ratio of the potential-energy inner product to the kinetic-energy inner product.

The formulation of Equation 5.180 can be applied to energy inner products developed directly from the kinetic and potential energies of the system as illustrated in the following example.

Example 5.27 The differential equation and boundary conditions used to determine the natural frequencies and boundary conditions for the torsional oscillations of an elastic shaft fixed at one end and with a thin disk and torsional spring attached to its other end are

$$\frac{\partial}{\partial x}\left(GJ\frac{\partial\theta}{\partial x}\right) = \rho J\frac{\partial^2\theta}{\partial t^2} \tag{a}$$

$$\theta(0,t) = 0 \tag{b}$$

$$-GJ(L)\frac{\partial\theta}{\partial x}(L,t) - k_t\theta(L,t) = I\frac{\partial^2\theta}{\partial t^2}(L,t) \tag{c}$$

Nondimensional variables are introduced as

$$x^* = \frac{x}{L} \tag{d}$$

$$t^* = t\sqrt{\frac{G_0}{\rho_0 L^2}} \tag{e}$$

$$\alpha(x^*) = \frac{GJ(x)}{G_0 J(L)} \tag{f}$$

$$\beta(x^*) = \frac{\rho J(x)}{\rho_0 J(L)} \tag{g}$$

The nondimensional formulation of Equation a, Equation b, and Equation c is

$$\frac{\partial}{\partial x}\left(\alpha\frac{\partial\theta}{\partial x}\right) = \beta\frac{\partial^2\theta}{\partial t^2} \tag{h}$$

$$\theta(0,t) = 0 \tag{i}$$

$$\frac{\partial\theta}{\partial x}(1,t) + \kappa\theta(1,t) = -\mu\frac{\partial^2\theta}{\partial t^2} \tag{j}$$

where $\kappa = k_t L/J(L)G$ and $\mu = I_D/\rho J(L)L$. Use of the normal-mode solution, $\theta(x,t) = \Theta(x)e^{i\omega t}$, in Equation h, Equation i, and Equation j leads to

$$\frac{d}{dx}\left(\alpha\frac{d\Theta}{dx}\right) + \beta\omega^2\Theta = 0 \tag{k}$$

$$\Theta(0) = 0 \tag{l}$$

$$\frac{d\Theta}{dx}(1) + \kappa\Theta(1) = \mu\omega^2\Theta(1) \tag{m}$$

Equation k is in the form of a Sturm-Liouville operator with $p(x) = -\alpha(x)$, $q(x) = 0$, and $r(x) = \beta(x)$. Consider the energy inner product,

$$(f,g)_K = \int_0^1 [\mathbf{L}f(x)]g(x)dx$$

$$= -\int_0^1 \frac{d}{dx}(\alpha(x)\frac{df}{dx})\frac{dg}{dx}dx \tag{n}$$

Application of integration by parts to Equation n leads to

$$(f,g)_K = \int_0^1 \alpha(x)\frac{df}{dx}\frac{dg}{dx} - \alpha(1)\frac{df}{dx}(1)g(1) \tag{o}$$

By definition $\alpha(1)=1$. Substitution of Equation m into Equation o leads to

$$(f,g)_K = \int_0^1 \alpha(x)\frac{df}{dx}\frac{dg}{dx} + \kappa f(1)g(1) - \mu\omega^2 f(1)g(1) \tag{p}$$

Using these inner products, Rayleigh's quotient becomes

$$R(f) = \frac{\int_0^1 \alpha(x)\left(\frac{df}{dx}\right)^2 dx + \kappa[f(1)]^2 - \mu\omega^2[f(1)]^2}{\int_0^1 \beta(x)[f(x)]^2 dx} \tag{q}$$

Equation q can be rearranged to give

$$\int_0^1 \alpha(x)\left(\frac{df}{dx}\right)^2 dx + \kappa[f(1)]^2 - \mu\omega^2[f(1)]^2 - R(f)\int_0^1 \beta(x)[f(x)]^2 dx = 0 \tag{r}$$

Note that if f is an eigenvector, $R(f)=\omega^2$, and Equation r can be written as

$$\int_0^1 \alpha(x)\left(\frac{df}{dx}\right)^2 dx + \kappa[f(1)]^2 - R(f)\left\{\int_0^1 \beta(x)[f(x)]^2 dx + \mu[f(1)]^2\right\} = 0 \tag{s}$$

Equation s is minimized as in Section 5.13. The definition of the potential-energy inner product used in Equation 5.180 is

$$(f,g)_K = \int_0^1 \alpha(x)\frac{df}{dx}\frac{dg}{dx}dx + \kappa f(1)g(1) \tag{t}$$

while the kinetic-energy inner product is defined as

$$(f,g)_M = \int_0^1 \beta(x)f(x)g(x)dx + \mu f(1)g(1) \tag{u}$$

Consider a shaft made from a uniform material with varying cross-section such that $\alpha(x)=\beta(x)=(1-\delta x)^2$. Consider a shaft with $\delta=0.2$, $\kappa=1.5$, and $u=1.0$. (a) Use a Rayleigh-Ritz method to determine an approximation for the five lowest natural frequencies and mode shapes using the five lowest mode shapes for a uniform shaft as basis functions. (b) Use the finite-element method to approximate the five lowest natural frequencies and mode shapes.

Solution (a) The five lowest natural frequencies and mode shapes of the uniform shaft are obtained by solving the problem,

$$\frac{d^2\Theta}{dx^2}+\omega^2\Theta=0 \tag{v}$$

$$\Theta(0)=0 \tag{w}$$

$$\frac{d\Theta}{dx}(1)+1.5\Theta(1)=\omega^2\Theta(x) \tag{x}$$

The solution of Equation v subject to Equation w is

$$\Theta(x)=C\sin(\omega x) \tag{y}$$

Application of Equation x to Equation y leads to

$$\tan(\omega)=\frac{\omega}{\omega^2-1.5} \tag{z}$$

There are an infinite, but countable, number of natural frequencies which satisfy Equation z; the first five are $\omega_1=1.345$, $\omega_2=3.461$, $\omega_3=6.443$, $\omega_4=9.531$, $\omega_5=12.646$. The eigenvectors of the form of Equation j are normalized by requiring

$$\int_0^1 [C_i\sin(\omega_i x)]^2\,dx+\omega^2[C_i\sin(\omega_i)]^2=1 \tag{aa}$$

The first five normalized mode shapes are

$$\begin{aligned}
\Theta_1(x)&=0.885\sin(1.345x)\\
\Theta_2(x)&=1.342\sin(3.461x)\\
\Theta_3(x)&=1.396\sin(6.443x)\\
\Theta_4(x)&=1.406\sin(9.531x)\\
\Theta_5(x)&=1.410\sin(12.646x)
\end{aligned} \tag{bb}$$

The MATHCAD file shown below provides the solution to part (a).

Rayleigh Ritz approximations for Example 5.27 (a)

Basis functions

$$\varphi 1(x) := 0.885 \cdot \sin(1.345 \cdot x)$$
$$\varphi 2(x) := 1.342 \cdot \sin(3.461 \cdot x)$$
$$\varphi 3(x) := 1.396 \cdot \sin(6.443 \cdot x)$$
$$\varphi 4(x) := 1.406 \cdot \sin(9.531 \cdot x)$$
$$\varphi 5(x) := 1.410 \cdot \sin(12.646 \cdot x)$$

$$\varphi(x) := \begin{pmatrix} \varphi 1(x) \\ \varphi 2(x) \\ \varphi 3(x) \\ \varphi 4(x) \\ \varphi 5(x) \end{pmatrix}$$

Parameters

$\mu := 1$ Mass ratio

$\kappa := 1.5$ Stiffness ratio

$\delta := .2$ Taper ratio

$\alpha(x) := (1 - \delta \cdot x)^2$ Nondimensional cross section area

Mass and stiffness matrix calculations

$$i := 1, 2..5$$

$$j := 1, 2..5$$

$$M_{i,j} := \int_0^1 \varphi(x)_i \cdot \varphi(x)_j \alpha(x) dx + \mu \cdot \varphi(1)_i \cdot \varphi(1)_j$$

$$M = \begin{pmatrix} 0.985 & -0.07 & 0.072 & -0.044 & 0.035 \\ -0.07 & 0.858 & 0.035 & 0.024 & -0.012 \\ 0.072 & 0.035 & 0.826 & 0.059 & 0.011 \\ -0.044 & 0.024 & 0.059 & 0.819 & 0.066 \\ 0.035 & -0.012 & 0.011 & 0.066 & 0.817 \end{pmatrix}$$

$$K_{i,j} := \int_0^1 \alpha(x) \cdot \left(\frac{d}{dx}\varphi(x)_i\right) \cdot \left(\frac{d}{dx}\varphi(x)_j\right) dx + \kappa \cdot \varphi(1)_i \cdot \varphi(1)_j$$

$$K = \begin{pmatrix} 1.832 & 0.179 & 0.197 & 0.02 & 0.097 \\ 0.179 & 9.665 & 1.768 & 0.129 & 0.434 \\ 0.197 & 1.768 & 33.627 & 4.691 & 0.203 \\ 0.02 & 0.129 & 4.691 & 73.718 & 9.02 \\ 0.097 & 0.434 & 0.203 & 9.02 & 129.973 \end{pmatrix}$$

Natural frequency calculations

$D := M^{-1} \cdot K$

$$D = \begin{pmatrix} 1.885 & 0.832 & -2.651 & 3.851 & -5.436 \\ 0.358 & 11.288 & 0.117 & -1.993 & 2.33 \\ 0.05 & 1.607 & 40.747 & -1.088 & -1.314 \\ 0.108 & -0.303 & 2.664 & 90.076 & -2.113 \\ 0.033 & 0.659 & -0.391 & 3.582 & 159.55 \end{pmatrix}$$

$\lambda := \text{eigenvals}(D)$

$$\lambda = \begin{pmatrix} 159.453 \\ 90.141 \\ 1.847 \\ 11.299 \\ 40.807 \end{pmatrix} \quad \text{Eigenvalues of D}$$

$\omega := \lambda^{0.5}$

$$\omega = \begin{pmatrix} 12.627 \\ 9.494 \\ 1.359 \\ 3.361 \\ 6.388 \end{pmatrix} \quad \text{Natural frequencies}$$

$W := \text{eigenvecs}(D)$

$$W = \begin{pmatrix} 0.035 & -0.047 & 0.999 & -0.108 & 0.074 \\ -0.016 & 0.027 & -0.038 & -0.993 & -7.047 \times 10^{-3} \\ 0.011 & 0.021 & 2.566 \times 10^{-4} & 0.054 & -0.996 \\ 0.031 & -0.997 & -1.366 \times 10^{-3} & -5.376 \times 10^{-3} & 0.053 \\ -0.999 & 0.051 & -1.766 \times 10^{-5} & 4.709 \times 10^{-3} & -4.87 \times 10^{-3} \end{pmatrix}$$

$Q := W^T \cdot M \cdot W$

$$Q = \begin{pmatrix} 0.81 & 0 & 2.028 \times 10^{-15} & 0 & 0 \\ 0 & 0.804 & 0 & 0 & 0 \\ 2.04 \times 10^{-15} & 0 & 0.991 & 7.399 \times 10^{-15} & 0 \\ 0 & 0 & 7.398 \times 10^{-15} & 0.84 & 0 \\ 0 & 0 & 0 & 0 & 0.81 \end{pmatrix}$$

Normalized eigenvectors

$$w1(x) := \frac{1}{(Q_{1,1})^{0.5}} \cdot \sum_{j=1}^{5} (W_{j,1} \cdot \varphi(x)_j)$$

$$w2(x) := \frac{1}{(Q_{2,2})^{0.5}} \cdot \sum_{j=1}^{5} (W_{j,2} \cdot \varphi(x)_j)$$

$$w3(x) := \frac{1}{(Q_{3,3})^{0.5}} \cdot \sum_{j=1}^{5} (W_{j,3} \cdot \varphi(x)_j)$$

$$w4(x) := \frac{1}{(Q_{4,4})^{0.5}} \cdot \sum_{j=1}^{5} (W_{j,4} \cdot \varphi(x)_j)$$

$$w5(x) := \frac{1}{(Q_{5,5})^{0.5}} \cdot \sum_{j=1}^{5} (W_{j,5} \cdot \varphi(x)_j)$$

Mode shape plots

$x := 0, .01 .. 1$

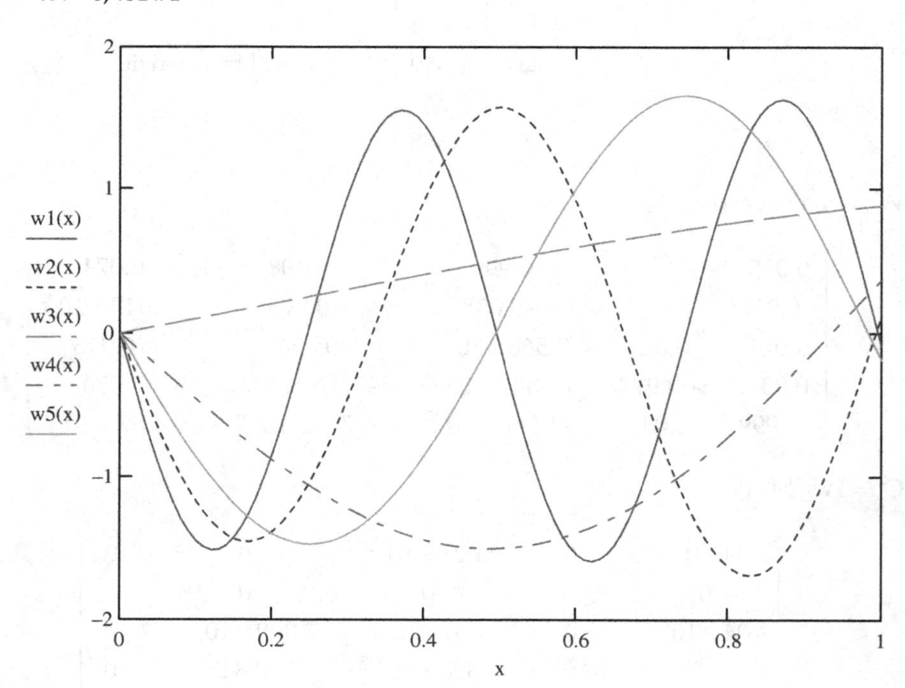

(b) Piecewise linear functions which satisfy the geometric boundary condition at $x=0$ and are continuous are defined as basis elements. The finite-element method is applied in the same fashion as the Rayleigh-Ritz method to approximate natural frequencies and mode shapes for this vibration problem. A mass matrix is defined from the kinetic-energy inner products of the basis elements. A stiffness matrix is defined from the potential-energy inner products of the basis elements. The square roots of the eigenvalues of $\mathbf{M}^{-1}\mathbf{K}$ are the approximations. The mode-shape approximations are obtained from the eigenvectors, as illustrated in the MATHCAD file shown below. This file also demonstrates the orthogonality of the finite-element approximations for the mode shapes.

Example 5.28 (b) Finite Element approximation for Natural Frequencies and Mode Shapes for Torsional Oscillations of Non uniform shaft Five elements of Equal Length from $x=0$ to $x=1$

Basis functions

$$\phi0(x) := (1 - 5 \cdot x)(\Phi(x) - \Phi(x - 0.2))$$
$$\phi1(x) := 5 \cdot x \cdot (\Phi(x) - \Phi(x - 0.2)) + (2 - 5 \cdot x) \cdot (\Phi(x - 0.2) - \Phi(x - 0.4))$$
$$\phi2(x) := \phi1(x - 0.2)$$
$$\phi3(x) := \phi1(x - 0.4)$$
$$\phi4(x) := \phi1(x - 0.6)$$
$$\phi5(x) := (5 \cdot x - 4) \cdot \Phi(x - 0.8)$$

$$x := 0, .01 .. 1$$

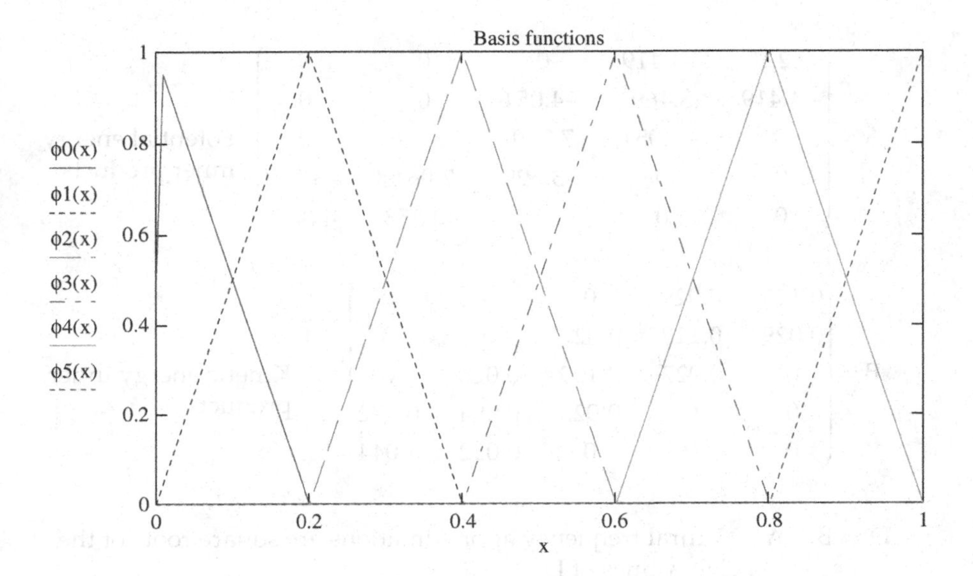

$$u(x) := \begin{pmatrix} \phi1(x) \\ \phi2(x) \\ \phi3(x) \\ \phi4(x) \\ \phi5(x) \end{pmatrix}$$ Column vector of basis functions to facilitate calculation of inner products

Non-uniform area

$\alpha(x) := (1 - 0.2 \cdot x)^2$ Non-uniform area of shaft

$\mu := 1$ Ratio of moment of inertia of disk to moment of inertia of shaft

$\kappa := 1.5$ Ratio of torsional stiffness of discrete spring to torsional stiffness of shaft

Energy inner products

$$i := 1..5$$

$$j := 1..5$$

$$B_{i,j} := \int_0^1 \alpha(x) \cdot u(x)_i \cdot u(x)_j dx + \mu \cdot u(1)_i \cdot u(1)_j$$

$$A_{i,j} := \int_0^1 \alpha(x) \cdot \left(\frac{d}{dx} u(x)_i\right) \cdot \left(\frac{d}{dx} u(x)_j\right) dx + \kappa \cdot u(1)_i \cdot u(1)_j$$

$$A = \begin{pmatrix} 9.221 & -4.419 & 0 & 0 & 0 \\ -4.419 & 8.469 & -4.051 & 0 & 0 \\ 0 & -4.051 & 7.749 & -3.699 & 0 \\ 0 & 0 & -3.699 & 7.061 & -3.353 \\ 0 & 0 & 0 & -3.353 & 4.867 \end{pmatrix}$$ Potential energy inner products

$$B = \begin{pmatrix} 0.123 & 0.029 & 0 & 0 & 0 \\ 0.029 & 0.113 & 0.027 & 0 & 0 \\ 0 & 0.027 & 0.103 & 0.025 & 0 \\ 0 & 0 & 0.025 & 0.094 & 0.022 \\ 0 & 0 & 0 & 0.022 & 1.044 \end{pmatrix}$$ Kinetic energy inner products

$D := B^{-1} \cdot A$ Natural frequency approximations are square roots of the eigenvalues of D

$\lambda := \text{eigenvals}(D)$

$$\lambda = \begin{pmatrix} 228.788 \\ 117.581 \\ 46.351 \\ 11.665 \\ 1.864 \end{pmatrix}$$ Eigenvalues of D

$$\text{wfe} : \lambda^{0.5}$$

$$\text{wfe} = \begin{pmatrix} 15.126 \\ 10.843 \\ 6.808 \\ 3.415 \\ 1.365 \end{pmatrix}$$

$Q := \text{eigenvals}(D)$

$$Q = \begin{pmatrix} -0.344 & 0.558 & -0.566 & 0.377 & 0.146 \\ 0.583 & -0.37 & -0.345 & 0.617 & 0.293 \\ -0.614 & -0.363 & 0.407 & 0.599 & 0.431 \\ 0.405 & 0.647 & 0.626 & 0.308 & 0.549 \\ -0.015 & -0.033 & -0.063 & -0.152 & 0.638 \end{pmatrix}$$ Eigenvectors of D will provide approximation to mode shapes

$Ax := \text{eigenvec}(D, \lambda_5)$

$$Ax = \begin{pmatrix} 0.146 \\ 0.293 \\ 0.431 \\ 0.549 \\ 0.638 \end{pmatrix}$$ Noting that lowest natural frequency corresponds to $\lambda 5$, the mode shape is determined using its corresponding eigenvector

$$z1 := \frac{1}{\sqrt{Ax^T \cdot B \cdot Ax}}$$ Normalization with respect to the kinetic energy inner product

$z1 = 1.386$

$$A1 = \begin{pmatrix} 0.202 \\ 0.406 \\ 0.597 \\ 0.76 \\ 0.884 \end{pmatrix}$$

$$w1(x) \sum_{i=1}^{5} (A1_i \cdot u(x)_i) \quad \text{Approximation for lowest mode shape}$$

Repeating the above procedure to determine approximations to mode shape of higher modes
$A \times 2 := \text{eigenvec}(D, \lambda_4)$

$$A \times 2 = \begin{pmatrix} -0.377 \\ -0.617 \\ -0.599 \\ -0.308 \\ 0.152 \end{pmatrix}$$

$$z1 := \frac{1}{\sqrt{A \times 2^T \cdot B \cdot A \times 2}} \qquad z1 = 2.416$$

$A2 := z1 \cdot A \times 2$

$$w2(x) := \sum_{i=1}^{5} (A2_i \cdot u(x)_i)$$

$A \times 3 := \text{eigenvec}(D, \lambda_3)$

$$A \times 3 = \begin{pmatrix} -0.566 \\ -0.345 \\ 0.407 \\ 0.626 \\ -0.063 \end{pmatrix}$$

$$z1 := \frac{1}{\sqrt{A \times 3^T \cdot B \cdot A \times 3}} \qquad z1 = 2.822$$

$A3 := z1 \cdot A \times 3$

$$wB(x) := \sum_{i=1}^{5} (A3_i \cdot u(x)_i)$$

$A \times 4 := \text{eigenvec}(D, \lambda_2)$

$$A \times 4 = \begin{pmatrix} -0.558 \\ 0.37 \\ 0.363 \\ -0.647 \\ 0.033 \end{pmatrix}$$

$$z1 := \frac{1}{\sqrt{A\times 4^{T}\cdot B\cdot A\times 4}} \quad z1 = 3.325$$

$A4 := z1 \cdot A\times 4$

$$w4(x) := \sum_{i=1}^{5}(A4_i \cdot u(x)_i)$$

$A\times 5 := \text{eigenvec}(D, \lambda_1)$

$$A\times 5 = \begin{pmatrix} 0.344 \\ -0.583 \\ 0.614 \\ -0.405 \\ 0.015 \end{pmatrix}$$

$$z1 := \frac{1}{\sqrt{A\times 5^{T}\cdot B\cdot A\times 5}} \quad z1 = 3.957$$

$A5 := z1 \cdot A\times 5$

$$w5(x) := \sum_{i=1}^{5}(A5_i \cdot u(x)_i)$$

$$x := 0, .01 .. 1$$

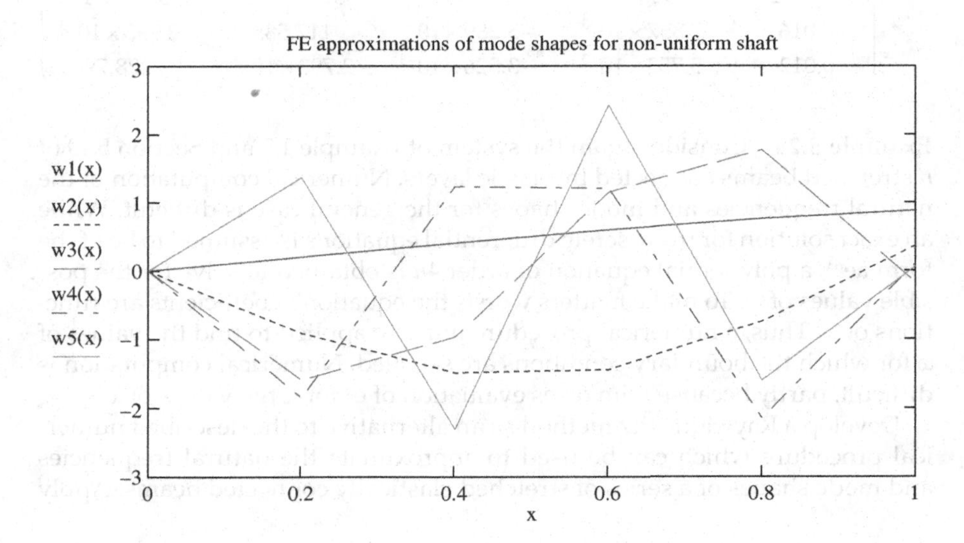

$$\mathrm{w(x)} := \begin{pmatrix} \mathrm{w1(x)} \\ \mathrm{w2(x)} \\ \mathrm{w3(x)} \\ \mathrm{w4(x)} \\ \mathrm{w5(x)} \end{pmatrix}$$

Orthogonality check

$$\mathrm{i} := 1..5$$

$$\mathrm{j} := 1..5$$

$$V_{i,j} := \int_0^1 \alpha(x)\cdot w(x)_i \cdot w(x)_j dx + \mu \cdot w(1)_i \cdot w(1)_j$$

$$V = \begin{pmatrix} 1 & 5.027\times10^{-6} & -1.125\times10^{-6} & -1.119\times10^{-6} & -5.613\times10^{-6} \\ 5.027\times10^{-6} & 1 & -1.788\times10^{-6} & 5.653\times10^{-7} & 2.948\times10^{-6} \\ -1.125\times10^{-6} & -1.788\times10^{-6} & 1 & -8.474\times10^{-6} & -7.281\times10^{-6} \\ -1.119\times10^{-6} & 5.653\times10^{-7} & -8.474\times10^{-6} & 1 & 9.96\times10^{-6} \\ -5.613\times10^{-6} & 2.948\times10^{-6} & -7.281\times10^{-6} & 9.96\times10^{-6} & 1 \end{pmatrix}$$

$$K_{i,j} := \int_0^1 \alpha(x)\cdot\left(\frac{d}{dx}w(x)_i\right)\cdot\left(\frac{d}{dx}w(x)_j\right)dx + \kappa \cdot w(1)_i \cdot w(1)_j$$

$$K = \begin{pmatrix} 1.848 & 2.036\times10^{-3} & -0.012 & 0.016 & 0.012 \\ 2.036\times10^{-3} & 11.67 & -6.945\times10^{-3} & 7.857\times10^{-3} & 5.753\times10^{-3} \\ -0.012 & -6.945\times10^{-3} & 46.357 & -5.289\times10^{-3} & -3.526\times10^{-3} \\ 0.016 & 7.857\times10^{-3} & -5.289\times10^{-3} & 117.585 & 2.793\times10^{-3} \\ 0.012 & 5.753\times10^{-3} & -3.526\times10^{-3} & 2.793\times10^{-3} & 228.79 \end{pmatrix}$$

Example 5.29 Consider again the system of Example 1.3 and Section 5.10 of n stretched beams connected by elastic layers. Numerical computation of the natural frequencies and mode shapes for the general case is difficult. While an exact solution for the discrete differential equations is assumed to be of the form $\mathbf{a}e^{\alpha x}$, a polynomial equation of order $4n$ is obtained to solve for the possible values of α. To make matters worse, the equation's coefficients are functions of ω. Thus, a numerical procedure must be applied to find the values of ω for which the boundary conditions are satisfied. Numerical computation is difficult, partly because it involves evaluation of e^x for large values of x.

Develop a Rayleigh-Ritz method as an alternative to the described numerical procedure which can be used to approximate the natural frequencies and mode shapes of a series of stretched elastically connected beams. Apply the procedure to the model of a triple-walled nanotube and plot the natural frequencies as a function of ε.

Solution Natural-frequency and mode-shape calculations for elastically connected unstretched Euler-Bernoulli beams are illustrated in Example 5.17. The mode shapes satisfy the same boundary conditions as the mode shapes for the stretched beams, except for the case when one end is free. Thus an appropriate choice of basis functions for a Rayleigh-Ritz approximation of the frequencies of a set of elastically connected stretched beams are the mode shapes for the corresponding unstretched beams.

Using the notation of Section 5.10, the first m sets of natural frequencies of the unstretched beams are defined as $\omega_{k,j}$ for $k=1,2,\ldots,m$ and $j=1,2,\ldots,n$. The corresponding normalized mode shapes are $\mathbf{a}_{\mathbf{k},\mathbf{j}}\ \phi_k(x)$. The energy inner products are defined by

$$(\mathbf{u},\mathbf{v})_{\mathrm{M}}=\int_0^1 \mathbf{v}^{\mathrm{T}}\mathbf{M}\mathbf{u}dx \tag{a}$$

$$(\mathbf{u},\mathbf{v})_{\mathrm{V}}=\int_0^1 \mathbf{v}^{\mathrm{T}}(\mathbf{K}_{\mathrm{a}}+\mathbf{K}_{\mathrm{b}}+\mathbf{K}_{\mathrm{c}})\mathbf{u}dx \tag{b}$$

An advantage of using mode shapes of the unstretched beams as basis functions for a Rayleigh-Ritz approximations for the response of the stretched beam is that the mass matrix for both problems is the same. Therefore the basis functions satisfy appropriate orthonormality conditions with respect to the kinetic-energy inner product,

$$(\mathbf{a}_{\mathbf{k},\mathbf{j}}\phi_k,\mathbf{a}_{\mathbf{p},\mathbf{q}}\phi_q)_{\mathrm{M}}=\delta_{j,p}\delta_{k,q} \tag{c}$$

In addition, the mode shapes for the unstretched beams are mutually orthogonal with respect to the potential-energy inner product for the unstretched beams,

$$\begin{aligned}(\mathbf{a}_{k,j}\phi_k,\mathbf{a}_{p,q}\phi_q)_L&=\int_0^1 \mathbf{a}_{p,q}^T\ \phi_p[(\mathbf{K}_b+\mathbf{K}_c)\mathbf{a}_{k,j}\phi_k]dx\\&=\omega_{k,j}^2\delta_{j,p}\delta_{k,q}\end{aligned} \tag{d}$$

Thus,

$$(\mathbf{a}_{k,j}\phi_k,\mathbf{a}_{p,q}\phi_q)_v=\omega_{k,j}^2\delta_{j,p}\delta_{k,q}+\int_0^1 \mathbf{a}_{p,q}^T\ \phi_p[\mathbf{K}_a(\mathbf{a}_{k,j}\phi_k)]dx \tag{e}$$

Reconsider the set of three fixed-fixed Euler Bernoulli beams of Example 5.17. The first five sets of intramodal frequencies are given in Table 5.6, while the corresponding normalized mode shapes are given in Table 5.7. These fifteen modes will be used as basis functions in the Rayleigh-Ritz approximations for the natural frequencies and mode shapes of a set of three elastically connected stretched fixed-fixed beams:

$$\begin{aligned} &\mathbf{u}_1 = a_{1,1}\phi_1 \quad \mathbf{u}_2 = a_{1,2}\phi_1 \quad \mathbf{u}_3 = a_{1,3}\phi_1 \\ &\mathbf{u}_4 = a_{2,1}\phi_2 \quad \mathbf{u}_5 = a_{2,2}\phi_2 \quad \mathbf{u}_6 = a_{2,3}\phi_2 \\ &\mathbf{u}_7 = a_{3,1}\phi_3 \quad \mathbf{u}_8 = a_{3,2}\phi_3 \quad \mathbf{u}_9 = a_{3,3}\phi_3 \\ &\mathbf{u}_{10} = a_{4,1}\phi_4 \quad \mathbf{u}_{11} = a_{4,2}\phi_4 \quad \mathbf{u}_{12} = a_{4,3}\phi_4 \\ &\mathbf{u}_{13} = a_{5,1}\phi_5 \quad \mathbf{u}_{14} = a_{5,2}\phi_5 \quad \mathbf{u}_{15} = a_{5,3}\phi_5 \end{aligned} \tag{f}$$

The Rayleigh-Ritz approximation is of the form

$$\mathbf{w}(x) = \sum_{i=1}^{15} \alpha_i \mathbf{u}_i \tag{g}$$

In view of the orthonormality condition of Equation c which is satisfied by the basis functions, the matrix **B** of Equation 5.173 and Equation 5.175 is the 15×15 identity matrix. The matrix **A** is the matrix of potential-energy inner products which are calculated using Equation d. Note that

$$\int_0^1 \mathbf{a}_{p,q}^T \ \phi_p[\mathbf{K}_a(\mathbf{a}_{k,j}\phi_k)]dx = -\varepsilon \int_0^1 \mathbf{a}_{p,q}^T \ \mathbf{a}_{k,j}\phi_p(x)\frac{d^2\phi_k}{dx^2}dx \tag{h}$$

The **A** matrix for this example is listed in Table 5.8. The natural-frequency approximations are the square roots of the eigenvalues of the matrix **A**. The natural frequencies for $\varepsilon = 50$ are listed in Table 5.9.
The Rayleigh-Ritz approximations for the mode shapes can be calculated by

$$\mathbf{w}_i = \sum_{j=1}^{15} p_{i,j}\mathbf{u}_i(x) \tag{i}$$

The mode-shape approximations for four modes are illustrated in Figure 5.19.

5.15 Green's functions

Let L be a linear second-order differential operator defined on a domain S_D, a subspace of $C^2[a, b]$. The operator is in the self-adjoint form of Equation 5.18.

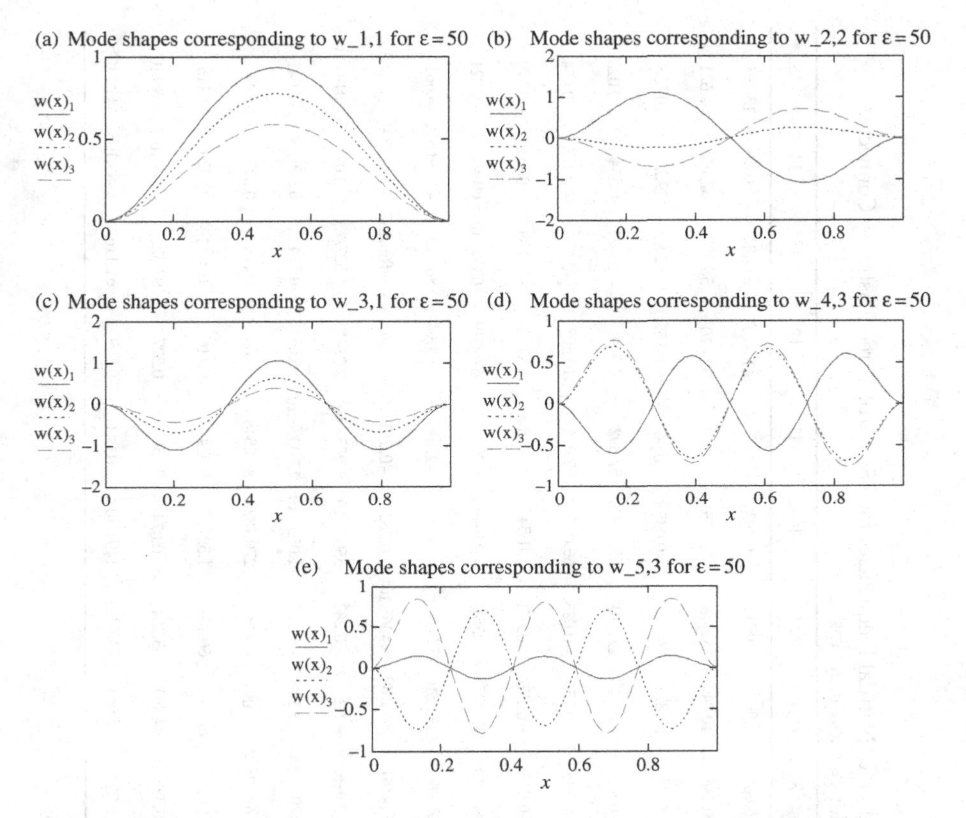

Figure 5.19 Mode-shape approximations obtained using the Rayleigh-Ritz method for a set of three elastically connected beams, (a) $w_{1,1}$, (b) $w_{2,2}$, (c) $w_{3,1}$, (d) $w_{4,3}$, and (e) $w_{5,3}$.

The Sturm-Liouville problem corresponding to the operator is $\mathbf{L}y=\lambda y$. Let λ_i $i=1,2,\ldots$ be the eigenvalues of the Sturm-Liouville problem with corresponding normalized eigenvectors $y_i(x)$. The eigenvectors are orthonormal with respect to the inner product $(f,g)_r=\int_z^b f(x)g(x)r(x)dx$.

Consider the nonhomogeneous problem,

$$Lw=f \tag{5.181}$$

where w is in S_D. The expansion theorem implies that there exist coefficients c_i $i=1,2,\ldots$. such that

$$w(x)=\sum_{i=1}^{\infty} c_i y_i(x) \tag{5.182}$$

It is assumed that $f(x)$ is of a form such that it has a convergent eigenvector expansion of the form

Table 5.8 The **A** matrix for Rayleigh-Ritz Approximation of Natural Frequencies for a Series of Three Elastically Connected Euler-Bernoulli Beams

	1	2	3	4	5	6	7	8	9	10	11	12	13	14	15
1	376×10^3	26.492	7.253	0.07	0.014	28×10^4	75.602	2.057	1.974	−0.193	0.058	41×10^3	61.436	33.098	13.291
2	26.492	299×10^4	81.654	0.024	0.071	0.011	00.116	40.849	31.643	−0.155	−0.145	32×10^3	58.932	62.122	52.621
3	7.253	81.654	238×10^4	168×10^3	0.014	0.06	29.604	86.871	06.777	−0.035	−0.06	−0.157	74.213	19.275	207.38
4	0.07	0.024	168×10^3	296×10^8	75.837	36.859	−0.856	−0.047	22×10^3	65.019	57.892	4.374	−0.765	0.334	−0.028
5	0.014	0.071	0.014	75.837	833×10^4	96.453	−0.385	−0.754	−0.085	66.075	92.826	10.508	−0.666	−0.476	−0.134
6	28.10^4	0.011	0.06	36.859	96.453	794×10^4	−0.055	−0.175	−0.674	101.54	85.948	15.401	−0.188	−0.318	−0.573
7	75.602	00.116	29.604	−0.856	−0.385	−0.055	856×10^4	37.597	33.381	−2.996	0.202	67×10^3	033×10^8	60.883	−9.221
8	2.057	40.849	86.871	−0.047	−0.754	−0.175	37.597	383×10^4	77.789	−1.3	−2.106	−0.094	86.043	44.461	16.661
9	1.974	−31.643	−306.777	-1.922×10^{-3}	−0.085	−0.674	133.381	577.789	5.436×10^4	−0.282	−0.615	−2.063	−166.487	−333.284	−674.831
10	−0.193	−0.155	−0.035	−665.019	−466.075	−101.54	−2.996	−1.3	−0.282	5502×10^4	1.055×10^3	270.482	1.527	−0.074	2.049×10^{-3}
11	0.058	−0.145	−0.06	157.892	−492.826	−185.948	0.202	−2.106	−0.615	1.055×10^3	17.008×10^4	853.548	0.374	0.956	−0.041
12	1.241×10^{-3}	6.732×10^{-3}	−0.157	4.374	10.508	−515.401	3.67×10^{-3}	−0.094	−2.063	270.482	853.548	9.166×10^4	0.116	0.327	0.899
13	61.436	58.932	74.213	−0.765	−0.666	−0.188	033.10^3	86.043	66.487	1.527	0.374	0.116	104×10^5	93.326	61.116
14	33.098	62.122	19.275	0.334	−0.476	−0.318	60.883	44.461	33.284	−0.074	0.956	0.327	93.326	409×10^5	098×10^8
15	13.291	52.621	207.38	−0.028	0.134	−0.573	−9.221	16.661	74.831	149×10^3	−0.041	0.899	61.116	098×10^8	641×10^5

Table 5.9 Rayleigh-Ritz Approximations for Natural Frequencies of Three Elastically Connected Stretched Beams with $\varepsilon = 50$

k,j	$\omega_{k,j}$
1,1	53.54
1,2	113.95
1,3	179.92
2,1	90.90
2,2	135.40
2,3	194.78
3,1	153.20
3,2	184.12
3,3	233.18
4,1	234.44
4,2	264.81
4,3	302.83
5,1	332.28
5,2	375.28
5,3	405.14

$$f(x) = \sum_{i=1}^{\infty} (f, y_i)_r y_i(x) \tag{5.183}$$

Substituting Equation 5.183 and Equation 5.182 into Equation 5.181 leads to

$$\mathbf{L}\left(\sum_{i=1}^{\infty} c_i y_i\right) = \sum_{i=1}^{\infty} (f, y_i)_r y_i \tag{5.184}$$

Using the linearity properties of $\mathbf{L}$ and noting that $\mathbf{L}y_i = \lambda_i y_i$, Equation 5.184 can be rearranged to give

$$\sum_{i=1}^{\infty} [\lambda_i c_i - (f, y_i)_r] y_i \tag{5.185}$$

The eigenvectors are linearly independent. Therefore,

$$c_i = \frac{1}{\lambda_i}(f, y_i)_r \tag{5.186}$$

Substituting Equation 5.186 into Equation 5.182, the nonhomogeneous solution can be expressed as

$$w(x) = \sum_{i=1}^{\infty} \frac{1}{\lambda_i} (f, y_i)_r y_i$$

$$= \sum_{i=1}^{\infty} \frac{1}{\lambda_i} \int_a^b f(\xi) y_i(\xi) y_i(x) r(\xi) d\xi \tag{5.187}$$

The series in Equation 5.187 is convergent. Therefore, the order of summation and integration may be changed, leading to

$$w(x) = \int_a^b K(x,\xi) f(\xi) r(\xi) d\xi \tag{5.188}$$

where the Green's function is

$$K(x,\xi) = \sum_{i=1}^{\infty} \frac{1}{\lambda_i} y_i(x) y_i(\xi) \tag{5.190}$$

The above shows the existence of a Green's function for a self-adjoint operator. Given the Green's function, Equation 5.188 is used to determine $w(x)$. In this sense, the integral operator,

$$\mathbf{G}f = \int_a^b K(x,\xi) f(\xi) r(\xi) d\xi \tag{5.190}$$

is the inverse operator of $\mathbf{L}$, because $\mathbf{L}(\mathbf{G}f) = \mathbf{L}w = f$ and $\mathbf{G}(\mathbf{L}w) = \mathbf{G}f = w$.

The general definition of a Green's function for a linear differential operator $\mathbf{L}$ is a function of ξ and x such that if $\mathbf{L}w = f$, then

$$w = \int_a^b K(x,\xi) f(\xi) r(\xi) d\xi \tag{5.191}$$

Equation 5.189 shows that the Green's function does not exist if the operator is positive semi-definite, in which case the smallest eigenvalue is zero. This is consistent with the concept that the Green's function represents an inverse operator for $\mathbf{L}$. If 0 is an eigenvalue of $\mathbf{L}$, then there exists a nontrivial $\mathbf{w}$ such that $\mathbf{Lw} = \mathbf{0}$. If an inverse $\mathbf{L}^{-1}$ exists, then $\mathbf{L}^{-1}(\mathbf{0}) = \mathbf{w}$. However, if $\mathbf{L}^{-1}$ is linear, then $\mathbf{L}^{-1}(\mathbf{0}) = \mathbf{0}$, which contradicts the assumption of a nontrivial $\mathbf{w}$. Also recall that if 0 is an eigenvalue of **L**, then **f** must satisfy a solvability condition for a solution of **Lw** = **f** to exist.

Suppose that the Green's function exists and is defined by Equation 5.190. Consider the conditions under which **G** is self-adjoint. To this end, consider

$$
\begin{aligned}
(\mathbf{G}f, g)_r &= \int_a^b (\mathbf{G}f) g(x) r(x) dx \\
&= \int_a^b \left[\int_a^b K(x,\xi) f(\xi) r(\xi) d\xi \right] g(x) r(x) dx \\
&= \int_a^b \int_a^b K(x,\xi) f(\xi) r(\xi) g(x) r(x) d\xi dx
\end{aligned}
\tag{5.192}
$$

Interchanging the order of integration in Equation 5.193 leads to

$$
(\mathbf{G}f, g)_r = \int_a^b \left[\int_a^b K(x,\xi) g(x) r(x) dx \right] f(\xi) r(\xi) d\xi \tag{5.193}
$$

If $K(x,\xi) = K(\xi,x)$, Equation 5.193 becomes

$$
\begin{aligned}
(\mathbf{G}f, g)_r &= \int_a^b \left[\int_a^b K(\xi,x) g(x) r(x) dx \right] f(\xi) r(\xi) d\xi \\
&= \int_a^b (\mathbf{G}g) f(\xi) r(\xi) d\xi \\
&= (f, \mathbf{G}g)_r
\end{aligned}
\tag{5.194}
$$

Equation 5.194 shows that the operator **G** is self-adjoint with respect to the inner product $(f, g)_r$ if the Green's function is symmetric, $K(x, \xi) = K(\xi, x)$.

The above discussion provides some insight into the Green's function, but does not provide a method for determining the Green's function. This follows from a physical understanding of the Green's function. To this end, consider the deflection of a beam pinned at both ends, as illustrated in Figure 5.20. The differential equation governing the deflection of the beam due to a distributed load $f(x)$ is

$$
EI\frac{d^4w}{dx^4} = f(x) \tag{5.195}
$$

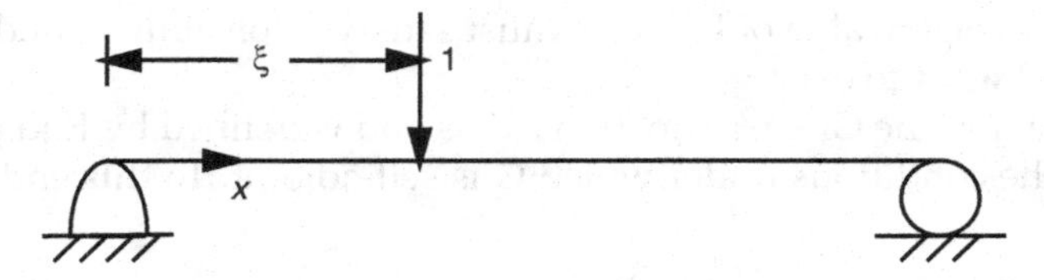

Figure 5.20 The kernel for the Green's function which is used to determine the deflection of the beam due to a distributed load is determined as the deflection of the beam at a distance x from its left support due to a concentrated unit load a distance ξ from its left support.

The boundary conditions accompanying Equation 5.195 are

$$w(0) = 0 \tag{5.196a}$$

$$\frac{d^2w}{dx^2}(0) = 0 \tag{5.196b}$$

$$w(L) = 0 \tag{5.196c}$$

$$\frac{d^2w}{dx^2}(L) = 0 \tag{5.196d}$$

Suppose that the load is simply a unit concentrated load applied at a distance ξ from the left support. The concentrated load can be mathematically described by $f(x) = \delta(x - \xi)$, where $\delta(x)$ is the Dirac delta function, also called the unit impulse function. The mathematical problem governing the deflection is

$$EI\frac{d^4w}{dx^4} = \delta(x - \xi) \tag{5.197}$$

subject to the boundary conditions of Equations 5.196. The solution can be obtained by direct integration of Equation 5.197 or by a number of other methods, including the moment-area method. Using direct integration to solve Equation 5.197, it should be noted that d^3w/dx^3 is discontinuous at $x = \xi$. Physically, the shear force developed in the cross-section of the beam is discontinuous across the concentrated load (Figure 5.21), and the shear force is proportional to d^3w/dx^3. Mathematically, the integral of the Dirac delta function is $\int_0^x \delta(\tau - \xi)d\tau = u(x - \xi)$, where $u(x - \xi)$ is the unit step function which is 0 for $x < \xi$ and 1 for $x > \xi$. Repeated integration leads to a solution of Equation 5.197 as

$$q(x,\xi) = \frac{1}{EI}\left[\frac{1}{6}(x - \xi)^3 u(x - \xi) + C_1\frac{x^3}{6} + C_2\frac{x^2}{2} + C_3x + C^4\right] \tag{5.198}$$

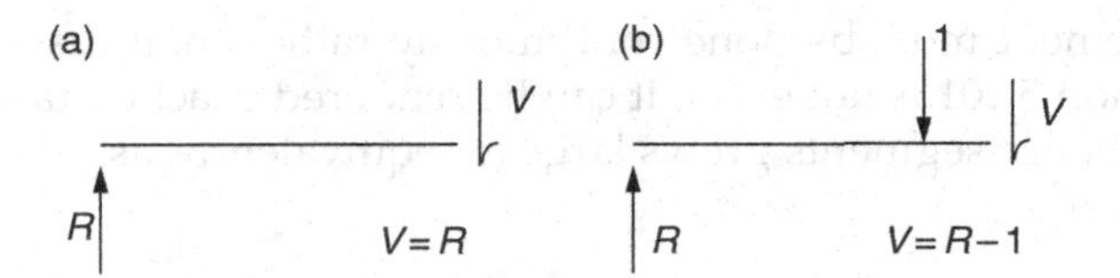

Figure 5.21 The internal shear force in the beam is discontinuous at the location where the concentrated load is applied.

Application of the boundary conditions, Equation 5.196

$$q(x,\xi)=\frac{1}{EI}\left[\frac{1}{6}(x-\xi)^3u(x-\xi)+\left(\frac{\xi}{L}-1\right)\frac{x^3}{6}+\frac{\xi L}{6}\left(1-\frac{\xi}{L}\right)\left(2-\frac{\xi}{L}\right)x\right] \tag{5.199}$$

The value of $q(a, b)$ represents the deflection at $x=a$ due to a concentrated unit load at $x=b$. Conversely $q(b, a)$ represents the deflection at $x=b$ due to a unit concentrated load applied at $x=a$. The deflection obtained if a unit concentrated load is applied at $x=a$ and then a unit concentrated load applied at $x=b$ is the same as if the order of loading were reversed. The total work done by application of the loads in the former case is $U_{a\to b}=1/2[q(a,a)]+q(a,b)+1/2[q(b,b)]$, while the total work done by application of the load first at b and then at a is $U_{b\to a}=1/2[q(b,b)]+q(b,a)+1/2[q(a,a)]$. Since the work is the same independent of the order of loading, $q(a, b)=q(b, a)$. Since a and b are arbitrary, a generalization of this result is

$$q(x,\xi)=q(\xi,x) \tag{5.200}$$

The function $q(x, \xi)$ is called an influence function.

The deflection of $w(x)$, a beam due to a superposition of loads $f_1(x)+f_2(x)$ is $w(x)=w_1(x)+w_2(x)$, where $w_1(x)$ is the deflection due to the loading described only by $f_1(x)$, and $w_2(x)$ is the deflection due to the loading described only by $f_2(x)$. Consider a beam subject to a distributed load per length of $f(x)$. The beam is divided into n segments of equal length, $\Delta x=L/n$. Define $f_i(x)$ as the load applied over the interval $i\Delta x\le x\ (i+1)\Delta x$. The distributed load over this interval is equivalent to a concentrated load $\hat{f}_i=\int_{i\Delta x}^{(i+1)\Delta x} f(x)dx$ applied at $\xi_i=i\Delta x+\int_{i\Delta x}^{(i+1)\Delta x}(x-i\Delta x)f(x)dx$. Using the concept of influence functions, the deflection along the length of the beam due to the distributed load applied over this interval is $w_i^{\,n}(x)=\hat{f}_i q(x,\xi_i)$. Using the superposition principle the deflection of the beam due to the entire distributed loading is

$$w^n(x)=\sum_{i=1}^{n}\hat{f}_i q(x,\xi_i) \tag{5.201}$$

A distributed load may be replaced by a concentrated load equal to its resultant at the centroid of the load distribution for the purposes of calculating internal forces in beams and reactions. However, to determine the deflection, this replacement must be done continuously rather than discretely. Therefore, Equation 5.201 is not exact. It can be rendered exact by taking the limit as the number of segments grows large (or equivalently, as $\Delta x \to 0$):

$$w(x) = \lim_{\Delta x \to 0} \sum_{i=1}^{n} \hat{f}_i q(x, \xi_i)$$
$$= \int_0^L f(\xi) q(x, \xi) d\xi \tag{5.202}$$

Equation 5.202 is an integral representation of the solution in the form of a Green's function operator. The influence function $q(x, \xi)$ is the Green's function for the problem described by Equation 5.195 and Equation 5.196. The Green's function for this problem satisfies the following properties:

- $q(x, \xi)$ satisfies all boundary conditions.
- $q(x, \xi)$ is a solution of the differential equation $Lw = \delta(x - \xi)$ for all x, $0 \leq x \leq \xi$, and all ξ, $\xi < x < L$.
- $q(x, \xi)$, $\partial q/\partial x(x, \xi)$, and $\partial^2 q/\partial x^2(x, \xi)$ are all continuous for all x, $0 \leq x \leq L$.
- $\partial^3 q/\partial x^3$ has a jump discontinuity at $x = \xi$. The value of the discontinuity is such that

$$EI \frac{\partial^3 q}{\partial x^3}(\xi^+, \xi) - EI \frac{\partial^3 q}{\partial x^3}(\xi^-, \xi) = -1 \tag{5.203}$$

Example 5.30 The thin variable-area rod shown in Figure 5.22 is subject to an internal heat generation which varies across its length. The rod is of length L, has a fixed temperature imposed at one end, and is insulated at the other end. The heat transfer coefficient between the rod and the ambient medium is small. Defining nondimensional variables by $x^* = x/L$, $\theta = T - T_0/T_0$, $\alpha(x^*) = A(x)/A_0$, and $f(x^*) = u(x)/u_{max}$, where $u(x)$ is the rate of internal heat

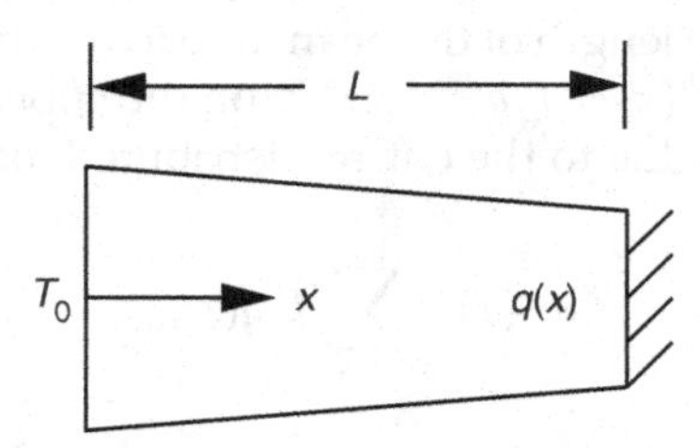

Figure 5.22 A Green's function is used to determine the temperature distribution in the variable-area rod.

generation across the entire cross-section and u_{max} its maximum value, the formulation of the problem governing the temperature distribution in the bar is

$$-\beta\frac{d}{dx}\left[\alpha(x)\frac{d\theta}{dx}\right]=f(x) \tag{a}$$

$$\theta(0)=0 \tag{b}$$

$$\frac{d\theta}{dx}(1)=0 \tag{c}$$

with $\beta=kA_0T_0/L^2u_{max}$. Determine the Green's function for the system with $\alpha(x)=1+\varepsilon x$ and use the Green's function to determine $\theta(x)$ if $f(x)=x(1-x)$.

Solution The Green's function is the solution of

$$-\beta\frac{d}{dx}\left[(1+\varepsilon x)\frac{dq}{dx}\right]=\delta(x-\xi) \tag{d}$$

subject to Equation b and Equation c. integration of Equation d gives

$$\beta(1+\varepsilon x)\frac{dq}{dx}=-u(x-\xi)+C_1 \tag{e}$$

$$\frac{dq}{dx}=-\frac{u(x-\xi)}{\beta(1+\varepsilon x)}+\frac{C_1}{\beta(1+\varepsilon x)} \tag{f}$$

Application of Equation c to Equation f gives $C_1=1$, leading to

$$\frac{dq}{dx}=\frac{1}{\beta(1+\varepsilon x)}-\frac{u(x-\xi)}{\beta(1+\varepsilon x)} \tag{g}$$

Integration of Equation g results in

$$q(x,\xi)=\frac{1}{\beta\varepsilon}\left[\ln(1+\varepsilon x)-u(x-\xi)\ln\left(\frac{1+\varepsilon x}{1+\varepsilon\xi}\right)\right]+C_2 \tag{h}$$

Application of Equation b to Equation h leads to $C_2=0$ and the Green's function of

$$q(x,\xi)=\frac{1}{\beta\varepsilon}\left[\ln(1+\varepsilon x)-u(x-\xi)\ln\left(\frac{1+\varepsilon x}{1+\varepsilon\xi}\right)\right] \tag{i}$$

Evaluation of the unit step functions of Equation g and Equation i leads to

$$\frac{dq}{dx} = \begin{cases} \dfrac{1}{\beta(1+\varepsilon x)} & x<\xi \\ 0 & x>\xi \end{cases} \tag{j}$$

$$q(x,\xi) = \begin{cases} \dfrac{1}{\beta\varepsilon}\ln(1+\varepsilon x) & x<\xi \\ \dfrac{1}{\beta\varepsilon}\ln(1+\varepsilon x) & x>\xi \end{cases} \tag{k}$$

Equation j and Equation k show that since the end where $x=1$ is insulated and a unit heat source is applied at $x=\xi$, the rate of heat transfer is zero for $x > \xi$, leading to a constant temperature in this region. The rate of heat transfer is discontinuous at $x=\xi$, while the temperature is continuous at this point.

The solution of Equation a for any $f(x)$ is of the form

$$\theta(x) = \int_0^1 f(\xi)q(x,\xi)d\xi \tag{l}$$

Substitution of the given $f(x)$ into Equation l leads to

$$\begin{aligned} \theta(x) &= \frac{1}{\beta\varepsilon}\int_0^1 \xi(1-\xi)\left[\ln(1+\varepsilon x) - u(x-\xi)\ln\left(\frac{1+\varepsilon x}{1+\varepsilon\xi}\right)\right]d\xi \\ &= \frac{\ln(1+\varepsilon x)}{\beta\varepsilon}\left[\int_0^1 \xi(1-\xi)d\xi + \int_x^1 \xi(1-\xi)\ln(1+\varepsilon\xi)d\xi\right] \\ &= \frac{\ln(1+\varepsilon x)}{36\beta\varepsilon^4}\Big\{12\varepsilon^3 + 6\left(5\varepsilon^3+18\varepsilon^2+15\varepsilon+8\right)\ln(1+\varepsilon) \\ &\quad -(13\varepsilon^3+66\varepsilon^2+89\varepsilon+40) - 6\left[2(1+\varepsilon x)^3 + 3(1+\varepsilon x)^2 + (1+\varepsilon)(1+\varepsilon x)\right] \\ &\quad +4(1+\varepsilon x)^3 + 9(1+\varepsilon x)^2 + (1+\varepsilon)(1+\varepsilon x)\Big\} \end{aligned} \tag{m}$$

The Green's function of Example 5.29 satisfies the following conditions:

- $q(x, \xi)$ satisfies all the boundary conditions.
- $q(x, \xi)$ is a solution of the differential equation $Lw=\delta(x-\xi)$ for all x, $0\leq x\leq \xi$ and all ξ, $\xi < x < 1$.

- $q(x, \xi)$ is continuous for all x, $0 \leq x \leq 1$.
- $\partial q/\partial x$ has a jump discontinuity at $x = \xi$. The value of the discontinuity is such $\partial q/\partial x(\xi^+, \xi) - \partial q/\partial x(\xi^-, \xi) = -1/\beta(1 + \varepsilon\xi)$.

The Green's function for a problem $\mathbf{L}w = f$ subject to appropriate boundary conditions is the function $q(x, \xi)$, which is the solution of $\mathbf{L}q = \delta(x - \xi)$ that satisfies all boundary conditions.

The above examples specify the conditions which a Green's function must satisfy and provide a rubric for the construction of the Green's function. Consider a linear differential operator $\mathbf{L}w = a_n(x)(d^n w/dx^n) + a_{n-1}(x)(d^{n-1}w/dx^{n-1}) + \cdots + a_1(dw/dx) + a_0 w$. The Green's function satisfies the following conditions:

- $q(x, \xi)$ satisfies $\mathbf{L}q = \delta(x - \xi)$ everywhere except at $x = \xi$.
- $q(x, \xi)$ satisfies all boundary conditions.
- $q(x, \xi)$, $\partial q/\partial x, \ldots, \partial^{n-2}q/\partial x^{n-2}$ are continuous for all x.
- $\partial^{n-1}q/\partial x^{n-1}$ is discontinuous at $x = \xi$, at which it has a jump discontinuity

$$\frac{\partial^{n-1}q}{\partial x^{n-1}}(\xi^+, \xi) - \frac{\partial^{n-1}q}{\partial x^{n-1}}(\xi^-, \xi) = -\frac{1}{a_n(\xi)} \tag{5.204}$$

The Green's function can be constructed for any linear differential operator which does not have an eigenvalue of zero, in which case $\mathbf{L}w = 0$ has a nontrivial solution. If $\mathbf{L}$ is self-adjoint, the Green's function is symmetric, and if the Green's function is symmetric, the Green's function operator is self-adjoint.

Example 5.31 Set up the Green's function solution for Example 5.29, assuming the heat transfer coefficient between the surface and the ambient is a finite value h.

Solution In terms of the nondimensional variables defined in Example 5.29, the nondimensional problem governing the temperature distribution in the surface is

$$-\beta \frac{d}{dx}\left[(1 + \varepsilon x)\frac{d\theta}{dx}\right] + \mu\theta = f(x) \tag{a}$$

$$\theta(0) = 0 \tag{b}$$

$$\frac{d\theta}{dx}(1) = 0 \tag{c}$$

where $\mu = hPu_{\max}T_0$.

Since the Green's function satisfies the differential equation $\mathbf{L}w = \delta(x - \xi)$ for all x except $x = \xi$, it is a homogeneous solution for $0 \leq x \leq \xi$ and for $\xi < x \leq 1$. Defining $z = 1 + \varepsilon x$, the homogeneous equation is

$$\frac{d}{dz}\left(z\frac{d\theta}{dz}\right) - \nu^2\theta = 0 \tag{d}$$

with $\nu^2 = \sqrt{\mu/\beta\varepsilon^2}$. The general solution of Equation d is

$$\theta(z) = C_1 I_0\left(2\nu z^{1/2}\right) + C_2 K_0\left(2\nu z^{1/2}\right) \tag{e}$$

Thus, the Green's function is of the form

$$q(x,\xi) = \begin{cases} C_1 I_0\left(2\nu(1+\varepsilon x)^{1/2}\right) + C_2 K_0\left(2\nu(1+\varepsilon x)^{1/2}\right) & 0 \le x < \xi \\ C_3 I_0\left(2\nu(1+\varepsilon x)^{1/2}\right) + C_4 K_0\left(2\nu(1+\varepsilon x)^{1/2}\right) & \xi < x \le 1 \end{cases} \tag{f}$$

Requiring the Green's function to satisfy the boundary conditions of Equation b and Equation c leads to

$$q(x,\xi) = \begin{cases} C_1\left[I_0\left(2\nu(1+\varepsilon x)^{1/2}\right) - \dfrac{I_0(2\nu)}{K_0(2\nu)} K_0\left(2\nu(1+\varepsilon x)^{1/2}\right)\right] & 0 \le x < \xi \\ C_3\left[I_0\left(2\nu(1+\varepsilon x)^{1/2}\right) + \dfrac{I_1\left(2\nu(1+\varepsilon)^{1/2}\right)}{K_1\left(2\nu(1+\varepsilon)^{1/2}\right)} K_0\left(2\nu(1+\varepsilon x)^{1/2}\right)\right] & \xi < x \le 1 \end{cases} \tag{f}$$

The solution is continuous at $x = \xi$. Thus,

$$C_1 = C_3 H = C_3 \frac{\left[I_0\left(2\nu(1+\varepsilon\xi)^{1/2}\right) + \dfrac{I_1\left(2\nu(1+\varepsilon)^{1/2}\right)}{K_1\left(2\nu(1+\varepsilon)^{1/2}\right)} K_0\left(2\nu(1+\varepsilon\xi)^{1/2}\right)\right]}{\left[I_0\left(2\nu(1+\varepsilon\xi)^{1/2}\right) - \dfrac{I_0(2\nu)}{K_0(2\nu)} K_0\left(2\nu(1+\varepsilon\xi)^{1/2}\right)\right]} \tag{g}$$

The form of C_3 is obtained by requiring $\partial q/\partial x(\xi^+,\xi) - \partial q/\partial \xi(\xi^-,\xi) = 1/\beta(1+\varepsilon\xi)$. Subsequent algebra leads to

$$C_3 = \frac{(1+\varepsilon\xi)^{1/2}}{\beta\nu\left[I_1\left(2\nu(1+\varepsilon\xi)^{1/2}\right)(1-H) + K_1\left(2\nu(1+\varepsilon\xi)^{1/2}\right)(Q+HR)\right]} \tag{h}$$

where $Q = I_1(2\nu(1+\varepsilon)^{1/2})/K_1(2\nu(1+\varepsilon)^{1/2})$ and $R = I_0(2\nu)/K_0(2\nu)$. The solution of the differential equation is

$$\theta(x) = \int_0^1 q(x,\xi) f(\xi) d\xi \tag{i}$$

Green's functions can be used to convert differential eigenvalue problems to eigenvalue problems for integral equations. Consider again the transverse deflections of the simply supported beam shown in Figure 5.20. Suppose that the beam is vibrating with a transverse deflection $w(x, t)$. The transverse load is due to the inertia of the beam, $f(x) = \rho A(\partial^2 w/\partial t^2)$. Then, from Equation 5.202,

$$w(x,t) = \int_0^L \rho A \frac{\partial^2 w}{\partial t^2}(\xi,t) q(x,\xi) d\xi \tag{5.205}$$

where the Green's function $q(x, \xi)$ is given in Equation 5.199. Substitution of a normal-mode solution $w(x, t) = W(x)e^{i\omega t}$ into Equation 5.205 leads to

$$W(x) = -\omega^2 \int_0^L \rho A W(\xi) q(x,\xi) d\xi \tag{5.206}$$

Noting that the kinetic-energy inner product for the beam is defined as $(f,g)_M = \int_0^L \rho A f(x) g(x) dx$, Equation 5.206 can be rewritten as

$$W(x) = -\omega^2 (W(\xi), q(x,\xi))_M \tag{5.207}$$

Dividing Equation 5.207 by ω^2 leads to

$$GW = \frac{1}{\omega^2} W \tag{5.208}$$

Equation 5.208 implies that the natural frequencies are the reciprocals of the square roots of the Green's function operator and that the mode shapes $W(x)$ are the corresponding eigenvectors. Since the Green's function is symmetric, the operator is self-adjoint with respect to the kinetic-energy inner product. This implies that the mode shapes corresponding to distinct natural frequencies are mutually orthogonal with respect to the kinetic-energy inner product.

The integral-equation formulation of the eigenvalue problem for the natural frequencies is not easier to solve than the differential equation formulation.

Problems

5.1. The differential equations governing the free vibrations of the system shown in Figure P5.1 are

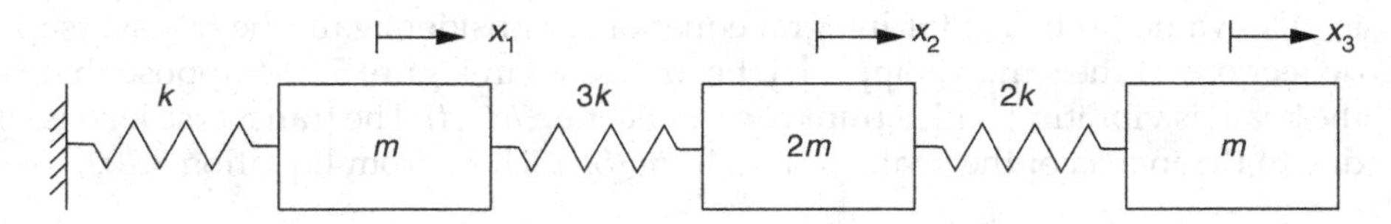

Figure P5.1 Three-degree-of-freedom system of Problems 5.1 and 5.2.

$$\begin{bmatrix} m & 0 & 0 \\ 0 & 2m & 0 \\ 0 & 0 & m \end{bmatrix}\begin{bmatrix} \ddot{x}_1 \\ \ddot{x}_2 \\ \ddot{x}_3 \end{bmatrix} + \begin{bmatrix} 4k & -3k & 0 \\ -3k & 5k & -2k \\ 0 & -2k & 2k \end{bmatrix}\begin{bmatrix} x_1 \\ x_2 \\ x_3 \end{bmatrix} = \begin{bmatrix} 0 \\ 0 \\ 0 \end{bmatrix} \tag{a}$$

Recall that the natural frequencies of a discrete system are the square roots of the eigenvalues of $\mathbf{M}^{-1}\mathbf{K}$ and that the mode shapes are the corresponding eigenvectors (Example 5.2). (a) Determine the natural frequencies for the system; (b) determine the normalized mode shapes for the system.

5.2. The initial conditions for the system of Figure P5.1 and Problem 5.1 are

$$\begin{bmatrix} x_1(0) \\ x_2(x) \\ x_3(x) \end{bmatrix} = \begin{bmatrix} 0 \\ 0 \\ \delta \end{bmatrix} \text{ and } \begin{bmatrix} \dot{x}_1(0) \\ \dot{x}_2(x) \\ \dot{x}_3(x) \end{bmatrix} = \begin{bmatrix} 0 \\ 0 \\ 0 \end{bmatrix} \tag{b}$$

Let ω_1, ω_2 and ω_3 be the natural frequencies of the system of Problem 5.1, and let $\mathbf{X}_1$, $\mathbf{X}_2$, and $\mathbf{X}_3$ be their corresponding normalized mode shapes. Let $\mathbf{x}$ represent the solution of Equation a of Problem 5.1 subject to the initial conditions of Equation b. At any time, $\mathbf{x}$ can be expanded in terms of the normalized mode shapes:

$$\mathbf{x}(t) = \alpha_1(t)\mathbf{X}_1 + \alpha_2(t)\mathbf{X}_2 + \alpha_3(t)\mathbf{X}_3 \tag{c}$$

a. Substitute Equation c into Equation a of Problem 5.1 and take the standard inner product of both sides with respect to $\mathbf{X}_j$ for an arbitrary j. Use the properties of mode-shape orthogonality to derive uncoupled differential equations for $\alpha_i(t)$.
b. Substitute Equation c into the initial conditions of Equation c to derive initial conditions for $\alpha_i(t)$.
c. Solve the differential equations to obtain $\alpha_i(t)$.
d. Determine $\mathbf{x}(t)$ using Equation c.

5.3. The differential equations governing the motion of the system shown in Figure P5.3 are

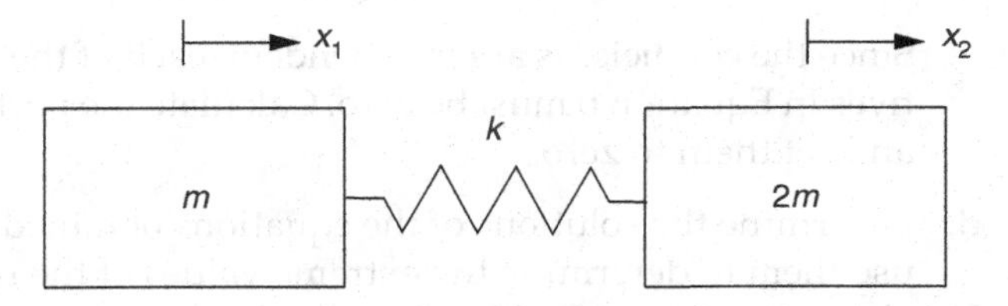

Figure P5.3 Two-degree-of-freedom unrestrained system of Problem 5.3.

$$\begin{bmatrix} m & 0 \\ 0 & 2m \end{bmatrix}\begin{bmatrix} \ddot{x}_1 \\ \ddot{x}_2 \end{bmatrix} + \begin{bmatrix} k & -k \\ -k & k \end{bmatrix}\begin{bmatrix} x_1 \\ x_2 \end{bmatrix} = \begin{bmatrix} 0 \\ 0 \end{bmatrix}$$

Determine the natural frequencies and normalized mode shapes for the system.

5.4. The stress tensor at a point in a solid is

$$\mathbf{S} = \begin{pmatrix} 200 & 150 & 100 \\ 150 & -200 & 300 \\ 100 & 300 & 100 \end{pmatrix} \text{kPa}$$

Determine the principal stresses and unit vectors normal to the places on which they act.

5.5. Recall that the stress vector acting on a plane whose unit normal is $\mathbf{n}$ can be calculated as $\mathbf{Sn}$. The normal stress acting on a plane, σ, is the component of the stress vector normal to the plane,

$$\sigma = (\mathbf{Sn}, \mathbf{n}) \tag{a}$$

where the inner product is the standard inner product (the dot product) on R^3. Prove that the maximum normal stress at a point is the largest principal stress and that the minimum normal stress is the smallest principal stress by following these steps. The expansion theorem implies that the unit normal can be written as a linear combination of the unit normals to the principal planes.

a. Assume a linear combination for the unit normal to a plane, $\mathbf{n} = \alpha_1\mathbf{n}_1 + \alpha_2\mathbf{n}_2 + \alpha_3\mathbf{n}_3$, noting that since the vector is a unit vector, the sum of the squares of the coefficients in the linear combination is one. Substitute the expansion into Equation a.
b. Use the properties of inner products and the orthonormality of the unit vectors normal to the principal planes to reduce the resulting equation.
c. The resulting expression is a function of the expansion coefficients. The normal stress is stationary if

$$d\sigma = 0 = \frac{\partial \sigma}{\partial \alpha_1} d\alpha_1 + \frac{\partial \sigma}{\partial \alpha_2} d\alpha_2 + \frac{\partial \sigma}{\partial \alpha_3} d\alpha_3 \quad \text{(b)}$$

Since the coefficients are independent, each of the partial derivatives in Equation b must be zero. Calculate the partial derivatives and set them to zero.

d. Determine the solutions of the equations obtained in part (c) and use them to determine the extreme values of the normal stress.

5.6. Using the notation of Problem 5.5, the shear stress vector acting on a plane whose unit normal is **n** is calculated as

$$\tau = \mathbf{Sn} - \sigma\mathbf{n} \quad \text{(c)}$$

The magnitude of the shear stress is

$$\|\tau\| = (\tau, \tau)^{\frac{1}{2}} = (\mathbf{Sn} - \sigma\mathbf{n}, \mathbf{Sn} - \sigma\mathbf{n})^{\frac{1}{2}} \quad \text{(d)}$$

Expand the unit normal in terms of the unit vectors to the principal planes.

a. Use the expansion in Equation d to prove that the maximum normal stress acts on a plane which makes an angle of $\pi/4$ to the plane of the largest normal stress.
b. Calculate the maximum shear stress and a unit vector normal to the plane on which it acts to obtain the state of stress described by the stress tensor of Equation a of Problem 5.4.

5.7. The stress vector defining the state of stress at a point is

$$\mathbf{S} = \begin{pmatrix} 100 & -50 & -100 \\ -50 & -100 & 200 \\ -100 & 200 & 200 \end{pmatrix} \text{kPa} \quad \text{(a)}$$

The principal stresses and unit vectors normal to the principal planes are

$$\sigma_1 = 350 \text{ kPa} \qquad \mathbf{n}_1 = \begin{bmatrix} -0.4082 \\ 0.4082 \\ 0.8165 \end{bmatrix}$$

$$\sigma_2 = 50 \text{ kPa} \qquad \mathbf{n}_2 = \begin{bmatrix} 0.9129 \\ 0.1826 \\ 03651 \end{bmatrix} \quad \text{(b)}$$

$$\sigma_2 = -200 \text{ kPa} \quad \mathbf{n}_3 = \begin{bmatrix} 0 \\ 0.8944 \\ -04472 \end{bmatrix}$$

A small change in loading leads to a stress vector of

$$\mathbf{S} = \begin{pmatrix} 100+10\varepsilon & -50+2\varepsilon & -100 \\ -50+2\varepsilon & -100+\varepsilon & 200 \\ -100 & 200 & 200 \end{pmatrix} \text{ kPa} \tag{c}$$

where ε is a small dimensionless parameter. The principal stresses for the new state of stress and the normal vectors can be expanded as

$$\hat{\sigma}_i = \sigma_i + \varepsilon\mu_i \tag{d}$$

$$\hat{\mathbf{n}}_\mathbf{i} = \mathbf{n}_\mathbf{i} + \varepsilon\mathbf{m}_\mathbf{i} \tag{e}$$

Determine the perturbations in principal stresses, μ_i, i = 1, 2, 3.

5.8. The differential equations governing the motion of the system shown in Figure P5.8 are

$$\begin{bmatrix} m & 0 & 0 & 0 \\ 0 & 2m & 0 & 0 \\ 0 & 0 & m & 0 \\ 0 & 0 & 0 & m \end{bmatrix} \begin{bmatrix} \ddot{x}_1 \\ \ddot{x}_2 \\ \ddot{x}_3 \\ \ddot{x}_4 \end{bmatrix} + \begin{bmatrix} 3k & -k & 0 & 0 \\ -k & 2k & -k & 0 \\ 0 & -k & 3k & -2k \\ 0 & 0 & -2k & 2k \end{bmatrix} \begin{bmatrix} x_1 \\ x_2 \\ x_3 \\ x_4 \end{bmatrix} = \begin{bmatrix} 0 \\ 0 \\ 0 \\ 0 \end{bmatrix} \tag{a}$$

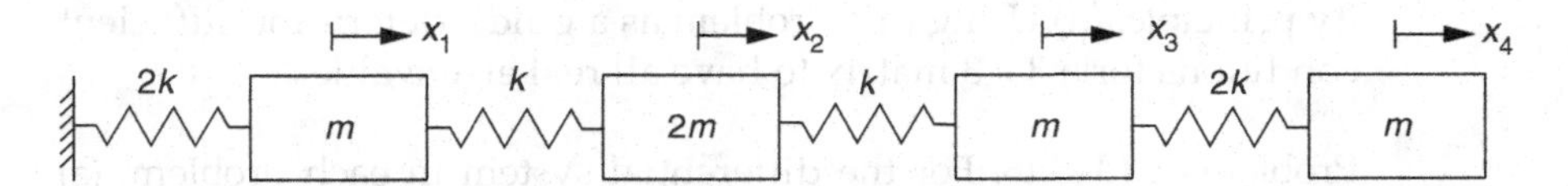

Figure P5.8 Four-degree-of-freedom of Problem 5.8.

The lowest natural frequency for the system is $0.4344\sqrt{k/m}$, which has a corresponding normalized eigenvector of $\mathbf{X} = \sqrt{1/m}$. [0.1397 0.4005 0.5541 0.5938].

a. Determine the perturbation in the lowest natural frequency if the particle of mass $2m$ is adjusted to $2m + \varepsilon m$, where ε is a small independent dimensionless parameter.
b. Determine the perturbation in the lowest natural frequency if the stiffness of the spring connecting the leftmost particle of mass m with the particle of mass $2m$ changes to a stiffness of $k - \varepsilon k$, where ε is a small dimensionless parameter.

5.9. a. Determine all eigenvalues and eigenvectors for the matrix

$$\mathbf{A} = \begin{bmatrix} 3 & -1 & 0 \\ -1 & 2 & -1 \\ 0 & -1 & 2 \end{bmatrix}$$

b. Normalize the eigenvectors and then demonstrate orthonormality of eigenvectors.

5.10. a. Determine all eigenvalues and eigenvectors of the matrix $\mathbf{A}=\begin{bmatrix}1 & 6\\ 3 & 4\end{bmatrix}$.

b. Determine all eigenvalues and eigenvectors of $\mathbf{A}^T$.

c. Demonstrate the principle of biorthogonality of the eigenvectors of $\mathbf{A}$ and $\mathbf{A}^T$.

5.11. A valid inner product defined for R^2 is $(\mathbf{x}, \mathbf{y})=x_1y_1+2x_2y_2$. (a) Show that the matrix $\mathbf{A}$ in Problem 5.10 is self-adjoint with respect to this inner product. (b) Demonstrate orthogonality of the eigenvectors of $\mathbf{A}$. (c) Since $\mathbf{A}$ is self-adjoint with respect to a valid inner product, its eigenvalues are all real. Use this idea to deduce sufficient conditions for ensuring that a 2×2 matrix has all real eigenvalues. Hint: Is there any inner product for which $\begin{bmatrix}1 & 6\\ -3 & 4\end{bmatrix}$ self-adjoint?

5.12. A valid inner product defined for R^3 is $(\mathbf{x}, \mathbf{y})=x_1y_1+x_2y_2+3x_3y_3$. (a) Give an example of a 3×3 matrix $\mathbf{A}$ which is self-adjoint with respect to this inner product. (b) Determine the eigenvalues and eigenvectors of this matrix and demonstrate orthogonality of the eigenvectors. (c) Determine the eigenvectors of $\mathbf{A}^T$. Note that $\mathbf{A}^T=\mathbf{A}^*$ with respect to the standard inner product for R^3. Demonstrate the biorthogonality principle. (d) Using this problem as a guide, determine sufficient conditions for a 3×3 matrix to have all real eigenvalues.

Problems 5.13–5.16. For the differential system in each problem, (a) specify an inner product for which the system is self-adjoint; (b) determine all eigenvalues and eigenvectors for the system; (c) normalize the eigenvectors; and (d) demonstrate orthogonality of the eigenvectors.

5.13. $\dfrac{d^2y}{dx^2}+\lambda y=0$

$$\frac{dy}{dx}(0)=0$$

$$y(1)+2y(1)=0\,.$$

5.14. $\dfrac{d^2y}{dx^2}+6\dfrac{dy}{dx}+\lambda y=0$

$$y(0)=0$$

$$\frac{dy}{dx}(1)=0\,.$$

5.15. $x^2 \frac{d^2y}{dx^2} + 4x\frac{dy}{dx} + \lambda y = 0$

$y(1) = 0$

$\frac{dy}{dx}(2) = 0.$

5.16. $\frac{d^2y}{dx^2} + \lambda y = 0$

$\frac{dy}{dx}(0) = 0$

$\frac{dy}{dx}(1) = 0$

5.17. $x^3 \frac{d^2y}{dx^2} + 4x^2 \frac{dy}{dx} + \lambda y = 0.$

Problems 5.18–5.23. refer to the following problem: The differential equation for determination of the natural frequencies and mode shapes for the longitudinal motion of a nonuniform bar is

$$\frac{d}{dx}\left[\alpha(x)\frac{dy}{dx}\right] + \omega^2\alpha(x)y = 0 \tag{a}$$

5.18. Find the first five natural frequencies and normalized mode shapes for a uniform bar, $\alpha(x) = 1$, which is fixed at both ends, $y(0) = 0$ and $y(1) = 0$.
5.19. Find the first five natural frequencies and normalized mode shapes for a uniform bar, $\alpha(x) = 1$, which is fixed at $x = 0$ and has a linear spring attached at $x = 1$ such that the boundary conditions are $y(0) = 0$ and $dy/dx(1) + 0.25y(1) = 0$.
5.20. Find the first five natural frequencies and normalized mode shapes for a uniform bar, $\alpha(x) = 1$, which is fixed at $x = 0$ and has a discrete particle and a linear spring attached at $x = 1$ such that the boundary conditions are $y(0) = 0$ and $dy/dx(1) + 0.25y(1) = 0.5\omega^2 y(1)$.
5.21. Find the first five natural frequencies and normalized mode shapes for a nonuniform bar with $\alpha(x) = 1 + 0.5x$ which is fixed at both ends, $y(0) = 0$ and $y(1) = 0$.
5.22. Find the first five natural frequencies and normalized mode shapes for a nonuniform bar with $\alpha(x) = (1\text{–}0.5x)^2$ which is fixed at $x = 0$ and free at $x = 1$ such that $y(0) = 0$ and $dy/dx(1) = 0$.

5.23. Find the first five natural frequencies and normalized mode shapes for a nonuniform bar with $\alpha(x)=(1-0.5x)^2$ which is fixed at $x=0$ and has a discrete particle and a linear spring attached at $x=1$ such that the boundary conditions are $y(0)=0$ and $\alpha(1)dy/dx(1)+0.25y(1)=0.5\omega^2y(1)$.

Problems 5.24–5.29 refer to the following general problem: the natural frequencies ω and mode shapes $w(x)$ of a uniform stretched beam on an elastic foundation are governed by the differential equation

$$\frac{d^4w}{dx^4}-\varepsilon\frac{d^2w}{dx^2}+\eta w-\omega^2 w=0 \tag{a}$$

5.24. Determine the first five natural frequencies and mode shapes for a beam with $\varepsilon=0$ and $\eta=1$ which is pinned at $x=0$ and $x=1$ such that

$$y(0)=0,\frac{d^2y}{dx^2}(0)=0,y(1)=0,\frac{d^2y}{dx^2}(1)=0.$$

5.25. Determine the first five natural frequencies and mode shapes for a beam with $\varepsilon=2$ and $\eta=0$ which is pinned at $x=0$ and $x=1$ such that

$$y(0)=0,\frac{d^2y}{dx^2}(0)=0,y(1)=0,\frac{d^2y}{dx^2}(1)=0.$$

5.26. Determine the first five natural frequencies and mode shapes for a beam with $\varepsilon=2$ and $\eta=1$ which is pinned at $x=0$ and fixed at $x=1$ such that

$$y(0)=0,\frac{d^2y}{dx^2}(0)=0,y(1)=0,\frac{dy}{dx}(1)=0.$$

5.27. Determine the first five natural frequencies and mode shapes for a beam with $\varepsilon=0$ and $\eta=0$ which is fixed at $x=0$ and has a linear spring attached at $x=1$ such that

$$y(0)=0,\frac{dy}{dx}(0)=0,\frac{d^2y}{dx^2}(1)=0,\frac{d^3y}{dx^3}(1)-0.5y(1)=0.$$

5.28. Determine the first five natural frequencies and mode shapes for a beam with $\varepsilon=0.5$ and $\eta=1$ which is fixed at $x=0$ and free at $x=1$ such that

$$y(0)=0,\frac{dy}{dx}(0)=0,\frac{d^2y}{dx^2}(1)=0,\frac{d^3y}{dx^3}(1)+0.5\frac{dy}{dx}(1)=0.$$

5.29. Determine the first five natural frequencies and mode shapes for a beam with $\varepsilon=0$ and $\eta=1$ which is free at $x=0$ and has an attached mass at $x=1$ such that

$$y(0)=0, \frac{dy}{dx}(0)=0, \frac{d^2y}{dx^2}(1)=0, \frac{d^3y}{dx^3}(1)=0.5\omega^2 y(1)=0.$$

5.30. One end of the nonuniform bar shown in Figure P5.30 is fixed, while its other end is attached to a spring whose other end is attached to a particle. The differential equations governing the displacement of the bar, $u(x,t)$, and the displacement of the particle, $y(t)$, are

$$\frac{\partial}{\partial x}\left(EA\frac{\partial u}{\partial x}\right)+\rho A u(x)=0 \quad \text{(a)}$$

$$u(0,t)=0 \quad \text{(b)}$$

$$EA(L)\frac{\partial u}{\partial x}(L,t)-k[y(t)-u(L,t)]=0 \quad \text{(c)}$$

$$m\frac{d^2y}{dt^2}+k[y(t)-u(L,t)]=0 \quad \text{(d)}$$

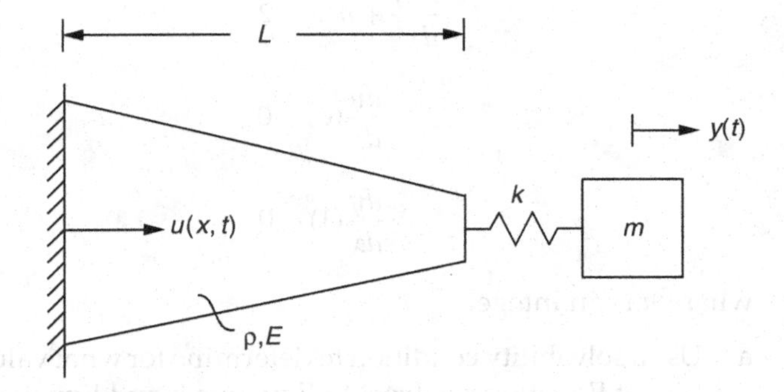

Figure P5.30 System for Problem 5.30.

a. Nondimensionalize the problem through introduction of appropriate nondimensional variables.
b. Introduce the normal-mode solution,

$$u(x,t)=U(x)e^{i\omega t} \quad \text{(e)}$$

$$y(t)=Ye^{i\omega t} \quad \text{(f)}$$

into the nondimensional equation and determine an eigenvalue-eigenvector problem whose solution leads to the natural frequencies and mode shapes.

c. Define the vector space V as the set of vectors of the form $g = \begin{bmatrix} f(x) \\ a \end{bmatrix}$, where $f(x)$ is in the subspace of $C^2[0,1]$ of all vectors $f(x)$ such that $f(0)=0$ and $du/dx(1)+\eta u(1)=0$, where η is an appropriate nondimensional parameter. Define an operator L whose domain is V by

$$\mathbf{L}\begin{bmatrix} f(x) \\ a \end{bmatrix} = \begin{bmatrix} -\frac{1}{\alpha(x)}\frac{d}{dx}\left(\alpha\frac{df}{dx}\right) \\ \mu\alpha f(1)-\nu \end{bmatrix} \tag{g}$$

where μ and ν are appropriate nondimensional constants. Show that the eigenvalue problem is of the form

$$\mathbf{Lg}=\omega^2\mathbf{g} \tag{h}$$

d. Determine an appropriate inner product for which L is self-adjoint.
e. Show that L is positive definite with respect to this inner product.
f. Determine the first five natural frequencies and mode shapes for a uniform bar, assuming that all nondimensional constants are one.
g. Repeat part (f) if $\alpha(x)=(1–0.1x)^3$.

5.31. Consider the boundary-value problem

$$\frac{d^2y}{dx^2}+n^2y=2x \tag{a}$$

$$\frac{dy}{dx}(0)=0 \tag{b}$$

$$\frac{dy}{dx}(1)=0 \tag{c}$$

where n is an integer.

a. Use a solvability condition to determine for what values of n a solution of Equation a subject to Equation b and Equation c exists.
b. Solve Equation a subject to Equation b and Equation c to confirm the results of part (a).
c. Determine $y(x)$ when a solution exists.

5.32. Show the algebraic steps necessary to obtain Equation 5.56 from Equation 5.55.

5.33. Use the modal-analysis method derived in Example 5.9 to determine the response of a two-degree-of-freedom mechanical system whose motion is described by the differential equation

$$\begin{bmatrix} m & 0 \\ 0 & 2m \end{bmatrix}\begin{bmatrix} \ddot{x}_1 \\ \ddot{x}_2 \end{bmatrix}+\begin{bmatrix} 2k & -k \\ -k & k \end{bmatrix}\begin{bmatrix} x_1 \\ x_2 \end{bmatrix}=\begin{bmatrix} 0 \\ F_0\sin(\omega t) \end{bmatrix}$$

and with all initial conditions equal to zero.

5.34. Determine the constraints on the constants a_1, b_1, a_2, b_2, α_1, β_1, γ_1, α_2, β_2 and γ_2 as defined in Equation 5.71 such that the differential operator of Equation 5.66 is positive definite with respect to the inner product defined in Equation 5.67.

Problems 5.35–5.44 refer to the problem of finding the eigenvalues and eigenvectors of the system

$$-\frac{d}{dx}\left(x^3\frac{dy}{dx}\right) = \lambda xy \qquad \text{(a)}$$

$$y(0) = 0 \qquad \text{(b)}$$

$$y(1) + 2y'(1) = 0 \qquad \text{(c)}$$

5.35. Using the exact solution, determine the five lowest eigenvalues and their corresponding eigenvectors.
5.36. Choose two polynomials of order three or less which satisfy the boundary conditions. Use Rayleigh's quotient to approximate the lowest eigenvalue.
5.37. Choose the elements of a basis for the subspace of $C^2[0,1]$ defined by the intersection of $P^4[0,1]$ with the set of all functions satisfying boundary conditions (b) and (c). Use these functions as a basis for a Rayleigh-Ritz approximation of the lowest eigenvalues of the system.
5.38. Choose the elements of a basis for the subspace of $C^2[0,1]$ defined by the intersection of $P^4[0,1]$ with the set of all functions satisfying boundary condition (b) (the geometric boundary condition). Use these functions as a basis for a Rayleigh-Ritz approximation of the lowest eigenvalues of the system.
5.39. Use a finite-element method to approximate the lowest eigenvalues and eigenvectors for the system. Use five equally spaced elements.

Problems 5.40–5.57 refer to the system of Problems 5.35–5.44, but with boundary condition (c) replaced by

$$y(1) + 2y'(1) = 0.5\lambda y(1) \qquad \text{(d)}$$

5.40. Develop the appropriate form of the inner product for orthogonality of the eigenvectors.
5.41. Use the exact solution to determine the five lowest eigenvectors.
5.42. Choose two polynomials of order three or less which satisfy the boundary conditions. Use Rayleigh's quotient to approximate the lowest eigenvalue.
5.43. Choose the elements of a basis for the subspace of $C^2[0,1]$ defined by the intersection of $P^4[0,1]$ with the set of all functions satisfying

boundary condition (b) (the geometric boundary condition). Use these functions as a basis for a Rayleigh-Ritz approximation of the lowest eigenvalues of the system.

5.44. Use the finite-element method to approximate the natural frequencies and mode shapes. Use five equally spaced elements.

5.45. Consider the eigenvalue problem,

$$(1-x^2)\frac{d^2y}{dx^2}-2x\frac{dy}{dx}+30y=\frac{\lambda}{1-x^2}y$$

which is to be solved on the interval $-1\le x\le 1$. (a) Determine the self-adjoint form of the differential operator. (b) What boundary conditions must be specified at $x=-1$ and $x=1$? (c) Determine the inner product for eigenvector orthogonality. (d) Determine the eigenvalues and eigenvectors for this problem (Hint: $30=(5)(6)$).

5.46. Prove that, if subject to appropriate boundary conditions, a sixth-order differential operator of the form

$$Ly=\frac{d^3}{dx^3}\left(f(x)\frac{d^3y}{dx^3}\right)+\frac{d^2}{dx^2}\left[g(x)\frac{d^2y}{dx^2}\right]+\frac{d}{dx}\left(h(x)\frac{dy}{dx}\right)+q(x)y$$

is self-adjoint with respect to the standard inner product on $C^6[0,1]$.

5.47. Under what conditions on $f(x)$, $g(x)$, $h(x)$, and $q(x)$ is the sixth-order operator of Problem 5.46 positive definite with respect to the standard inner product on $C^6[0,1]$?

5.48. Determine the exact value of the critical buckling length for the column of Example 5.22.

5.49. Use the Rayleigh-Ritz method to approximate the critical buckling length of a fixed-free column. Use a basis of polynomials of order six or less, which satisfy the boundary conditions.

5.50. Use the Rayleigh-Ritz method to approximate the critical rotational speed of a fixed-free column. Use a basis of polynomials of order six or less, which satisfy the boundary conditions.

5.51. Use the finite-element method to approximate the critical buckling length of a fixed-free column. Use five equally spaced elements.

Chapter 6

Partial differential equations

6.1 Homogeneous partial differential equations

A partial differential equation is a differential equation with more than one independent variable. Although the independent variables can represent a variety of coordinates and scales, this study focuses on partial differential equations in which the independent variables are spatial coordinates and time.

Consider a problem in which it is desired to determine the variation of a dependent variable, say ϕ, in a region of space **R** which is bounded by a surface **S**. Let **r** be a position vector from the origin of the coordinate system to a particle or point in **R**. Most generally, **r** is a function of three independent spatial coordinates. The dependent variable may also be a function of time and designated as $\phi(\mathbf{r},t)$. The surface **S** is described by $g(\mathbf{r},t)=0$. The problem is an interior problem if g is bounded and the solution is to be obtained for position vectors defined within the interior of g, as illustrated in Figure 6.1a. The problem is an exterior problem if the solution is to be obtained for position vectors defined outside of g, as illustrated in Figure 6.1b.

The general form of a linear partial differential equation is

$$\mathbf{L}\phi + \mathbf{M}\phi + \mathbf{G}\phi = f(\mathbf{r},t) \tag{6.1}$$

where **L** is a linear operator involving derivatives with respect to spatial variables, **M** is a linear operator involving derivatives with respect to t, and **G** is a linear operator involving mixed partial derivatives involving time and spatial variables.

Determination of a solution of Equation 6.1 requires application of appropriate initial conditions and boundary conditions. The number of initial conditions required depends on the order of **M**. If **M** is first-order, then it is necessary to specify $\phi(\mathbf{r},0)$. If **M** is second-order it, is also necessary to specify $\partial\phi/\partial t(\mathbf{r},0)$. The number and type of boundary conditions depend on the order and form of **L**. The boundary conditions are prescribed on **S**. For interior problems, it is often necessary to specify that the dependent variable is bounded and single-valued at all points within $g(\mathbf{r},t)$. For exterior problems, it is also required that the dependent variable be single-valued, but conditions as $||\mathbf{r}|| \to \infty$ must be specified. Such condition, called radiation conditions, place restrictions on ϕ far away from $g(\mathbf{r},t)$. In some

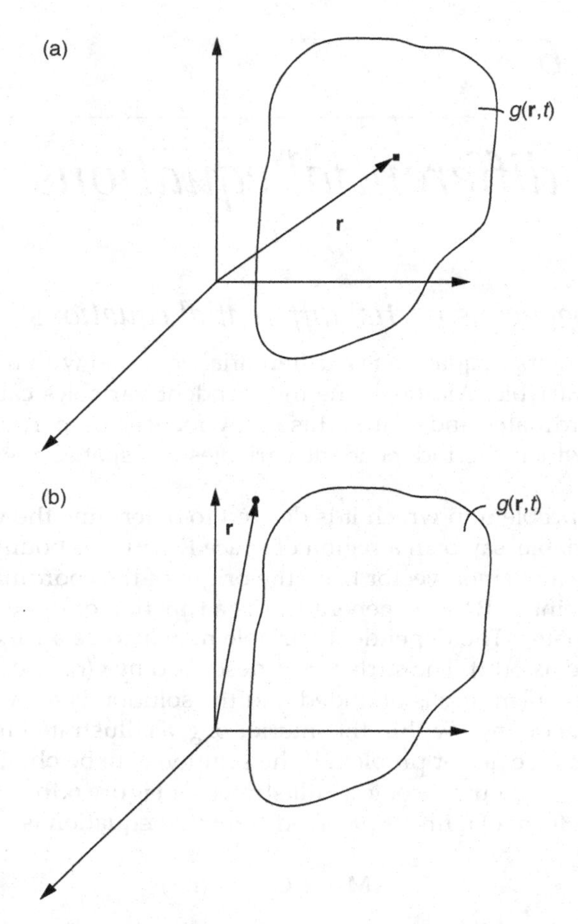

Figure 6.1 (a) An interior problem is one in which the solution is determined for every position vector **r** everywhere within the volume bounded by $g(\mathbf{r},t)$. (b) An exterior problem is one in which the solution is determined for position vectors **r** outside the closed surface defined by $g(\mathbf{r},t)$.

cases, the radiation condition is simply that ϕ approaches zero as $||\mathbf{r}|| \rightarrow \infty$. Solutions of the wave equation in exterior regions where reflection cannot occur must satisfy a condition that all waves are propagating away from the body.

Second-order linear partial differential equations are classified based on their form. The classification of a differential equation with variable coefficients may change with time or spatial position. The equations considered in this study maintain their classification for all **r** and t. Homogeneous second-order problems considered in this study are of the form of Equation 6.1 with $\mathbf{G} = 0$ and $\mathbf{L}\phi = -\nabla^2\phi$.

- The partial differential equation is elliptic if $\mathbf{M}=0$ and $\phi=\phi(\mathbf{r})$ and is independent of t. The resulting equation,

$$\nabla^2\phi = 0 \tag{6.2}$$

is called Laplace's equation. Laplace's equation governs problems such as steady-state heat transfer and potential flow of incompressible and inviscid fluids.

- The partial differential equation is parabolic if $\mathbf{M}\phi = p(\mathbf{r})\partial\phi/\partial t$. The resulting equation,

$$-\nabla^2\phi = p(\mathbf{r})\frac{\partial\phi}{\partial t} \tag{6.3}$$

is called the diffusion equation. The diffusion equation governs unsteady state conductive and convective heat and mass transfer problems.

- The partial differential equation is hyperbolic if $\mathbf{M}\phi = p(\mathbf{r})\partial^2\phi/\partial t^2$. The resulting equation,

$$\nabla^2\phi = p(\mathbf{r})\frac{\partial^2\phi}{\partial t^2} \tag{6.4}$$

is called the wave equation and governs propagation of waves in solids and fluids.

A unsteady state problem is homogeneous if $f(\mathbf{r},t)=0$ and all boundary conditions are homogeneous. A steady-state problem is homogeneous if $f(\mathbf{r})=0$ and if nonhomogeneous boundary conditions occur for only one independent variable. A method called separation of variables is applied in subsequent sections to solve homogeneous partial differential equations in bounded domains. Solutions of nonhomogeneous partial differential equations are approached using superposition of solutions and eigenvector expansion methods. Solutions are considered in Cartesian, cylindrical, and spherical coordinates.

Solutions in unbounded domains are usually more difficult to obtain than in interior regions. A problem in an exterior region is considered at the end of this chapter.

6.2 *Second-order steady-state problems, Laplace's equation*

Consider a partial differential equation derived from a mathematical model of a steady-state system; the response of the system is independent of time. The dependent variables are spatial coordinates defined by the position vector **r**, measured from the origin of a coordinate system. The dependent variable is $\phi(\mathbf{r})$. Assume that the partial differential equation governing $\phi(\mathbf{r})$ is of the form

$$\mathbf{L}\phi = 0 \tag{6.5}$$

where **L** is a linear operator. It is desired to determine the solution of Equation 6.5 in a region in space **R** which is bounded by a closed surface described by $g(\mathbf{r}) = 0$.

Determination of a solution of Equation 6.5 requires application of appropriate boundary conditions. Without loss of generality, assume $\mathbf{r} = x\mathbf{i} + y\mathbf{j} + z\mathbf{k}$ such that the solution of Equation 6.5 is expressed using Cartesian coordinates. The boundary of the region is described by $0 \le x \le a$, $0 \le y \le b$ and $0 \le z \le c$, as illustrated in Figure 6.2. It is desired to determine $\phi(x,y,z)$ everywhere in the parallelepiped. Nonhomogeneous boundary conditions are prescribed on boundary planes in one variable, say z. The number of boundary conditions required depends on the order of the operator **L**. Nonhomogeneous boundary conditions applied to other surfaces lead to a nonhomogeneous problem which is solved using a superposition method as described in Section 6.4. Solution of Equation 6.5 when **L** is a second-order differential operator requires one boundary condition defined on each plane that is part of the boundary of **R**. The boundary conditions on each plane are a linear

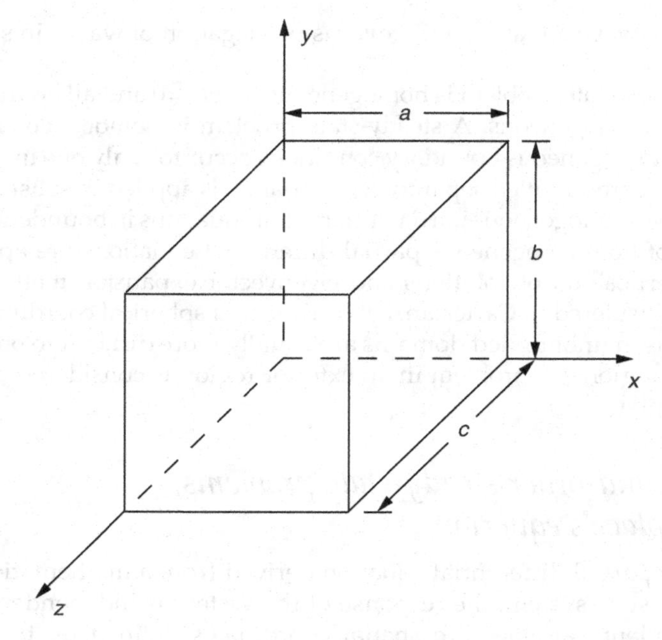

Figure 6.2 Laplace's equation is solved inside the cubic region.

combination of ϕ and $\partial\phi/\partial n$, where $\mathbf{n}$ is the unit normal to the boundary. In general,

$$\frac{\partial \phi}{\partial n} = (\nabla\phi)\cdot\mathbf{n} = \frac{(\nabla\phi)\cdot(\nabla g)}{|\nabla g|} \tag{6.6}$$

The nonhomogeneous boundary conditions for the problem under discussion are of the form

$$\phi(x,y,0) - \alpha_z \frac{\partial \phi}{\partial z}(x,y,0) = f(x,y) \tag{6.7}$$

$$\phi(x,y,c) + \beta_z \frac{\partial \phi}{\partial z}(x,y,c) = g(x,y) \tag{6.8}$$

Homogeneous boundary conditions are of the form

$$\phi(0,y,z) - \alpha_x \frac{\partial \phi}{\partial x}(0,y,z) = 0 \tag{6.9}$$

$$\phi(a,y,z) + \beta_x \frac{\partial \phi}{\partial x}(a,y,z) = 0 \tag{6.10}$$

$$\phi(x,0,z) - \alpha_y \frac{\partial \phi}{\partial y}(x,0,z) = 0 \tag{6.11}$$

$$\phi(x,b,z) + \beta_y \frac{\partial \phi}{\partial y}(x,b,z) = 0 \tag{6.12}$$

The method of separation of variables assumes the solution can be expressed as a product of the form

$$\phi(x,y,z) = \Phi(x,y)Z(z) \tag{6.13}$$

The operator $\mathbf{L}$ is said to be separable if

$$\mathbf{L}[\Phi(x,y)Z(z)] = Z(z)r(z)\mathbf{L}_{xy}[\Phi(x,y)] + \Phi(x,y)q(x,y)\mathbf{L}_z[Z(z)] \tag{6.14}$$

where $\mathbf{L}_{xy}$ is a partial differential operator whose domain is defined by the boundary conditions specified below, and $\mathbf{L}_z$ is an ordinary differential operator defined for $0 \le z \le c$. If $\mathbf{L}$ is separable, then substitution of Equation 6.13 into Equation 6.5 leads to

$$\frac{1}{\Phi(x,y)r(x,y)}\mathbf{L}_{xy}[\Phi(x,y)] = -\frac{1}{Z(z)q(z)}\mathbf{L}_{z}[Z(z)] \tag{6.15}$$

The left-hand side of Equation 6.15 is a function of x and y, but not of z. The right-hand side of Equation 6.15 is a function of z only. However, x, y, and z are independent variables. Both sides of Equation 6.15 can be equal for all values of x, y, and z if and only if each side is equal to the same constant; call it λ. Thus Equation 6.15 separates to

$$\mathbf{L}_{xy}[\Phi(x,y)] = \lambda r(x,y)\Phi(x,y) \tag{6.16}$$

$$\mathbf{L}_{z}[Z(z)] = -\lambda q(z)Z(z) \tag{6.17}$$

The argument used to separate Equation 6.15 into Equation 6.16 and Equation 6.17 is called the separation argument.

Substitution of Equation 6.13 into the boundary conditions, Equation 6.9, gives

$$\begin{gathered}\Phi(0,y)Z(z) - \alpha_x\frac{\partial\Phi}{\partial x}(0,y)Z(z) = 0\\ \left[\Phi(0,y) - \alpha_x\frac{\partial\Phi}{\partial x}(0,y)\right]Z(z) = 0\end{gathered} \tag{6.18}$$

Equation 6.18 must be satisfied for all z; thus

$$\Phi(0,y) - \alpha_x\frac{\partial\Phi}{\partial x}(0,y) = 0 \tag{6.19}$$

Substitution of Equation 6.13 into the remaining boundary conditions, Equation 6.10, Equation 6.11, and Equation 6.12, leads to

$$\Phi(a,y) + \beta_x\frac{\partial\Phi}{\partial x}(a,y) = 0 \tag{6.20}$$

$$\Phi(x,0) - \alpha_y\frac{\partial\Phi}{\partial y}(x,0) = 0 \tag{6.21}$$

$$\Phi(x,b) + \beta_y\frac{\partial\Phi}{\partial y}(x,b) = 0 \tag{6.22}$$

Equation 6.16, Equation 6.19, Equation 6.20, Equation 6.21, and Equation 6.22 constitute a boundary-value problem to solve for $\Phi(x,y)$. The system is homogeneous, and therefore a nontrivial solution exists only for certain values of λ. Thus Equation 6.16 and Equation 6.19, Equation 6.20, Equation 6.21, and Equation 6.22 constitute an eigenvalue problem.

Consider first the case in which $\mathbf{L}_{xy}$ is dependent on only one independent variable, say x, in which case $\mathbf{L}_x$ is an ordinary differential operator. If it is second-order, it can be written in a self-adjoint form. Let S_x be the domain of $\mathbf{L}_x$ specified by the boundary conditions, Equation 6.19 and Equation 6.20. Then $\mathbf{L}_x$ is self-adjoint with respect to the inner product $(f,g)_r = \int_0^a f(x)g(x)r(x)dx$ on S_x, where $r(x)$ is determined by writing the operator in its self-adjoint form. The system has an infinite, but countable, number of eigenvalues, $\lambda_1 \leq \lambda_2 \leq \lambda_3 \leq \ldots \leq \lambda_{k-1} \leq \lambda_k \leq \lambda_{k+1} \leq \ldots$, with an eigenvector corresponding to λ_k of $\Phi_k(x)$ for each k. The eigenvectors satisfy an orthogonality condition, $(\Phi_i(x), \Phi_j(x))_r = 0$ for $i \neq j$. The eigenvectors can be normalized by requiring that $(\Phi_i(x), \Phi_i(x))_r = 1$.

Let $Z_{k,1}(z)$ and $Z_{k,2}$ be two linearly independent solutions of Equation 6.16, which is assumed to be of second order, corresponding to $\lambda = \lambda_k$. Define

$$\phi_k(x,z) = (A_k Z_{k,1}(z) + B_k Z_{k,2}(z))\Phi_k(x) \tag{6.23}$$

Equation 6.23 is a solution of Equation 6.5 subject to the boundary conditions of Equation 6.9 and Equation 6.10. However, $\phi_k(x,y)$ does not satisfy the boundary conditions of Equation 6.7 and Equation 6.8. The most general solution of a linear homogenous system is a linear combination of all possible solutions. Thus the general solution of Equation 6.5, Equation 6.9, and Equation 6.10 is

$$\phi(x,z) = \sum_{k=1}^{\infty} (A_k Z_{k,1}(z) + B_k Z_{k,2}(z))\phi_k(x) \tag{6.24}$$

Equation 6.24 can also be viewed as an expansion for $\phi(x,z)$ in terms of the normalized eigenvectors of Equation 6.16, Equation 6.19, and Equation 6.20. Such an expansion is appropriate because for any z, $0 \leq z \leq c$, $\phi(x,z)$ is in S_x, and its eigenvector expansion is of the form $\phi(x,z) = \sum_{k=1}^{\infty} q_k(z)\phi_k(x)$. However, Equation 6.16 suggests that $q_k(z) = A_k Z_{k,1}(z) + B_k Z_{k,2}(z)$.

If the mathematical modeling of the problem is accurate, then $f(x)$ and $g(x)$ must satisfy boundary conditions 6.9 and 6.10. Thus $f(x)$ and $g(x)$ should be in S_x, in which case they each have an eigenvector expansion which converges pointwise to the functions. Thus

$$f(x) = \sum_{k=1}^{\infty} u_k \Phi_k(x) \tag{6.25}$$

$$g(x) = \sum_{k=1}^{\infty} v_k \Phi_k(x) \tag{6.26}$$

where

$$u_k = (f(x), \Phi_k(x))_r \tag{6.27}$$

$$v_k = (f(x), \Phi_k(x))_r \tag{6.28}$$

Substitution of Equation 6.24 into Equation 6.5, using the eigenvector expansion of Equation 6.25, leads to

$$\sum_{k=1}^{\infty}(A_k Z_{k,1}(0) + B_k Z_{k,2}(0))\Phi_k(x) - \alpha_z \sum_{k=1}^{\infty}(A_k Z'_{k,1}(0) + B_k Z'_{k,2}(0))\Phi_k(x)$$

$$= \sum_{k=1}^{\infty} u_k \Phi_k(x)$$

$$\sum_{k=1}^{\infty}\{A_k[Z_{k,1}(0) - \alpha_z Z'_{k,1}(0)] + B_k[Z_{k,2}(0) - \alpha_z Z'_{k,2}(0)] - u_k\}\Phi_k(x) = 0 \tag{6.29}$$

The eigenvectors are linearly independent, and therefore each coefficient in the linear combination of Equation 6.29 is identically zero, leading to

$$A_k[Z_{k,1}(0) - \alpha_z Z'_{k,1}(0)] + B_k[Z_{k,2}(0) - \alpha_z Z'_{k,2}(0)] = u_k \tag{6.30}$$

Similar substitution of Equation 6.24 into Equation 6.5 using Equation 6.28 leads to

$$A_k[Z_{k,1}(c) + \beta_z Z'_{k,1}(c)] + B_k[Z_{k,2}(c) + \beta_z Z'_{k,2}(c)] = v_k \tag{6.31}$$

Equation 6.30 and Equation 6.31 can be solved simultaneously for A_k and B_k.

The procedure described above is called separation of variables. The method is directly applicable when the partial differential operator is separable and when nonhomogeneous boundary conditions exist only on a portion of the boundary described by coordinate curves of one variable.

The following problem illustrates the application of the method of separation of variables to a two-dimensional problem governed by Laplace's equation.

Example 6.1 The thin rectangular slab, illustrated in Figure 6.3 is constrained to have a constant temperature T_1 on three sides and a different constant temperature T_2 on its fourth side. Determine the steady-state temperature distribution in the slab.

Solution The nondimensional partial differential equation governing the temperature distribution in the slab is

$$\frac{\partial^2 \Theta}{\partial x^2} + \frac{\partial^2 \Theta}{\partial y^2} = 0 \tag{a}$$

where

$$x^* = \frac{x}{a} \tag{b}$$

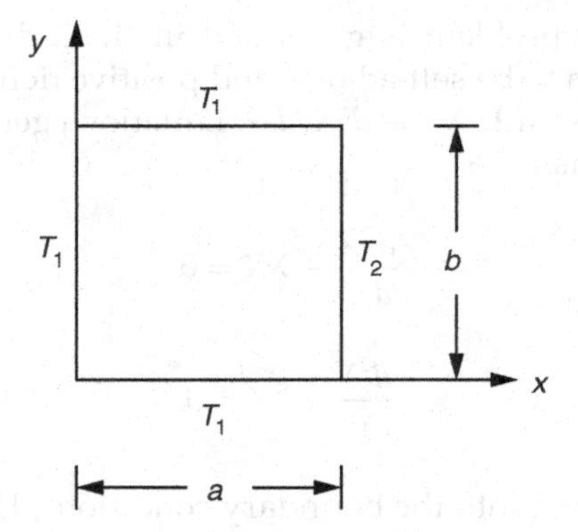

Figure 6.3 The temperature of three sides of the thin rectangular slab of Example 6.1 is maintained at T_1 while the constant temperature on the fourth side is T_2. A steady-state temperature distribution is assumed.

$$y^* = \frac{y}{a} \tag{c}$$

$$\Theta(x,y) = \frac{T - T_1}{T_2 - T_1} \tag{d}$$

Note that *'s have been dropped from nondimensional variables in Equation a. The nondimensional boundary conditions are

$$\Theta(0,y) = 0 \tag{e}$$

$$\Theta(1,y) = 0 \tag{f}$$

$$\Theta(x,0) = 0 \tag{g}$$

$$\Theta\left(x, \frac{b}{a}\right) = 1 \tag{h}$$

A product solution of Equation a is assumed to be of the form

$$\Theta(x,y) = X(x)Y(y) \tag{i}$$

Substitution of Equation i into Equation a, using the separation argument, leads to

$$-\frac{1}{X(x)}\frac{d^2X}{dx^2} = \frac{1}{Y(y)}\frac{d^2Y}{dy^2} = \lambda \tag{j}$$

Note that the boundary conditions for constant values of x are homogeneous, while the boundary condition at $y = b/a$ is nonhomogeneous. Thus the Sturm-Liouville problem is generated in the x-direction. That is, the operator $\mathbf{L}_x$ is chosen to be self-adjoint and positive definite on an appropriate S_x. Such is the case if $\mathbf{L}_x X = -d^2X/dx^2$. Equation j generates two ordinary differential equations:

$$\frac{d^2X}{dx^2} + \lambda X = 0 \tag{k}$$

$$\frac{d^2Y}{dy^2} - \lambda Y = 0 \tag{l}$$

Substituting Equation i into the boundary conditions, Equation e and Equation f, leads to

$$X(0) = 0 \tag{m}$$

$$X(1) = 0 \tag{n}$$

The eigenvalues and normalized eigenvectors obtained from solution of the eigenvalue problem of Equation k, Equation m, and Equation n are $\lambda_k = (k\pi)^2$ and $X_k(x) = \sqrt{2}\sin(k\pi x)$. The eigenvectors are orthogonal with respect to, and are normalized with respect to, the standard inner product on $C[0,1]$.

The general solution of Equation l corresponding to $\lambda_k = (k\pi)^2$ is

$$Y_k(x) = A_k \cosh(k\pi y) + B_k \sinh(k\pi y) \tag{o}$$

Thus the general solution to Equation a which satisfies boundary conditions Equation e and Equation f is

$$\Theta(x,y) = \sum_{k=1}^{\infty} [A_k \cosh(k\pi y) + B_k \sinh(k\pi y)]\sqrt{2}\sin(k\pi x) \tag{p}$$

Application of Equation g to Equation p results in

$$\Theta(x,0) = 0 = \sum_{k=1}^{\infty} \sqrt{2} A_k \sin(k\pi x) \tag{q}$$

Equation q is satisfied if and only if $A_k = 0$, which leads to

$$\Theta(x,y) = \sum_{k=1}^{\infty} \sqrt{2} B_k \sinh(k\pi y)\sin(k\pi x) \tag{r}$$

The remaining boundary condition, Equation h, is applied by expanding $g(x)=1$ in an eigenvector expansion. However, $g(x)=1$ is not in S_x, which would require that $g(0)=0$ and $g(1)=0$. A convergent eigenvector expansion is available, but it does not converge to $g(x)$ at $x=0$ and $x=1$. This situation is not a result of the mathematics, but rather of the physics of the modeling. The boundary conditions specify two values for Θ at the points $(0, b/a)$ and $(1, b/a)$. The boundary conditions at $x=0$ and $x=1$ specify $\Theta=0$, while the boundary condition at $y=b/a$ requires $\Theta=1$. The modeling is inaccurate, and therefore convergence to $g(x)$ is not anticipated at these points. Indeed, the right-hand side of Equation p converges to one value only when evaluated at these points.

The constant temperature on each boundary is an assumption made to simplify the problem. In actuality, conduction heat transfer causes the temperature on each side near the corners to vary for some distance away from the corners, as illustrated in Figure 6.4. The actual variation of temperature over the boundary is difficult to determine; it depends on factors not known from the problem statement, such as how the temperatures are maintained. If the actual temperatures are used as boundary conditions, then the boundary conditions on three faces are nonhomogeneous and the method of separation of variables cannot be directly applied.

Since the boundary conditions violate the laws of physics, the use of an eigenvector expansion for $g(x)$ which does not converge to $g(x)$ at $x=0$ and $x=1$ provides a reasonable approximation to the actual problem to be solved. To this end,

$$g(x)=1=\sum_{k=1}^{\infty} v_k\sqrt{2}\sin(k\pi x) \tag{s}$$

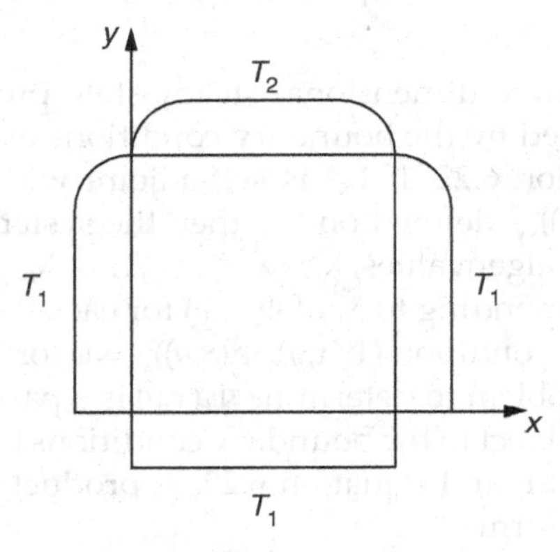

Figure 6.4 The mathematical model used in Example 6.1 violates the laws of physics because it suggests that the temperature is multivalued at the corners of the region.

where

$$v_k = \int_0^1 (1)\sqrt{2}\sin(k\pi x)dx$$

$$= \frac{\sqrt{2}}{k\pi}\left[1-(-1)^k\right]$$

$$= \begin{cases} 0 & k = 2,4,6,\ldots \\ \dfrac{2\sqrt{2}}{k\pi} & k = 1,3,5\ldots \end{cases} \tag{t}$$

Application of Equation h to Equation r using Equation t leads to

$$\sum_{k=1}^{\infty} v_k\sqrt{2}\sin(k\pi x) = \sum_{k=1}^{\infty}\sqrt{2}B_k \sinh\left(k\pi\frac{b}{a}\right)\sin(k\pi x) \tag{u}$$

Equating the coefficients of $\sin(k\pi x)$ from both sides of Equation u leads to

$$B_k = \frac{v_k}{\sinh\left(k\pi\frac{b}{a}\right)} \tag{v}$$

Substitution of Equation v into Equation r gives

$$\Theta(x,y) = \frac{4}{\pi}\sum_{k=1,3,5}^{\infty}\frac{\sinh(k\pi y)}{k\sinh\left(k\pi\frac{b}{a}\right)}\sin(k\pi x) \tag{w}$$

Now consider a three-dimensional steady-state problem. Let S_{xy} be the domain of $\mathbf{L}_{xy}$ specified by the boundary conditions of Equation 6.20, Equation 6.21, and Equation 6.22. If $\mathbf{L}_{xy}$ is self-adjoint with respect to an inner product, $(F(x,y), G(x,y))_{xy}$, defined on S_{xy}, then the system has an infinite, but countable, number of eigenvalues, $\lambda_1 \le \lambda_2 \le \lambda_3 \le \ldots \le \lambda_{k-1} \le \lambda_k \le \lambda_{k+1} \le \ldots$, with an eigenvector corresponding to λ_k of $\Phi_k(x,y)$ for each k. The eigenvectors satisfy an orthogonality condition $(\Phi_i(x,y), \Phi_j(x,y))_{xy} = 0$ for $i \ne j$.

The eigenvalue problem to determine $\Phi_k(x,y)$ is a partial differential equation, Equation 6.16, subject to the boundary conditions of Equation 6.19, Equation 6.20, Equation 6.21, and Equation 6.22. A product solution for $\Phi_k(x,y)$ is assumed to be of the form

$$\Phi(x,y) = X(x)Y(y) \tag{6.32}$$

The operator $\mathbf{L}_{xy}$ is separable if $\mathbf{L}_{xy}[u(x)v(y)] = g(y)v(y)\mathbf{L}_x u(x) + f(x)u(x)\mathbf{L}_y v(y)$ for all $u(x)$ and $v(x)$ in S_{xy}. If $\mathbf{L}_{xy}$ is separable, substitution of Equation 6.32 into Equation 6.16 leads to

$$\frac{1}{X(x)f(x)}\mathbf{L}_x[X(x)] - \lambda = \frac{1}{Y(x)g(y)}\mathbf{L}_y[Y(y)] \tag{6.33}$$

Use of the separation argument implies that both sides of Equation 6.33 are equal to the same constant; call it μ. Thus, Equation 6.33 separates into

$$\mathbf{L}_x X(x) - \lambda = \mu X(x) \tag{6.34}$$

$$\mathbf{L}_y Y(x) = \mu Y(y) \tag{6.35}$$

Substitution of Equation 6.32 into Equation 6.19, Equation 6.20, Equation 6.21, and Equation 6.22 leads to

$$X(0) - \alpha_x \frac{dX}{dx}(0) = 0 \tag{6.36}$$

$$X(a) + \beta_x \frac{dX}{dx}(a) = 0 \tag{6.37}$$

$$Y(0) - \alpha_y \frac{dY}{dy}(0) = 0 \tag{6.38}$$

$$Y(b) + \beta_y \frac{d\Phi}{dy}(b) = 0 \tag{6.39}$$

Equation 6.35, Equation 6.38, and Equation 6.39 constitute an eigenvalue problem for $Y(y)$ on a domain S_y. Its solution leads to an infinite number of eigenvalues μ_ℓ for $\ell = 1,2,\ldots$ such that $\mu_1 \leq \mu_2 \leq \ldots \leq \mu_{\ell-1} \leq \mu_\ell \leq \mu_{\ell+1} \leq \ldots$ Let $Y_\ell(y)$ be the eigenvector corresponding to μ_ℓ normalized with respect to an inner product $(f(y),g(y))_y$ defined on S_y.

For each ℓ, Equation 6.34, Equation 6.36, and Equation 6.37 constitute an eigenvalue problem. There are an infinite, but countable, number of eigenvalues, $\ell_{\ell,m}$ $m = 1,2,\ldots$, for each problem, such that $\lambda_{\ell,1} \leq \lambda_{\ell,2} \leq \ldots \leq \lambda_{\ell,m-1} \leq \lambda_{\ell,m} \leq \lambda_{\ell,m+1} \leq \ldots$ Let $X_{\ell,m}$ be the eigenvector corresponding to $\lambda_{\ell,m}$ which has been normalized with respect to an inner product $(p(x),s(x))_x$ defined on S_x.

The eigenvalues of $\mathbf{L}_x$ are also the eigenvalues of $\mathbf{L}_{xy}$. Thus $\mathbf{L}_{xy}$ has a doubly infinite number of eigenvalues. The eigenvectors of $\mathbf{L}_{xy}$ are

$$\Phi_{\ell,m}(x,y) = X_{\ell,m}(x)Y_\ell(y) \tag{6.40}$$

Equation 6.17 can be written for each $\lambda_{\ell,m}$ $\ell=1,2,\ldots$ and $m=1,2,\ldots$:

$$\mathbf{L}_{\mathbf{z}}Z_{\ell,m}(z)=-\lambda_{\ell,m}Z_{\ell,m}(z) \tag{6.41}$$

Equation 6.41 is a linear second-order ordinary differential equation whose general solution is written as

$$Z_{\ell,m}(z)=A_{\ell,m}Z_{\ell,m,1}(z)+B_{\ell,m}Z_{\ell,m,2}(z) \tag{6.42}$$

such that a solution of Equation 6.5 subject to Equation 6.7, Equation 6.8, Equation 6.9, and Equation 6.10 is

$$\phi_{\ell,m}(x,y,z)=[A_{\ell,m}Z_{\ell,m,1}(z)+B_{\ell,m}Z_{\ell,m,2}(z)]X_{\ell,m}(x)Y_{\ell}(y) \tag{6.43}$$

The most general solution of a linear homogeneous problem is a linear combination of all possible solutions,

$$\phi(x,y,z)=\sum_{\ell=1}^{\infty}\sum_{m=1}^{\infty}[A_{\ell,m}Z_{\ell,m,1}(z)+B_{\ell,m}Z_{\ell,m,2}(z)]X_{\ell,m}(x)Y_{\ell}(y) \tag{6.44}$$

Recall that $\mathbf{L}_{xy}$ is self-adjoint with respect to an inner product on S_{xy}. Thus any element in S_{xy} has an expansion in terms of eigenvectors of $\mathbf{L}_{xy}$. Assume that $f(x,y)$ is in S_{xy}. Its eigenvector expansion is

$$f(x,y)=\sum_{\ell=1}^{\infty}\sum_{m=1}^{\infty}u_{\ell,m}X_{\ell,m}(x)Y_{\ell}(y) \tag{6.45}$$

where the expansion coefficients are

$$u_{\ell,k}=(f(x,y),X_{\ell,m}(x)Y_{\ell}(y))_{xy} \tag{6.46}$$

The function $g(x,y)$ in Equation 6.8 has an eigenvector expansion of the form

$$g(x,y)=\sum_{\ell=1}^{\infty}\sum_{m=1}^{\infty}v_{\ell,m}X_{\ell,m}(x)Y_{\ell}(y) \tag{6.47}$$

Equation 6.44 can be substituted into the boundary condition, Equation 6.7, with $f(x)$ replaced by Equation 6.45, leading to

$$\begin{aligned}&\sum_{\ell=1}^{\infty}\sum_{m=1}^{\infty}[A_{\ell,m}Z_{\ell,m,1}(0)+B_{\ell,m}Z_{\ell,m,2}(0)]X_{\ell,m}(x)Y_{\ell}(y)\\&-\alpha_z\sum_{\ell=1}^{\infty}\sum_{m=1}^{\infty}[A_{\ell,m}Z'_{\ell,m,1}(0)+B_{\ell,m}Z'_{\ell,m,2}(0)]X_{\ell,m}(x)Y_{\ell}(y)\\&=\sum_{\ell=1}^{\infty}\sum_{m=1}^{\infty}u_{\ell,m}X_{\ell,m}(x)Y_{\ell}(y)\end{aligned} \tag{6.48}$$

Combining the summations in Equation 6.48, and noting the linear independence of the eigenvectors of $\mathbf{L}_{xy}$, leads to

$$[Z_{\ell,m,1}(0)-\alpha_z Z'_{\ell,m,1}(0)]A_{\ell,m}+[Z_{\ell,m,2}(0)-\alpha_z Z'_{\ell,m,2}(0)]B_{\ell,m}=u_{\ell,m} \tag{6.49}$$

A similar procedure used with Equation 6.8 gives

$$[Z_{\ell,m,1}(c)+\beta_z Z'_{\ell,m,1}(c)]A_{\ell,m}+[Z_{\ell,m,2}(c)+\beta_z Z'_{\ell,m,2}(c)]B_{\ell,m}=v_{\ell,m} \tag{6.50}$$

Equation 6.49 and Equation 6.50 can then be solved simultaneously for $A_{\ell,m}$ and $B_{\ell,m}$.

Example 6.2 The solid rectangular block shown in Figure 6.5 is in a medium of temperature T_∞. The heat transfer coefficient between the block and the medium is h. Defining

$$x^* = \frac{x}{a} \quad y^* = \frac{y}{a} \quad z^* = \frac{z}{a} \quad \text{and} \quad \Theta = \frac{T-T_\infty}{T_1-T_\infty} \tag{a}$$

the problem governing the steady nondimensional temperature distribution in the block is

$$\frac{\partial^2\Theta}{\partial x^2}+\frac{\partial^2\Theta}{\partial y^2}+\frac{\partial^2\Theta}{\partial z^2}=0 \tag{b}$$

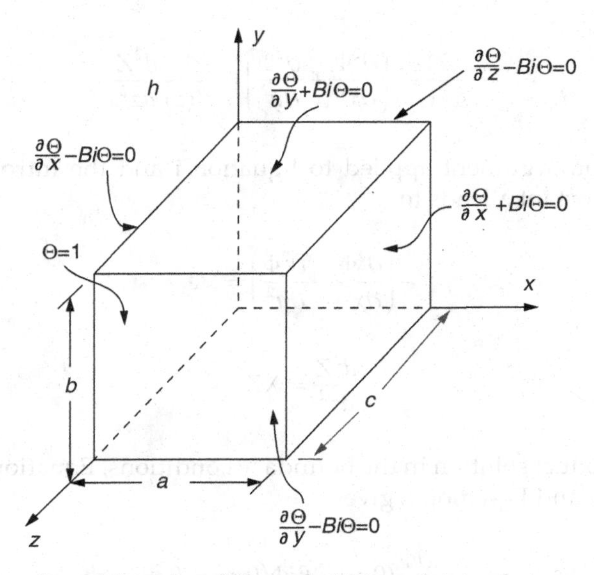

Figure 6.5 One face of the cube of Example 6.2 is maintained at a constant temperature while the remainder of the surface is subject to heat transfer by convection with the ambient.

$$\frac{\partial \Theta}{\partial x}(0,y,z) - Bi\Theta(0,y,z) = 0 \tag{c}$$

$$\frac{\partial \Theta}{\partial x}(1,y,z) + Bi\Theta(1,y,z) = 0 \tag{d}$$

$$\frac{\partial \Theta}{\partial y}(x,0,z) - Bi\Theta(x,0,z) = 0 \tag{e}$$

$$\frac{\partial \Theta}{\partial y}\left(x,\frac{b}{a},z\right) + Bi\Theta\left(x,\frac{b}{a},z\right) = 0 \tag{f}$$

$$\frac{\partial \Theta}{\partial z}(x,y,0) - Bi\Theta(x,y,0) = 0 \tag{g}$$

$$\Theta\left(x,y,\frac{c}{a}\right) = 1 \tag{h}$$

where $Bi = ha/k$ is the Biot number. Determine the steady-state temperature distribution in the block.

Solution The nonhomogeneous boundary condition is on a face of constant z, and therefore the appropriate form of the initial product solution is $\Theta(x,y,z) = \Phi(x,y)Z(z)$, which when substituted into Equation b, leads to

$$-\frac{1}{\Phi(x,y)}\left(\frac{\partial^2 \Phi}{\partial x^2} + \frac{\partial^2 \Phi}{\partial y^2}\right) = \frac{1}{Z(z)}\frac{d^2 Z}{dz^2} \tag{i}$$

The separation argument applied to Equation i and the introduction of a separation constant λ leads to

$$-\left(\frac{\partial^2 \Phi}{\partial x^2} + \frac{\partial^2 \Phi}{\partial y^2}\right) = \lambda\Phi \tag{j}$$

$$\frac{d^2 Z}{dz^2} = \lambda Z \tag{k}$$

Use of the product solution in the boundary conditions, Equation c, Equation d, Equation e, and Equation f, gives

$$\frac{\partial \Phi}{\partial x}(0,y) - Bi\Phi(0,y) = 0 \tag{l}$$

$$\frac{\partial \Phi}{\partial x}(1,y) + Bi\Phi(1,y) = 0 \tag{m}$$

$$\frac{\partial \Phi}{\partial y}(x,0) - Bi\Phi\,(x,0) = 0 \tag{n}$$

$$\frac{\partial \Phi}{\partial y}\left(x,\frac{b}{a}\right) + Bi\Phi\left(x,\frac{b}{a}\right) = 0 \tag{o}$$

Equation j, Equation l, Equation m, Equation n, and Equation o constitute an eigenvalue problem. The operator defined by $\mathbf{L}_{xy}\Phi = -(\partial^2\Phi/\partial x^2 + \partial^2\Phi/\partial y^2)$ is positive definite and self-adjoint on S_{xy} as defined by Equation l, Equation m, Equation n, and Equation o with respect to the inner product

$$(f(x,y),g(x,y))_{xy} = \int_0^1 \int_0^{\frac{b}{a}} f(x,y)g(x,y)dydx \tag{p}$$

Thus, 5 all values of λ are real and positive, and eigenvectors corresponding to distinct eigenvalues are orthogonal with respect to the inner product defined in Equation p.

A product solution for $\Phi(x,y)$ is assumed to be of the form $\Phi(x,y) = X(x)Y(y)$, which when substituted into Equation j, leads to

$$\frac{1}{X(x)}\frac{d^2X}{dx^2} + \lambda = -\frac{1}{Y}\frac{d^2Y}{dy^2} \tag{q}$$

Application of the separation argument to Equation q and introduction of a separation constant μ leads to

$$\frac{d^2X}{dx^2} + (\lambda - \mu)X = 0 \tag{r}$$

$$\frac{d^2Y}{dx^2} + \mu Y = 0 \tag{s}$$

Substitution of the product solution into the boundary conditions, Equation l, Equation m, Equation n, and Equation o, gives

$$\frac{dX}{dx}(0) - BiX(0) = 0 \tag{t}$$

$$\frac{dX}{dx}(1) + BiX(1) = 0 \tag{u}$$

$$\frac{dY}{dy}(0) - BiY(0) = 0 \tag{v}$$

$$\frac{dY}{dy}\left(\frac{b}{a}\right) + BiY\left(\frac{b}{a}\right) = 0 \tag{w}$$

The operator defined by $\mathbf{L}_y Y = -d^2Y/dy^2$ is positive definite and self-adjoint on S_y as defined by the boundary conditions of Equation v and Equation w with respect to

$$(f(y), g(y))_y = \int_0^{\frac{b}{a}} f(y)g(y)dy \tag{x}$$

which is the standard inner product on $C\,[0, b/a]$. Thus all values of μ are real and positive, and eigenvectors corresponding to distinct eigenvalues of $\mathbf{L}_y$ are orthogonal with respect to the inner product in Equation w.

The solution of Equation s is

$$Y(y) = C_1 \cos\left(\sqrt{\mu}y\right) + C_2 \sin\left(\sqrt{\mu}y\right) \tag{y}$$

Application of Equation v to Equation y leads to

$$C_2 = \frac{Bi}{\sqrt{\mu}} C_1 \tag{z}$$

Subsequent application of Equation t results in

$$\tan\left(\sqrt{\mu}\frac{b}{a}\right) = \frac{2Bi}{(Bi)^2 - \mu} \tag{aa}$$

Equation aa has an infinite, but countable, number of solutions. The first five solutions for $b/a = 2$ and $Bi = 4$ are given in Table 6.1. The eigenvector corresponding to an eigenvalue μ_ℓ is

Table 6.1 Eigenvalues for Example 6.2

m	1	2	3	4	5
μ_m	0.054	3.439	14.88	37.248	60.289
β_m	0.22	34.521	86.716	156.487	245.647
$\lambda_{1,m}$	0.27	34.575	86.770	156.541	245.702
$\lambda_{2,m}$	3.659	37.960	90.155	159.926	249.186
$\lambda_{3,m}$	15.10	49.401	101.696	171.367	260.54
$\lambda_{4,m}$	37.468	71.769	123.964	193.735	282.895
$\lambda_{5,m}$	61.109	94.810	147.005	216.716	306.136

$$Y_\ell(y) = C_\ell\left[\cos\left(\sqrt{\mu_\ell}y\right) + \frac{Bi}{\mu_\ell}\sin\left(\sqrt{\mu_\ell}y\right)\right] \tag{bb}$$

Normalization of the eigenvector with respect to the inner product in Equation x leads to

$$C_\ell = \left[\frac{4\mu_\ell}{2\mu_\ell + 2Bi(1+Bi) + \frac{1}{\sqrt{\mu_\ell}}(\mu_\ell - Bi^2)\sin\left(2\sqrt{\mu_\ell}\frac{b}{a}\right) - 2Bi\cos\left(2\sqrt{\mu_\ell}\frac{b}{a}\right)}\right]^{\frac{1}{2}} \tag{cc}$$

Equation r has nontrivial solutions for each μ_ℓ corresponding to eigenvalues $\lambda_{\ell,m}$ $m = 1,2,\ldots$ which are obtained from Equation r, Equation t, and Equation u. Note that the solution of this system is obtained by comparison with the system of Equation s, Equation u, and Equation v. The systems are the same if in Equation s, Equation u, and Equation v, y is replaced by x, b/a is replaced by 1, and μ is replaced by $\lambda - \mu$. Thus the eigenvalues are obtained by solving

$$\tan\left(\sqrt{\lambda - \mu_\ell}\right) = \frac{2Bi}{(Bi)^2 - (\lambda - \mu_\ell)} \tag{dd}$$

There are an infinite, but countable, number of solutions of Equation dd for each ℓ. Note that Equation dd may be written as

$$\tan\left(\sqrt{\beta}\right) = \frac{2Bi}{(Bi)^2 - \beta} \tag{ee}$$

where $\beta = \lambda - \mu_\ell$. The solutions of Equation ee, β_m $m = 1,2,\ldots$ are independent of μ_ℓ.

The corresponding eigenvectors are of the form

$$X_m(x) = D_m\left[\cos\left(\sqrt{\beta_m}x\right) + \frac{Bi}{\beta_m}\sin\left(\sqrt{\beta_m}x\right)\right] \tag{ff}$$

Equation ff can be normalized by choosing

$$D_m = \left[\frac{4\beta_m}{2\beta_m + 2Bi(1+Bi) + \frac{1}{\sqrt{\beta_m}}(\beta_m - Bi^2)\sin\left(2\sqrt{\beta_m}\right) - 2Bi\cos\left(2\sqrt{\beta_m}\right)}\right]^{\frac{1}{2}} \tag{gg}$$

The eigenvalues of $\mathbf{L}_{xy}$ are given by

$$\lambda_{\ell,m} = \mu_\ell + \beta_m \tag{hh}$$

where the μ_ℓ are the solutions of Equation aa and the β_m are the solutions of Equation ee. The normalized eigenvectors are

$$\Phi_{\ell,m} = X_m(x)Y_\ell(y)$$
$$= C_\ell D_m\left[\cos\left(\sqrt{\beta_m}\,x\right)+\frac{Bi}{\beta_m}\sin\left(\sqrt{\beta_m}\,x\right)\right]\left[\cos\left(\sqrt{\mu_\ell}\,y\right)+\frac{Bi}{\mu_\ell}\sin\left(\sqrt{\mu_\ell}\,y\right)\right] \tag{ii}$$

For each $\lambda_{\ell,m}$, the general solution of Equation k is

$$Z_{\ell,m}(z) = A_{\ell,m}\cosh\left(\sqrt{\lambda_{\ell,m}}\,z\right)+B_{\ell,m}\sinh\left(\sqrt{\lambda_{\ell,m}}\,z\right) \tag{jj}$$

The general solution of Equation b is

$$\Theta(x,y,z) = \sum_{\ell=1}^{\infty}\sum_{m=1}^{\infty}\left\{C_\ell D_m\left[\cos\left(\sqrt{\beta_m}\,x\right)+\frac{Bi}{\beta_m}\sin\left(\sqrt{\beta_m}\,x\right)\right]\right.$$
$$\times\left[\cos\left(\sqrt{\mu_\ell}\,y\right)+\frac{Bi}{\mu_\ell}\sin\left(\sqrt{\mu_\ell}\,y\right)\right]$$
$$\left.\times\left[A_{\ell,m}\cosh\left(\sqrt{\lambda_{\ell,m}}\,z\right)+B_{\ell,m}\sinh\left(\sqrt{\lambda_{\ell,m}}\,z\right)\right]\right\} \tag{kk}$$

Application of Equation g to Equation kk gives

$$0 = \frac{\partial\Theta}{\partial z}(x,y,0) - Bi\Theta(x,y,0)$$
$$= \sum_{\ell=1}^{\infty}\sum_{m=1}^{\infty}C_\ell D_m\sqrt{\lambda_{\ell,m}}\,B_{\ell,m}\left[\cos\left(\sqrt{\beta_m}\,x\right)+\frac{Bi}{\beta_m}\sin\left(\sqrt{\beta_m}\,x\right)\right]$$
$$\times\left[\cos\left(\sqrt{\mu_\ell}\,y\right)+\frac{Bi}{\mu_\ell}\sin\left(\sqrt{\mu_\ell}\,y\right)\right]$$
$$-Bi\sum_{\ell=1}^{\infty}\sum_{m=1}^{\infty}C_\ell D_m A_{\ell,m}\left[\cos\left(\sqrt{\beta_m}\,x\right)+\frac{Bi}{\beta_m}\sin\left(\sqrt{\beta_m}\,x\right)\right]$$
$$\times\left[\cos\left(\sqrt{\mu_\ell}\,y\right)+\frac{Bi}{\mu_\ell}\sin\left(\sqrt{\mu_\ell}\,y\right)\right]$$
$$= \sum_{\ell=1}^{\infty}\sum_{m=1}^{\infty}C_\ell D_m\left(\sqrt{\lambda_{\ell,m}}\,B_{\ell,m} - BiA_{\ell,m}\right)\left[\cos\left(\sqrt{\beta_m}\,x\right)+\frac{Bi}{\beta_m}\sin\left(\sqrt{\beta_m}\,x\right)\right]$$
$$\times\left[\cos\left(\sqrt{\mu_\ell}\,y\right)+\frac{Bi}{\mu_\ell}\sin\left(\sqrt{\mu_\ell}\,y\right)\right] \tag{ll}$$

Equation ll is satisfied if and only if

$$A_{\ell,m} = \frac{\sqrt{\lambda_{\ell,m}}}{Bi} B_{\ell,m} \tag{mm}$$

Application of Equation h to Equation kk using Equation mm leads to

$$\Theta\left(x,y,\frac{c}{a}\right) = 1 = \sum_{\ell=1}^{\infty}\sum_{m=1}^{\infty}\left\{C_\ell D_m\left[\cos\left(\sqrt{\beta_m}x\right) + \frac{Bi}{\beta_m}\sin\left(\sqrt{\beta_m}x\right)\right]\right.$$
$$\times\left[\cos\left(\sqrt{\mu_\ell}y\right) + \frac{Bi}{\mu_\ell}\sin\left(\sqrt{\mu_\ell}y\right)\right]$$
$$\left.\times B_{\ell,m}\left[\frac{\sqrt{\lambda_{\ell,m}}}{Bi}\cosh\left(\sqrt{\lambda_{\ell,m}}\frac{c}{a}\right) + \sinh\left(\sqrt{\lambda_{\ell,m}}\frac{c}{a}\right)\right]\right\} \tag{nn}$$

Equation nn is satisfied by expanding $f(x,y)$ in an expansion in terms of the eigenvectors of $\mathbf{L}_{xy}$,

$$1 = \sum_{\ell=1}^{\infty}\sum_{m=1}^{\infty} v_{\ell,k}\left\{C_\ell D_m\left[\cos\left(\sqrt{\beta_m}x\right) + \frac{Bi}{\beta_m}\sin\left(\sqrt{\beta_m}x\right)\right]\right.$$
$$\left.\times\left[\cos\left(\sqrt{\mu_\ell}y\right) + \frac{Bi}{\mu_\ell}\sin\left(\sqrt{\mu_\ell}y\right)\right]\right\} \tag{oo}$$

where

$$v_{\ell,m} = \int_0^1\int_0^{\frac{b}{a}} (1)C_\ell D_m\left[\cos\left(\sqrt{\beta_m}x\right) + \frac{Bi}{\beta_m}\sin\left(\sqrt{\beta_m}x\right)\right]$$
$$\times\left[\cos\left(\sqrt{\mu_\ell}y\right) + \frac{Bi}{\mu_\ell}\sin\left(\sqrt{\mu_\ell}y\right)\right]dydx$$
$$= \frac{C_\ell D_m}{\sqrt{\mu_\ell\beta_m}}\left[\sin\left(\sqrt{\beta_m}\right) - \frac{Bi}{\beta_m}\cos\left(\sqrt{\beta_m}\right) + \frac{Bi}{\beta_m}\right]$$
$$\times\left[\sin\left(\sqrt{\mu_\ell}\frac{b}{a}\right) - \frac{Bi}{\mu_\ell}\cos\left(\sqrt{\mu_\ell}\frac{b}{a}\right) + \frac{Bi}{\mu_\ell}\right] \tag{pp}$$

Equation nn is satisfied if and only if

$$B_{\ell,m} = \frac{v_{\ell,m}}{\frac{\sqrt{\lambda_{\ell,m}}}{Bi}\cosh\left(\sqrt{\lambda_{\ell,m}}\frac{c}{a}\right) + \sinh\left(\sqrt{\lambda_{\ell,m}}\frac{c}{a}\right)} \tag{qq}$$

The solution of Equation b, Equation c, Equation d, Equation e, Equation f, Equation g, and Equation h is

$$\Theta(x,y,z)=\sum_{\ell=1}^{\infty}\sum_{m=1}^{\infty}\left\{B_{\ell,m}C_{\ell}D_m\left[\cos\left(\sqrt{\beta_m}x\right)+\frac{Bi}{\beta_m}\sin\left(\sqrt{\beta_m}x\right)\right]\right.$$

$$\times\left[\cos\left(\sqrt{\mu_\ell}y\right)+\frac{Bi}{\mu_\ell}\sin\left(\sqrt{\mu_\ell}y\right)\right]$$

$$\left.\times\left[\frac{\sqrt{\lambda_{\ell,m}}}{Bi}\cosh\left(\sqrt{\lambda_{\ell,m}}z\right)+\sinh\left(\sqrt{\lambda_{\ell,m}}z\right)\right]\right\} \tag{rr}$$

The eigenvectors of $\mathbf{L}_{xy}$ are products of the eigenvectors of two eigenvalue problems. Each eigenvalue problem is self-adjoint, and therefore each set of eigenvectors satisfies its own orthonormality conditions,

$$\left(Y_p,Y_q\right)_y=\delta_{p,q} \tag{6.51}$$

$$\left(X_{\ell,r},X_{\ell,s}\right)_x=\delta_{r,s} \tag{6.52}$$

Mode-shape orthonormality of eigenvectors of $\mathbf{L}_{xy}$ requires that

$$\left(Y_pX_{p,r},Y_qX_{q,s}\right)_{xy}=\delta_{p,q}\delta_{r,s} \tag{6.53}$$

The inner product used in Equation 6.53 is of the form

$$(f(x,y),g(x,y))_{xy}=\int_0^{b/a}\int_0^{b/a}f(x,y)g(x,y)h(x,y)dxdy \tag{6.54}$$

When $\mathbf{L}_{xy}$ is separable, $h(x,y)=h_x(x)h_y(y)$. Then

$$\left(Y_pX_{p,r},Y_qX_{q,s}\right)_{xy}=\int_0^1\int_0^{b/a}Y_p(y)X_{p,r}(x)Y_q(y)X_{q,s}(x)h_x(x)h_y(y)dydx$$

$$=\left[\int_0^1 X_{p,r}(x)X_{q,s}(x)h_x(x)dx\right]\left[\int_0^{b/a}Y_p(y)Y_q(y)h_y(y)dy\right]$$

$$=\left(X_{p,r},X_{q,s}\right)_x\left(Y_p,Y_q\right)_y \tag{6.55}$$

The y-subscripted inner product in Equation 6.55 is zero unless $p=q$. If $p=q$, the x-subscripted inner product is zero unless $r=s$. Thus Equation 6.55 agrees with Equation 6.53. In addition,

$$(Y_p X_{p,r}, Y_p X_{p,r})_{xy} = (X_{p,r}, X_{p,r})_x (Y_p, Y_p)_y \tag{6.56}$$

Equation 6.56 shows that if the x- and y-eigenvectors are normalized with respect to their individually subscripted inner products, then they are normalized with respect to the xy-subscripted inner product.

6.3 Time-dependent problems: Initial value problems

Consider the case when the operator of Equation 6.1 involves derivatives with respect to time, $\mathbf{M} \neq 0$. The diffusion equation and the wave equation are examples of such partial differential equations. Initial-value problems for partial differential equations occur when a system has an initial state other then its equilibrium state and when the response varies with spatial coordinates. The response of the system is then time-dependent.

It is assumed in this section that the partial differential equations, boundary conditions, and initial conditions are all nondimensional and that all boundary conditions are homogeneous. The operator $\mathbf{L}$ is assumed to be separable in time such that

$$\mathbf{L}[f(\mathbf{r})g(t)] = s(t)g(t)\mathbf{L}_\mathrm{r}[f(\mathbf{r})] + h(\mathbf{r})f(\mathbf{r})\mathbf{L}_\mathrm{t}[g(t)] \tag{6.57}$$

where $\mathbf{L}_\mathrm{r}$ is an operator involving only spatial derivatives, $\mathbf{L}_\mathrm{t}$ is an operator involving only derivatives with respect to time, and $s(t)$ and $f(\mathbf{r})$ are functions determined from the separation procedure.

Consider a problem of the form

$$\mathbf{L}\phi(\mathbf{r},t) = 0 \tag{6.58}$$

where $\mathbf{L}$ is an operator separable in time. The region in which ϕ is defined is bounded by a surface described by $f(\mathbf{r}) = 0$. The boundary conditions are of the form

$$\phi(\mathbf{r},t) - \alpha(\mathbf{r})\frac{\partial \phi}{\partial \mathbf{n}} = 0 \tag{6.59}$$

everywhere on S, the surface of the region. The problem has an appropriate number of initial conditions, at least one of which is nonhomogeneous. A problem whose canonical form is the diffusion equation has one initial condition, while a problem whose canonical form is the wave equation requires two initial conditions.

Since **L** is separable in time, the method of separation of variables is used, assuming a product solution of

$$\phi(\mathbf{r},t) = \Phi(\mathbf{r})T(t) \tag{6.60}$$

Substitution of Equation 6.60 into Equation 6.58, while applying the condition for separation, Equation 6.57, leads to

$$s(t)T(t)\mathbf{L}_r[\Phi(\mathbf{r})] + h(\mathbf{r})\Phi(\mathbf{r})\mathbf{L}_t[T(t)] = 0 \tag{6.61}$$

Equation 6.61 can be rearranged to yield

$$\frac{1}{h(\mathbf{r})\Phi(\mathbf{r})}\mathbf{L}_r[\Phi(\mathbf{r})] = -\frac{1}{s(t)T(t)}\mathbf{L}_t[T(t)] \tag{6.62}$$

The usual separation argument is applied with separation parameter λ, leading to

$$\mathbf{L}_r[\Phi(\mathbf{r})] = \lambda h(\mathbf{r})\Phi(\mathbf{r}) \tag{6.63}$$

$$\mathbf{L}_t[T(t)] = -\lambda s(t)T(t) \tag{6.64}$$

Use of the product solution in the boundary condition, Equation 6.59, leads to

$$\Phi(\mathbf{r},t) - \alpha(\mathbf{r})\frac{\partial\Phi}{\partial\mathbf{n}} = 0 \tag{6.65}$$

on S.

Equation 6.63 and Equation 6.64 form an eigenvalue-eigenvector problem. If $\mathbf{L}_r$ is self-adjoint and positive definite with respect to the inner product

$$(f(\mathbf{r}), g(\mathbf{r}))_V = \int_V f(\mathbf{r})g(\mathbf{r})h(\mathbf{r})dV \tag{6.66}$$

then all eigenvalues are real and positive, and the eigenvectors corresponding to distinct eigenvalues are orthogonal with respect to the inner product defined in Equation 6.66.

The solution proceeds in a manner similar to that for steady-state problems. Once all of the eigenvalues and eigenvectors of $\mathbf{L}_r$ have been determined, a solution of Equation 6.64 is obtained for each eigenvalue. Each resulting product solution of the form of Equation 6.60 satisfies the partial differential equation and all boundary conditions. The most general solution is a linear combination of all product solutions. Eigenvector expansions of nonhomogeneous terms in the initial conditions are developed and used to impose the initial conditions.

Example 6.3 A bar of thickness L is at a uniform temperature T_0 when it is suddenly submerged in a bath of temperature T_∞. Determine the unsteady state temperature distribution in the bar when: (a) the heat transfer coefficient is small and convection heat transfer is neglected; (b) the heat transfer coefficient is of an intermediate value and heat convection is of the same order of magnitude as conduction at the ends of the bar, but the cross-sectional area is large compared to the perimeter times the length of the bar, such that convection over the length of the bar can be neglected; (c) the heat transfer coefficient is large enough such that convection must be included over the length of the bar, but a one-dimensional assumption is used; and (d) the thickness is comparable to the length, meaning that a two dimensional model is required.

Solution Define $x^* = x/L$, $t^* = \rho c_p L^2/k$, and $\Theta = (T - \mathrm{T}_\infty)/(\mathrm{T}_0 - \mathrm{T}_\infty)$. (a) If the heat transfer coefficient is small, the temperature at the ends is the same as the temperature of the bath. The problem governing the nondimensional temperature distribution is

$$\frac{\partial \Theta}{\partial t} = \frac{\partial^2 \Theta}{\partial x^2} \tag{a}$$

$$\Theta(0,t) = 0 \tag{b}$$

$$\Theta(1,t) = 0 \tag{c}$$

$$\Theta(x,0) = 1 \tag{d}$$

A product solution of Equation a is assumed to be of the form

$$\Theta(x,t) = X(x)T(t) \tag{e}$$

which, when substituted into Equation a, leads to

$$-\frac{1}{X}\frac{d^2X}{dx^2} = -\frac{1}{T}\frac{dT}{dt} \tag{f}$$

The separation argument is applied to Equation f. Defining λ as the separation constant, the separated differential equations are

$$\frac{d^2X}{dx^2} + \lambda X \tag{g}$$

$$\frac{dT}{dt}+\lambda T=0 \tag{h}$$

Substitution of Equation e into the boundary conditions, Equation b and Equation c, leads to

$$X(0)=0 \tag{i}$$

$$X(1)=0 \tag{j}$$

The eigenvalues and normalized eigenvectors for the system comprised of Equation g subject to Equation i and Equation j are

$$\lambda_\ell=(\ell\pi)^2 \quad \ell=1,2,... \tag{k}$$

$$X_\ell(x)=\sqrt{2}\sin(\ell\pi x) \tag{l}$$

The corresponding solutions of Equation h are

$$T_\ell(t)=A_\ell e^{-(\ell\pi)^2 t} \tag{m}$$

The general solution becomes

$$\Theta(x,t)=\sum_{\ell=1}^{\infty}\sqrt{2}A_\ell e^{-(\ell\pi)^2 t}\sin(\ell\pi x) \tag{n}$$

The eigenvector expansion of the nonhomogeneous initial condition is

$$1=\sum_{\ell=1}^{\infty}v_\ell\sqrt{2}\sin(\ell\pi x) \tag{o}$$

where

$$\begin{aligned} v_k&=\left(1,\sqrt{2}\sin(\ell\pi x)\right)\\ &=\int_0^1\sqrt{2}\sin(\ell\pi x)dx\\ &=\frac{\sqrt{2}}{\ell\pi}\left[1-(-1)^\ell\right]\\ &=\frac{2\sqrt{2}}{\ell\pi}\begin{cases}0 & \ell=2,4,6,...\\ 1 & \ell=1,3,5,...\end{cases} \end{aligned} \tag{p}$$

Application of the initial condition, Equation d, to Equation n using Equation o and Equation p leads to $A_\ell = v_\ell$. The time-dependent temperature distribution in the plate is

$$\Theta(t) = \frac{4}{\pi}\sum_{\ell=1}^{\infty}\frac{1}{\ell}e^{-(\ell\pi)^2 t}\sin(\ell\pi x) \tag{q}$$

The expansion of Equation q converges to the nondimensional temperature distribution everywhere except at $x=0$ and $x=1$. This, of course, is because $f(x)=1$ is not in the domain of the operator defined in Equation h when subject to the boundary conditions of Equation i and Equation j. This analysis is consistent with the mathematical modeling, but it violates physical principles. The modeling suggests that the temperature is discontinuous at the ends of the bar at $t=0$. Before submergence in the bath, the bar has a uniform temperature T_0, but the modeling suggests that immediately after submergence, the temperature at each end is T_∞. The more precise physical model is that used in part (b), in which heat transfer occurs through convection at the ends of the bar and a transient temperature response occurs at the ends of the bar.

(b) The appropriate nondimensional boundary conditions for the model as stated are

$$\frac{\partial \Theta}{\partial x}(0,t) - Bi\Theta(0,t) = 0 \tag{r}$$

$$\frac{\partial \Theta}{\partial x}(1,t) + Bi\Theta(1,t) = 0 \tag{s}$$

The problem governed by Equation g, Equation r, and Equation s is the same as that obtained in the solution of Example 6.2. The eigenvalues are the positive solutions of

$$\tan\left(\sqrt{\lambda}\right) = \frac{2Bi}{(Bi)^2 - \lambda} \tag{t}$$

The normalized eigenvector corresponding to an eigenvalue λ_ℓ is

$$X_\ell(x) = D_\ell\left[\cos\left(\sqrt{\lambda_\ell}x\right) + \frac{Bi}{\lambda_\ell}\sin\left(\sqrt{\lambda_\ell}x\right)\right] \tag{u}$$

where

$$D_\ell = \left[\frac{4\lambda_\ell}{2\lambda_\ell + 2Bi(1+Bi) + \frac{1}{\sqrt{\lambda_\ell}}(\lambda_\ell - Bi^2)\sin\left(2\sqrt{\lambda_\ell}\right) - 2Bi\cos\left(2\sqrt{\lambda_\ell}\right)}\right]^{\frac{1}{2}} \tag{v}$$

The solution of Equation h corresponding to an eigenvalue λ_ℓ is

$$T_\ell(t) = A_\ell e^{-\lambda_\ell t} \tag{w}$$

The general solution is

$$\Theta(x,t) = \sum_{\ell=1}^{\infty} A_\ell e^{-\lambda_\ell t} D_\ell \left[\cos\left(\sqrt{\lambda_\ell}\, x\right) + \frac{Bi}{\lambda_\ell}\sin\left(\sqrt{\lambda_\ell}\, x\right)\right] \tag{x}$$

The eigenvector expansion for the nonhomogeneous function in the boundary condition of Equation d is

$$1 = \sum_{\ell=1}^{\infty} v_\ell D_\ell \left[\cos\left(\sqrt{\lambda_\ell}\, x\right) + \frac{Bi}{\lambda_\ell}\sin\left(\sqrt{\lambda_\ell}\, x\right)\right] \tag{y}$$

where

$$v_\ell = \int_0^1 D_\ell \left[\cos\left(\sqrt{\lambda_\ell}\, x\right) + \frac{Bi}{\lambda_\ell}\sin\left(\sqrt{\lambda_\ell}\, x\right)\right] dx$$

$$= \frac{D_\ell}{\sqrt{\lambda_\ell}}\left[\sin\left(\sqrt{\lambda_\ell}\right) - \frac{Bi}{\lambda_\ell}\cos\left(\sqrt{\lambda_\ell}\right) + \frac{Bi}{\beta_\ell}\right] \tag{z}$$

Application of the initial condition, Equation d, to Equation x using Equation y and Equation z, leads to $A_\ell = v_\ell$. The solution for the temperature distribution is

$$\Theta(x,t) = \sum_{\ell=1}^{\infty} v_\ell e^{-\lambda_\ell t} D_\ell \left[\cos\left(\sqrt{\lambda_\ell}\, x\right) + \frac{Bi}{\lambda_\ell}\sin\left(\sqrt{\lambda_\ell}\, x\right)\right] \tag{aa}$$

Evaluation of Equation aa at $x = 0$ leads to

$$\Theta(0,t) = \sum_{\ell=1}^{\infty} v_\ell e^{-\lambda_\ell t} D_\ell \tag{bb}$$

The function $f(x) = 1$ is still not in the domain of the Sturm-Liouville problem, but the solution of Equation aa leads to a time-dependent temperature at each end.

(c) Let w be the width of the plate and p its thickness. The area over which heat conduction occurs is $A = wp$, while the perimeter for convection is $P = 2(w + p)$. A second Biot number is defined as $Bi_2 = hPL/kA$. The nondimensional governing partial differential equation when convective heat transfer across the surface is included is

$$\frac{\partial \Theta}{\partial t} = \frac{\partial^2 \Theta}{\partial x^2} - Bi_2 \Theta \tag{cc}$$

The boundary conditions are those of Equation r and Equation s and the initial condition is that of Equation d. A product-solution assumption, $\Theta(x,t) = X(x)T(t)$, when substituted into Equation cc, leads to

$$-\frac{1}{X(x)}\frac{d^2X}{dt^2} = -\frac{1}{T}\left(\frac{dT}{dt} + Bi_2\right) \tag{dd}$$

Application of the usual separation argument to Equation dd, using λ as the separation parameter, leads to

$$\frac{d^2X}{dx^2} + \lambda X = 0 \tag{ee}$$

$$\frac{dT}{dt} + (Bi_2 + \lambda)T = 0 \tag{ff}$$

The problem for $X(x)$ is the same as that of part (b) and thus has the same solution. The only difference in this problem is the solution for Equation ff, which for a specific λ_ℓ is

$$T_\ell(t) = A_\ell e^{-(\lambda_\ell + Bi_2)t} \tag{gg}$$

The temperature distribution is

$$\Theta(x,t) = e^{-Bi_2 t} \sum_{\ell=1}^{\infty} v_\ell e^{-\lambda_\ell t} D_\ell \left[\cos\left(\sqrt{\lambda_\ell}\, x\right) + \frac{Bi}{\lambda_\ell} \sin\left(\sqrt{\lambda_\ell}\, x\right)\right] \tag{hh}$$

where all terms are as specified in the solution to part (b).

(d) The temperature distribution is two-dimensional, $\Theta(x,y,t)$. The nondimensional governing equation is

$$\frac{\partial \Theta}{\partial t} = \frac{\partial^2 \Theta}{\partial x^2} + \frac{\partial^2 \Theta}{\partial y^2} \tag{ii}$$

The boundary conditions for conduction heat transfer over the entire surface are

$$\frac{\partial \Theta}{\partial x}(0,y,t) - Bi\Theta(1,y,t) = 0 \tag{jj}$$

$$\frac{\partial \Theta}{\partial x}(1,y,t) + Bi\Theta(1,y,t) = 0 \tag{kk}$$

$$\frac{\partial \Theta}{\partial y}(x,0,t) - Bi\Theta(x,0,t) = 0 \tag{ll}$$

$$\frac{\partial \Theta}{\partial y}\left(x,\frac{p}{L}y,t\right) + Bi\Theta\left(x,\frac{p}{L}y,t\right) = 0 \tag{mm}$$

The initial condition is that of Equation d, $\Theta(x,y,0) = 1$.

A product solution of Equation ii is assumed to be of the form

$$\Theta(x,y,t) = \Phi(x,y)T(t) \tag{nn}$$

which when substituted into Equation ii, leads to

$$-\frac{1}{\Phi}\left(\frac{\partial^2 \Phi}{\partial x^2} + \frac{\partial^2 \Phi}{\partial y^2}\right) = -\frac{1}{T}\frac{\partial T}{\partial t} \tag{oo}$$

Application of the usual separation argument to Equation oo and defining a separation parameter λ leads to

$$\frac{\partial^2 \Phi}{\partial x^2} + \frac{\partial^2 \Phi}{\partial y^2} = -\lambda\Phi \tag{pp}$$

$$\frac{dT}{dt} + \lambda T = 0 \tag{qq}$$

Use of the product solution in the boundary conditions, Equation jj, Equation kk, Equation ll, and Equation mm, gives

$$\frac{\partial \Phi}{\partial x}(0,y) - Bi\Phi(1,y) = 0 \tag{rr}$$

$$\frac{\partial \Phi}{\partial x}(1,y) + Bi\Phi(1,y) = 0 \tag{ss}$$

$$\frac{\partial \Phi}{\partial y}(x,0) - Bi\Phi(x,0) = 0 \tag{tt}$$

$$\frac{\partial \Phi}{\partial y}\left(x,\frac{p}{L}y\right) + Bi\Phi\left(x,\frac{p}{L}y\right) = 0 \tag{uu}$$

A product solution is assumed for Equation pp subject to Equation rr, Equation ss, Equation tt, and Equation uu as follows:

$$\Phi(x,y) = X(x)Y(y) \tag{vv}$$

Substitution of Equation vv into Equation pp, Equation rr, Equation ss, Equation tt, and Equation uu leads to problems similar to those solved in Example 6.2. The solution, Equation h of Example 6.2, is repeated below:

$$\begin{aligned}\Phi_{\ell,m} &= X_m(x)Y_\ell(y)\\ &= C_\ell D_m\left[\cos\left(\sqrt{\beta_m}\,x\right)+\frac{Bi}{\beta_m}\sin\left(\sqrt{\beta_m}\,x\right)\right]\left[\cos\left(\sqrt{\mu_\ell}\,y\right)+\frac{Bi}{\mu_\ell}\sin\left(\sqrt{\mu_\ell}\,y\right)\right]\end{aligned} \tag{ww}$$

Equations defining the parameters in Equation ww are given as Equation z, Equation aa, Equation bb, Equation cc, Equation dd, Equation ee, Equation ff, and Equation gg of Example 6.2.

The details of the remainder of the solution are similar to those in part (a). The solution of Equation qq is Equation w for a specific eigenvalue, $\lambda_{\ell,m}$. The general solution of the partial differential equation is

$$\begin{aligned}\Theta(x,y,t) = \sum_{\ell=1}^{\infty}\sum_{m=1}^{\infty} C_\ell D_m A_\ell e^{-\lambda_{\ell,m}t}\left[\cos\left(\sqrt{\beta_m}\,x\right)+\frac{Bi}{\beta_m}\sin\left(\sqrt{\beta_m}\,x\right)\right]\\ \times\left[\cos\left(\sqrt{\mu_\ell}\,y\right)+\frac{Bi}{\mu_\ell}\sin\left(\sqrt{\mu_\ell}\,y\right)\right]\end{aligned} \tag{xx}$$

Application of the initial condition leads to $A_\ell = v_\ell$, where, from Equation oo of Example 6.2,

$$\begin{aligned}v_{\ell,m} = \frac{C_\ell D_m}{\sqrt{\mu_\ell \beta_m}}\left[\sin\left(\sqrt{\beta_m}\right)-\frac{Bi}{\beta_m}\cos\left(\sqrt{\beta_m}\right)+\frac{Bi}{\beta_m}\right]\\ \times\left[\sin\left(\sqrt{\mu_\ell}\,\frac{p}{L}\right)-\frac{Bi}{\mu_\ell}\cos\left(\sqrt{\mu_\ell}\,\frac{p}{L}\right)+\frac{Bi}{\mu_\ell}\right]\end{aligned} \tag{yy}$$

Example 6.4 The nondimensional problem governing the displacement, $u(x,t)$, of the elastic bar shown in Figure 6.6 is

$$\frac{\partial}{\partial x}\left[\alpha(x)\frac{\partial u}{\partial x}\right] = \alpha(x)\frac{\partial^2 u}{\partial x^2} \tag{a}$$

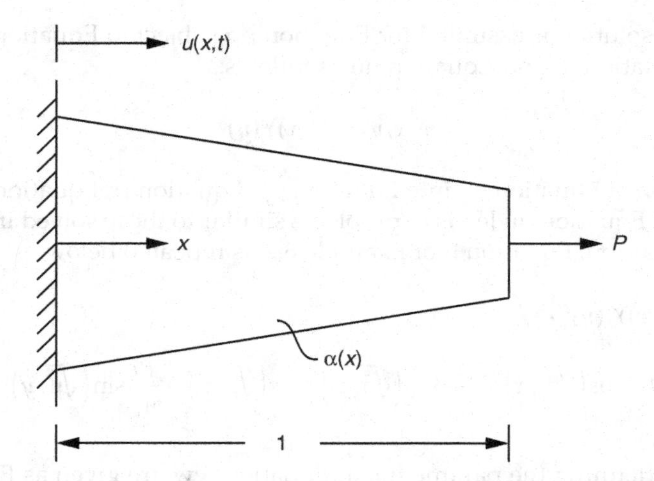

Figure 6.6 The force P is removed from the end of the elastic bar at $t=0$. The resulting motion is obtained through solution of the wave equation with a nonhomogeneous initial condition.

The bar is fixed at $x=0$ and free at $x=1$, leading to the boundary conditions,

$$u(0,t)=0 \tag{b}$$

$$\frac{\partial u}{\partial x}(1,t)=0 \tag{c}$$

Before $t=0$, the right end of the bar is subject to a force P which leads to a static displacement in the bar, given by

$$u(x,0)=\frac{P}{A_0E_0}\int_0^x \frac{1}{\alpha(\xi)}d\xi \tag{d}$$

The force is removed at $t=0$, resulting in subsequent motion of the bar. Since the bar is at rest at $t=0$,

$$\frac{\partial u}{\partial t}(x,0)=0 \tag{e}$$

Determine the response if (a) $\alpha(x)=1$ and (b) $\alpha(x)=(1-0.1x)^2$.

Solution (a) For $\alpha=1$, the partial differential equation and the initial condition, Equation d, become

$$\frac{\partial^2 u}{\partial x^2}=\frac{\partial^2 u}{\partial t^2} \tag{f}$$

$$u(x,0) = \frac{Px}{A_0 E_0} \tag{g}$$

A product solution of Equation a can be assumed to be of the form $u(x,t) = X(x)T(t)$, which when substituted into Equation f, leads to

$$-\frac{1}{X(x)}\frac{d^2X}{dx^2} = -\frac{1}{T(t)}\frac{d^2T}{dt^2} \tag{h}$$

Applying the usual separation argument to Equation h and choosing λ as the separation constant leads to

$$\frac{d^2X}{dx^2} + \lambda X = 0 \tag{i}$$

$$\frac{d^2T}{dt^2} + \lambda T = 0 \tag{j}$$

Application of the product solution to the boundary conditions, Equation b and Equation c, results in

$$X(0) = 0 \tag{k}$$

$$\frac{dX}{dx}(1) = 0 \tag{l}$$

The eigenvalues and normalized eigenvectors of the system of Equation i, Equation k, and Equation l are

$$\lambda_\ell = \left[\frac{(2\ell-1)\pi}{2}\right]^2 \tag{m}$$

$$X_\ell(x) = \sqrt{2}\sin\left[\frac{(2\ell-1)\pi}{2}x\right] \tag{n}$$

For a specific λ_ℓ, the general solution of Equation j is

$$T_\ell(t) = A_\ell \cos\left[\frac{(2\ell-1)\pi}{2}t\right] + B_\ell \sin\left[\frac{(2\ell-1)\pi}{2}t\right] \tag{o}$$

A solution which satisfies Equation a, Equation b, and Equation c is

$$u(x,t) = \sum_{\ell=1}^{\infty}\left\{A_\ell \cos\left[\frac{(2\ell-1)\pi}{2}t\right] + B_\ell \sin\left[\frac{(2\ell-1)\pi}{2}t\right]\right\}\sqrt{2}\sin\left[\frac{(2\ell-1)\pi}{2}x\right] \tag{p}$$

Application of the initial condition, Equation e, to Equation p requires $B_\ell = 0$.

The initial condition, Equation g, is imposed through an eigenvector expansion,

$$\frac{Px}{A_0E_0} = \sum_{\ell=1}^{\infty} v_\ell \sqrt{2} \sin\left[\frac{(2\ell-1)\pi}{2}x\right] \tag{q}$$

where

$$\begin{aligned} v_\ell &= \left(\frac{Px}{A_0E_0}, \sqrt{2}\sin\left[\frac{(2\ell-1)\pi}{2}x\right]\right) \\ &= \int_0^1 \left(\frac{Px}{A_0E_0}\right)\sqrt{2}\sin\left[\frac{(2\ell-1)\pi}{2}x\right]dx \\ &= \sqrt{2}\frac{4P(-1)^{\ell+1}}{[(2\ell-1)\pi]^2 A_0E_0} \end{aligned} \tag{r}$$

Application of Equation g to Equation p leads to $A_\ell = v_\ell$. Then, substituting Equation r into Equation p,

$$u(x,t) = \frac{8P}{\pi^2 A_0 E_0} \sum_{\ell=1}^{\infty} \frac{(-1)^{\ell+1}}{(2\ell-1)^2} \sin\left[\frac{(2\ell-1)\pi}{2}x\right]\cos\left[\frac{(2\ell-1)\pi}{2}t\right] \tag{s}$$

(b) Using $\alpha(x) = (1-0.1x)^2$, Equation a and Equation d become

$$\frac{\partial}{\partial x}\left[(1-0.1x)^2\frac{\partial u}{\partial x}\right] = (1-0.1x)^2\frac{\partial^2 u}{\partial t^2} \tag{t}$$

$$u(x,0) = \frac{Px}{A_0E_0(1-0.1x)} \tag{u}$$

Substitution of the product solution $u(x,t) = X(x)T(t)$ into Equation t leads to

$$-\frac{1}{(1-0.1x)^2 X(x)}\frac{d}{dx}\left[(1-0.1x)^2\frac{dX}{dx}\right] = -\frac{1}{T(t)}\frac{d^2T}{dt^2} \tag{v}$$

Applying the usual separation argument to Equation v and defining the separation parameter as λ results in

$$-\frac{1}{(1-0.1x)^2}\frac{d}{dx}\left[(1-0.1x)^2\frac{dX}{dx}\right]=\lambda X \tag{w}$$

$$\frac{d^2T}{dx^2}+\lambda T=0 \tag{x}$$

Equation w with boundary conditions Equation k and Equation l constitutes a Sturm-Liouville problem with $p(x)=-(1-0.1x)^2$, $q(x)=0$ and $r(x)=(1-0.1x)^2$. The results of Section 5.4 prove that $\mathbf{L}_x$ is self-adjoint and positive definite with respect to the inner product,

$$(f(x),g(x))_r=\int_0^1 f(x)g(x)(1-0.1x)^2\,dx \tag{y}$$

The solution of Equation w is aided by a change in independent variable; let

$$z=1-0.1x \tag{z}$$

Substitution of Equation z into Equation w leads to

$$\frac{d}{dz}\left(z^2\frac{dX}{dz}\right)+100\lambda z^2X=0 \tag{aa}$$

Comparison with Equation 3.108 shows that Equation aa has a Bessel-function solution of the form of Equation 3.110 with $r=2, s=2$, and $b=10\sqrt{\lambda}$. The resulting general solution of Equation aa is

$$X(z)=C_1z^{-\frac{1}{2}}J_{-\frac{1}{2}}\left(10\sqrt{\lambda}z\right)+C_2z^{-\frac{1}{2}}Y_{-\frac{1}{2}}\left(10\sqrt{\lambda}z\right) \tag{bb}$$

Equation bb may also be written in terms of spherical Bessel functions as

$$X(z)=C_1j_0\left(10\sqrt{\lambda}z\right)+C_2y_0\left(10\sqrt{\lambda}z\right) \tag{cc}$$

Substitution of Equation z into Equation cc leads to

$$X(x)=C_1j_0\left(10\sqrt{\lambda}(1-0.1x)\right)+C_2y_0\left(10\sqrt{\lambda}(1-0.1x)\right) \tag{dd}$$

Application of Equation k to Equation dd gives

$$C_2 = -\frac{j_0\left(10\sqrt{\lambda}\right)}{y_0\left(10\sqrt{\lambda}\right)} C_1 \tag{ee}$$

Substitution of Equation l into Equation dd using Equation ee leads to a transcendental equation for λ,

$$j_0'\left(9\sqrt{\lambda}\right) y_0\left(10\sqrt{\lambda}\right) - y_0'\left(9\sqrt{\lambda}\right) j_0\left(10\sqrt{\lambda}\right) = 0 \tag{ff}$$

Equation ff has an infinite number of solutions which are indexed by λ_ℓ, $\ell = 1, 2, \ldots$. The eigenvector corresponding to λ_ℓ is

$$X_\ell(x) = C_\ell \left[j_0\left(10\sqrt{\lambda_\ell}(1-0.1x)\right) - \frac{j_0\left(10\sqrt{\lambda_\ell}\right)}{y_0\left(10\sqrt{\lambda_\ell}\right)} y_0\left(10\sqrt{\lambda_\ell}(1-0.1x)\right) \right] \tag{gg}$$

The eigenvectors can be normalized by requiring $(X_\ell, X_\ell) = 1$ or

$$\int_0^1 C_\ell^2 \left[j_0\left(10\sqrt{\lambda_\ell}(1-0.1x)\right) - \frac{j_0\left(10\sqrt{\lambda_\ell}\right)}{y_0\left(10\sqrt{\lambda_\ell}\right)} y_0\left(10\sqrt{\lambda_\ell}(1-0.1x)\right) \right]^2 (1-0.1x)^2 \, dx = 1 \tag{hh}$$

Closed-form solutions of the integrals in Equation hh are not known. For a specific λ_ℓ, the solution of Equation x is

$$T_\ell(t) = A_\ell \cos\left(\sqrt{\lambda_\ell} t\right) + B_\ell \sin\left(\sqrt{\lambda_\ell} t\right) \tag{ii}$$

The remainder of the solution is similar to that for part (a). The general solution is a linear combination of all product solutions. Application of Equation e leads to $B_\ell = 0$. Application of Equation u leads to $A_\ell = v_\ell$, where

$$\begin{aligned} v_\ell &= \left(\frac{Px}{A_0 E_0 (1-0.1x)}, X_\ell \right)_r \\ &= \int_0^1 \frac{Px}{A_0 E_0} C_\ell \left[j_0\left(10\sqrt{\lambda_\ell}(1-0.1x)\right) - \frac{j_0\left(10\sqrt{\lambda_\ell}\right)}{y_0\left(10\sqrt{\lambda_\ell}\right)} y_0\left(10\sqrt{\lambda_\ell}(1-0.1x)\right) \right] \\ &\quad \times (1-0.1x) dx \end{aligned} \tag{jj}$$

6.4 Nonhomogeneous partial differential equations

Direct application of separation of variables requires that the partial differential operator be separable and that a sufficient number of boundary conditions be homogeneous to generate eigenvalue problems. If the problem is unsteady state, eigenvalue problems must be generated in all spatial directions. If the problem is steady, eigenvalue problems must be generated in all but one of the spatial directions.

Nonhomogeneous problems occur when the governing partial differential equation is nonhomogeneous or when there are too many nonhomogeneous boundary conditions to apply separation of variables directly. Two methods are presented in this section to handle nonhomogeneous terms. A superposition method can often be used for linear problems in which the solution is a superposition of solutions, each of which is obtained through application of separation of variables or through solution of an ordinary differential equation. Nonhomogeneous problems which do not lend themselves to superposition can be solved using an eigenvector expansion method in which a solution is assumed to be an eigenvector expansion of solutions of the corresponding homogeneous problem obtained by removing the nonhomogeneous terms. These approaches are illustrated through the following examples.

Example 6.5 The steady nondimensional temperature distribution in the slab shown in Figure 6.7 is governed by the two-dimensional Laplace's equation,

$$\frac{\partial^2 \theta}{\partial x^2} + \frac{\partial^2 \theta}{\partial y^2} = 0 \tag{a}$$

The boundary conditions corresponding to Figure 6.7 are

$$\theta(0, y) = 0 \tag{b}$$

$$\theta(1, y) = \delta_1 \tag{c}$$

$$\theta(x, 0) = 0 \tag{d}$$

$$\theta(x, \alpha) = \delta_2 \tag{e}$$

Separation of variables cannot be directly applied to solve Equation a subject to the boundary conditions of Equation b, Equation c, Equation d, and Equation e because the system has nonhomogeneous boundary conditions for constant values of both x and y. Consider the superposition,

$$\theta(x, y) = \theta_1(x, y) + \theta_2(x, y) \tag{f}$$

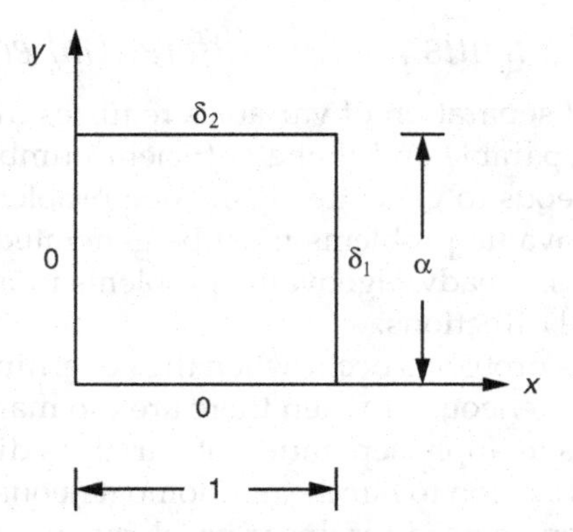

Figure 6.7 The slab has nonhomogeneous boundary conditions on its sides, described by constant values of two independent variables. Separation of variables is not directly applicable to Laplace's equation subject to these boundary conditions. Therefore, a superposition method is used.

where $\theta_1(x,y)$ satisfies

$$\frac{\partial^2\theta_1}{\partial x^2}+\frac{\partial^2\theta_1}{\partial y^2}=0 \tag{g}$$

$$\theta_1(0,y)=0 \tag{h}$$

$$\theta_1(1,y)=\delta_1 \tag{i}$$

$$\theta_1(x,0)=0 \tag{j}$$

$$\theta_1(x,\alpha)=0 \tag{k}$$

and $\theta_2(x,y)$ satisfies

$$\frac{\partial^2\theta_2}{\partial x^2}+\frac{\partial^2\theta_2}{\partial y^2}=0 \tag{l}$$

$$\theta_2(0,y)=0 \tag{m}$$

$$\theta_2(1,y)=0 \tag{n}$$

$$\theta_2(x,0)=0 \tag{o}$$

$$\theta_2(x,\alpha)=\delta_2 \tag{p}$$

It is easy to verify by substitution that $\theta(x,y)$ given by Equation f satisfies Equation a, Equation b, Equation c, Equation d, and Equation e.

The method of separation of variables is used to solve for $\theta_1(x,y)$ as

$$\theta_1(x,y) = \frac{4\delta_1}{\pi}\sum_{n=1,3,5}^{\infty}\frac{\sin\left(\frac{n\pi}{\alpha}y\right)\sinh\left(\frac{n\pi}{\alpha}x\right)}{n\sinh(n\pi)} \tag{q}$$

The solution for $\theta_2(x,y)$ is obtained by interchanging the roles of x and y, using δ_2 in place of δ_1, replacing α by 1, and noting that $\sinh(n\pi)$ should be replaced by $\sinh(n\pi a)$. Thus,

$$\theta_2(x,y) = \frac{4\delta_2}{\pi}\sum_{n=1,3,5}^{\infty}\frac{\sin(n\pi x)\sinh(n\pi y)}{n\sinh(n\pi\alpha)} \tag{r}$$

The superposition formula, Equation f of Example 6.5, is obvious, and the choice of problems defining the superposition components is obvious.

Example 6.6 The slab shown in Figure 6.8 is subject to an internal heat generation which is a function of x only. The partial differential equation governing the nondimensional temperature distribution is

$$\frac{\partial^2\theta}{\partial x^2} + \frac{\partial^2\theta}{\partial y^2} = cf(x) \tag{a}$$

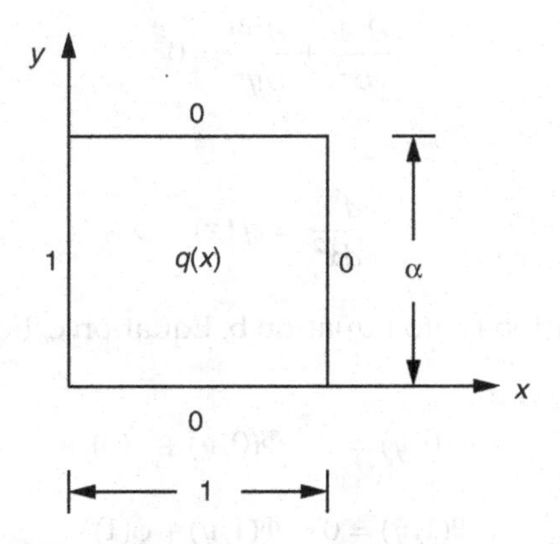

Figure 6.8 Internal heat generation leads to a nonhomogeneous partial differential equation. Since the nonhomogeneous term is a function of x only, a superposition of a function of x only and a separate function of x and y is developed.

The boundary conditions corresponding to the slab shown in Figure 6.7 are

$$\theta(0,y)=1 \tag{b}$$

$$\theta(1,y)=0 \tag{c}$$

$$\theta(x,0)=0 \tag{d}$$

$$\theta(x,\alpha)=0 \tag{e}$$

Determine a superposition formula to solve Equation a subject to Equation b, Equation c, Equation d, and Equation e. The problems defining each element of the superposition formula must be directly solvable by separation of variables or be represented by ordinary differential equations.

Solution Since the internal heat generation is a function of x only, it is reasonable to assume that it leads to a term dependent only on x, suggesting a superposition formula of the form

$$\theta(x,y)=\Phi(x,y)+\phi(x) \tag{f}$$

Substitution of Equation f into Equation a leads to

$$\frac{\partial^2\Phi}{\partial x^2}+\frac{\partial^2\Phi}{\partial y^2}+\frac{d^2\phi}{dx^2}=cf(x) \tag{g}$$

It is desired to obtain a homogeneous problem for $\Phi(x,y)$. Choosing

$$\frac{\partial^2\Phi}{\partial x^2}+\frac{\partial^2\Phi}{\partial y^2}=0 \tag{h}$$

leads to

$$\frac{d^2\phi}{dx^2}=cf(x) \tag{i}$$

Substitution of Equation f into Equation b, Equation c, Equation d, and Equation e gives

$$\theta(0,y)=1=\Phi(0,y)+\phi(0) \tag{j}$$

$$\theta(1,y)=0=\Phi(1,y)+\phi(1) \tag{k}$$

$$\theta(x,0)=0=\Phi(x,0)+\phi(x) \tag{l}$$

$$\theta(x,\alpha)=0=\Phi(x,\alpha)+\phi(x) \tag{m}$$

Equation l and Equation m lead to nonhomogeneous boundary conditions at constant values of y for $\Phi(x,y)$:

$$\Phi(x,0) = -\phi(x) \tag{n}$$

$$\Phi(x,\alpha) = -\phi(x) \tag{o}$$

The problem for $\Phi(x,y)$ can be solved directly by separation of variables only if the boundary conditions defining its problem at constant values of x are homogeneous,

$$\Phi(0,y) = 0 \tag{p}$$

$$\Phi(1,y) = 0 \tag{q}$$

The choices of Equation p and Equation q, when used in Equation k and Equation l, lead to

$$\phi(0) = 1 \tag{r}$$

$$\phi(1) = 0 \tag{s}$$

The problem governing $\phi(x)$ is the ordinary differential equation of Equation i subject to the boundary conditions of Equation r and Equation s. The differential equation is easily solved for a specific $f(x)$ by integrating both sides with respect to x twice and then solving for the constants of integration through application of Equation r and Equation s. The problem governing $\Phi(x,y)$ is the partial differential equation of Equation h subject to the boundary conditions of Equation n, Equation o, Equation p, and Equation q. The problem is directly solvable by separation of variables with the eigenvalue problem in the x-direction. Boundary conditions p and q are applied using eigenvector expansions of $\phi(x)$.

Example 6.7 The steady nondimensional temperature distribution of the slab shown in Figure 6.9 is governed by the problem,

$$\frac{\partial^2 \theta_s}{\partial x^2} + \frac{\partial^2 \theta_s}{\partial y^2} = 0 \tag{a}$$

$$\theta_s(0,y) = 0 \tag{b}$$

$$\frac{\partial \theta_s}{\partial x}(1,y) + Bi\theta_s(1,y) = 0 \tag{c}$$

$$\theta_s(x,0) = 0 \tag{d}$$

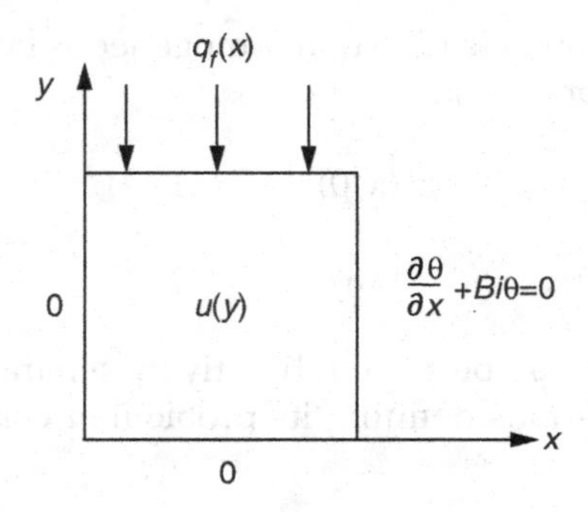

Figure 6.9 The slab has a steady-state temperature distribution when the internal heat generation is initiated. Since the internal heat generation is independent of time, a different steady state will be achieved. Therefore, an initial attempt at a superposition involves a transient temperature distribution and a steady-state temperature distribution. Nonhomogeneous boundary conditions make several additional superpositions necessary.

$$\frac{\partial \theta_s}{\partial y}(x,\alpha) = c_1 q_f(x) \tag{e}$$

where $q_f(x)$ is an imposed heat flux. The problem described by Equation a, Equation b, Equation c, Equation d, and Equation e can be solved directly by separation of variables. The eigenvalue problem is in the x-direction, and the boundary condition at $x=\alpha$ is imposed using an eigenfunction expansion for $q_f(x)$.

At $t=0$, an internal heat source begins to function. The heat added is constant with x, but a function of y. The initiation of the heat generation leads to a transient temperature distribution, $\Theta(x,y,t)$, in the slab. The unsteady state problem is governed by

$$\frac{\partial^2 \Theta}{\partial x^2} + \frac{\partial^2 \Theta}{\partial y^2} = \frac{\partial \Theta}{\partial t} - c_2 u(y) \tag{f}$$

$$\Theta(x,y,0) = \theta_s(x,y) \tag{g}$$

$$\Theta(0,y,t) = 0 \tag{h}$$

$$\frac{\partial \Theta}{\partial x}(1,y,t) + Bi\Theta(1,y,t) = 0 \tag{i}$$

$$\Theta(x,0,t) = 0 \tag{j}$$

$$\frac{\partial \Theta}{\partial y}(x,\alpha,t) = c_1 q_f(x) \tag{k}$$

Determine a superposition formula which can be used to solve the problem governed by Equation f, Equation g, Equation h, Equation i, Equation j, and Equation k, such that the problem governing each term in the superposition can be solved directly by separation of variables or is an ordinary differential equation. Specify the differential equations and all initial and boundary conditions governing each problem.

Solution The internal heat generation is constant with time. A new steady state will eventually be achieved. The initial transient solution will die out, leaving this new steady state. A reasonable superposition is thus

$$\Theta(x,y,t) = \Phi(x,y,t) + S(x,y) \tag{l}$$

where $\Phi(x,y,t)$ is the transient response and $S(x,y)$ is the eventual steady-state response.

Substitution of Equation l into Equation f leads to

$$\frac{\partial^2 \Phi}{\partial x^2} + \frac{\partial^2 \Phi}{\partial y^2} + \frac{\partial^2 S}{\partial x^2} + \frac{\partial^2 S}{\partial y^2} = \frac{\partial \Phi}{\partial t} - c_2 u(y) \tag{m}$$

The transient response is a function of three independent variables, whereas the steady-state response is a function of two independent variables. For simplicity, it is therefore desirable to choose a homogeneous problem to given in the transient response. The separable partial differential equation governing $\Phi(x,y)$ is chosen as

$$\frac{\partial^2 \Phi}{\partial x^2} + \frac{\partial^2 \Phi}{\partial y^2} = \frac{\partial \Phi}{\partial t} \tag{n}$$

Subtract. Equation n from Equation n leads to

$$\frac{\partial^2 S}{\partial x^2} + \frac{\partial^2 S}{\partial y^2} = -c_2 u(y) \tag{o}$$

Substitution of the superposition assumption, Equation l, into the initial condition, Equation g, leads to

$$\theta_s(x,y) = \Phi(x,y,0) + S(x,y) \tag{p}$$

Equation p is satisfied by choosing

$$\Phi(x,y,0) = \theta_s(x,y) - S(x,y) \tag{q}$$

Substitution of Equation l into the boundary conditions, Equation h, Equation i, Equation j, and Equation k, results in

$$\Theta(0,y,t)=0=\Phi(0,y,t)+S(0,y) \tag{r}$$

$$\frac{\partial\Theta}{\partial x}(1,y,t)+Bi\Theta(1,y,t)=0=\frac{\partial\Phi}{\partial x}(1,y,t)+Bi\Phi(1,y,t)+\frac{\partial S}{\partial x}(1,y)+BiS(1,y) \tag{s}$$

$$\Theta(x,0,t)=0=\Phi(x,0,t)+S(x,0) \tag{t}$$

$$\frac{\partial\Theta}{\partial y}(x,\alpha,t)=c_1q_f(x)=\Phi(x,\alpha,t)+S(x,\alpha) \tag{u}$$

The problem for $\Phi(x,y,t)$ can be solved directly by separation of variables only if all boundary conditions which it satisfies are homogeneous. To this end, boundary conditions are chosen from Equation r, Equation s, Equation t, and Equation u such that

$$\Phi(0,y,t)=0 \tag{v}$$

$$\frac{\partial\Phi}{\partial x}(1,y,t)+Bi\Phi(1,y,t)=0 \tag{w}$$

$$\Phi(x,0,t)=0 \tag{x}$$

$$\frac{\partial\Phi}{\partial x}(x,\alpha,t)=0 \tag{y}$$

The boundary conditions which must be satisfied by $S(x,y)$ are therefore

$$S(0,y)=0 \tag{z}$$

$$\frac{\partial S}{\partial x}(1,y)+BiS(1,y)=0 \tag{aa}$$

$$S(x,0)=0 \tag{bb}$$

$$\frac{\partial S}{\partial x}(x,\alpha)=c_1q_f(x) \tag{cc}$$

The problem for $\Phi(x,y,t)$ defined by Equation n, Equation q, Equation v, Equation w, Equation x, and Equation y can be directly solved using separation of variables once $S(x,y)$ is known. The problem defining $S(x,y)$, Equation o, Equation z, Equation aa, Equation bb, and Equation cc, is nonhomogeneous. The governing partial differential equation has a nonhomogeneous term which is a function of y only. This suggests a superposition formula of the form

$$S(x,y)=A(x,y)+B(y) \tag{dd}$$

Substitution of Equation dd into Equation o and requiring a homogeneous partial differential equation to be satisfied by $A(x,y)$ leads to

$$\frac{\partial^2 A}{\partial x^2} + \frac{\partial^2 A}{\partial y^2} = 0 \tag{ee}$$

$$\frac{d^2 B}{dy^2} = -c_2 u(y) \tag{ff}$$

The boundary conditions for $S(x,y)$, Equation z, Equation aa, Equation bb, and Equation cc, are satisfied when the superposition of Equation dd is used by choosing

$$A(0,y) = -B(y) \tag{gg}$$

$$\frac{\partial A}{\partial x}(1,y) + BiA(1,y) = -BiB(y) \tag{hh}$$

$$A(x,0) = 0 \tag{ii}$$

$$\frac{\partial \mathrm{A}}{\partial y}(x,\alpha) = cq_f(x) \tag{jj}$$

$$\mathrm{B}(0) = 0 \tag{kk}$$

$$\frac{dB}{dy}(\alpha) = 0 \tag{ll}$$

$B(y)$ is easily obtained by solving the ordinary differential equation, Equation ff, subject to the boundary conditions, Equation kk and Equation ll. However, the problem governing $A(x,y)$, Equation ee, Equation gg, Equation hh, Equation ii, Equation jj, Equation kk, and Equation ll, has nonhomogeneous boundary conditions at constant values of x and a constant value of y and is itself nonhomogeneous. This final nonhomogeneity is relieved by a simple superposition similar to that used in Example 16.1,

$$A(x,y) = C(x,y) + D(x,y) \tag{mm}$$

where

$$\frac{\partial^2 C}{\partial x^2} + \frac{\partial^2 C}{\partial y^2} = 0 \tag{nn}$$

$$C(0,y) = -B(y) \tag{oo}$$

$$\frac{\partial C}{\partial x}(1,y) + BiC(1,y) = -BiB(y) \tag{pp}$$

$$C(x,0)=0 \tag{qq}$$

$$\frac{\partial C}{\partial y}(x,\alpha)=0 \tag{rr}$$

and

$$\frac{\partial^2 D}{\partial x^2}+\frac{\partial^2 D}{\partial y^2}=0 \tag{ss}$$

$$D(0,y)=0 \tag{tt}$$

$$\frac{\partial D}{\partial x}(1,y)+BiD(1,y)=0 \tag{uu}$$

$$D(x,0)=0 \tag{vv}$$

$$\frac{\partial D}{\partial y}(x,\alpha)=c_2q_f(x) \tag{ww}$$

The final superposition formula is

$$\Theta(x,y,t)=\Phi(x,y,t)+B(y)+C(x,y)+D(x,y) \tag{xx}$$

The technique used in the previous example, using a superposition of a transient solution and a steady-state solution, is applicable in problems where an initial steady state is disturbed due to a change in conditions and a new steady state is expected to develop after the transient response decays. The thin slab shown in Figure 6.10 is at a uniform temperature when it is heated from below, leading to a time-dependent temperature in the slab. The resulting problem is nonhomogeneous due to a nonhomogeneous boundary condition. However, if all other sides are insulated, there is no surface for heat transfer with the ambient medium, and instead of reaching a steady state, the temperature in the slab will increase without bound. In this case, a superposition assuming that the temperature is the sum of a transient

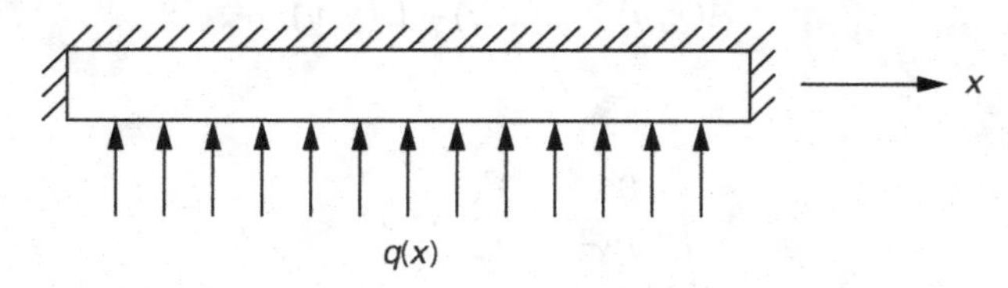

Figure 6.10 The bar has a uniform temperature when the heat flux is applied. All other sides are insulated, and therefore a new steady state is not attained. A superposition formula must include a function of time only.

response and an eventual steady state will not work. If one were attempted, it would become apparent that the nonhomogeneous boundary condition cannot be satisfied.

The difficulty in attaining a new steady state can also be explained by noting that the eigenvalue problem in the spatial direction obtained when solving for the transient response is only non-negative definite. The system has a zero eigenvalue due to the imposition of derivative boundary conditions at each boundary. Therefore, when the general solution is formed from a superposition of eigenvectors, the transient term corresponding to the zero eigenvalue does not approach zero for large t.

A successful superposition formula for the system shown in Figure 6.10 includes a function of time which is allowed to grow without bound,

$$\theta(x,t) = \Phi(x,t) + \phi(x) + \sigma(t) \tag{6.67}$$

Superposition methods are not successful when the nonhomogeneous term is a function of time or a function of all spatial variables. In such cases, the nonhomogeneity can be addressed using eigenvector expansions. The following two examples illustrate this concept.

Example 6.8 The thin slab shown in Figure 6.11 is subject to an internal heat generation which is a function of x and y. The problem is formulated as

$$\frac{\partial^2\theta}{\partial x^2} + \frac{\partial^2\theta}{\partial y^2} = f(x,y) \tag{a}$$

$$\theta(0,y) = 0 \tag{b}$$

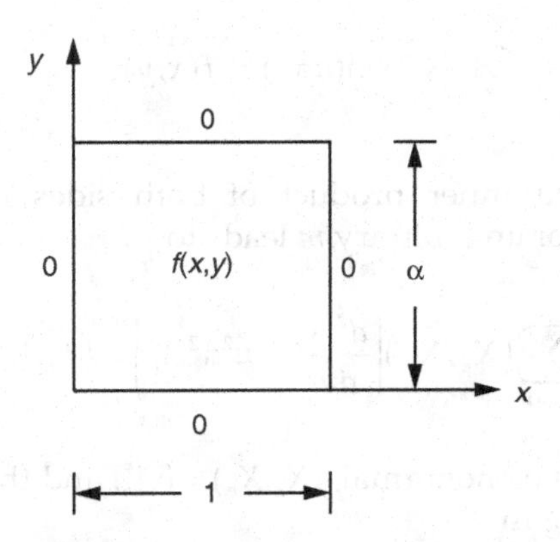

Figure 6.11 Internal heat generation leads to a nonhomogeneous differential equation. The nonhomogeneity is a function of both spatial variables, and therefore a superposition in which the temperature is expanded in a series of eigenvectors of the corresponding homogeneous problem is used. Orthogonality of the eigenvectors is used to uncouple the differential equations for the coefficients in the expansion.

$$\theta(1,y)=0 \tag{c}$$

$$\theta(x,0)=0 \tag{d}$$

$$\frac{\partial\theta}{\partial y}(x,\alpha)=0 \tag{e}$$

Determine the solution of Equation a, Equation b, Equation c, Equation d, and Equation e.

Solution Consider the eigenvalue problem, $LX=-d^2X/dx^2=\lambda X$, with $X(0)=0$ and $X(1)=1$. This is the eigenvalue problem obtained using a separation of variables method a neutral if Equation a is homogeneous and at least one of Equation d and Equation e is nonhomogeneous. The eigenvalues and normalized eigenvectors are $\lambda_n=n^2\pi^2$ and $X_n(x)=\sqrt{2}\sin(n\pi x)$. The eigenvectors are orthonormal with respect to the standard inner product on C[0,1].

For any y, $\theta(x,y)$ is in the domain of L. Thus it has an eigenvector expansion of the form

$$\theta(x,y)=\sum_{n=1}^{\infty}A_n(y)\sqrt{2}\sin(n\pi x) \tag{f}$$

Substitution of Equation f into Equation a leads to

$$\sum_{n=1}^{\infty}A_n(y)\left(-n^2\pi^2\right)\sqrt{2}\sin(n\pi x)+\sum_{n=1}^{\infty}\frac{d^2A_n}{dy^2}\sqrt{2}\sin(n\pi x)=f(x,y)$$

$$\sum_{n=1}^{\infty}\left(\frac{d^2A_n}{dy^2}-n^2\pi^2A_n\right)\sqrt{2}\sin(n\pi x)=f(x,y) \tag{g}$$

Taking the standard inner product of both sides of Equation g with $X_m(x)=\sqrt{2}\sin(m\pi x)$ for an arbitrary m leads to

$$\sum_{n=1}^{\infty}(X_n,X_m)\left(\frac{d^2A_n}{dy^2}-n^2\pi^2A_n\right)=(f,A_m) \tag{h}$$

The eigenvectors are orthonormal, $(X_n,X_m)=\delta_{n,m}$, and the sum collapses to a single term, resulting in

$$\frac{d^2A_m}{dy^2}-m^2\pi^2A_m=g_m(y)\quad m=1,2,... \tag{i}$$

where

$$g_m(y) = (f, X_m)$$
$$= \int_0^1 f(x,y)\sqrt{2}\sin(m\pi x)dx \tag{j}$$

Use of the eigenvector expansion leads to a set of uncoupled differential equations whose solutions are the expansion coefficients.

Substitution of Equation f into Equation d leads to

$$\theta(x,0) = 0 = \sum_{n=1}^{\infty} A_n(0)\sqrt{2}\sin(n\pi y) \tag{k}$$

The eigenvectors are linearly independent, and therefore a linear combination can be set equal to zero only if each coefficient is zero. Thus, Equation k is valid only if

$$A_n(0) = 0 \quad n = 1,2,\ldots \tag{l}$$

Using a similar argument, Equation e is satisfied only if

$$\frac{dA_n}{dy}(\alpha) = 0 \quad n = 1,2,\ldots \tag{m}$$

The coefficients in the eigenvector expansion are obtained by solving Equation g subject to Equation l and Equation m. The homogeneous solution of Equation g is

$$A_{m,h} = C_1\cosh(m\pi y) + C_2\sinh(m\pi y) \tag{n}$$

The particular solution is obtained using variation of parameters as

$$A_{m,p}(y) = -\cosh(m\pi y)\int_0^y \sinh(m\pi\tau)g_m(\tau)d\tau$$
$$+ \sinh(m\pi y)\int_0^y \cosh(m\pi\tau)g_m(\tau)d\tau$$
$$= \int_0^y \sinh[m\pi(y-\tau)]g_m(\tau)d\tau \tag{o}$$

The solution for $A_m(y)$ obtained by adding Equation n and Equation o and applying the initial conditions is

$$A_m = -\frac{\sinh(m\pi y)}{\cosh(m\pi\alpha)}\int_0^{\alpha} \cosh[m\pi(\alpha-\tau)]g_m(\tau)d\tau$$

$$+\int_0^{y} \sinh[m\pi(y-\tau)]g_m(\tau)d\tau \tag{p}$$

The solution for $\theta(x,y)$ is obtained by substitution of Equation j into Equation p and the result into Equation f:

$$\theta(x,y) = 2\sum_{n=1}^{\infty}\sin(n\pi x)\left\{-\frac{\sinh(m\pi y)}{\cosh(m\pi\alpha)}\right.$$

$$\times\int_0^{\alpha}\int_0^{1}\cosh[m\pi(\alpha-\tau)]\sin(n\pi\beta)f(\beta,\tau)d\beta d\tau$$

$$\left.+\int_0^{y}\int_0^{1}\sinh[n\pi(y-\tau)]\sin(n\pi\beta)f(\beta,\tau)d\beta d\tau\right\} \tag{q}$$

Example 6.9 Figure 6.12 illustrates a uniform circular shaft, fixed at one end, with a thin disk attached at its other end. The system is at rest in equilibrium when a time-dependent torque is applied to the thin disk. The nondimensional problem governing the resulting torsional oscillations of the shaft and disk is

$$\frac{\partial^2\theta}{\partial x^2} = \frac{\partial^2\theta}{\partial t^2} \tag{a}$$

$$\theta(0,t) = 0 \tag{b}$$

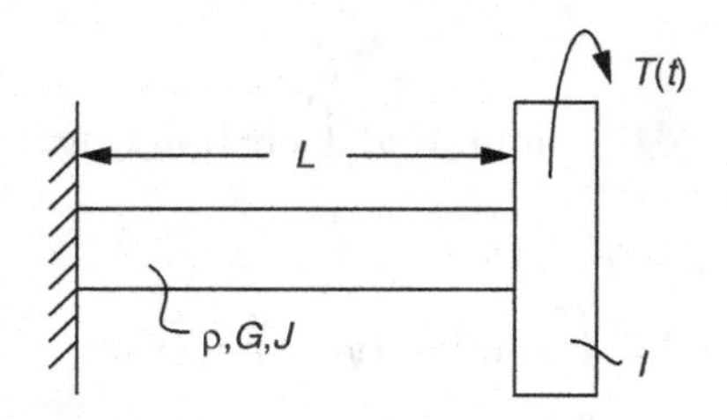

Figure 6.12 The applied torque leads to a time-dependent term in the boundary condition specified at the end of the shaft. An eigenvector expansion is used to determine the time-dependent angular displacement of the shaft.

$$\frac{\partial\theta}{\partial x}(1,t)+\mu\frac{\partial^2\theta}{\partial t^2}=\Lambda f(t) \tag{c}$$

$$\theta(x,0)=0 \tag{d}$$

$$\frac{\partial\theta}{\partial t}(x,0)=0 \tag{e}$$

where $\beta = I/\rho JL$ and $\Lambda = T_0L/JG$, and where T_0 is a characteristic value of the applied torque. The nondimensionalization of the problem without the applied torque is developed in Example 5.10.

The problem specified by Equation a, Equation b, Equation c, Equation d, and Equation e cannot be solved by separation of variables because the nonhomogeneous term is in a boundary condition rather than an initial condition. Develop a modal analysis solution in which the forced response is obtained through an eigenvector expansion of mode shapes for the free response.

Solution The free response of the system is due to a nonzero initial condition with no subsequent energy input to the system. The free response can be obtained using separation of variables assuming $\theta(x,t) = X(x)T(t)$. The resulting eigenvalue problem defining $X(x)$ is

$$\frac{d^2X}{dx^2}=-\lambda X \tag{f}$$

$$X(0)=0 \tag{g}$$

$$\frac{dX}{dx}(1)=-\mu\lambda X(1) \tag{h}$$

The solution of the eigenvalue problem specified in Equation f, Equation g, and Equation h is obtained in Example 5.13. The eigenvalue λ_k is the kth positive solution of

$$\tan\left(\sqrt{\lambda}\right)=\frac{1}{\mu\sqrt{\lambda}} \tag{i}$$

Its corresponding eigenvector is

$$X_k(x)=C_k\sin\left(\sqrt{\lambda_k}\,x\right) \tag{j}$$

It is shown in Example 5.13 that the eigenvectors are orthogonal with respect to the inner product defined by

$$(f,g)=\int_0^1 f(x)g(x)dx+\mu f(1)g(1) \tag{k}$$

Normalizing with respect to the inner product of Equation k leads to a set of orthonormal eigenvectors defined by

$$X_k(x) = \sqrt{\frac{2}{1+\mu\left[\sin\left(\sqrt{\lambda_k}\right)\right]^2}}\sin\left(\sqrt{\lambda_k}\,x\right) \quad k = 1,2,... \tag{l}$$

The solution of Equation a, Equation b, Equation c, Equation d, and Equation e, at every instant of time, satisfies the same boundary conditions as the eigenvectors obtained from Equation f, Equation g, and Equation h. Since $\theta(x,t)$ is in the domain of the operator L as defined in Equation f, Equation g, and Equation h, application of the expansion theorem implies that there exist time-dependent coefficients, $A_k(t)$, such that

$$\theta(x,t) = \sum_{k=1}^{\infty} A_k(t) X_k(x) \tag{m}$$

Substitution of Equation m into Equation a gives

$$\sum_{k=1}^{\infty} A_k \frac{d^2 X_k}{dx^2} = \sum_{k=1}^{\infty} \frac{d^2 A_k}{dt^2} X_k \tag{n}$$

Substituting Equation f into Equation n and rearranging results yields

$$\sum_{k=1}^{\infty}\left(\frac{d^2 A_k}{dt^2} + \lambda_k A_k\right) X_k = 0 \tag{o}$$

Substitution of Equation m into the nonhomogeneous boundary condition, Equation c, leads to

$$\sum_{k=1}^{\infty} A_k(t)\frac{dX_k}{dx}(1) + \mu \sum_{k=1}^{\infty} \frac{d^2 A_k}{dt^2} X_k(x) = \Lambda f(t) \tag{p}$$

Substitution of Equation h into Equation p leads to

$$\mu \sum_{k=1}^{\infty}\left(\frac{d^2 A_k}{dt^2} + \lambda_k A_k\right) X_k(1) = \Lambda f(t) \tag{q}$$

Multiplying Equation o by $X_m(x)$ and integrating from 0 to 1 leads to

$$\sum_{k=1}^{\infty}\left(\frac{d^2 A_k}{dt^2} + \lambda_k A_k\right)\int_0^1 X_k(x) X_m(x)\,dx = 0 \tag{r}$$

Mode-shape orthonormality with respect to the inner product of Equation i implies

$$\int_0^1 X_k(x)X_m(x)dx = \delta_{k,m} - \mu X_k(1)X_m(1) \tag{s}$$

which when substituted into Equation r, leads to

$$\frac{d^2A_m}{dt^2} + \lambda_m A_m = \mu\sum_{k=1}^{\infty}\left(\frac{d^2A_k}{dt^2} + \lambda_k A_k\right)X_k(1)X_m(1) \tag{t}$$

Substitution of Equation q into Equation t gives

$$\frac{d^2A_m}{dt^2} + \lambda_m A_m = X_m(1)\Lambda f(t) \tag{u}$$

where $X_m(1)$ is evaluated from Equation l, leading to

$$\frac{d^2A_m}{dt^2} + \lambda_m A_m = \Lambda \sin\left(\sqrt{\lambda_m}\right)\sqrt{\frac{2}{1+\mu\left[\sin\left(\sqrt{\lambda_m}\right)\right]^2}}f(t) \tag{v}$$

Initial conditions for Equation v are obtained from substitution of Equation m into Equation d and Equation e, leading to

$$A_m(0) = 0 \tag{w}$$

$$\frac{dA_m}{dt}(0) = 0 \tag{x}$$

Suppose $f(t) = \sin(\omega t)$. The solution of Equation v subject to the initial conditions of Equation w and Equation x is

6.5 *Problems in cylindrical coordinates*

Cylindrical coordinates are illustrated in Figure 6.13. The coordinates r, θ and z are used to define a point in the cylinder, or even exterior to the cylinder. The coordinate z is measured from a defined point along the axis of the cylinder. The coordinate r is the distance from the axis of the cylinder. The coordinate θ is measured circumferentially counterclockwise from a defined plane which includes the z-axis. To describe all points on the surface and interior of a cylinder of length L and radius R $0 \le z \le L$, $0 \le r \le R$ and $0 \le \theta \le 2\pi$.

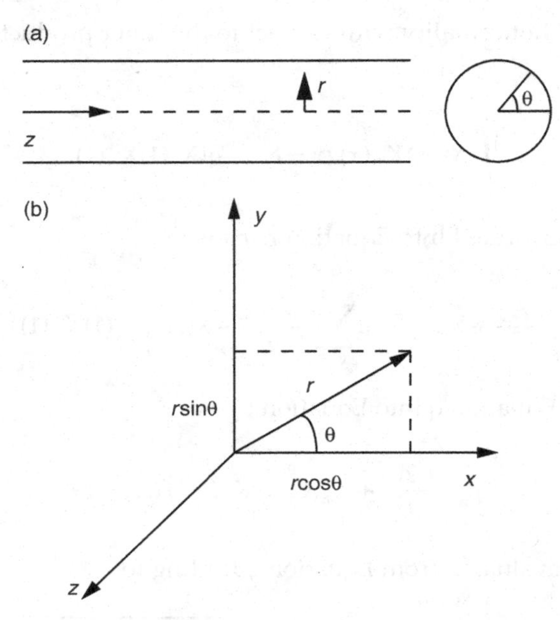

Figure 6.13 (a) The cylindrical coordinate system is described by an axial coordinate, z, a radial coordinate, r, and a circumferential coordinate, θ. (b) Cylindrical coordinates can be converted to Cartesian coordinates and vice versa.

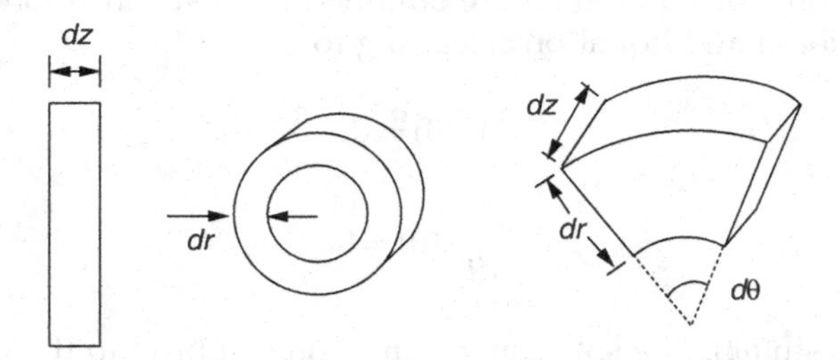

Figure 6.14 A differential volume of a cylinder is formed by taking a slice perpendicular to the axis of the cylinder, taking an annular ring from the slice, and then cutting the ring along an arc.

A differential volume within the cylinder is illustrated in Figure 6.14. The region is formed by first taking a slice of the cylinder of thickness dz, with the slice cut perpendicular to the axis of the cylinder with faces parallel. Then an annular ring is cut from the slice. The inner radius of the ring is r and its outer radius $r+dr$. Finally, a sector of the ring is cut, defining a plug of volume dV. The angle of the sector is $d\theta$ with its right face along the radius defined by the angle θ. The area of the face of the sector is $dA=rdrd\theta$, while the volume of the plug is $dV=rdrd\theta dz$.

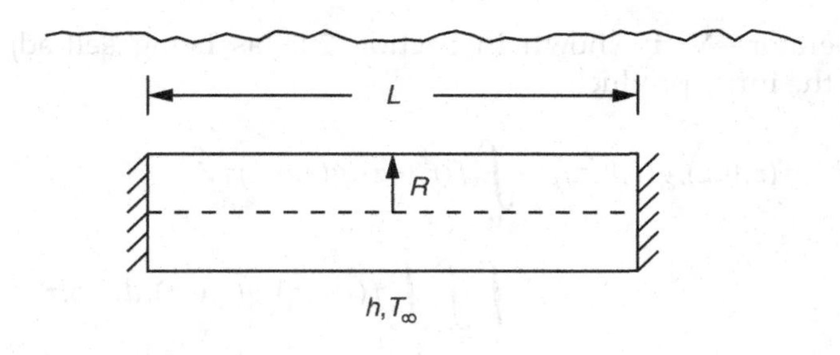

Figure 6.15 The cylinder maintains a steady-state temperature distribution $f(r,\theta,z)$ when it is immersed in a medium with an ambient temperature T_∞. The resulting unsteady state temperature distribution is obtained by solving the diffusion equation in cylindrical coordinates.

Cylindrical coordinates are related to Cartesian coordinates through

$$x = r\cos\theta \tag{6.68}$$

$$y = r\sin\theta \tag{6.69}$$

$$z = z \tag{6.70}$$

Differential equations written in cylindrical coordinates can be derived either by applying appropriate conservation laws to differential volumes, as in Example 1.4, or in the case of systems in which governing equations are written in vector operators, simply by using the following identities to convert the operators to cylindrical coordinates. The gradient in cylindrical coordinates is

$$\nabla\phi = \frac{\partial\phi}{\partial r}\mathbf{e}_r + \frac{1}{r}\frac{\partial\phi}{\partial\theta}\mathbf{e}_\theta + \frac{\partial\phi}{\partial z}\mathbf{e}_z \tag{6.71}$$

The divergence operator written in cylindrical coordinates is

$$\nabla\cdot\mathbf{v} = \frac{1}{r}\frac{\partial}{\partial r}(rv_r) + \frac{1}{r}\frac{\partial v_\theta}{\partial\theta} + \frac{\partial v_z}{\partial z} \tag{6.72}$$

The Laplacian operator in cylindrical coordinates is

$$\nabla^2\phi = \frac{1}{r}\frac{\partial}{\partial r}\left(r\frac{\partial\phi}{\partial r}\right) + \frac{1}{r^2}\frac{\partial^2\phi}{\partial\theta^2} + \frac{\partial^2\phi}{\partial z^2} \tag{6.73}$$

Laplace's equation, the diffusion equation, and the wave equation are all separable in cylindrical coordinates. Nonhomogeneous equations can be solved using the methods of Section 6.4.

The operator $-\nabla^2$ is shown in Section 2.11 as being self-adjoint with respect to the inner product,

$$\left(f(r,\theta,z),g(r,\theta,z)\right)=\int_V f(r,\theta,z),g(r,\theta,z)dV$$

$$=\int\int\int f(r,\theta,z),g(r,\theta,z)rdrd\theta dz \tag{6.74}$$

The inner product for a full cylinder is

$$\left(f(r,\theta,z),g(r,\theta,z)\right)=\int_0^L\int_0^{2\pi}\int_0^R f(r,\theta,z),g(r,\theta,z)rdrd\theta dz \tag{6.75}$$

If the region over which the problem to be solved is a partial cylinder, then the limits of integration of Equation 6.75 must be adjusted accordingly.

It is also shown in Section 2.11 that $-\nabla^2\phi$ is positive definite with respect to the inner product of Equation 6.75 if ϕ or a linear combination of ϕ and $\nabla\phi\cdot n$ are specified over a portion of the surface. If $\nabla\phi\cdot n$ is specified everywhere on the surface, the operator is non-negative definite.

Example 6.10 The cylinder of Figure 6.15 has a temperature distribution $T_0 f(r,\theta,z)$ when it is immersed in a bath of temperature T_∞. The ends of the cylinder are insulated, while its heat is transfered through its circumference with a heat transfer coefficient h. The following nondimensional variables are introduced:

$$r^*=\frac{r}{R} \tag{a}$$

$$z^*=\frac{z}{R} \tag{b}$$

$$t^*=\frac{\rho ct}{kR^2} \tag{c}$$

$$\Lambda(r^*,\theta,z^*,t^*)=\frac{T(r,\theta,z,t)-T_\infty}{T_0-T_\infty} \tag{d}$$

where ρ is the mass density of the cylinder, c is its specific heat and k is its thermal conductivity.

It is not necessary to nondimensionalize the circumferential coordinate θ because it is measured in radians which are inherently nondimensional. Using the same length scale to nondimensionalize both r and z enables the nondimensional form of the Laplacian to be written as in Equation 6.73, understanding that r and z are nondimensional. Either L or R can be used to define the nondimensional variables. The use of R may be preferable because the ends at constant values of z are insulated, whereas the outer circumference of the cylinder is subject to convection heat transfer from the bath. In this case, the Biot number is a more accurate measure of the ratio of convective heat transfer to the rate of conductive heat transfer.

The mathematical form of the problem in terms of nondimensional variables with the *'s dropped is the partial differential equation,

$$\nabla^2\Lambda = \frac{\partial \Lambda}{\partial t} \tag{e}$$

subject to the boundary conditions

$$\frac{\partial \Lambda}{\partial r}(1,\theta,z,t) + Bi\Lambda(1,\theta,z,t) = 0 \tag{f}$$

$$\frac{\partial \Lambda}{\partial z}(r,\theta,0,t) = 0 \tag{g}$$

$$\frac{\partial \Lambda}{\partial z}(r,\theta,\alpha,t) = 0 \tag{h}$$

where $\alpha = L/R$. The initial condition is

$$\Lambda(r,\theta,z,0) = f(r,\theta,z) \tag{i}$$

In addition to satisfying Equation f, Equation g, Equation h, and Equation i, the temperature must be bounded and single-valued everywhere in the cylinder. Determine $\Lambda(r,\theta,z,t)$.

Solution The governing partial differential equation is homogeneous and separable. The boundary conditions are all homogeneous. Thus, separation of variables is applicable. To this end, assume a product solution of the form

$$\Lambda(r,\theta,z,t) = U(r,\theta,z)T(t) \tag{j}$$

Substitution of Equation j into Equation e gives

$$T(t)\nabla^2 U = U\frac{dT}{dt}$$

$$\frac{1}{U}\nabla^2 U = \frac{1}{T}\frac{dT}{dt} \tag{k}$$

Applying the usual separation argument to Equation k and defining the separation constant as $-\lambda$ leads to

$$\frac{dT}{dt} + \lambda T = 0 \tag{l}$$

and

$$\nabla^2 U = -\lambda U$$

$$\frac{1}{r}\frac{\partial}{\partial r}\left(r\frac{\partial U}{\partial r}\right) + \frac{1}{r^2}\frac{\partial^2 U}{\partial \theta^2} + \frac{\partial^2 U}{\partial z^2} = -\lambda U \tag{m}$$

Substitution of Equation l into the boundary conditions, Equation f, Equation g, and Equation h, results in

$$\frac{\partial U}{\partial r}(1,\theta,z) + BiU(1,\theta,z) = 0 \tag{n}$$

$$\frac{\partial U}{\partial z}(r,\theta,0) = 0 \tag{o}$$

$$\frac{\partial U}{\partial z}(r,\theta,\alpha) = 0 \tag{p}$$

Equation m is a separable partial differential equation and is subject to homogeneous boundary conditions. Assume a product solution of Equation m of the form

$$U(r,\theta,z) = R(r)\Theta(\theta)Z(z) \tag{q}$$

Substitution of Equation q into Equation m gives

$$\frac{\Theta Z}{r}\frac{d}{dr}\left(r\frac{dR}{dr}\right) + \frac{RZ}{r^2}\frac{d^2\Theta}{d\theta^2} + R\Theta\frac{d^2 Z}{dz^2} = -\lambda R\Theta Z \tag{r}$$

Equation r can be multiplied by $r^2/R\Theta Z$ and rearranged, resulting in

$$\frac{r}{R}\frac{d}{dr}\left(r\frac{dR}{dr}\right) + \frac{r^2}{Z}\frac{d^2 Z}{dz^2} + \lambda r^2 = -\frac{1}{\Theta}\frac{d^2\Theta}{d\theta^2} \tag{s}$$

The usual separation argument is applied to Equation s with a separation constant of μ. The resulting equations are

$$\frac{d^2\Theta}{d\theta^2} + \mu\Theta = 0 \tag{t}$$

and

$$\frac{r}{R}\frac{d}{dr}\left(r\frac{dR}{dr}\right)+\frac{r^2}{Z}\frac{d^2Z}{dz^2}+\left(\lambda r^2-\mu\right)=0 \tag{u}$$

Equation u can be divided by r^2 and rearranged to yield

$$\frac{1}{rR}\frac{d}{dr}\left(r\frac{dR}{dr}\right)-\frac{\mu}{r^2}=-\frac{1}{Z}\frac{d^2Z}{dz^2}-\lambda \tag{v}$$

The usual separation argument is used for Equation v with a separation constant of $-\kappa$, resulting in

$$\frac{d^2Z}{dz^2}+(\lambda-\kappa)Z=0 \tag{w}$$

and

$$r\frac{d}{dr}\left(r\frac{dR}{dr}\right)+\left(\kappa r^2-\mu\right)R=0 \tag{x}$$

Application of Equation q to Equation n, Equation o, and Equation p leads to

$$\frac{dR}{dr}(1)+BiR(1)=0 \tag{y}$$

$$\frac{dZ}{dz}(0)=0 \tag{z}$$

$$\frac{dZ}{dz}(\alpha)=0 \tag{aa}.$$

Separation of variables has been successfully applied. The values of the separation constant λ are the eigenvalues of $-\nabla^2 U$ subject to the boundary conditions, Equation n, Equation o, and Equation p, and the conditions that U remain finite and single-valued at every point in the cylinder. The eigenvalue problem for $-\nabla^2 U$ can itself be solved using separation of variables. The values of the separation constant μ are the eigenvalues of $-d^2\Theta/d\theta^2$ subject only to the single-valuedness condition. This is a non-negative definite Sturm-Liouville problem, which was solved in Example 5.6 Thus there are an infinite, but countable, number of values of μ: $\mu_0,\mu_1,\mu_2,\ldots,\mu_{n-1},\mu_n,\mu_{n+1},\ldots$ There are then an infinite number of equations of the form of Equation x to solve for $R(r)$ which is indexed using n. The values of κ are the eigenvalues of the operator

$-1/r[d(rdR_n)/dr]+\mu_n/r^2R_n$ subject to Equation y and the requirement that $R(r)$ be finite for all r. For each n, this is a Sturm-Liouville problem leading to an infinite number of eigenvalues, $\kappa_{n,1},\kappa_{n,2},\ldots,\kappa_{n,m-1},\kappa_{n,m},\kappa_{n,m+1},\ldots$ with corresponding eigenvectors. The values of $\beta=\lambda-\kappa$ are the eigenvalues of $-d^2Z/dz^2$, subject to the boundary conditions, Equation z and Equation aa. Since this is also a Sturm-Liouville problem, there are an infinite number of eigenvalues $\beta_0,\beta_1,\beta_2,\ldots,\beta_{p-1},\beta_p,\beta_{p+1},\ldots$ with corresponding eigenvectors. Thus the eigenvalues of $-\nabla^2U$ are of the form

$$\lambda_{n,m,p}=\beta_p+\kappa_{n,m} \tag{bb}$$

for $n,p=0,1,2,\ldots$ and $m=1,2,\ldots$ The details of the solutions follow.

Let $U_{n,m,p}(r,\theta,z)=R_{n,m}\Theta_nZ_p$ be an eigenvector of $-\nabla^2$ corresponding to an eigenvalue $\lambda_{n,m,p}$. The eigenvector is normalized using the inner product of Equation 6.75. Thus the normalization condition is

$$\int_0^{\alpha}\int_0^{2\pi}\int_0^{1}\left[R_{n,m}(r)\Theta_n(\theta)Z_p(z)\right]^2 rdrd\theta dz=1 \tag{cc}$$

Equation cc can be rewritten as

$$\left\{\int_0^1\left[R_{n,m}(r)\right]^2 rdr\right\}\left\{\int_0^{2\pi}\left[\Theta_n(\theta)\right]^2 d\theta\right\}\left\{\int_0^{\alpha}\left[Z_p(z)\right]^2 dz\right\}=1 \tag{dd}$$

Equation dd shows that if each term in the product is normalized individually, then the eigenvector $U_{n,m,p}(r,\theta,z)=R_{n,m}\Theta_nZ_p$ is normalized. This is the approach taken below.

The solution of Equation t is

$$\Theta(\theta)=C_1\cos\left(\sqrt{\mu}\theta\right)+C_2\sin\left(\sqrt{\mu}\theta\right) \tag{ee}$$

The two sets of coordinates, (r,θ,z) and $(r,\theta+2\pi,z)$, represent the same point. Single-valuedness of the temperature requires $\Theta(\theta)=\Theta(\theta+2\pi)$. Single-valuedness of the rate of heat transfer, $q_\theta=\nabla T\cdot e_\theta=(1/r)\partial T/\partial\theta$, requires $d\Theta/d\theta(\theta)=d\Theta/d\theta(\theta+2\pi)$. Thus the eigenvector and its derivative must be periodic of period 2π. These are the conditions considered in Example 5.6. The eigenvalues for the periodic response are $\mu_n=n^2$ $n=0,1,2,\ldots$ The normalized eigenvector corresponding to $\mu=0$ is

$$\Theta_0(\theta)=\frac{1}{\sqrt{2\pi}} \tag{ff}$$

An eigenvalue $\mu_n = n^2$ $n = 1,2,\ldots$ has two linearly independent eigenvectors. These normalized eigenvectors are

$$\Theta_{1n} = \frac{1}{\sqrt{\pi}}\cos(n\theta) \tag{gg}$$

$$\Theta_{2n} = \frac{1}{\sqrt{\pi}}\sin(n\theta) \tag{hh}$$

For $\mu_n = n^2$, Equation x becomes

$$r\frac{d}{dr}\left(r\frac{dR_n}{dr}\right) + \left(\kappa r^2 - n^2\right)R_n = 0 \tag{ii}$$

Equation ii can be recognized as Bessel's equation of order n with a general solution of

$$R_n(r) = D_{1n}J_n\left(\sqrt{\kappa}r\right) + D_{2n}Y_n\left(\sqrt{\kappa}r\right) \tag{jj}$$

$R_n(0)$ is finite only at $n=0$ if $D_{2n}=0$, leading to $R_n(r) = D_{1n}J_n(\sqrt{\kappa}r)$. Application of Equation y leads to the transcendental equation whose solutions define the eigenvalues by

$$\sqrt{\kappa}J_n'\left(\sqrt{\kappa}\right) + BiJ_n\left(\sqrt{\kappa}\right) = 0 \tag{kk}$$

For each n, Equation kk has an infinite number of solutions indexed as $\kappa_{n,1} < \kappa_{n,2} < \ldots < \kappa_{n,m-1} < \kappa_{n,m} < \kappa_{n,m+1} < \ldots$ The corresponding normalized eigenvectors are

$$R_{n,m}(r) = D_{n,m}J_n\left(\sqrt{\kappa_{n,m}}r\right) \tag{ll}$$

where

$$D_{n,m} = \frac{1}{\left\{\int_0^1 \left[J_n\left(\sqrt{\kappa_{n,m}}r\right)\right]^2 r\,dr\right\}^{\frac{1}{2}}}$$

$$= \left\{\left[\frac{1}{2} + \left(\frac{Bi}{\kappa_{n,m}}\right)^2\right] J_n(\kappa_{n,m})^2\right\}^{-\frac{1}{2}} \tag{mm}$$

The eigenvalues of the problems defined by Equation w, Equation z, and Equation aa are

$$\beta_p = \left(\frac{p\pi}{\alpha}\right)^2 \quad p = 0,1,2,... \tag{nn}$$

The corresponding normalized eigenvectors are

$$Z_0(z) = \sqrt{\frac{1}{\alpha}} \tag{oo}$$

$$Z_p(z) = \sqrt{\frac{2}{\alpha}}\cos\left(\frac{p\pi}{\alpha}z\right) \tag{pp}$$

Note that the problem defining $Z(z)$ is non-negative definite because its lowest eigenvalue is zero.

Summarizing, the eigenvalues for U are

$$\lambda_{n,m,p} = \left(\frac{p\pi}{\alpha}\right)^2 + \kappa_{n,m} \tag{qq}$$

where $\kappa_{n.m}$ is the mth positive solution of Equation kk for each n. The ranges of the indices in Equation qq are $n, p = 0,1,2,\ldots$ and $m = 1,2,\ldots$ The corresponding eigenvectors are

$$U_{0,m,0} = \sqrt{\frac{1}{2\pi\alpha}} D_{0,m} J_0\left(\sqrt{\kappa_{0,m}}\, r\right) \quad m = 1,2,... \tag{rr1}$$

$$U_{0,m,p} = \sqrt{\frac{1}{\pi\alpha}} D_{0,m} J_0\left(\sqrt{\kappa_{0,m}}\, r\right)\cos\left(\frac{p\pi}{\alpha}z\right) \quad m,p = 1,2,... \tag{rr2}$$

$$U_{n,m,0,1} = \sqrt{\frac{1}{\pi\alpha}} D_{n,m} J_n\left(\sqrt{\kappa_{n,m}}\, r\right)\cos(n\theta) \quad n,m = 1,2,... \tag{rr3}$$

$$U_{n,m,0,2} = \sqrt{\frac{1}{\pi\alpha}} D_{n,m} J_n\left(\sqrt{\kappa_{n,m}}\, r\right)\sin(n\theta) \quad n,m = 1,2,... \tag{rr3}$$

$$U_{n,m,p,1} = \sqrt{\frac{2}{\pi\alpha}} D_{n,m} J_n\left(\sqrt{\kappa_{n,m}}\, r\right)\cos(n\theta)\cos\left(\frac{p\pi}{\alpha}z\right) \quad n,m,p = 1,2,... \tag{rr5}$$

$$U_{n,m,p,2} = \sqrt{\frac{2}{\pi\alpha}} D_{n,m} J_n\left(\sqrt{\kappa_{n,m}}\, r\right)\sin(n\theta)\cos\left(\frac{p\pi}{\alpha}z\right) \quad n,m,p = 1,2,... \tag{rr6}$$

The solution of Equation l leads to

$$T_{n,m,p}(t) = A_{n,m,p}e^{-\lambda_{n,m,p}t} \tag{ss}$$

The general solution is a linear combination of all homogeneous solutions

$$\begin{aligned}\Lambda(r,\theta,z,t) = &\sum_{m=1}^{\infty}\left[A_{0,m,0}e^{-\lambda_{0,m,0}t}U_{0,m,0}(r,\theta,z) + \sum_{p=1}^{\infty}A_{0,m,p}e^{-\lambda_{0,m,p}t}U_{0,m,p}(r,\theta,z)\right] \\ &+\sum_{n=1}^{\infty}\left\{\sum_{m=1}^{\infty}e^{-\lambda_{n,m,0}t}[A_{n,m,0}U_{n,m,0,1}(r,\theta,z) + B_{n,m,0}U_{n,m,0,2}(r,\theta,z)]\right. \\ &\left.+\sum_{m=1}^{\infty}\sum_{p=1}^{\infty}e^{-\lambda_{n,m,p}t}[A_{n,m,p}U_{n,m,p,1}(r,\theta,z) + B_{n,m,p}U_{n,m,p,2}(r,\theta,z)]\right\}\end{aligned} \tag{tt}$$

Application of the initial condition, Equation i, to Equation tt leads to

$$\begin{aligned}f(r,\theta,z) = &\sum_{m=1}^{\infty}\left[A_{0,m,0}U_{0,m,0}(r,\theta,z) + \sum_{p=1}^{\infty}A_{0,m,p}U_{0,m,p}(r,\theta,z)\right] \\ &+\sum_{n=1}^{\infty}\left\{\sum_{m=1}^{\infty}[A_{n,m,0}U_{n,m,0,1}(r,\theta,z) + B_{n,m,0}U_{n,m,0,2}(r,\theta,z)]\right. \\ &\left.+\sum_{m=1}^{\infty}\sum_{p=1}^{\infty}[A_{n,m,p}U_{n,m,p,1}(r,\theta,z) + B_{n,m,p}U_{n,m,p,2}(r,\theta,z)]\right\}\end{aligned} \tag{uu}$$

The coefficients in the summations on the right-hand side of Equation uu are the coefficients in the eigenvector expansion for $f(r,\theta,z)$. For example,

$$A_{n,m,p,1} = \int_0^{\alpha}\int_0^{2\pi}\int_0^1 \sqrt{\frac{2}{\pi\alpha}}D_{n,m}J_n\left(\sqrt{\kappa_{0,m}}r\right)\cos(n\theta)\cos\left(\frac{p\pi}{\alpha}\right)f(r,\theta,z)rdrd\theta dz \tag{vv}$$

Example 6.11 A sector of the membrane shown in Figure 6.16 has been made from fibers of a nanomaterial. The natural frequencies of the membrane when it is clamped over its entire perimeter are to be measured and the data used to predict the properties of the material. The differential equation governing the vibrations of the membrane is

$$T\nabla^2 w = \rho\frac{\partial^2 w}{\partial t^2} \tag{a}$$

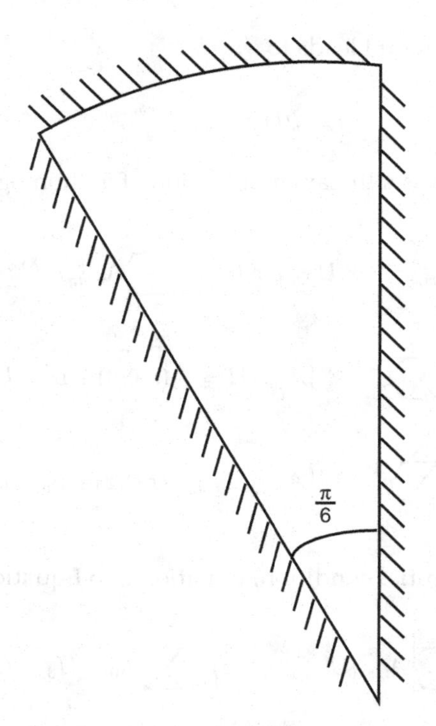

Figure 6.16 The natural frequencies of the membrane of Example 6.11 are obtained using separation of variables.

where $w(r,\theta,t)$ is the displacement of the membrane, T is the tension in the membrane, and ρ is its mass density. The boundary conditions are

$$w(R,\theta,t)=0 \tag{b}$$

$$w(r,0,t)=0 \tag{c}$$

$$w\left(r,\frac{\pi}{6},t\right)=0 \tag{d}$$

Nondimensional variables defined by $r^*=r/R, w^*=w/R$, and $t^*=t\sqrt{T/\rho R^2}$ can be substituted into Equation a, Equation b, Equation c, and Equation d, leading to

$$\nabla^2 w=\frac{\partial^2 w}{\partial t^2} \tag{e}$$

$$w(1,\theta,t)=0 \tag{f}$$

where *'s have been dropped from nondimensional variables. The boundary conditions represented by Equation c and Equation d have the same equations in nondimensional form. Determine the natural frequencies and normalized mode shapes of the membrane.

Solution A product solution of Equation e is assumed to be of the form

$$w(r,\theta,t)=Q(t)U(r,\theta) \tag{g}$$

Substitution of Equation g into Equation e, rearranging, and using the usual separation argument leads to

$$\frac{d^2Q}{dt^2}+\lambda Q=0 \tag{h}$$

$$-\nabla^2 U=\lambda U \tag{i}$$

where λ is a separation constant. Noting that the general solution of Equation h is $Q=C_1\cos(\sqrt{\lambda}t)+C_2\sin(\sqrt{\lambda}t)$ it is clear that the natural frequencies are the square roots of the values of the separation constants, which are the eigenvalues of $-\nabla^2 Q$ subject to $Q(1,\theta)=0$, $Q(r,0)=0$ and $Q(r,\pi/6)=0$.

To solve Equation i, a product solution is assumed of the form

$$Q(r,\theta)=R(r)\Theta(\theta) \tag{j}$$

Substitution of Equation j into Equation i leads to

$$\frac{1}{r}\frac{d}{dr}\left(r\frac{dR}{dr}\right)\Theta(\theta)+\frac{1}{r^2}R(t)\frac{d^2\Theta}{d\theta^2}+\lambda R\Theta=0 \tag{k}$$

Equation k can be rearranged to yield

$$\frac{r}{R}\frac{d}{dr}\left(r\frac{dR}{dr}\right)+\frac{\lambda r^2}{R}=-\frac{1}{\Theta}\frac{d^2\Theta}{dr^2} \tag{l}$$

The usual separation argument is used on Equation l with a separation constant μ. The resulting differential equations are

$$\frac{d^2\Theta}{d\theta^2}+\mu\Theta=0 \tag{m}$$

$$r\frac{d}{dr}\left(r\frac{dR}{dr}\right)+\left(\lambda r^2-\mu\right)R=0 \tag{n}$$

Use of the product solution in the boundary conditions leads to

$$R(1) = 0 \tag{o}$$

$$\Theta(0) = 0 \tag{p}$$

$$\Theta\left(\frac{\pi}{6}\right) = 0 \tag{q}$$

The general solution for Equation m is

$$\Theta(\theta) = C_1 \cos\left(\sqrt{\mu}\theta\right) + C_2 \sin\left(\sqrt{\mu}\theta\right) \tag{r}$$

Application of Equation p to Equation q leads to $C_1 = 0$. Application of Equation q to Equation r with $C_1 = 0$ leads to

$$\sin\left(\sqrt{\mu}\frac{\pi}{6}\right) = 0 \tag{s}$$

Equation r is satisfied for

$$\mu_n = 36n^2 \quad n = 1,2,... \tag{t}$$

The eigenvectors can be normalized by requiring that $\int_0^{\pi/6} C_2^2[\sin(6n\theta)]^2 \, d\theta = 1$ which leads $C_2 = \sqrt{12/\pi}$. The normalized eigenvectors are

$$\Theta(\theta) = \sqrt{\frac{12}{\pi}} \sin(6n\theta) \tag{u}$$

Substitution of Equation t into Equation n leads to

$$r\frac{d}{dr}\left(r\frac{dR_n}{dr}\right) + \left(\lambda r^2 - 36n^2\right)R_n = 0 \tag{v}$$

The solution of Equation v is

$$R_n(r) = D_{1n}J_{6n}\left(\sqrt{\lambda}r\right) + D_{2n}Y_{6n}\left(\sqrt{\lambda}r\right) \tag{w}$$

The solution is finite at $r = 0$ only if $D_{2n} = 0$. Subsequent application of Equation o to Equation w leads to

$$J_{6n}\left(\sqrt{\lambda}\right) = 0 \tag{x}$$

There are an infinite, but countable, number of solutions of Equation x for each n. These are indexed as $\lambda_{n,m}$, which is the mth solution of Equation x for a specific n. The eigenvectors are normalized by choosing

$$D_{1,n} = \left[\int_0^1 \left[J_{6n}\left(\sqrt{\lambda_{n,m}}r\right)\right]^2 r dr\right]^{-\frac{1}{2}}$$

$$= \left\{\frac{2}{\left[\left[J_{6n}\left(\sqrt{\lambda_{n,m}}\right)\right]^2 - J_{6n-1}\left(\sqrt{\lambda_{n,m}}\right)J_{6n+1}\left(\sqrt{\lambda_{n,m}}\right)\right]}\right\}^{\frac{1}{2}} \tag{y}$$

It is clear from Equation h that the natural frequencies are the square roots of the eigenvalues of U, $\omega_{m,n} = \sqrt{\lambda_{m,n}}$. The first three natural frequencies for the first five values of n are given in Table 6.2.

Table 6.2 Natural Frequencies of membrane of Example 6.11

n	1	2	3
$\omega_{n,1}$	9.936	16.698	23.257
$\omega_{n,2}$	17.004	20.79	27.698
$\omega_{n,3}$	20.321	24.495	31.65
$\omega_{n,4}$	23.586	24.495	35.375
$\omega_{n,5}$	26.82	31.46	38.965

Example 6.12 The cylinder shown in Figure 6.17 is at a uniform temperature T_0 when a uniform heat flux is suddenly applied to the right end. The circumference of the cylinder is maintained at T_0, while the left end of the cylinder is insulated. Determine the resulting unsteady state temperature distribution in the cylinder.

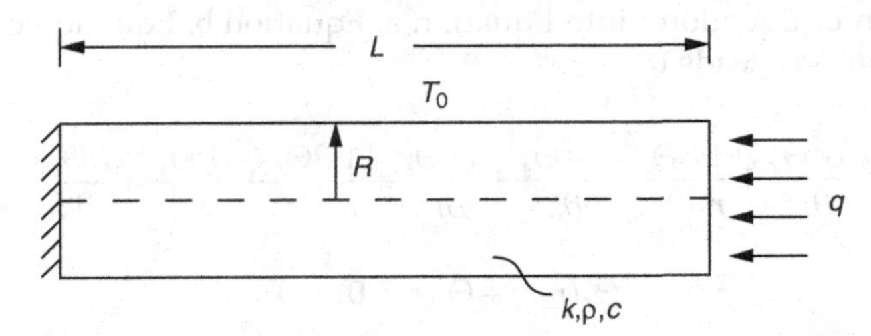

Figure 6.17 The cylinder of Example 6.12 is at a uniform temperature when a heat flux begins at its left end. The resulting problem for the unsteady-state temperature distribution is nonhomogeneous, and therefore a superposition solution is used.

Solution Since the initial temperature is uniform and the applied heat flux is uniform over the end of the cylinder, the temperature is constant circumferentially for every z and every r and thus independent of θ. The mathematical formulation for the nondimensional temperature distribution, $\Theta(r,z,t)=(T(r,z,t)-T_0)/T_0$, in terms of the nondimensional varibles $r^*=r/L$, $z^*=z/L$ and $t^*=kt/\rho cL^2$ is

$$\frac{\partial^2\Theta}{\partial r^2}+\frac{1}{r}\frac{\partial\Theta}{\partial r}+\frac{\partial^2\Theta}{\partial z^2}=\frac{\partial\Theta}{\partial t} \tag{a}$$

$$\Theta(r,z,0)=0 \tag{b}$$

$$\Theta\left(\frac{R}{L},z,t\right)=0 \tag{c}$$

$$\frac{\partial\Theta}{\partial z}(r,0,t)=0 \tag{d}$$

$$\frac{\partial\Theta}{\partial z}(r,1,t)=\alpha \tag{e}$$

where $\alpha=qL/kT_0$

The boundary condition at $z=1$, Equation e, is nonhomogeneous, and therefore the system of Equation a, Equation b, Equation c, Equation d, and Equation e is nonhomogeneous. Heat can escape from the cylinder through its circumference, and therefore the temperature should approach a steady-state distribution. This suggests that the temperature is a superposition of an unsteady state or transient solution and a steady-state solution of the form

$$\Theta(r,z,t)=\Theta_s(r,z)+\Theta_t(r,z,t) \tag{f}$$

Substitution of Equation f into Equation a, Equation b, Equation c, Equation d, and Equation e leads to

$$\frac{\partial^2\Theta_s}{\partial r^2}+\frac{1}{r}\frac{\partial\Theta_s}{\partial r}+\frac{\partial^2\Theta_s}{\partial z^2}+\frac{\partial^2\Theta_t}{\partial r^2}+\frac{1}{r}\frac{\partial\Theta_t}{\partial r}+\frac{\partial^2\Theta_t}{\partial z^2}=\frac{\partial\Theta_t}{\partial t} \tag{g}$$

$$\Theta_s(r,z)+\Theta_t(r,z,0)=0 \tag{h}$$

$$\Theta_s\left(\frac{R}{L},z\right)+\Theta_s\left(\frac{R}{L},z,t\right)=0 \tag{i}$$

$$\frac{\partial \Theta_s}{\partial z}(r,0)+\frac{\partial \Theta_t}{\partial z}(r,0,t)=0 \tag{j}$$

$$\frac{\partial \Theta_s}{\partial z}(r,1)+\frac{\partial \Theta_t}{\partial z}(r,1,t)=\alpha \tag{k}$$

The goal of the superposition is to choose a problem for $\Theta_t(r,z,t)$ which can be solved directly using separation of variables. The problem must consist of a homogeneous partial differential equation, a nonhomogeneous initial condition, and homogeneous boundary conditions. The following choice of problem for $\Theta_t(r,z,t)$ meets these requirements:

$$\frac{\partial^2 \Theta_t}{\partial r^2}+\frac{1}{r}\frac{\partial \Theta_t}{\partial r}+\frac{\partial^2 \Theta_t}{\partial z^2}=\frac{\partial \Theta_t}{\partial t} \tag{l}$$

$$\Theta_t(r,z,0)=-\Theta_s(r,z) \tag{m}$$

$$\Theta_t\left(\frac{R}{L},z,t\right)=0 \tag{n}$$

$$\frac{\partial \Theta_t}{\partial z}(r,0,t)=0 \tag{o}$$

$$\frac{\partial \Theta_t}{\partial z}(r,1,t)=0 \tag{p}$$

Substitution of Equation l, Equation m, Equation n, Equation o, and Equation p into Equation g, Equation h, Equation i, Equation j, Equation k, and Equation l leads to formulating the problem to be solved for the steady-state response as

$$\frac{\partial^2 \Theta_s}{\partial r^2}+\frac{1}{r}\frac{\partial \Theta_s}{\partial r}+\frac{\partial^2 \Theta_s}{\partial z^2}=0 \tag{q}$$

$$\Theta_s\left(\frac{R}{L},z,t\right)=0 \tag{r}$$

$$\frac{\partial \Theta_s}{\partial z}(r,0,t)=0 \tag{s}$$

$$\frac{\partial \Theta_s}{\partial z}(r,1,t)=\alpha \tag{t}$$

The steady-state solution can be obtained by solve Equation q subject to Equation r, Equation s, and Equation t, direct application of separation of variables.

A product solution of Equation q is assumed as

$$\Theta_s(r,z) = R(r)Z(z) \tag{u}$$

Substituting Equation u into Equation q, rearranging, and using the usual separation argument, noting that since the nonhomogeneous boundary condition is applied at a constant value of z, the Sturm-Liouville problem must be in the r-coordinate, leads to

$$\frac{d^2Z}{dz^2} - \lambda Z = 0 \tag{v}$$

$$r^2\frac{d^2R}{dr^2} + r\frac{dR}{dr} + \lambda r^2 R = 0 \tag{w}$$

$$R\left(\frac{R}{L}\right) = 0 \tag{x}$$

The nontrivial solutions of Equation w and Equation x subject to the constraint that the temperature remains finite at $r=0$ are

$$R_n(r) = A_n J_0\left(\sqrt{\lambda_n} r\right) \tag{y}$$

where the eigenvalue λ_n is the nth solution of

$$J_0\left(\sqrt{\lambda}\frac{R}{L}\right) = 0 \tag{z}$$

and

$$A_n = \left[\int_0^{\frac{R}{L}} \left[J_0\left(\sqrt{\lambda_n} r\right)\right]^2 r dr\right]^{-\frac{1}{2}}$$

$$= \left[\frac{\sqrt{\lambda_n} L}{R J_1\left(\sqrt{\lambda_n}\frac{R}{L}\right)}\right]^{\frac{1}{2}} \tag{aa}$$

The solution of Equation v which satisfies the boundary condition at $z=0$ is

$$Z_n(z) = C_n \cosh\left(\sqrt{\lambda_n} z\right) \tag{bb}$$

The boundary condition at $z=1$ is satisfied using an eigenvector expansion for $f(r)=\alpha$, leading to

$$\begin{aligned} C_n &= \frac{1}{\sqrt{\lambda_n}\sinh\left(\sqrt{\lambda_n}\right)}(\alpha, R_n(r))_r \\ &= \frac{1}{\sqrt{\lambda_n}\sinh\left(\sqrt{\lambda_n}\right)}\int_0^{\frac{R}{L}} \alpha A_n J_0\left(\sqrt{\lambda_n} r\right) r dr \\ &= \frac{\alpha R A_n J_1\left(\sqrt{\lambda_n}\frac{R}{L}\right)}{L\lambda_n \sinh\left(\sqrt{\lambda_n}\right)} \end{aligned} \tag{cc}$$

The steady-state solution is

$$\Theta_s(r,z) = \sum_{n=1}^{\infty} C_n \cosh\left(\sqrt{\lambda_n} z\right) A_n J_0\left(\sqrt{\lambda_n} r\right) \tag{dd}$$

The transient response is obtained through solution of Equation l, Equation m, Equation n, Equation o, and Equation p using separation of variables and using Equation dd to impose the initial condition. To this end, a product solution of Equation l is assumed to be of the form

$$\Theta_t(r,z,t) = \hat{R}(r)\hat{Z}(z)\hat{T}(t) \tag{ee}$$

Substituting Equation ee into Equation l, rearranging, and using the usual separation arguments leads to

$$\frac{d\hat{T}}{dt} + \mu\hat{T} = 0 \tag{ff}$$

$$r^2\frac{d^2\hat{R}}{dr^2} + r\frac{d\hat{R}}{dr} + \kappa r^2\hat{R} = 0 \tag{gg}$$

$$\frac{d^2\hat{Z}}{dz^2} + (\mu - \kappa)\hat{Z} = 0 \tag{hh}$$

It should be noted that the problem defining $\hat{R}(r)$ is the same as that defining $R(r)$, and therefore $\hat{R}_n(r) = R_n(r)$, which is stated in Equation y, Equation z, and Equation aa. The corresponding normalized solution for $\hat{Z}_m(z)$ which satisfies Equation hh and $d\hat{Z}/dz\,(0) = d\hat{Z}/dz\,(1) = 0$ is

$$\hat{Z}_0(z) = 1 \tag{ii}$$

$$\hat{Z}_m(z) = \frac{1}{\sqrt{2}}\cos(m\pi z) \quad m = 1,2,3,... \tag{jj}$$

The eigenvalues are of the form

$$\mu_{n,m} = \lambda_n + m^2\pi^2 \tag{kk}$$

The solutions of Equation ff are of the form

$$\hat{T}_{n,m}(t) = F_{n,m}e^{-\mu_{n,m}t} \tag{ll}$$

The general form of the transient solution is the sum of all possible solutions, which, using Equation aa, Equation jj, and Equation ll, is

$$\Theta_t(r,z,t) = \sum_{n=1}^{\infty}\left[A_n J_0\left(\sqrt{\lambda_n}r\right) \times \left(F_{n,0}e^{-\lambda_n t} + \sum_{m=1}^{\infty} F_{n,m}\frac{1}{\sqrt{2}}\cos(m\pi z)e^{-\mu_{n,m}t}\right)\right] \tag{mm}$$

Application of the initial condition leads to

$$-\sum_{i=1}^{\infty} C_i \cosh\left(\sqrt{\lambda_i}z\right)A_i J_0\left(\sqrt{\lambda_i}r\right) = \sum_{n=1}^{\infty}\left[A_n J_0\left(\sqrt{\lambda_n}r\right) \times \left(F_{n,0} + \sum_{m=1}^{\infty} F_{n,m}\frac{1}{\sqrt{2}}\cos(m\pi z)\right)\right] \tag{nn}$$

The left-hand side of Equation nn has an eigenvector expansion in terms of the eigenvectors obtained for the transient solution, which leads to

$$F_{n,0} = -\left(\sum_{i=1}^{\infty} C_i \cosh\left(\sqrt{\lambda_i}z\right)A_i J_0\left(\sqrt{\lambda_i}r\right), A_n J_0\left(\sqrt{\lambda_n}r\right)\right)$$

$$= -\sum_{i=1}^{\infty} A_i A_n C_i \int_0^1 \int_0^{\frac{R}{L}} \cosh\left(\sqrt{\lambda_i}z\right) J_0\left(\sqrt{\lambda_i}r\right) J_0\left(\sqrt{\lambda_n}r\right) r\,dr\,dz \tag{oo}$$

Application of orthonormality of eigenvectors in the r-direction, $\int_0^{R/L} A_i A_n J_0(\sqrt{\lambda_i}r)J_0(\sqrt{\lambda_n}r)rdr = \delta_{i,n}$, reduces Equation oo to

$$F_{n,0} = -C_n \int_0^1 \sinh\left(\sqrt{\lambda_n}z\right)dz$$

$$= \frac{C_n}{\sqrt{\lambda_n}}\left[1 - \cosh\sqrt{\lambda_n}\right] \tag{pp}$$

Using a similar analysis,

$$F_{n,m} = -\left(\sum_{j=1}^{\infty} C_i \cosh\left(\sqrt{\lambda_i}z\right) A_i J_0\left(\sqrt{\lambda_i}r\right), \frac{A_n}{\sqrt{2}} J_0\left(\sqrt{\lambda_n}r\right)\cos(m\pi z)\right)$$

$$= -\sum_{i=1}^{\infty} \frac{A_i A_n C_i}{\sqrt{2}} \int_0^1 \int_0^{\frac{R}{L}} \cosh\left(\sqrt{\lambda_i}z\right)\cos(m\pi z) J_0\left(\sqrt{\lambda_i}r\right) J_0\left(\sqrt{\lambda_n}r\right) rdrdz$$

$$= -\frac{C_n}{\sqrt{2}} \int_0^1 \cosh\left(\sqrt{\lambda_n}z\right)\cos(m\pi z)dx$$

$$= \frac{(-1)^{m+1} C_n \sinh\sqrt{\lambda_n}}{\lambda_n + m^2\pi^2} \tag{qq}$$

6.6 *Problems in spherical coordinates*

The spherical coordinate system is illustrated in Figure 6.18. The coordinates r, θ and ϕ are used to locate a point uniquely in the sphere. The radial coordinate r is the distance from the center of the sphere to a point such that $0 \le r \le R$. The circumferential coordinate θ is the angle measured counterclockwise within a plane passing through the center of the sphere from a reference line to the projection of the radius to the point on the plane such that $0 \le \theta \le 2\pi$. The azimuthal coordinate ϕ is the angle made by the projection to the radius. The azimuthal coordinate may be defined by either $-\pi/2 \le \phi \le \pi/2$ or $0 \le \phi \le \pi$.

Spherical coordinates can be converted to Cartesian coordinates by

$$x = r\cos\phi\cos\theta \tag{6.76}$$

$$y = r\cos\phi\sin\theta \tag{6.77}$$

$$z = r\sin\phi \tag{6.78}$$

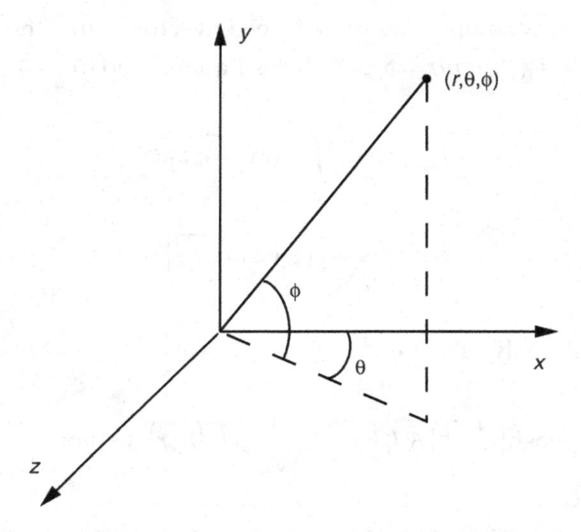

Figure 6.18 The coordinates of a point in a spherical coordinate system are represented in terms of a radial coordinate, r, a circumferential coordinate, θ, and an azimuthal coordinate, ϕ.

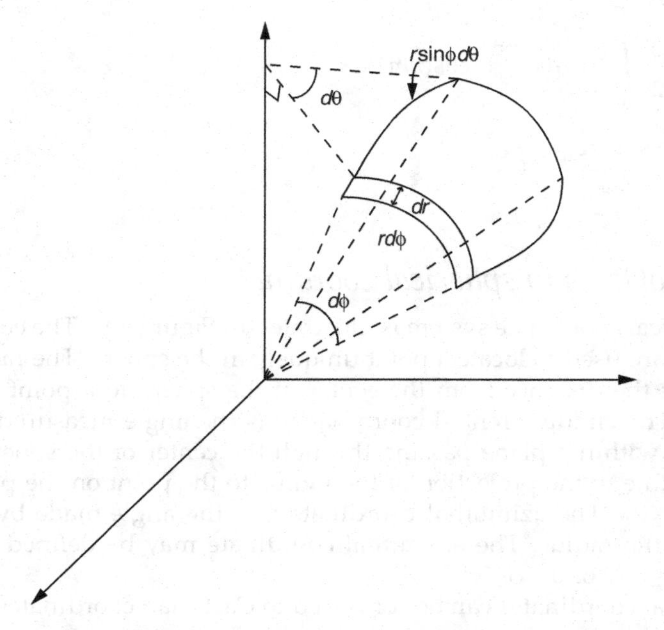

Figure 6.19 The volume of a differential element in a spherical coordinate system is $dV = r^2 \sin\phi dr d\theta d\phi$.

A differential volume within the sphere is illustrated in Figure 6.19. The volume of this element is

$$dV = r^2 \sin\phi dr d\theta d\phi \tag{6.79}$$

Vector operators written in spherical coordinates are

$$\nabla\Lambda=\frac{\partial\Lambda}{\partial r}\mathbf{e}_r+\frac{1}{r\sin\phi}\frac{\partial\Lambda}{\partial\theta}\mathbf{e}_\theta+\frac{1}{r}\frac{\partial\Lambda}{\partial\phi}\mathbf{e}_\phi \tag{6.80}$$

$$\nabla\cdot\mathbf{G}=\frac{1}{r^2}\frac{\partial}{\partial r}\left(r^2G_r\right)+\frac{1}{r\sin\phi}\frac{\partial G_\theta}{\partial\theta}+\frac{1}{r\sin\phi}\frac{\partial}{\partial\phi}\left(\sin\phi G_\phi\right) \tag{6.81}$$

$$\nabla^2\Lambda=\frac{1}{r^2}\frac{\partial}{\partial r}\left(r^2\frac{\partial\Lambda}{\partial r}\right)+\frac{1}{r^2\sin^2\phi}\frac{\partial^2\Lambda}{\partial\theta^2}+\frac{1}{r^2\sin\phi}\frac{\partial}{\partial\phi}\left(\sin\phi\frac{\partial\Lambda}{\partial\phi}\right) \tag{6.82}$$

Laplace's equation, the diffusion equation, and the wave equation are all separable in spherical coordinates.

The operator $-\nabla^2$ was shown in Example 2.31 to be self-adjoint with respect to the inner product of Equation a of that example. When the boundary of the region over which $-\nabla^2$ is bounded by $r=1$ with $0\leq\theta\leq 2\pi$ and $0\leq\phi\leq\pi$, the inner product is defined as

$$(f(r,\theta,\phi),g(r,\theta,\phi))=\int_0^{\pi}\int_0^{2\pi}\int_0^{1} f(r,\theta,\phi)g(r,\theta,\phi)r^2\sin\phi\, dr\, d\theta\, d\phi \tag{6.83}$$

Example 6.13 The surface of the sphere shown in Figure 6.20 is maintained at a constant temperature distribution, $f(\theta,\phi)$. Determine the steady-state nondimensional temperature, $\Lambda(r,\theta,\phi)$. Assume that the radial coordinate is nondimensional with $r^*=r/R$.

Solution Laplace's equation in spherical coordinates reduces to

$$\frac{\partial}{\partial r}\left(r^2\frac{\partial\Lambda}{\partial r}\right)+\frac{1}{\sin^2\phi}\frac{\partial^2\Lambda}{\partial\theta^2}+\frac{1}{\sin\phi}\frac{\partial}{\partial\phi}\left(\sin\phi\frac{\partial\Lambda}{\partial\phi}\right)=0 \tag{a}$$

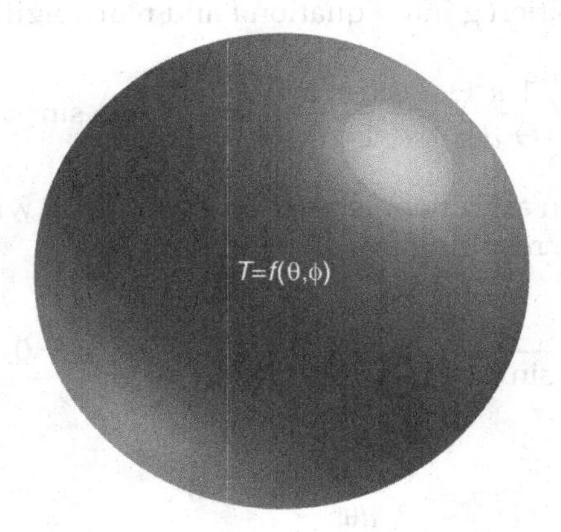

Figure 6.20 The surface of the sphere is maintained at a constant temperature distribution.

The nonhomogeneous boundary condition is at $r=1$. Therefore, a product solution can be assumed to be of the form

$$\Lambda(r,\theta,\phi) = R(r)U(\theta,\phi) \tag{b}$$

Substitution of Equation b into Equation a leads to

$$U\frac{d}{dr}\left(r^2\frac{dR}{dr}\right) + \frac{R}{\sin^2\phi}\frac{\partial^2 U}{\partial\theta^2} + \frac{R}{\sin\phi}\frac{\partial}{\partial\phi}\left(\sin\phi\frac{\partial U}{\partial\phi}\right) = 0 \tag{c}$$

Dividing Equation c by RU and rearranging leads to

$$\frac{1}{R}\frac{d}{dr}\left(r^2\frac{dR}{dr}\right) = -\frac{1}{U\sin^2\phi}\frac{\partial^2 U}{\partial\theta^2} - \frac{1}{U\sin\phi}\frac{\partial}{\partial\phi}\left(\sin\phi\frac{\partial U}{\partial\phi}\right) \tag{d}$$

Applying the usual separation argument to Equation d and defining the separation constant as λ leads to

$$r^2\frac{d^2R}{dr^2} + r\frac{dR}{dr} - \lambda R = 0 \tag{e}$$

$$-\frac{1}{\sin^2\phi}\frac{\partial^2 U}{\partial\theta^2} - \frac{1}{\sin\phi}\frac{\partial}{\partial\phi}\left(\sin\phi\frac{\partial U}{\partial\phi}\right) = \lambda U \tag{f}$$

It is noted that Equation e is a Cauchy-Euler equation (Section 3.5).

A product solution for U is assumed to be of the form

$$U(\theta,\phi) = \Theta(\theta)\Phi(\phi) \tag{g}$$

Substitution of Equation g into Equation f and rearranging gives

$$-\frac{1}{\Theta}\frac{d^2\Theta}{d\theta^2} = \frac{\sin\phi}{\Phi}\frac{d}{d\phi}\left(\sin\phi\frac{d\Phi}{d\phi}\right) - \lambda\sin^2\phi \tag{h}$$

The usual separation argument is used on Equation h, with μ defined as the separation constant, resulting in

$$\frac{1}{\sin\phi}\frac{d}{d\phi}\left(\sin\phi\frac{d\Phi}{d\phi}\right) + \left(\lambda - \frac{\mu}{\sin^2\phi}\right)\Phi = 0 \tag{i}$$

$$\frac{d^2\Theta}{d\theta^2} + \mu\Theta = 0 \tag{j}$$

Solutions of Equation e, Equation i, and Equation j are required to be finite and single-valued at all points in the sphere.

The coordinates $(r,\theta+2n\pi,\phi)$ for any integer value of n refer to the same point as the coordinates (r,θ,ϕ) Thus, for the temperature and rate of heat transfer to be single-valued, $\Theta(\theta)$ and $d\Theta/d\theta$ must be periodic of period 2π. Equation j with the periodicity conditions is the eigenvalue problem which defines the trigonometric Fourier series. Its eigenvalues, obtained in Section 5.5 and used in Section 6.2, are $\mu_n=n^2$ for $n=0,1,2,\ldots$. The normalized eigenvector corresponding to $n=0$ is

$$\Theta_0=\frac{1}{\sqrt{2\pi}} \tag{k}$$

Each eigenvalue for $n>0$ has two linearly independent normalized eigenvectors given by

$$\Theta_{n,1}(\theta)=\frac{1}{\sqrt{\pi}}\cos(n\theta) \tag{l1}$$

$$\Theta_{n,2}(\theta)=\frac{1}{\sqrt{\pi}}\sin(n\theta) \tag{l2}$$

Equation i can be written as an eigenvalue problem for each value of n,

$$-\frac{1}{\sin\phi}\frac{d}{d\phi}\left(\sin\phi\frac{d\Phi_n}{d\phi}\right)+\frac{n^2}{\sin^2\phi}=\lambda\Phi_n \tag{m}$$

Equation n is of the form of a Sturm-Liouville problem with $p(\phi)=-\sin\phi$, $q(\phi)=n^2/\sin^2\phi$, and $r(\phi)=\sin\phi$. For each n, there are an infinite, but countable, number of eigenvalues, $\lambda_{n,m}$ for $m=1,2,\ldots$. The eigenvectors $\Phi_{n,m}$ and $\Phi_{n,p}$, respectively corresponding to $\lambda_{n,m}$ and $\lambda_{n,p}$, satisfy the orthogonality condition $(\Phi_{n,m},\Phi_{n,p})=0$, where

$$\left(\Phi_{n,m},\Phi_{n,p}\right)_\phi=\int_0^{\frac{\pi}{2}}\Phi_{n,m}(\phi)\Phi_{n,p}(\phi)\sin\phi d\phi \tag{n}$$

Since $0\le\phi\le\pi$ and $p(0)=p(\pi)=0$, then according to theorem 5.7 the eigenvectors must only satisfy the conditions that they are finite at $\phi=0$ and $\phi=\pi$.

Using the change of independent variable, $s=\cos\phi$, Equation n can be rewritten as

$$\frac{d}{ds}\left[\left(1-s^2\right)\frac{d\Phi_n}{ds}\right]+\left(\mu-\frac{n^2}{1-s^2}\right)\Phi_n=0 \tag{o}$$

Equation o is the associated Legendre's equation which is solved in Section 3.7. Unless $\mu = m(m+1)$, its solution is not finite at $s = \pm 1$, which correspond to $\phi = 0$ and $\phi = \pi$. For $\mu = m(m+1)$, the general solution of Equation o is

$$\Phi_{n,m} = C_1 P_m^n(s) + C_1 Q_m^n(s),$$

where $P_m^n(s)$ and $Q_m^n(s)$ are the associated Legendre functions of the first and second kinds of order m and index n. Noting that $Q_m^n(s)$ is not finite at $s = \pm 1$, $C_2 = 0$. Thus

$$\Phi_{n,m}(\phi) = C_{n,m} P_m^n(\cos\phi)\sin^n(\phi) \tag{p}$$

The eigenvectors can be normalized by choosing

$$C_{n,m} = \left\{ \int_0^{\pi} \left[P_m^n(\cos\phi) \right]^2 \sin^{2n+1}\phi d\phi \right\}^{-\frac{1}{2}} \tag{q}$$

The differential equation for $R(r)$ now becomes

$$r^2 \frac{d^2 R_m}{dr^2} + r \frac{dR_m}{dr} - m(m+1)R_m = 0 \tag{r}$$

As discussed in Section 3.5, solution of Equation r is assumed as $R_m = r^{\alpha}$, which when substituted into Equation r, leads to a quadratic equation whose roots are the values of α. The general solution is then obtained as

$$R_m(r) = A_m r^m + B_m r^{-(m+1)} \tag{s}$$

The solution must remain finite at the center of the sphere, $r = 0$, requiring $B_m = 0$.

Using Equation b, Equation g, Equation k, Equation l, Equation m, Equation p, and Equation s, the general solution of Equation a is

$$\Lambda(r,\theta,\phi) = \sum_{m=0}^{\infty} \left\{ A_{0,m} \frac{C_{0,m}}{\sqrt{2\pi}} P_m(\cos\phi) r^m \right.$$

$$\left. + \sum_{n=1}^{\infty} \frac{C_{n,m}}{\sqrt{\pi}} P_m^n(\cos\phi)\sin^n(\phi) r^m \left(A_{n,m}\cos\theta + B_{n,m}\sin\theta \right) \right. \tag{t}$$

The boundary condition $\Lambda(1,\theta,\phi) = f(\theta,\phi)$ is applied using an eigenvector expansion for $f(\theta,\phi)$. The constants in Equation t can be obtained as

$$A_{0,m} = \left(f, \frac{C_{0,m}}{\sqrt{2\pi}} P_m(\cos\phi) \right) = \frac{C_{0,m}}{\sqrt{2\pi}} \int_0^{2\pi} \int_0^{\pi} f(\theta,\phi) P_m(\cos\phi) \sin\phi d\phi d\theta \tag{u}$$

$$A_{n,m} = \left(f, \frac{C_{n,m}}{\sqrt{\pi}} P_m^n(\cos\phi)\sin^n(\phi)\cos\theta \right)$$

$$= \frac{C_{n,m}}{\sqrt{\pi}} \int_0^{2\pi} \int_0^{\pi} f(\theta,\phi) P_m^n(\cos\phi)\sin^n(\phi)\cos\theta \sin^{n+1}\phi d\phi d\theta \tag{v}$$

$$B_{n,m} = \left(f, \frac{C_{n,m}}{\sqrt{\pi}} P_m^n(\cos\phi)\sin^n(\phi)\sin\theta \right)$$

$$= \frac{C_{n,m}}{\sqrt{\pi}} \int_0^{2\pi} \int_0^{\pi} f(\theta,\phi) P_m(\cos\phi)\sin\theta \sin^{n+1}\phi d\phi d\theta \tag{w}$$

Example 6.14 The sphere of Example 6.13 is at a steady state with a temperature distribution given by Equation t, Equation u, Equation v, and Equation w when it is suddenly placed in a medium of temperature T_∞. The heat transfer coefficient between the sphere and the ambient is h. The nondimensional problem formulation for the resulting nondimensional temperature, $\Lambda(r,\theta,\phi,t)$, is

$$\nabla^2\Lambda = \frac{\partial\Lambda}{\partial t} \tag{a}$$

$$\frac{\partial\Lambda}{\partial r}(1,\theta,\phi,t) + Bi\Lambda(1,\theta,\phi,t) \tag{b}$$

$$\Lambda(r,\theta,\phi,0) = g(r,\theta,\phi) \tag{c}$$

where $g(r,\theta,\phi)$ is $\Lambda(1,\theta,\theta)$ of Example 6.13 adjusted for a change in the definition in nondimensional temperature. Determine $\Lambda(r,\theta,\phi,t)$.

Solution A product solution of Equation a can be assumed to be of the form

$$\Lambda(r,\theta,\phi,t) = T(t)U(r,\theta,\phi) \tag{d}$$

Substitution of Equation d into Equation a and Equation b, rearrangement of the resulting equation, and application of the usual separation argument with the separation parameter defined as λ leads to

$$\frac{dT}{dt} + \lambda T = 0 \tag{e}$$

$$-\nabla^2 U = \lambda U \tag{f}$$

$$\frac{\partial U}{\partial r}(1,\theta,\phi) + BiU(1,\theta,\phi) = 0 \tag{g}$$

The values of the separation constant are the values of the eigenvalues of $-\nabla^2 U$ subject to Equation g. The eigenvectors must be finite and single-valued at all points in the sphere. Eigenvectors corresponding to distinct eigenvectors are orthogonal with respect to the inner product,

$$(f(r,\theta,\phi), g(r,\theta,\phi)) = \int_0^{\pi}\int_0^{2\pi}\int_0^{1} f(r,\theta,\phi)g(r,\theta,\phi)r^2 \sin\phi \, dr d\theta d\phi \tag{h}$$

A product solution of Equation f is assumed to be of the form

$$U(r,\theta,\phi) = R(r)\Theta(\theta)\Phi(\phi) \tag{i}$$

Substituting Equation i into Equation f with $\nabla^2\Lambda$ obtained from Equation 6.82 gives

$$\frac{\Theta\Phi}{r^2}\frac{d}{dr}\left(r^2\frac{dR}{dr}\right) + \frac{R\Phi}{r^2\sin^2\phi}\frac{d^2\Theta}{d\theta^2} + \frac{R\Theta}{r^2\sin\phi}\frac{d}{d\phi}\left(\sin\phi\frac{d\Phi}{d\phi}\right) = -\lambda R\Theta\Phi \tag{j}$$

Multiplying Equation j by $r^2/R\Theta\Phi$ and rearranging leads to

$$-\frac{1}{R}\frac{d}{dr}\left(r^2\frac{dR}{dr}\right) - \lambda r^2 = \frac{1}{\sin^2\phi\Theta}\frac{d^2\Theta}{d\theta^2} + \frac{1}{\sin\phi\Phi}\frac{d}{d\phi}\left(\sin\phi\frac{d\Phi}{d\phi}\right) \tag{k}$$

Applying the usual separation argument to Equation k and denoting the separation constant as μ results in

$$\frac{d}{dr}\left(r^2\frac{dR}{dr}\right) + \left(\lambda r^2 - \mu\right)R \tag{l}$$

$$\frac{1}{\sin^2\phi\Theta}\frac{d^2\Theta}{d\theta^2} + \frac{1}{\sin\phi\Phi}\frac{d}{d\phi}\left(\sin\phi\frac{d\Phi}{d\phi}\right) = -\mu \tag{m}$$

Multiplying Equation m by $\sin^2\phi$ and rearranging leads to

$$-\frac{1}{\Theta}\frac{d^2\Theta}{d\theta^2} = \frac{\sin\phi}{\Phi}\frac{d}{d\phi}\left(\sin\phi\frac{d\Phi}{d\phi}\right) + \mu\sin^2\phi \tag{n}$$

Applying the usual separation argument to Equation n and denoting the separation constant as κ results in

$$\frac{d^2\Theta}{d\theta^2}+\kappa\Theta=0 \tag{o}$$

$$\frac{1}{\sin\phi}\frac{d}{d\phi}\left(\sin\phi\frac{d\Phi}{d\phi}\right)+\left(\mu-\frac{\kappa}{\sin^2\phi}\right)\Phi=0 \tag{p}$$

The values of the separation constant κ are the eigenvalues of Equation o subject to periodicity conditions. The eigenvalues are $\kappa_n=n^2$ and the corresponding normalized eigenvectors are those of Equation k. Equation l, and Equation m of Example 6.13.

Equation p is the same as that encountered in the steady-state problem of Example 6.13. The eigenvalues are of the form $\mu_m=m\ (m+1)$, and the normalized eigenvector is expressed in terms of the associated Legendre function of the first kind, as in Equation p of Example 6.13.

The self-adjoint form of equation l is

$$-\frac{1}{r^2}\frac{d}{dr}\left(r^2\frac{dR_m}{dr}\right)+\frac{m(m+1)}{r^2}R_m=\lambda R_m \tag{q}$$

Equation q is in proper Sturm-Liouville form with $p(r)=-r^2$, $q(r)=\mu/r^2$, and $r(r)=r^2$. For each m, its eigenvalues are infinite, but countable, $\lambda_{m,p}$ for $p=1,2,\ldots$. The orthogonality condition for $R_{m,p}$ and $R_{m,q}$ corresponding to distinct eigenvalues $\lambda_{m,p}$ and $\lambda_{m,q}$ is

$$\begin{aligned}0&=\left(R_{m,p}(r),R_{m,q}(r)\right)_r\\&=\int_0^1 R_{m,p}(r)R_{m,q}(r)r^2dr\end{aligned} \tag{r}$$

Equation q is the spherical Bessel's equation discussed in Section 3.6. Its general solution is

$$R_m(r)=D_m j_m\left(\sqrt{\lambda}r\right)+E_m y_m\left(\sqrt{\lambda}r\right) \tag{s}$$

where $j_m(r)$ and $y_m(r)$ are the spherical Bessel functions of the first and second kinds of order m and argument r. Noting that $y_m(0)$ is not finite at $r=0$ it is required that $E_m=0$ for the temperature to be finite at $r=0$. Application of the product solution to Equation g leads to $dR/dr(1) + BiR(1) = 0$, which when applied to Equation s, results in

$$\sqrt{\lambda}\, j_m'\left(\sqrt{\lambda}\right) + (Bi)\, j_m\left(\sqrt{\lambda}\right) = 0 \tag{t}$$

The eigenvalue $\lambda_{m,p}$ is the pth positive solution of Equation t. The resulting normalized eigenvectors are

$$R_{m,p}(r) = D_{m.p} j_m\left(\sqrt{\lambda_{m,p}}\, r\right) \tag{u}$$

where

$$D_{m,p} = \left[\int_0^1 \left[j_m\left(\sqrt{\lambda_{m,p}}\, r\right)\right]^2 r^2 dr\right]^{-\frac{1}{2}} \tag{v}$$

The closed-form evaluation of Equation v is in terms of confluent hypergeometric functions and is beyond the scope of this study.

The normalized eigenvectors of Equation f are

$$U_{0,m,p} = \frac{D_{m,p} C_{n,m}}{\sqrt{2\pi}} P_m(\cos\phi)\, j_m\left(\sqrt{\lambda_{m,p}}\, r\right) \tag{w}$$

$$U_{m,n,p,1} = \frac{D_{m,p} C_{n,m}}{\sqrt{\pi}} P_m^n(\cos\phi) \sin^n(\phi)\, j_m\left(\sqrt{\lambda_{m,p}}\, r\right) \cos\theta \tag{x}$$

$$U_{m,n,p,2} = \frac{D_{m,p} C_{n,m}}{\sqrt{\pi}} P_m^n(\cos\phi) \sin^n(\phi)\, j_m\left(\sqrt{\lambda_{m,p}}\, r\right) \sin\theta \tag{y}$$

Noting that the general solution of Equation e is $T_{n,m,p}(t) = A_{n,m,p} e^{-\lambda_{m,p} t}$, the solution of Equation a is

$$\Lambda(r,\theta,\phi,t) = \sum_{m=1}^{\infty} \sum_{p=1}^{\infty} e^{-\lambda_{m,p} t} \left\{ A_{0,m,p} U_{0,m,p}(r,\theta,\phi) + \sum_{n=1}^{\infty} \left[A_{n,m,p} U_{m,n,p,1}(r,\theta,\phi) + B_{n,m,p} U_{m,n,p,2}(r,\theta,\phi) \right] \right\} \tag{z}$$

Example 6.15 The surface of the sphere shown in Figure 6.20 is pulsating such that the distance from the center of the sphere to a point on the surface of the sphere is

$$r = 1 + w(\theta,\phi,t) \tag{a}$$

The pulsations of the sphere lead to the initiation and propagation of acoustic waves in the surrounding inviscid fluid. If $\Phi(r,\theta,\phi,t)$ is the velocity potential in the fluid which is related to the velocity vector by

$$\mathbf{v} = \nabla\Phi \tag{b}$$

then the waves are governed by the wave equation,

$$\nabla^2\Phi = \frac{\partial^2\Phi}{\partial t^2} \tag{c}$$

A boundary condition is that the normal component of velocity of the fluid at the surface of the sphere is equal to the velocity of the particle on the sphere at that point,

$$\frac{\partial\Phi}{\partial r}(1,\theta,\phi,t) = \frac{\partial w}{\partial t}(\theta,\phi,t) \tag{d}$$

Since there is no reflection, the waves are only outgoing.

(a) Determine $\Phi(r,t)$ if $w(t) = \sum_{p=0}^{\infty} A_p \sin(\omega_p t + \gamma_p)$; $w(t)$ is a periodic function with a trigonometric Fourier series representation.
(b) Determine $\Phi(r,\theta,\phi,t)$ if $w(\theta,\phi,t) = f(\theta,\phi)\sin(\omega t)$, where f is a continuous function over the surface of the sphere.

Solution (a) The boundary condition on the surface of the sphere is

$$\frac{\partial\Phi}{\partial r}(1,t) = \sum_{p=1}^{\infty} A_p \omega_p \cos(\omega_p t + \nu_p) \tag{e}$$

The pulsations of the body are such that it retains the shape of a sphere, but with changing radius. The pulsations are independent of θ and ϕ, and therefore the waves are uniformly propagated from the surface of the sphere and are independent of θ and ϕ. Equation c becomes

$$\frac{1}{r^2}\frac{\partial}{\partial t}\left(r^2\frac{\partial\Phi}{\partial r}\right) = \frac{\partial^2\Phi}{\partial t^2} \tag{f}$$

Then the boundary condition, Equation e, suggests a solution of the form

$$\Phi(r,t) = \sum_{p=1}^{\infty}\left[g_{1p}\sin(\omega_p t) + g_{2p}\cos(\omega_p t)\right] \tag{g}$$

Substitution of Equation g into Equation f leads to

$$\frac{1}{r^2}\frac{d}{dr}\left(r^2\frac{dg_{1p}}{dr}\right)+\omega_p^2 g_{1p}=0$$

$$r^2\frac{d^2g_{1p}}{dr^2}+2r\frac{dg_{1p}}{dr}+\omega_p^2 r^2 g_{1p}=0 \tag{h}$$

and

$$r^2\frac{d^2g_{2p}}{dr^2}+2r\frac{dg_{2p}}{dr}+\omega_p^2 r^2 g_{2p}=0 \tag{i}$$

The solutions of Equation h and Equation i are written using spherical Bessel functions as

$$g_{1p}(r)=C_{1p}j_0(\omega_p r)+C_{2p}y_0(\omega_p r) \tag{j}$$

$$g_{1p}(r)=C_{3p}j_0(\omega_p r)+C_{4p}y_0(\omega_p r) \tag{k}$$

Asymptotic relations for the spherical Bessel functions are

$$j_0(\omega r)\approx\frac{1}{\omega r}\sin(\omega r)+O\left(\frac{1}{r^2}\right)\quad \text{as } r\to\infty \tag{l}$$

$$y_0(\omega r)\approx-\frac{1}{\omega r}\cos(\omega r)+O\left(\frac{1}{r^2}\right)\quad \text{as } r\to\infty \tag{m}$$

Combining Equation g, Equation i, Equation l, and Equation m, the behavior of the waves for large r is

$$\begin{aligned}\Phi(r,t)\approx\sum_{p=1}^{\infty}\frac{1}{\omega_p r}&\{[C_{1p}\sin(\omega_p r)-C_{2p}\cos(\omega_p r)]\sin(\omega_p t)\\&+[C_{3p}\sin(\omega_p r)-C_{4p}\cos(\omega_p r)]\cos(\omega_p t)\}+0(\tfrac{1}{r^2})\end{aligned} \tag{n}$$

Use of trigonometric identities in Equation n leads to

$$\begin{aligned}\Phi(r,t)\approx\sum_{p=1}^{\infty}\frac{1}{2\omega_p r}&\{[C_{1p}\cos(\omega_p(t-r))-C_{1p}\cos(\omega_p(t+r))\\&-C_{2p}\sin(\omega_p(t+r))-C_{2p}\sin(\omega_p(t-r))]\\&+[C_{3p}\sin(\omega_p(t+r))+C_{3p}\sin(\omega_p(t-r))\\&-C_{4p}\cos(\omega_p(t-r))-C_{4p}\cos(\omega_p(t+t))]\}+0(\tfrac{1}{r^2})\end{aligned} \tag{o}$$

Waves are continuously generated from the surface of the pulsating sphere. A wave is generated at the surface at $t=0$. The wave propagates in the radial direction. A wave of the form $\cos(\omega(t-r))$ or $\sin(\omega(t-r))$ has a positive velocity and propagates in the positive radial direction (an outgoing wave), whereas a wave of the form $\cos(\omega(t+r))$ or $\sin(\omega(t+r))$ has a negative velocity and propagates in the negative radial direction (an incoming wave). However, the region is unbounded, and outgoing waves are not reflected. Since only outgoing waves are permitted in the solution,

$$C_{3p} - C_{2p} = 0 \tag{p}$$

$$C_{1p} + C_{4p} = 0 \tag{q}$$

The solution becomes

$$\Phi(r,t) = \sum_{p=1}^{\infty} \{[C_{1p} j_0(\omega_p r) + C_{2p} y_0(\omega_p r)]\sin(\omega_p t) + [C_{2p} j_0(\omega_p r) - C_{1p} y_0(\omega_p r)]\cos(\omega_p t)\} \tag{r}$$

Satisfaction of Equation e requires that

$$\sum_{p=1}^{\infty} \omega_p \{[C_{1p} j_0'(\omega_p) + C_{2p} y_0'(\omega_p)]\sin(\omega_p t) + [C_{2p} j_0'(\omega_p) - C_{1p} y_0'(\omega_p)]\cos(\omega_p t)\} = \sum_{p=1}^{\infty} A_p \omega_p \cos(\omega_p t + \nu_p) \tag{s}$$

The trigonometric terms on the right-hand side of Equation s can be expanded using a trigonometric identity; then coefficients of like terms are equated for each p, leading to

$$\left[C_{1p} j_0'(\omega_p) + C_{2p} y_0'(\omega_p)\right]\omega_p = -A_p \omega_p \sin \nu_p \tag{t}$$

$$\left[C_{2p} j_0'(\omega_p) - C_{1p} y_0'(\omega_p)\right]\omega_p = A_p \omega_p \cos \nu_p \tag{u}$$

Equation t and Equation u are solved simultaneously, leading to

$$C_{1p} = -A_p \frac{j_0'(\omega_p)\sin \nu_p + y_0'(\omega_p)\cos \nu_p}{\left[j_0'(\omega_p)\right]^2 + \left[y_0'(\omega_p)\right]^2} \tag{v}$$

$$C_{2p} = A_p \frac{j_0'(\omega_p)\cos \nu_p - y_0'(\omega_p)\sin \nu_p}{\left[j_0'(\omega_p)\right]^2 + \left[y_0'(\omega_p)\right]^2} \tag{w}$$

(b) The boundary condition on the surface of the sphere is

$$\frac{\partial \Phi}{\partial r}(1,\theta,\phi,t) = \omega f(\theta,\phi)\cos(\omega t) \tag{x}$$

The form of Equation x and the knowledge that waves must be outgoing only suggest a superposition solution of the form

$$\Phi(r,\theta,\phi,t) = A(r,\theta,\phi)\sin(\omega t) + B(r,\theta,\phi)\cos(\omega t) \tag{y}$$

Substitution into Equation c and choosing appropriate problems leads to

$$\nabla^2 A + \omega^2 A = 0 \tag{z}$$

$$\nabla^2 B + \omega^2 B = 0 \tag{aa}$$

A product solution of Equation z is assumed to be of the form

$$A(r,\theta,\Phi) = R(r)\Theta(\theta)\Psi(\phi) \tag{bb}$$

Substitution of Equation bb into Equation z leads to

$$\frac{\Theta\Psi}{r^2}\frac{d}{dr}\left(r^2\frac{dR}{dr}\right) + \frac{R\Psi}{r^2\sin^2\phi}\frac{d^2\Theta}{d\theta^2} + \frac{R\Theta}{r^2\sin\phi}\frac{d}{d\phi}\left(\sin\phi\frac{d\Psi}{d\phi}\right) + \omega^2 R\Theta\Psi = 0 \tag{cc}$$

Multiplying Equation cc by $r^2/R\Theta\Psi$ and rearranging leads to

$$-\frac{1}{R}\frac{d}{dr}\left(r^2\frac{dR}{dr}\right) + \omega^2 r^2 = \frac{1}{\sin^2\phi\Theta}\frac{d^2\Theta}{d\theta^2} + \frac{1}{\sin\phi\Phi}\frac{d}{d\phi}\left(\sin\phi\frac{d\Psi}{d\phi}\right) \tag{dd}$$

Applying the usual separation argument to Equation dd and denoting the separation constant as μ results in

$$\frac{d}{dr}\left(r^2\frac{dR}{dr}\right) + \left(\omega^2 r^2 - \mu\right)R \tag{ee}$$

$$\frac{1}{\sin^2\phi\Theta}\frac{d^2\Theta}{d\theta^2} + \frac{1}{\sin\phi\Phi}\frac{d}{d\phi}\left(\sin\phi\frac{d\Psi}{d\phi}\right) = -\mu \tag{ff}$$

Multiplying Equation ff by $\sin^2\phi$ and rearranging leads to

$$-\frac{1}{\Theta}\frac{d^2\Theta}{d\theta^2} = \frac{\sin\phi}{\Phi}\frac{d}{d\phi}\left(\sin\phi\frac{d\Psi}{d\phi}\right) + \mu\sin^2\phi \tag{gg}$$

Applying the usual separation argument to Equation gg and denoting the separation constant as κ results in

$$\frac{d^2\Theta}{d\theta^2}+\kappa\Theta=0 \tag{hh}$$

$$\frac{1}{\sin\phi}\frac{d}{d\phi}\left(\sin\phi\frac{d\Phi}{d\phi}\right)+\left(\mu-\frac{\kappa}{\sin^2\phi}\right)\Psi=0 \tag{ii}$$

The conditions satisfied by the solution of Equation hh are that the velocity potential and the velocity be single-valued everywhere within the acoustic field, which lead to the velocity potential and the θ-component of the velocity being periodic in θ with period 2π. The eigenvalues of Equation hh subject to the periodicity conditions are of the form $\kappa_n = n^2$ for $n=0,1,2,\ldots$.

The normalized eigenvector corresponding to $n=0$ is

$$\Theta_0=\frac{1}{\sqrt{2\pi}} \tag{jj}$$

Each eigenvalue for $n>0$ has two linearly independent normalized eigenvectors given by

$$\Theta_{n,1}(\theta)=\frac{1}{\sqrt{\pi}}\cos(n\theta) \tag{kk}$$

$$\Theta_{n,2}(\theta)=\frac{1}{\sqrt{\pi}}\sin(n\theta) \tag{ll}$$

The conditions satisfied by Equation ii are that the velocity must remain finite at every point in the field. Following the discussion of the solution of Equation n of Example 6.13, the eigenvalues of Equation ii subject to these conditions are of the form $\mu_m = m\,(m+1)$ and their eigenvectors are

$$\Psi_{n,m}(\phi)=C_{n,m}P_m^n(\cos\phi)\sin^n(\phi) \tag{mm}$$

where $C_{n,m}$ is chosen to normalize the eigenvector.

The general solution of Equation ee is then

$$R_m(r)=D_{1m}j_m(\omega r)+D_{2m}y_m(\omega r) \tag{nn}$$

The general solution of Equation z is

$$A_{0,m}(r,\theta,\phi)=\frac{C_{0,m}}{\sqrt{\pi}}\left[D_{1,m,0}j_m(\omega r)+D_{2,m,0}y_m(\omega r)\right]P_m(\cos\phi)\sin^n(\phi) \tag{oo}$$

$$A_{n,m}(r,\theta,\phi) = \frac{C_{n,m}}{\sqrt{2\pi}}\{[D_{1,m,1}j_m(\omega r) + D_{2,m,1}y_m(\omega r)]\cos(n\theta)$$

$$+[D_{1,m,2}j_m(\omega r) + D_{2,m,2}y_m(\omega r)]\sin(n\theta)\}P_m^n(\cos\phi)\sin^n(\phi) \qquad \text{(pp)}$$

The solution of Equation aa is similar to that of Equation oo and Equation pp, except with different arbitrary constants; call them E. Application of the superposition formula, Equation y, leads to

$$\Phi(r,\theta,\phi,t) = \sum_{m=0}^{\infty}\left[\frac{C_{0,m}}{\sqrt{\pi}}[D_{1,m,0}j_m(\omega r) + D_{2,m,0}y_m(\omega r)]P_m(\cos\phi)\right.$$

$$+\sum_{n=1}^{\infty}\frac{C_{n,m}}{\sqrt{2\pi}}\{[D_{1,m,1}j_m(\omega r) + D_{2,m,1}y_m(\omega r)]\cos(n\theta)$$

$$\left.+[D_{1,m,2}j_m(\omega r) + D_{2,m,2}y_m(\omega r)]\sin(n\theta)\}P_m^n(\cos\phi)\sin^n(\phi)\right]\sin(\omega t)$$

$$+\sum_{m=0}^{\infty}\left[\frac{C_{0,m}}{\sqrt{\pi}}[E_{1,m,0}j_m(\omega r) + E_{2,m,0}y_m(\omega r)]P_m(\cos\phi)\right.$$

$$+\sum_{n=1}^{\infty}\frac{C_{n,m}}{\sqrt{2\pi}}\{[E_{1,m,1}j_m(\omega r) + E_{2,m,1}y_m(\omega r)]\cos(n\theta)$$

$$\left.+[E_{1,m,2}j_m(\omega r) + E_{2,m,2}y_m(\omega r)]\sin(n\theta)\}P_m^n(\cos\phi)\sin^n(\phi)\right]\cos(\omega t)$$

(qq)

Asymptotic relations for the spherical Bessel functions are

$$j_m(\omega r) \approx \frac{1}{\omega r}\sin\left(\omega r - \frac{1}{2}m\pi\right) + O\left(\frac{1}{r^2}\right) \quad \text{as } r \to \infty \qquad \text{(rr)}$$

$$y_0(\omega r) \approx -\frac{1}{\omega r}\cos\left(\omega r - \frac{1}{2}m\pi\right) + O\left(\frac{1}{r^2}\right) \quad \text{as } r \to \infty \qquad \text{(ss)}$$

The solution includes only outgoing waves if (see part a)

$$D_{1,m,0} + E_{2,m,0} = 0 \qquad \text{(tt)}$$

$$E_{1,m,0} - D_{2,m,0} = 0 \qquad \text{(uu)}$$

$$D_{1,m,1} + E_{2,m,1} = 0 \qquad \text{(vv)}$$

$$E_{1,m,1} - D_{2,m,1} = 0 \tag{ww}$$

$$D_{1,m,2} + E_{2,m,2} = 0 \tag{xx}$$

$$E_{1,m,2} - D_{2,m,2} = 0 \tag{yy}$$

Application of the boundary condition on the surface of the sphere leads to

$$\sum_{m=0}^{\infty}\left[\frac{C_{0,m}}{\sqrt{\pi}}[D_{1,m,0}j'_m(\omega)+D_{2,m,0}y'_m(\omega)]P_m(\cos\phi)\right.$$

$$+\sum_{n=1}^{\infty}\frac{C_{n,m}}{\sqrt{2\pi}}\{[D_{1,m,1}j'_m(\omega)+D_{2,m,1}y'_m(\omega)]\cos(n\theta)$$

$$\left.+[D_{1,m,2}j'_m(\omega)+D_{2,m,2}y'_m(\omega)]\sin(n\theta)\}P_m^n(\cos\phi)\sin^n(\phi)\right]\sin(\omega t)$$

$$+\sum_{m=0}^{\infty}\left[\frac{C_{0,m}}{\sqrt{\pi}}[E_{1,m,0}j'_m(\omega)+E_{2,m,0}y'_m(\omega)]P_m(\cos\phi)\right.$$

$$+\sum_{n=1}^{\infty}\frac{C_{n,m}}{\sqrt{2\pi}}\{[E_{1,m,1}j'_m(\omega)+E_{2,m,1}y'_m(\omega)]\cos(n\theta)$$

$$\left.+[E_{1,m,2}j'_m(\omega)+E_{2,m,2}y'_m(\omega)]\sin(n\theta)\}P_m^n(\cos\phi)\sin^n(\phi)\right]\cos(\omega t)$$

$$= f(\theta,\phi)\cos(\omega t) \tag{zz}$$

Eigenvector expansions are used to obtain

$$f(r,\theta)=\sum_{m=0}^{\infty}\left\{F_mP_m(\cos\phi)+\sum_{n=1}^{\infty}[G_{m,n,1}\cos(n\theta)+G_{m,n,2}\sin(n\theta)]P_m^n(\cos\phi)\sin^n(\phi)\right\} \tag{aaa}$$

where

$$F_m=\frac{\int_0^{2\pi}\int_0^{\pi}f(\theta,\phi)P_m(\cos\phi)\sin\phi\, d\phi\, d\theta}{\int_0^{2\pi}\int_0^{\pi}[P_m(\cos\phi)]^2\sin\phi\, d\phi\, d\theta} \tag{bbb}$$

$$G_{m,n,1}=\frac{\int_0^{2\pi}\int_0^{\pi}f(\theta,\phi)\cos(n\theta)P_m^n(\cos\phi)\sin^{n+1}\phi\, d\phi\, d\theta}{\int_0^{2\pi}\int_0^{\pi}[\cos(n\theta)P_m(\cos\phi)\sin^n(\phi)]^2\sin\phi\, d\phi\, d\theta} \tag{ccc}$$

$$G_{m,n,2} = \frac{\int_0^{2\pi}\int_0^{\pi} f(\theta,\phi)\sin(n\theta)P_m^n(\cos\phi)\sin^{n+1}\phi d\phi d\theta}{\int_0^{2\pi}\int_0^{\pi}\left[\sin(n\theta)P_m(\cos\phi)\sin^n(\phi)\right]^2\sin\phi d\phi d\theta} \quad \text{(ddd)}$$

Equation zz and Equation aaa are used to obtain

$$\frac{C_{0,m}}{\sqrt{\pi}}\left[D_{1,m,0}j_m'(\omega) + D_{2,m,0}y_m'(\omega)\right] = F_m \quad \text{(eee)}$$

$$\frac{C_{0,m}}{\sqrt{\pi}}\left[E_{1,m,0}j_m'(\omega) + E_{2,m,0}y_m'(\omega)\right] = 0 \quad \text{(fff)}$$

$$\frac{C_{n,m}}{\sqrt{2\pi}}\left[D_{1,m,1}j_m'(\omega) + D_{2,m,1}y_m'(\omega)\right] = G_{m,n,1} \quad \text{(ggg)}$$

$$\frac{C_{0,m}}{\sqrt{\pi}}\left[E_{1,m,0}j_m'(\omega) + E_{2,m,0}y_m'(\omega)\right] = 0 \quad \text{(hhh)}$$

$$\frac{C_{n,m}}{\sqrt{2\pi}}\left[D_{1,m,2}j_m'(\omega) + D_{2,m,2}y_m'(\omega)\right] = G_{m,n,2} \quad \text{(iii)}$$

$$\frac{C_{n,m}}{\sqrt{2\pi}}\left[E_{1,m,2}j_m'(\omega) + E_{2,m,2}y_m'(\omega)\right] = 0 \quad \text{(jjj)}$$

Equation tt, Equation uu, Equation vv, Equation ww, Equation xx, and Equation yy and Equation eee, Equation fff, Equation ggg, and Equation hhh are solved simultaneously to obtain the constants.

Problems

6.1. The nondimensional natural frequencies ω and mode shapes $\phi(x,y)$ of a rectangular membrane can be determined from

$$\frac{\partial^2\phi}{\partial x^2} + \frac{\partial^2\phi}{\partial y^2} + \omega^2\phi = 0 \quad \text{(a)}$$

Determine the natural frequencies and normalized mode shapes for a membrane which is clamped along its entire edge. The appropriate boundary conditions are $\phi(0,y)=0$, $\phi(1,y)=0$, $\phi(x,0)=0$ and $\phi(x,\alpha)=0$ where, is the ratio of the lengths of the sides of the membrane.

6.2. Solve Problem 6.1 if the membrane is free along the edge $y=0$ such that its boundary condition $\partial\phi/\partial x(x,0)=0$

6.3. Determine the steady-state temperature distribution, $\Theta(x,y)$ in the thin square slab shown in Figure P6.3. The problem is governed by Laplace's equation subject to the boundary conditions $\Theta(0,y)=1$,

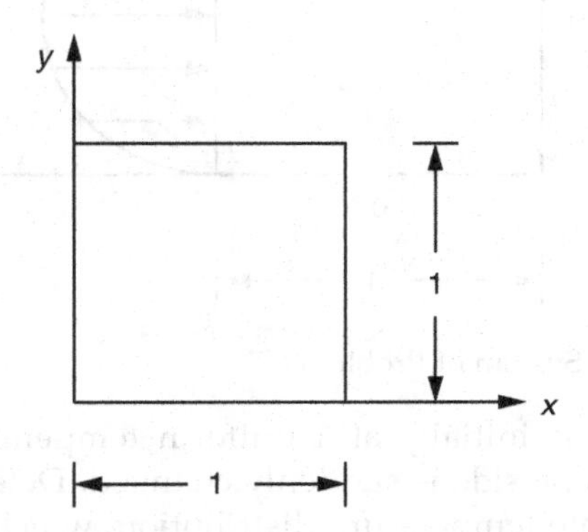

Figure P6.3 System of Problems 6.3–6.5.

$\partial\Theta/\partial x(1,y)=0$, $\Theta(x,0)=0$ and $\partial\Theta/\partial y(x,1)=0$.

6.4. Solve Problem 6.3 if the boundary conditions are $\Theta(0,y)=y(1-y)$, $\partial\Theta/\partial x\,(1,y)=0$, $\Theta(x,0)=0$ and $\partial\Theta/\partial y\,(x,1)=0$.

6.5. Solve Problem 6.3 if the boundary conditions are $\Theta(0,y)=y(1-y)$, $\partial\Theta/\partial x(1,y)=0$, $\Theta(x,0)=0$ and $\partial\Theta/\partial y(x,1)+1.2\Theta(1,y)=0$.

6.6. The unsteady state temperature distribution in an extended surface with a large heat transfer coefficient is governed by

$$\frac{\partial^2\Theta}{\partial x^2}-(Bi)\Theta=\frac{\partial\Theta}{\partial t} \tag{a}$$

The surface has an initial temperature distribution when the ambient temperature suddenly changes to Θ_1. The left end of the surface is maintained at the new temperature, while the right end of the surface is insulated. The boundary conditions are $\Theta(0,t)=0$ and $\partial\Theta/\partial x(1,t)=0$. Assume that the initial condition $\Theta(x,0)=\Theta_0\cos(\pi x)$.

6.7. The thin slab shown in Figure P6.7 initially is at a uniform temperature when a heat flux is applied to one side. Determine the resulting unsteady state temperature distribution, which is governed by the partial differential equation $\partial^2\Theta/\partial x^2+\partial^2\Theta/\partial y^2=\partial\Theta/\partial t$ subject to the boundary conditions $\Theta(0,x,t)=0$, $\partial\Theta/\partial x\,(1,y,t)=\sin(\pi y/\alpha)$, $\Theta(x,0,t)=0$ and $\Theta(x,\alpha,t)=0$ and the initial condition $\Theta(x,y,0)=0$.

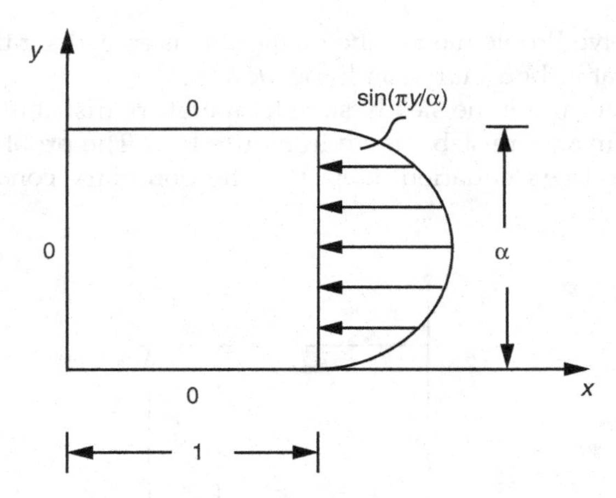

Figure P6.7 System of Problem 6.7.

6.8. A thin slab is initially at a uniform temperature when the temperature of one side is suddenly changed. Determine the resulting unsteady state temperature distribution, which is governed by the partial differential equation $\partial^2\Theta/\partial x^2 + \partial^2\Theta/\partial y^2 = \partial\Theta/\partial t$ subject to the boundary conditions $\Theta(0,x,t) = 1$, $\partial\Theta/\partial x\,(1,y,t) = 0$, $\Theta(x,0,t) = 0$ and $\Theta(x,\alpha,t) = 0$ and the initial $\Theta(x,y,0) = 0$.

6.9. A thin slab is initially at a temperature $f(x,y)$ when the temperature of two parallel sides is changed to the same temperature while the other two sides remain insulated. Determine the resulting unsteady state temperature distribution, which is governed by the partial differential equation $\partial^2\Theta/\partial x^2 + \partial^2\Theta/\partial y^2 = \partial\Theta/\partial t$ subject to the boundary conditions $\Theta(0,x,t) = 0$, $\Theta(1,y,t) = 0$, $\partial\Theta/\partial y\,(x,0,t) = 0$ and $\partial\Theta/\partial y\,(x,\alpha,t) = 0$ and the initial condition $\Theta(x,y,0) = f(x,y)$.

6.10. The thin shaft shown in Figure P6.10 is initially subject to a torque at its end which is suddenly removed. Determine the resulting response of the shaft, which is governed by the nondimensional partial differential equation $\partial^2\Theta/\partial x^2 = \partial^2\Theta/\partial t^2$ subject to the initial conditions $\Theta(0,t) = 0$, $\partial\Theta/\partial x(1,t) = 0$, and $\Theta(x,0) = Cx$, where C is a constant and $\partial\Theta/\partial t(x,0) = 0$.

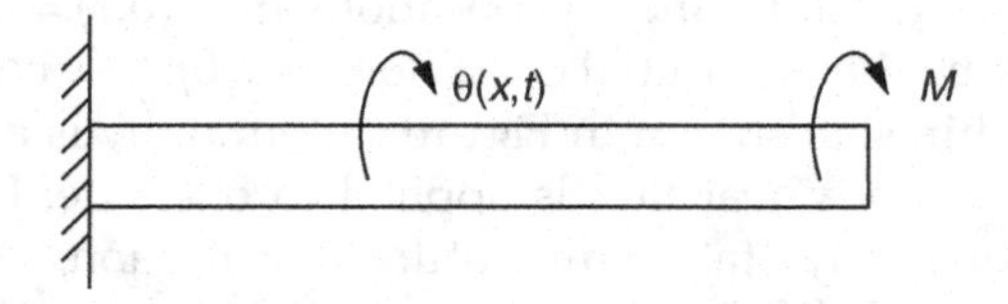

Figure P6.10 System of Problem 6.10.

6.11. The disk at the end of the thin shaft shown in Figure P6.11 is initially subject to a torque which is suddenly removed. Determine the resulting response of the shaft, which is governed by the non-dimensional partial differential equation $\partial^2\Theta/\partial x^2 = \partial^2\Theta/\partial t^2$ subject to the initial conditions $\Theta(0,t) = 0$, $\partial\Theta/\partial x(1,t) = -\mu\partial^2\Theta/\partial t^2(1,t)$ and $\Theta(x,0) = Cx$ and $\partial\Theta/\partial t(x,0) = 0$, where μ and C are constants.

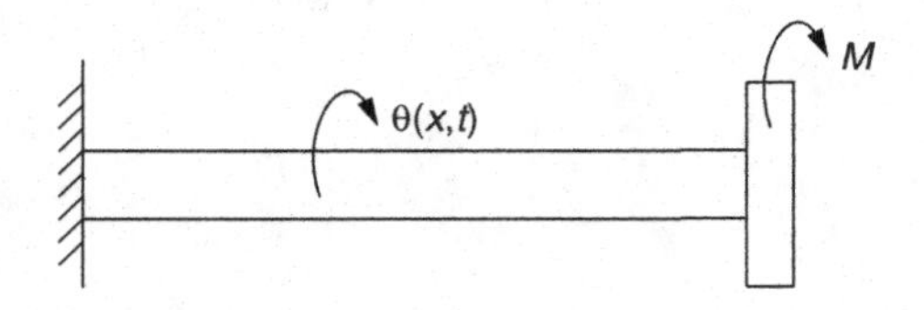

Figure P6.11 System of Problem 6.11.

Problems 6.12–6.16 refer to the system shown in Figure P6.12. The beam is subject to a concentrated load at its end, which leads to a static deflection of the form

$$w_0(x) = \alpha\left(\frac{x^2}{2} - \frac{x^3}{6}\right) \tag{a}$$

When the load is removed, the resulting vibrations are governed by

$$\frac{\partial^4 w}{\partial x^4} + \frac{\partial^2 w}{\partial t^2} = 0 \tag{b}$$

The boundary conditions are

$$w(0,t) = 0 \tag{c}$$

$$\frac{\partial w}{\partial x}(0,t) = 0 \tag{d}$$

$$\frac{\partial^3 w}{\partial x^3}(1,t) + \eta w(1,t) + \mu_1 \frac{\partial^2 w}{\alpha t^2}(1,t) \tag{e}$$

$$\frac{\partial^2 w}{\partial x^2}(1,t) + \mu_2 \frac{\partial^3 w}{\partial x \partial t^2} \tag{f}$$

The initial conditions are

$$w(x,0) = \alpha\left(\frac{x^2}{2} - \frac{x^3}{6}\right) \tag{g}$$

$$\frac{\partial w}{\partial t}(x,0) = 0 \tag{h}$$

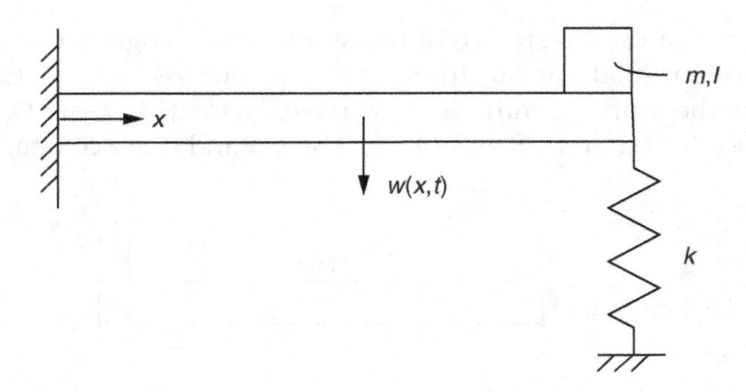

Figure P6.12 System of Problems 6.12–6.16.

6.12. Determine $w(x,t)$ if $\eta = \mu_1 = \mu_2 = 0$.
6.13. Determine $w(x,t)$ if $\eta = 2$ and $\mu_1 = \mu_2 = 0$.
6.14. Determine $w(x,t)$ if $\eta = \mu_1 = 0$ and $\mu_2 = 1$.
6.15. Determine $w(x,t)$ if $\mu_1 = 0.5$ and $\eta = \mu_2 = 0$.
6.16. Determine $w(x,t)$ if $\mu_1 = 0.5$, $\mu_1 = 0.5$ and $\eta = 2$.

Problems 6.17–6.24 refer to the nondimensional temperature distribution in a cylinder which, in general, is governed by

$$\frac{\partial^2 \Phi}{\partial r^2} + \frac{1}{r}\frac{\partial \Phi}{\partial r} + \frac{1}{r^2}\frac{\partial^2 \Phi}{\partial \theta^2} + \frac{1}{r^2}\frac{\partial^2 \Phi}{\partial z^2} = \frac{\partial \Phi}{\partial t} \tag{a}$$

Nondimensional variables are introduced such that $0 \leq z \leq 1$, $0 \leq r \leq R/L = \alpha$, and $0 \leq \theta \leq 2\pi$.

6.17. Determine the steady-state temperature distribution when the temperature at each end is determined by different constants and the circumference is insulated. The temperature is independent of θ. The boundary conditions for $\Phi(r,z)$ are $\partial\Phi/\partial r(\alpha,z) = 0$, $\Phi(r,0) = 0$, and $\Phi(r,1) = 1$.
6.18. Determine the steady-state temperature distribution in the cylinder when the temperature at each end is maintained at the same value and a heat flux $q(\theta,z)$ is applied to the circumference of the cylinder. The boundary conditions for $\Phi(r,\theta,z)$ are $\partial\Phi/\partial r(\alpha,\theta,z) = q(\theta,z)$, $\Phi(r,\theta,0) = 0$, and $\Phi(r,\theta,1) = 0$.
6.19. The cylinder is at a uniform temperature when the temperature of the surrounding medium is suddenly changed. Both ends of the cylinder are insulated, but heat is transferred by convection to the surrounding medium from the circumferential surface of the cylinder. The unsteady state temperature is independent of θ. The boundary conditions for $\Phi(r,z,t)$ are $\partial\Phi/\partial r(\alpha,z,t) + Bi\Phi(\alpha,z,t) = 0$, $\partial\Phi/\partial z(r,0,t) = 0$, and $\partial\Phi/\partial z(r,1,t) = 0$. The initial condition for the system is $\Phi(r,z,0) = 1$.
6.20. The cylinder has a steady-state temperature distribution of $f(r,\theta,z)$ when the temperature of the surrounding medium is suddenly

changed. Both ends of the cylinder are insulated, but heat is transferred by convection to the surrounding medium from the circumferential surface of the cylinder. The boundary conditions for $\Phi(r,\theta,1,t)$ are $\partial\Phi/\partial r(\alpha,\theta,z,t) + Bi\Phi(\alpha,\theta,z,t) = 0$, $\partial\Phi/\partial z(r,\theta,0,t) = 0$, and $\partial\Phi/\partial z(r,\theta,1,t) = 0$. The initial condition for the system is $\Phi(r,\theta,z,0) = f(r,\theta,z)$.

6.21. The cylinder has a steady-state temperature distribution of $f(r,\theta,z)$ when the left end is suddenly subject to a uniform heat flux. The right end of the cylinder is insulated, but heat is transferred by convection to the surrounding medium from the circumferential surface of the cylinder. The boundary conditions for is $\Phi(r,\theta,z,t)$ are $\partial\Phi/\partial r(\alpha,\theta,z,t) + Bi\Phi(\alpha,\theta,z,t) = 0$, $\partial\Phi/\partial z(r,\theta,0,t) = q$, and $\partial\Phi/\partial z(r,\theta,1,t) = 0$. The initial condition for the system is $\Phi(r,\theta,z,0) = f(r,\theta,z)$.

6.22. The cylinder has a uniform temperature when the circumferential surface is suddenly subject to a heat flux $q(z)$. Both ends of the cylinder are insulated. The boundary conditions for $\Phi(r,z,t)$ are $\partial\Phi/\partial r(\alpha,z,t) = q(z), \partial\Phi/\partial z(r,0,t) = 0$, and $\partial\Phi/\partial z(r,1,t) = 0$. The initial condition for the system is $\Phi(r,z,0) = 1$.

6.23. The cylinder has a steady-state temperature distribution $f(r,\theta,z)$ when an internal heat generation $u(z)$ begins in the cylinder. The right end of the cylinder is insulated, but heat is transferred by convection to the surrounding medium from the circumferential surface of the cylinder. The governing partial differential equation is modified to $\partial^2\Phi/\partial r^2 + (1/r)\partial\Phi/\partial r + (1/r^2)\partial^2\Phi/\partial\theta^2 + (1/r^2)\partial^2\Phi/\partial z^2 = \partial\Phi/\partial t + u(z)$. The boundary conditions for $\Phi(r,\theta,z,0)$ are $\partial\Phi/\partial r(\alpha,\theta,z,t) + Bi\Phi(\alpha,\theta,z,t) = 0$, $\partial\Phi/\partial z(r,\theta,0,t) = 0$, and $\partial\Phi/\partial z(r,\theta,1,t) = 0$. The initial condition for the system is $\Phi(r,\theta,z,0) = f(r,\theta,z)$.

6.24. The cylinder has the steady-state temperature distribution obtained in the solution of Problem 6.18 when an internal heat generation begins in the cylinder. The boundary conditions remain the same as in Problem 6.18, but the governing partial differential equation is modified to $\partial^2\Phi/\partial r^2 + (1/r)\partial\Phi/\partial r + (1/r^2)\partial^2\Phi/\partial\theta^2 + (1/r^2)\partial^2\Phi/\partial z^2 = \partial\Phi/\partial t + u(z)$. Solve the problem to determine $\Phi(r,\theta,z,t)$.

6.25. Determine the time-dependent response of the system of Example 6.4 when $\alpha(x) = (1 - 0.1x)^3$.

6.26. The disk at the end of the thin shaft shown in Figure P6.11 is at rest in its equilibrium position when a time-dependent uniformly distributed torque is applied across the length of the shaft. Determine the resulting response of the shaft, which is governed by the nondimensional partial differential equation $\partial^2\Theta/\partial x^2 = \partial^2\Theta/\partial t^2 + \phi x(1-x)\sin\omega t$ subject to the boundary conditions $\Theta(0,t) = 0$ and $\partial\Theta/\partial x(1,t) = -\mu[\partial^2\Theta/\partial t^2(1,t)]$ and the initial conditions $\Theta(x,0) = 0$ and $\partial\Theta/\partial t(x,0) = 0$, where μ and ϕ are constants.

6.27. Determine the steady-state temperature distribution in a sphere whose surface has a temperature distribution $f(\theta)$ which is independent of ϕ.

6.28. The curved surface of a hemisphere, as illustrated in Figure P6.28, is subject to a heat flux, $q(\theta,\phi)$. The flat surface is maintained at a constant temperature. Determine the steady-state temperature distribution in the sphere, $\Phi(r,\theta,\phi)$, using the boundary conditions $\partial\Phi/\partial r(1,\theta,\phi) = q(\theta,\phi)$ and $\Phi(r,\theta,\pi/2) = 0$.

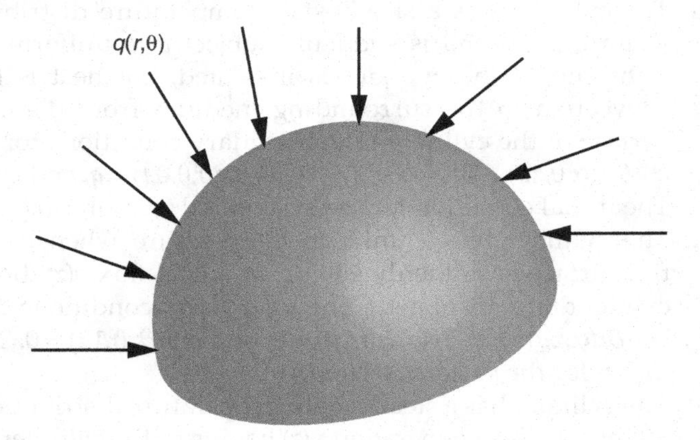

Figure P6.28 System of Problems 6.28 and 6.29.

6.29. Solve Problem 6.28 using the boundary conditions $\partial\Phi/\partial r(1,\theta,\phi) = q(\theta,\phi)$ and $\Phi(r,\pi,\phi) = 0$.

6.30. Determine the steady-state temperature distribution in the spherical shell shown in Figure 6.30 if the inner surface is insulated and the outer surface is maintained at a constant temperature, $f(\theta,\phi)$.

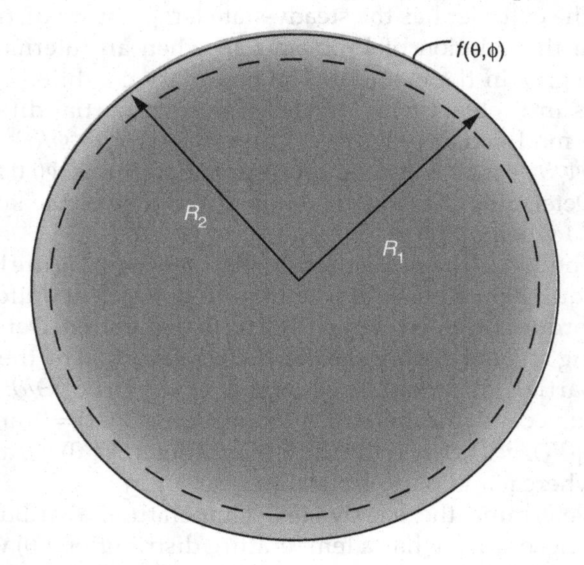

Figure P6.30 System of Problem 6.30.

The nondimensional boundary conditions are $\partial\Phi/\partial r(1,\theta,\phi)=0$ and $\Phi(R_2/R_1)=f(\theta,\phi)$.

6.31. The sphere shown in Figure P6.31 is maintained at a uniform temperature T_1 when the exterior is subject to a uniform heat flux q. Determine the resulting unsteady state temperature distribution in the sphere when (a) q is constant over the surface of the sphere and (b) $q=q(\theta,\phi)$

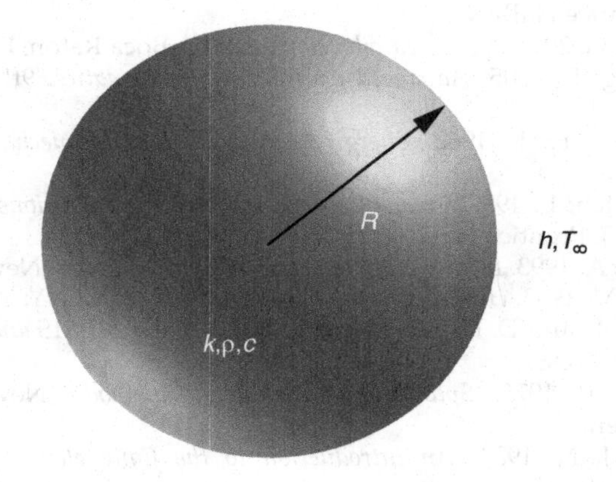

Figure P6.31 System of Problems 6.31 and 6.32.

6.32. The sphere shown in Figure P6.31 is maintained at a uniform temperature T_1 when an internal heat generation u is initiated in the sphere. Determine the resulting unsteady state temperature distribution in the sphere when (a) u is a constant throughout the sphere all (b) $u=u(r,\theta)$ and (c) $u=u(\mathrm{r},\theta,\phi)$.

6.33 Rework Example 6.8 by first considering the eigenvalue problem

$$-\frac{d^2Y}{dy^2}=\mu Y$$

$$Y(0)=0$$

$$Y'(\alpha)=0$$

Then expand $f(x,y)$ for each x in terms of the eigenvectors for this problem. Show that Equation q of Example 6.8 is attained.

References

1. Abramowitz, M., and I. Stegun. 1965. *Handbook of mathematical functions.* Mineola, NY: Dover.
2. Arpaci, V. 1966. *Conduction heat transfer.* Reading, MA: Addison Wesley.
3. Courant, R., and Hilbert, D. 1962. *Methods of mathematical physics.* New York, NY: Wiley-Interscience.

4. Franklin, J. N. 1968. *Matrix theory*. Englewood Cliffs, NJ: Prentice Hall.
5. Greenberg, M. 1998. *Advanced engineering mathematics*. 2nd ed. Englewood Cliffs, NJ: Prentice Hall.
6. Haberman, R. 2003. *Applied partial differential equations*. 4th ed. Englewood Cliffs, NJ: Prentice Hall.
7. Haug, E., and K. Choi. 1993. *Methods of engineering mathematics*. Englewood Cliffs, NJ: Prentice Hall.
8. Hildebrand, F. 1976. *Advanced calculus for applications*. 2nd ed. Englewood Cliffs, NJ: Prentice Hall.
9. Kelly, S. G. 2007. *Advanced vibration analysis*. Boca Raton, FL: CRC Press.
10. Kreyszig, E. 2005. *Advanced engineering mathematics*. 9th ed. New York, NY: Wiley.
11. Langhaar, H. L. 1962. *Energy methods in applied mechanics*. New York, NY: Wiley.
12. Meirovitch, L. 1997. *Principles and techniques of vibrations*. 1997. Upper Saddle River, NJ: Prentice Hall.
13. Nayfeh, A. 1993. *Introduction to perturbation techniques*. New York, NY: Wiley.
14. Ozisik, M. 1993. *Heat conduction*. 2nd ed. New York, NY: Wiley.
15. Pelesko, J., and D. H. Bernstein. 2003. *Modeling MEMS and NEMS*. Boca Raton, FL: CRC Press.
16. Prenter, P. 1975. *Splines and variational methods*. New York, NE: Wiley-Interscience.
17. Reddy, J. N. 1984. *An introduction to the finite element method*. New York: McGraw-Hill.
18. Strang, G., and G. J. Fix. 1973. *An analysis of the finite element method*. Englewood Cliffs, NJ: Prentice Hall.
19. Strang, G. 2005. *Linear algebra and its applications*. 4th ed. Belmont, CA: Brooks Cole.
20. Weinstock, R. 1952. *Calculus of variations with applications to physics and engineering*. New York, NY: McGraw-Hill.
21. White, F. 2005. *Viscous fluid flow*. 3rd ed. New York, NY: McGraw-Hill.
22. Wylie, C. R. 1975. *Advanced engineering mathematics*. New York, NY: McGraw-Hill.
23. Zhang, Z. 2007. *Nano/Microscale heat transfer*. New York, NY: McGraw-Hill.
24. Zienkiewicz, O. C., and R. I. Taylor. 1991. *The finite element method*. 4th ed. New York, NY: McGraw-Hill.

Index

Q

R

S

T

Printed in the United States
by Baker & Taylor Publisher Services

Printed in the United States
by Baker & Taylor Publisher Services